Issues in Maritime Cyber Security

ISSUES IN MARITIME CYBER SECURITY

EDITED BY

JOSEPH DiRENZO III,
NICOLE K. DRUMHILLER,
AND FRED S. ROBERTS

Westphalia Press
An Imprint of the Policy Studies Organization
Washington, DC
2017

Issues in Maritime Cyber Security
All Rights Reserved © 2017 by Policy Studies Organization

Westphalia Press
An imprint of Policy Studies Organization
1527 New Hampshire Ave., NW
Washington, D.C. 20036
info@ipsonet.org

ISBN-10: 1-63391-555-7
ISBN-13: 978-1-63391-555-8

Cover and interior design by Jeffrey Barnes
jbarnesbook.design

Daniel Gutierrez-Sandoval, Executive Director
PSO and Westphalia Press

Updated material and comments on this edition
can be found at the Westphalia Press website:
www.westphaliapress.org

CONTENTS

Preface, ix

Foreword
Robert Parker, xiii

Part I: Introductions/State of Cyber

1. Threats to Global Navigation
   David B. Moskoff and William G. Kaag 3
2. Cyber Security Considerations for the Maritime Environment
   Lily Ablon 17
3. U.S. Coast Guard Cyber Threat Information Sharing
   Jennifer M. Konon 25
4. Navigating the Cyber Threats to the U.S. Maritime
   Transportation System
   Jim Twist, Blake Rhoades, and Ernest Wong 57
5. Cyber Seaworthiness: A Call to Action
   Kevin S. Cook and David L. Nichols 81

Part II: Policy

6. Cyber Risks in the Maritime Transportation System
   Charles D. Michel, Paul F. Thomas, and Andrew E. Tucci 89
7. Barriers Against Cyber Security Threats: Using Barrier Management
   to Visualize and Administer Cyber Security Countermeasures
   Pål B. Kristoffersen 103
8. Cyber Warfare and Maritime Security: A Call for
   International Regulation
   Laura Sturdevant 115
9. Commercial Maritime Cyber Security: Risk and Vulnerability:
   What We Did not Know, We Did not Know
   Bruce Clark 129

10. Maritime Cyber Security: The Unavoidable Wave of Change
    Kate B. Belmont 181

11. Developing Risk-Based Performance Standards for Cyber Security in the Maritime Transportation System
    Kimberly Young-McLear 191

12. Cyber Security in Maritime Domain: A Risk Management Perspective
    Unal Tatar and Adrian Gheorghe 201

13. An Analysis of Response Strategies Used to Mitigate Disruptions to the Maritime Transportation System Caused by a Cyber Incident
    Joseph Couch 211

14. Cyber Security Industrial Controls
    Jim Cooper 221

15. Basic Cyber Security is Easy (So Why is Implementing it So Hard?)
    Jerry Doherty 231

16. Cyber Security and the Maritime Transportation System: Closing the Financial and Regulatory Incentive Gap
    Chris Conley 237

17. Information Sharing for Maritime Cyber Risk Management
    Dennis Egan, Darby Hering, Paul Kantor,
    Christie Nelson, and Fred Roberts 271

18. How Do We Promote the Use of Sound Cyber Risk Management Principles?
    Eduardo V. Martinez, Jessica L. Adkisson, Jacob A. Babb,
    Eric S. Casida, Frank K. Hooton, Gabriel Nunez,
    Jeremy Ouittschreiber, and Joshua S. Weishbecker 303

Part III: Technical

19. Economic Consequence Analysis of Maritime Cyber Threats
    Adam Rose 321

20. Securing the Integrity of Your Control System
    Mate J. Csorba and Nicolai Husteli 357

21. Man and Machine: How Visual Analytics Can Enable Insights in Maritime Cyber Security
    Sungahn Ko, Abish Malik, Guizhen Wang, and David S. Ebert ... 375

22. GPS & Shipping: Countering the Threat of Interference
    Jeff D. Coffed and Joe Rolli 397

23. GPS Jamming and Spoofing: Maritime's Biggest Cyber Threat
    Dana Goward 407

24. Windows on Submarines: Cyber Vulnerabilities and
    Opportunities in the Maritime Domain
    Erik Gartzke and Jon Lindsay 417

25. Cyber Security Resiliency in the Maritime Sector: A Systems
    Approach to Analyzing Gaps in the NIST Framework
    Kimberley Young-McLear and John Fossaceca 433

26. Game Theoretic Defense for Maritime Security
    Sara McCarthy, Arunesh Sinha and Milind Tambe 449

27. Maritime Cyber Security: What about Digital Forensics?
    Scott Blough and Gordan Crews 483

28. Maritime Cyber Threat Intelligence
    Tom Gresham 499

29. The Need for a User-Friendly Cyber Security Vulnerability
    Tool for Coast Guard Marine Inspectors
    *Mark Behne, Benjamin Chapman, Michael Clancy,
    Koachar Mohammad, and Kimberley Young-McLear* 515

Part IV: The Way Forward

30. The Next Great Battlefront
    Gerald Feltman 527

31. The Combined Joint Operations from the Sea Centre of
    Excellence is Way Ahead for Maritime Cyber Security
    Ovidiu Marius Portase 537

32. Toward a Maritime Cyber Security Compliance Regime
    *Mark R. Heckman, John McCready, David Mayhew,
    and Winnie L. Callahan* 543

33. Evergreen Cyber Project
    Eric Popiel 569

PREFACE

This book, and its focus on the cyber systems that encompass the Maritime Transportation System (MTS) was born out of the most basic of academic frameworks—a massive and unmet gap in the body of knowledge regarding a system that impacts every continent in the world and the vast majority of its people either directly or indirectly. The facts are simple: the world relies on maritime commerce to move exceptionally large portions of goods, services, and people at levels unprecedented in its collective history. Shipping by sea is not only economical, especially with large heavy items, but also affords access where other modes of transportation might not be as available. Using the United States as an example, "by value, vessels carry 53 and 38 percent of U.S. imports and exports, respectively—the largest share of any mode" (Chambers and Liu 2016). Additionally, in just the United States alone there are over 200 ferry operators, over 12 million recreation boats, and over 360 ports (Chambers and Liu 2016; MarEx 2016). The Maritime Administration (MARAD) added that there are "25,000 miles of navigable channels, 238 locks at 192 locations and over 3,700 marine terminals" (Maritime Administration 2016). The reliance on cyber networks, and the infrastructure they control is growing dramatically within the MTS—from the complex programs managing the loading and unloading of containers to waiting trucks, to the global navigation systems onboard the vessels that bring food grown in the mid-west to China, to the hydraulic valves designed to protect spills into waterways that are located and controlled by cyber systems within chemical, water/wastewater, or petroleum plants. Indeed, our ports are increasingly automated, even our largest vessels operate with tiny crews (and could one day soon operate themselves), and oil rigs, cargo handling systems, and other components of the MTS are managed by a vast array of software and related hardware. The impact of the cyber element on the international Maritime Transportation System is significant. Yet, with the clear advantages this brings, come vulnerabilities, challenges, and indeed the possibility of dramatic disruptions with cascading effects from cyber attacks and cyber accidents. The complexity of the "problem" is daunting and failure to solve it even more significant than at any other time in the history of maritime transportation. As the human–machine interface becomes more pronounced, and capabilities such as artificial intelligence develop for the benefit of mankind, the need for focused, recurring, and dedicated study to issues related to the cyber systems that encompass the world's Maritime Transportation System is more important than ever.

The idea for this book started in March of 2015 when Rutgers University's Department of Homeland Security Center of Excellence—CCICADA—The Command, Control and Interoperability Center for Advanced Data Analysis and American Military University co-hosted the first academic learning seminar and symposium on cyber issues within the Maritime Transportation System. Initially the hosts anticipated 20–25 attendees. By the time the event had begun over 125 stakeholders came including senior federal, state, and local officials, and the audience ballooned to over 300 when counting those who participated through live streaming of the event. When a review of the literature was conducted to offer a baseline of the current body of knowledge, less than 20 peer reviewed published works were found. The March 2015 event generated research questions and interest from a variety of stakeholders. This interest jump-started the now-rapid development of research on "maritime cyber security" and led to a second focused symposium in June 2015. The event was held at the California Maritime Academy's new Richmond, CA Maritime Security Center. This event drew even more interest from industry which further underscored its importance. The March event also featured six research questions posed by then-VADM Chuck Michel of the US Coast Guard and aimed at contributing to the body of knowledge in this field. Those questions led to several universities working with CCICADA in a University Maritime Cyber Initiative and that work was highlighted at the June Cal Maritime event and at subsequent maritime cyber security working sessions at the Port of Long Beach, the University of Southern California, and the University of North Carolina during 2016. Some of the work coming out of the Maritime Cyber University initiative appears in four papers in this book.

The editors took on the challenge of developing an academic work product that would not only start to address the gap in the body of knowledge but facilitate more in-depth research. Some of the key questions involve appropriate metrics for cyber health of a vessel, the role of the Coast Guard in maritime cyber security, and key ways for information to be shared within the private sector and between the private sector and government. One area that has only begun to be discussed is the role the insurance industries will play in framing cyber security and cyber risk within the Maritime Transportation System. Other key areas in need of future research include how to best address the human elements within this area, especially when it comes to maintaining basic cyber hygiene practices and the recognition that this issue requires recognition and buy-in from all members operating within the MTS.

PREFACE

The three of us are fortunate in that the vast majority of subject matter experts we asked to provide input did so; in fact, we were amazed by the overall response. From industry experts, to the academic community, from an international perspective, to the legal profession, we covered a wide range of topics. But even now, as we reflect on the essays that have made their way into this book, we continue to find areas within the world of cyber in the MTS that need further exploration. This is truly a multidisciplinary issue—and this effort scratches only the very tip of the knowledge iceberg. This is perhaps the most over-arching reason we put this book together. Every waterfall starts with a single drop of water, and as such, we hope that this book will ultimately move others to consider the important issues contained herein and continue to move the body of knowledge forward. Plans are already in the works to frame a follow-up volume... like the new release of a smart phone, the next step in advancement is already underway. Indeed, some papers in this book stem from 2015 and so will already require updating/follow-up in this fast-moving environment.

We would like to thank the leaders and organizations who have made this work possible. The Department of Homeland Security Office of University Programs has supported the maritime cyber security work of the CCICADA Center that led to this volume, under Grant 2009-ST-061-CCI002-06. Rutgers University, American Military University (AMU), The California Maritime Academy, University of Southern California (USC), and University of North Carolina—Chapel Hill contributed by hosting relevant meetings, and Rutgers, AMU, USC, and University of San Diego supported Coast Guard RDC—University research initiatives. We would like to thank the people who played a special role in putting the topic of maritime cyber security on the map and assisted us with advice, guidance, and planning for this book, in particular VADM Rob Parker, USCG (Ret), Captain Michael Dickey, USCG, Captain Andrew Tucci, USCG, Captain Bruce Clark, USCG (Ret), Captain Peter Crain, CN, (Ret), Captain David Moskoff, USMMS, David Boyd, USCG, Kate B. Belmont, Blank Rome LLP.

All essays in this book have been reviewed. We would like to thank the many people who acted as reviewers: Nicole Beebe, PhD, University of Texas—San Antonio, David Boyd, USCG, Professor Scott Blough, Tiffin University, Captain Bruce Clark, USCG (Ret), Gordon Crews, PhD, Tiffin University, RADM Bob Day, USCG (Ret.), Dennis Egan, PhD, Rutgers University, Captain Dana Goward, USCG (Ret), Tom Gresham, Port of San Diego, Dr. Jack McCready, USCG (Ret), Captain David Moskoff, USMMS, Captain Dermot Mulholland, RCN, VADM Rob Parker, USCG (Ret), CPT Blake Rhoades, USA, Captain James Spotts, USCG (Ret), Captain Andrew Tucci, USCG, Clay Wilson, American Military University (Ret), LTC Ernie Wong, USA.

Dr. Joseph DiRenzo III
American Military University

Dr. Nicole K. Drumhiller
American Public University System

Dr. Fred S. Roberts
Rutgers University

February 2017

REFERENCES

Chambers, Matthew, and Mindy Liu. 2016. "Bureau of Transportation Statistics." https://www.rita.dot.gov/bts/sites/rita.dot.gov.bts/files/publications/by_the_numbers/maritime_trade_and_transportation/index.html (accessed December 20).

MarEx. 2016. "A Nation with 360 Ports." *The Maritime Executive*, April 14. http://www.maritime-executive.com/features/a-nation-with-360-ports (accessed January 8, 2017).

Maritime Administration (MARAD). 2016. "Marine Transportation System (MTS)." https://www.marad.dot.gov/ports/marine-transportation-system-mts/ (accessed December 20).

FOREWORD

Maritime Cyber Security. It is a term of art that does not exactly roll off the tongue. Is there such a thing as Land Cyber security? Air? Space? Some say it is a misnomer. Some say the idea does not sufficiently encompass the challenges of cyber threats and vulnerabilities in the domain. Still others say it is just a continuation of "nothing new under the sun" when it comes to addressing the attendant challenges of cyber in the global maritime commons.

In this maritime domain that is steeped in autonomy and millennia of tradition, the idea that a ship and its crew, passengers, and cargo could be at risk or put thousands of others unwittingly at risk because of cyber intrusions or accidents was slow to take hold. The happier news about the long traditions of the sea is that it includes a predisposition for innovation and a bias for action once awareness of a problem is realized. This suggests that awareness of the problem is a key factor for the maritime domain and cyber challenges. I am excited to see that the maritime community and its supporters are now rapidly gaining awareness and taking increasingly innovative action against this burgeoning threat. Whether it is enough, soon enough, and will be sufficiently timely on a continuing basis is yet to be known.

Just 6 years ago I found a dearth of awareness or concern for this topic as I travelled and talked about it throughout the country and internationally. At that time, I was the U.S. Coast Guard's Atlantic Area Commander; a position that comes with insufficient resources to address all mandated missions and no shortage of challenges to address increasing, and increasingly complex, threats and vulnerabilities. I brought with me a newfound appreciation and concern about cyber awareness, safety, and security that I had acquired in my previous job as the Director of Operations at U.S. Southern Command. During my transition between jobs, the U.S. Coast Guard was dealing with the explosion with loss of 11 lives on the DEEPWATER HORIZON oil rig and the ensuing massive and unrelenting oil spill. This chaotic and all-consuming activity would go on literally for years. Acting strategically and deliberately on a set of threats and vulnerabilities that had not had catastrophic consequences (yet) under this operational shadow made awareness and action of this "new" cyber challenge difficult on the best of days. I am happy to say this is now rapidly changing, although not quickly enough to outpace the metastasizing growth of the threat.

Adding to the challenge is the complex global nature of the marine transportation system and its many inter-modal connected parts, the explosive growth and increased reliance on technology in the maritime sector, and the raw asymmetric vulnerabilities this combination brings. I am encouraged by the progress made during and since this seminal Maritime Cyber Security Symposium in March 2015, and cautiously optimistic about the efforts to mitigate threats and vulnerabilities by our government, the private sector, academia, and other researchers to keep the global economy and its related stability on a relatively even keel.

In reviewing the articles and the often robust discussions that followed, several observations stood out. I found unanimous agreement among those who are even remotely aware of cyber challenges in the maritime domain, that there is an issue that needs to be addressed. Within that audience, there is a wide variance in what the issues are and how they should be addressed, but clearly there are issues that have compelled many to action. Nearly all wanted to categorize the issues as safety or security, evolutionary or revolutionary, or some other such label that would place it in a familiar structure to either set strategy, tactics or policy, or otherwise inform and organize operational activity. Not surprisingly, where the advocate sat had a major impact on how they saw each of these issues and categorizations. Upon reflection, each is correct for the contributions of their represented group, and each will benefit from a more common and appropriate lexicon that is specific to cyber and its implications in the maritime domain. To get to that common ground it is useful to understand both the existing structures and tensions of the operational environment and its governance (both government and private enterprise) and the nature of the cyber challenge in the maritime domain.

Also clear to most casual cyber observers, this is a problem that can strike seemingly from nowhere, with no perceptible notice, requiring near-simultaneous response to defend or protect, and many participants with varying expertise to keep systems, data and our way of life safe and secure. This also means that cyber systems and the physical internet of things attached to them can be put at risk by a wide variety of actors and vectors. At first blush it would seem that ships and maritime facilities enjoy an odd advantage in this regard since one of their perennial modern challenges is lack of connectivity or bandwidth. Dig a little deeper and you can find a false sense of security that can arise from this wrong sense of limited vulnerability. When coupled with a dynamic and rapidly metastasizing threat and, an increasingly complex operating environment and smaller profit margins this can become a very real and unattended danger to the most isolated participant.

We have learned much about cyber security and its implications in the maritime domain, and in so doing have realized how much more we need to know and do. Conversations have so far revolved around: whether existing plans, policies, protocols, and procedures can or should apply to things cyber in the maritime world; what "resilience" means vis-à-vis maritime cyber; how to prevent cyber events impacting the domain including people, the MTS itself, the environment, and property or facilities; and how much and what type of investment yields the best bang for the buck for risk mitigation. Again, there is a lot of discussion and activity, but the question remains whether it will be enough, soon enough, and decisive enough to provide whatever level of resilience is determined to be necessary or sufficient.

Awareness is a key factor as mentioned above. Our awareness to date has brought us to understand that the characteristics of cyber include: extremely (almost incomprehensively) compressed timeline; high complexity; very dynamic; potentially disproportionate scale of action to impact; mostly anonymous; a utility versus vulnerability paradox; easily spoofed legitimacy; equally devastating disruptions from a technical error or criminal or terrorist act; importance of strong information sharing, often in real time; perceived cost-benefit gain from automation; oft under-informed trust or reliance (precision does not always equate to accuracy); and a realization that the challenge is not unique to, or contained by, the maritime environment. As our awareness matures, collectively and individually, this list should appropriately morph and change to match the threat. This creates great challenge for large bureaucracies with less than nimble processes, which can lead to suboptimal decision and policy making and therefore increase complexity of an already very complex challenge. At least rudimentary awareness and understanding of this threat should be a minimal threshold for leaders across the maritime community.

Policy making to address a challenge with the above described characteristics is a particularly difficult and dynamic problem. Many actors from government, academia, private sector, and various adversaries make almost any solution suboptimal by definition. This further complicates the risk calculus of an already complex scenario. Just within the Department of Homeland Security you have NPPD, USCERT, USCG, S&T, Secret Service and others. Add Department of Justice for Investigations and prosecutions, DoD to defend the nation's cyber space, IMO to help normalize regulations for the global maritime commons and many other transportation, energy, commerce, and labor constituencies, not to mention individual countries, regional cooperatives, and alliances. Hard just got harder, and it needs to happen faster and more often unless we find a more agile and adaptive way to make policy and decisions regarding maritime cyber. The underlying debate here at the time of writing is whether existing policy and decision-making structures can or should scale to meet this challenge or if a new structure is called for. No clear winners have emerged in this debate in this author's opinion. What is clear is that we need to think a lot and potentially a lot differently, then act toward a set of solutions that help us achieve cyber resilience in the maritime domain. One good example of this is included in the article on "Cyber Seaworthiness."

Help to answer some of these challenges comes in many forms and fora. Unsurprisingly, each tends to take on the characteristics reflected by the lens through which they see this part of the universe. An industry perspective tends to see risk in terms of corporate bottom line and liability first. Government leans toward whatever the current perception is of the public good. Academia tends to focus on pedagogical lanes. Adversaries need not look far or wide to find easy gaps and access points in this landscape if we do not accelerate our collective teamwork in the direction of solving cyber challenges, current and future.

Research into this arena is blossoming at extraordinary rates. Even with that, it may not be equal to the challenges of the threats and vulnerabilities yet. The landscape for research opportunities and better awareness or understanding of threats and vulnerabilities is rich. Threat vectors can be internal to systems, vessels, facilities, organizations, or countries. They can be caused by connection or adjacency to external threats from any of the above. They can be external, whether persistent or wildly stochastic, from industrial espionage, criminal activity, malicious activity, negligence, or spurious signals caused by environmental conditions. Vulnerabilities exist in control systems, imbedded systems, navigation, and timing systems. These same systems can be secondarily impacted by simple power loss or radio frequency or electromagnetic interference. Further, the MTS is subject to legal vulnerabilities where laws and regulations do not adequately enable commerce and protect its benefactors in a properly risk mitigated way. There is also the persistent threat of human vulnerabilities, either as an accident, unwitting proxy or malicious or criminal act.

Consequences of compromised vulnerabilities can lead to loss of life, economic loss, and environmental damage in the maritime domain. An insidious characteristic of cyber is its ability to disproportionally affect a large number of connected things simultaneously, potentially across the entire global commons. Sorting through the highest risk scenarios has become a more arduous task, and as the threats and responses evolve and mature by various actors, complexity increases at a rate well above linear expansion. Many are trying to answer this call, whether to be competitive in the global market or shepherd the public good. I applaud all and wish you well. I especially applaud you if you are among the 2% that ever read this far into a foreword and have not yet lost interest and skipped ahead (Thank you for your time and attention!).

Additionally, we would offer some advice on places to focus at this relatively early stage of attack on the problem set. What is needed first is improved awareness, especially by leadership both in and out of government, across the global maritime commons. With all its complexities, this problem may best be addressed to an audience with more to do and less time and resources to do it, through the use of dynamic visual analytics. Initial awareness is a good starting point, but a continued understanding, appreciation, and tracking of the problem set by monitoring in a way that can be rapidly assimilated is key to long-term resilience and near-term deterrence alike. A common taxonomy through standards and language will likewise be helpful.

Teamwork... The more complex and dynamic and broad in scope the problem, the more necessary it becomes. This problem calls for visible leadership support and high expectations (with commensurate follow up and testing to ensure expectations are met) for cyber awareness and a responsive cyber culture. This is more of an "all hands" evolution than any of the challenges I faced in 35 years in the business and it is more persistent and burgeoning. Just one person's bad cyber hygiene can open a door for a really bad actor in ways that have not existed before, and that can be harder to detect and defend when the time factor is so radically compressed. The lines that encompass the team are more expansive than before, drawing in governments, corporations and private citizens as never before. High assurance ways to collaborate and "proof" electronic identities and entities will be helpful here to build and maintain the critical trust needed to address these issues in a lasting way to keep commerce flowing and keep our populations safe and secure.

The formation of these expanded teams seems to call for a different tool set as well, whether based on existing tools (policies, protocols, tactics, techniques, and procedures) or a completely different set altogether. As a 35-year operator in this domain I would offer that simpler is often better, especially when it needs to be fast and adapt to a dynamic threat and environment. The NIST cyber security framework, roadmap and reference list are reasonable starting points, but fall short of answering the mail specific to this domain for this threat stream. Rules will almost surely be complex in such conditions, but the framework and concepts that underlie them should enable and encourage agile and nimble response; tactically, operationally and strategically.

Additional research, with accelerated and more open collaboration, can help inform more agile planning and investment decisions from the local to the global level, as well as operational and tactical tools at the operational and tactical level, be it private, corporate, or government. Policies and procedures will need to be more agile and adaptive throughout the maritime community. Engaging think tanks and the academic community to partner in this fight will reduce blind spots and accelerate our collective learning while growing a new generation of awareness.

As we simultaneously build and fly this plane, we must also remain vigilant, whether it is day to day with cyber hygiene, awareness, incident detection and repair, vulnerability scans, and reviews or in system design or type approval for high assurance systems such as navigation, control, or safety systems. We must leverage that which is common, and not necessarily unique to the maritime environment, in the solutions offered by others more experienced across all domains. Exercise, auditing, and testing of our cyber readiness or cyber seaworthiness will also help us sustain appropriate levels of cyber resilience in the maritime domain.

Whether a term of art, a misnomer or just a transient label, the issue we now know as maritime cyber security and safety is a challenge that has so far outpaced governance solutions. There is cause for optimism in the maritime sector at the same time there is great need for increased and more agile awareness, teamwork, and action against this dynamic threat. Together we stand a far better chance of defeating and defending these threats and we commend you for your contributions and the future actions you will take to help address and mitigate this burgeoning challenge. Again, I am excited to see that the maritime community and its supporters are now rapidly gaining awareness and taking increasingly innovative action against this burgeoning threat. Whether it is enough, soon enough, and will be sufficiently timely on a continuing basis is yet to be known.

The majority of the works that follow were part of the Maritime Cyber Security Symposium held at Rutgers University and co-sponsored by American Military University in March 2015. Others came as we discussed the idea of developing this volume. The challenge of publishing this first compendium kept pushing the publication date to the right on the calendar, which in turn opened the door to more engaging thoughts, which again pushed the date to the right, and so on. As such, we offer you this snapshot in time with full assurance that it will be incomplete, and hopeful that it will serve to accelerate learning, awareness, research, policy making and operations in the cyber world that simultaneously enables and threatens the maritime domain, marine transportation system, and its many users and benefactors, as well as many who will live full lives never appreciating that 90% of everything in their daily lives would not be the same without benefit of a safe and secure marine transportation system.

Rob Parker
U.S. Coast Guard (Ret)

PART I:

Introductions/State of Cyber

CHAPTER 1:
THREATS TO GLOBAL NAVIGATION

David B. Moskoff
U.S. Merchant Marine Academy

William G. Kaag
U.S. Navy (Ret.)

Abstract

This paper examines vulnerabilities of Global Navigation Satellite Systems (GNSS) and threats posed to these systems. Also considered are differences between maritime cyber threat issues such as radio frequency (RF) signal denial compared with traditional cyber threats. Various options available to the maritime community to mitigate these threats are discussed, such as crew training, stakeholder education, improved maritime equipment, and technological advances, including eLoran.

If this event had been a GPS failure instead of a GLONASS failure ... the entire world would have plunged into a catastrophe.
 —Nunzio Gambale, CEO of Locata, after the 11-hour outage of GLONASS April 2014

Introduction

Originally developed to guide Allied convoys safely across the Atlantic, the use of synchronized low frequency radio signals as a navigational aid revolutionized modern maritime navigation in the 1940s. Faced with operating ships and aircraft over vast areas, researchers pioneered the use of radio signals to aid navigation in regions where poor weather conditions made traditional methods—

such as dead reckoning and celestial navigation—exceptionally difficult. This system was eventually named LOng RAnge Navigation (LORAN). When in range of three or more shore-based transmitters, LORAN receivers placed on-board ships and aircraft allowed operators to fix their location within minutes regardless of the weather. The original system, known as LORAN-A, and its eventual replacement, LORAN-C, were operated by the U.S. Coast Guard and other nations until 2010. The U.S. portions of the system were phased out in favor of the satellite-based Global Positioning System (GPS) which became operational in July of 1995. Figure 1 shows basic GNSS operation including GPS. The latest LORAN Position Navigation and Timing (PNT) system known as "eLoran" is currently in use or under consideration in several countries. Among the entities that historically operated and maintained these terrestrial loran systems are national Coast Guards or defense organizations. Eventually, Loran C systems throughout the world are expected to be replaced by eLoran or a similar complementary system.

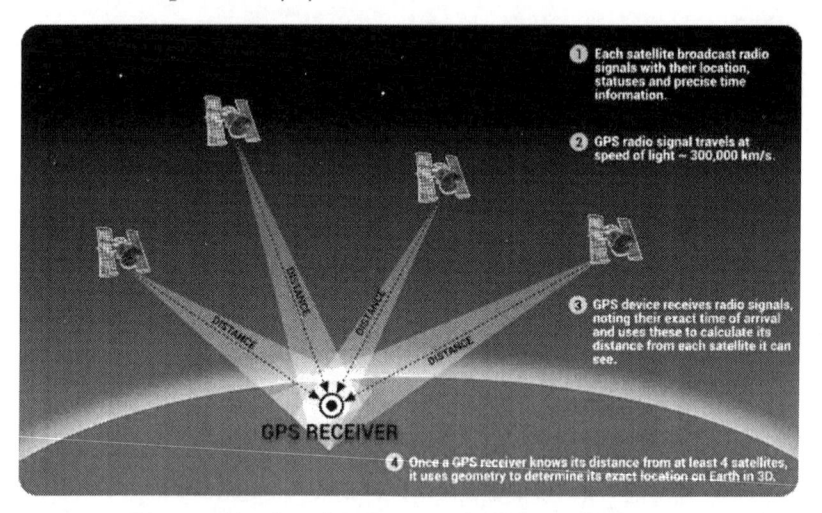

Figure 1: How satellite-based PNT systems operate—GPS as an example. (image courtesy of itsabouttimebook.com 2016 at http://itsabouttimebook.com/how-gps-works/time)

The impact of GPS on the commercial transportation industry has been enormous. Everything that moves—ships, cars, trains, aircraft, and even farm equipment—is now navigated by GPS, or a similar GNSS system. "Companies worldwide use GPS to timestamp business transactions, maintain records, and ensure traceability. Major financial institutions use GPS to" synchronize their computer networks around the world (National Coordination Office for Space-Based Positioning, Navigation, and Timing 2014). Large and small businesses now use "automated systems that can track, update, and manage multiple transactions made by a global network of customers" (National Coordination Office for Space-Based Positioning, Navigation, and Timing 2014). These systems require accurate timing information often to nanosecond levels available through GNSS such as GPS (National Coordination Office for Space-Based Positioning, Navigation, and Timing 2014).

Reliance of the Maritime Industry on GNSS

The commercial maritime industry has become especially reliant on GNSS technology. The paper chart, which has been used on the bridge of most ships in one form or another for the past several hundred years, is being rapidly replaced by electronic charts—also called eCharts. eCharts provide a continuous, real-time plot of the true and relative movements of both the vessel and nearby objects often using radar images and automatic information system (AIS) transponder signatures superimposed on the electronic chart. Figure 2 is an example of a nautical eChart. Most merchant marine academies continue to teach their cadets skillsets such as how to fix a vessel's position using terrestrial and celestial bearings. However, these techniques are less often used in the modern shipping industry, which continues to move irreversibly toward the use of fully integrated electronic bridges.[1] Yet, in the event of GNSS compromise, these basic seamanship skills may be necessary to counter a cyber attack to provide the resiliency necessary to this vital transportation system.

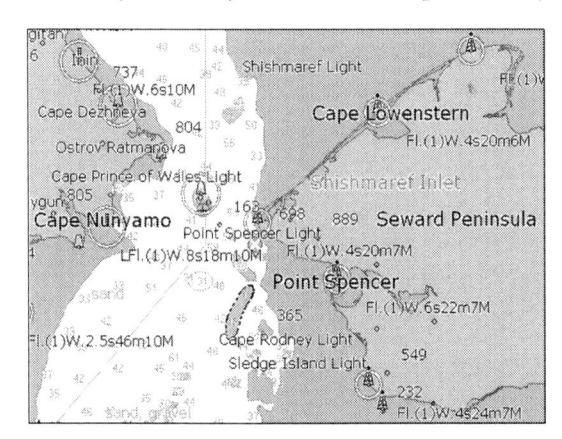

Figure 2: Sample eChart. (Ship Technology Global 2014)

---

1 An Integrated Bridge System is a combination of systems which are interconnected in order to allow centralized access to sensor information and command/control from workstations with the aim of increasing safe and efficient ship's management by suitably qualified personnel (International Maritime Organization 2016).

Several other satellite-based PNT systems are also in operation. In 1995, the same year that GPS became operational, the Russian Federation announced deployment of GLONASS. This system has been hampered by uneven funding and suffered a well-publicized 11-hour service outage in April 2014, among other failures. In May of 2016, a GLONASS expert provided a presentation on the details of this outage at the GNSS Conference in Croatia. In Asia, China has deployed its COMPASS (also known as Bei Dou) satellite navigation system. The system currently provides only regional coverage, however China has announced plans to provide global coverage by the year 2020. In Europe, the European Space Agency continues development of the GALILEO satellite navigation system. When complete, GALILEO will provide low-precision PNT services to the general public, while high-precision services will be available for a fee to commercial and military subscribers.

GNSS Signals

Signals produced by PNT satellite systems range between 1162 and 1610 MHz. Figure 3 depicts Lower and Upper L-Band frequency distribution of the four major GNSS. U.S. GPS emits two types of signals: (1) a Course Acquisition Code (referred to as the C/A code), which is broadcast on a single frequency and available free to all users; and (2) a second signal (referred to as the P(Y) code), which is broadcast on a separate encrypted frequency available only to the military. These two signals, C/A and P(Y) are equally accurate. However, the availability of the second signal on a different frequency allows the military to compensate for naturally occurring interference within the ionosphere, resulting in a more accurate fix and greater system resiliency.

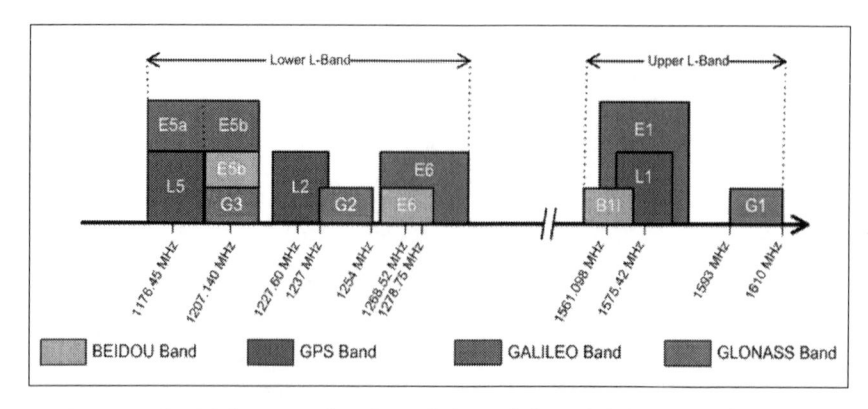

Figure 3: GNSS frequency bands, including all four global systems for 2017. (image courtesy of ExploreLabs.com 2015 at http://www.explorelabs.com/blog/designing-a-gps-receiver/)

It is important to note that GNSS pulses are extremely weak. GPS signals have been compared with the light emitted by a "40 Watt light bulb as seen from 11,000 miles away (17,700 km)."[2] As such GNSS signals are vulnerable to:

1) *Jamming and Interference.* The broadcast of a stronger signal that intentionally or unintentionally blocks or impacts a GNSS satellite signal.

2) *Spoofing.* The broadcast of a false GNSS signal, but at a slightly greater power. This deceives the GNSS receiver into locking onto the spoofed signal. Once the receiver has locked onto the stronger spoofed signal, the false signal gradually phases out of sync with the actual GNSS signal, causing the receiver to report false PNT data (information generated by the spoofer). This incremental phase out makes a spoofing attack very difficult to detect (The Mitre Corporation, 2014).

3) *Meaconing.* The intentional delay and rebroadcast of a GNSS signal intended to introduce error to receivers.

4) *Extreme Space Weather (ESW).* Solar activity such as solar flares, coronal mass ejections, high-speed solar wind, and the impact of energetic particles on the earth's ionosphere.

5) *Other Vulnerabilities.* Kinetic or laser attacks to the satellite constellations or collisions with space debris are a few of other known susceptibilities of GNSS.

Shipboard Systems Affected by the Loss of GNSS Signals

A significant proportion of navigation equipment on the bridge of a modern ocean-going commercial vessel or offshore energy platform will likely be affected by the loss of GNSS signals. Various shipboard equipment and maritime aids which might suffer impacts are identified in Figure 4.

For components listed in Figure 4, the loss of GNSS may not prevent the component from functioning through an alternate sensor input. However, tests conducted by the General Lighthouse Authorities (GLA) of the United Kingdom and Ireland in 2008 showed how easily error messages and auditory warnings prompted by the loss of GPS can easily distract (and overwhelm) a vessel's bridge team (Grant et al. 2008). This can be especially dangerous for vessels operating in confined waterways, near shallow areas, or maneuvering in higher traffic densities.

---

[2] Daniels, Charlie. 2014. Senior National Policy Analyst with Overlook Systems Technologies. (W. G. Kaag, Interviewer)

ISSUES IN MARITIME CYBER SECURITY

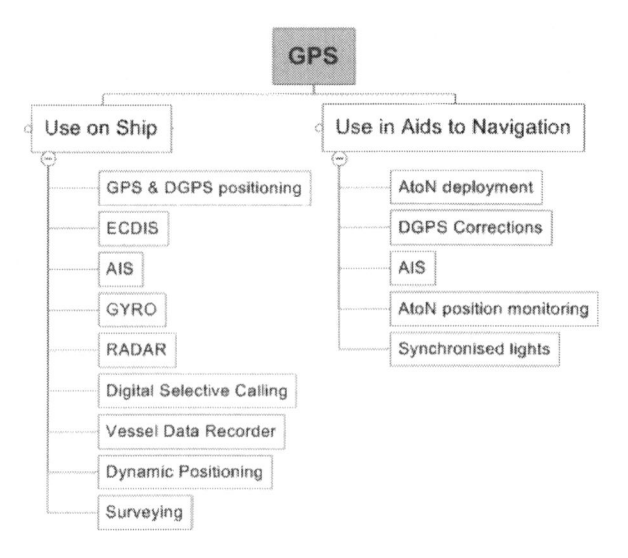

```
                          GPS
               ┌───────────┴───────────┐
         Use on Ship              Use in Aids to Navigation
               │                           │
         GPS & DGPS positioning        AtoN deployment
         ECDIS                         DGPS Corrections
         AIS                           AIS
         GYRO                          AtoN position monitoring
         RADAR                         Synchronised lights
         Digital Selective Calling
         Vessel Data Recorder
         Dynamic Positioning
         Surveying
```

Figure 4: Maritime navigation equipment that uses GPS as a data input. (Grant et al. 2008)

These vulnerabilities are not unique to the maritime industry. A number of other industries are also at risk. For instance, the aviation and financial industries are heavily dependent on properly functioning PNT systems and would be affected in varying degrees by a cyber attack on GNSS. However, largely unique to the maritime industry is that much of marine environment information transfer is via radio frequency (RF) and not a dedicated hard-line network or directional microwave dish. A good example of this type of transfer is positioning by satellite systems. Data being sent to and from shipboard computers along with other shipboard technology is cyber; therefore, interference with the data flow constitutes a cyber threat.

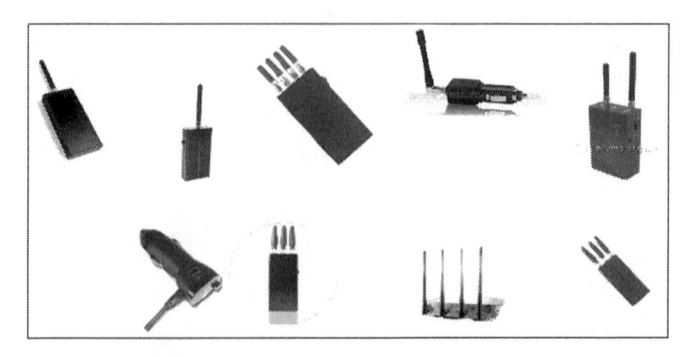

Figure 5: Small jammers that can be purchased via the Internet. Source: U.S. Government

CHAPTER 1

GNSS Jamming Equipment

With some exceptions, use of GNSS jammers is generally illegal in the U.S., Canada and Europe. Despite this, jammers of various sizes and power ratings as illustrated in Figure 5 can be purchased via the internet. These small handheld jammers are extremely difficult for law enforcement officials to locate and suppress because they can be used intermittently, disguised or hidden easily, are highly mobile, and if necessary disposed of quickly by perpetrators.

As discussed by Jones in 2011, advanced GPS receivers are more resistant to jamming than conventional designs. Receivers equipped with nulling antennas[3] are more resistant to jamming than receivers without them (Jones 2011). In Figure 6, the purple dashed horizontal line indicates receiver tolerance for obtaining the coarse acquisition (C/A) code.

As described earlier, the course/acquisition (C/A) code is the worldwide PNT signal recognized by all civilian GPS receivers; a civilian GPS receiver must first acquire (or capture) and track the C/A signal to obtain navigational coordinates. A typical GPS receiver can successfully acquire and remain locked onto the C/A signal as long as the jamming environment is below the C/A code acquisition threshold of 27 dB for 100 km. Equally important, a minimal 1-Watt interference signal at a range of 100 km can prevent a typical C/A receiver from acquiring the GPS signal (Jones 2011).

[Michael Jones, 2011]:

"Theoretically, at least, a 10-milliwatt jammer will prevent a receiver from acquiring the C/A code at a distance of 10 kilometers, and a receiver already tracking the C/A code will lose lock about a kilometer from the jammer."

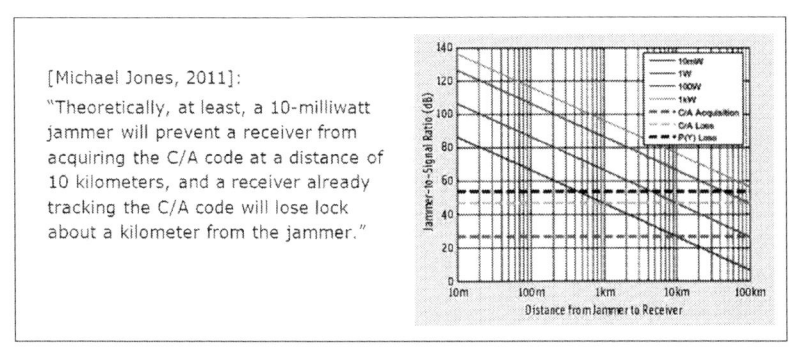

Figure 6: GPS jamming environment as a function of interference power and distance from jammer to the GPS receiver. The environment is given for four levels of interference power from 10 mW to 1kW (Jones 2011)

Figure 7 shows the area affected by a GPS jammer during tests conducted at Bridlington, U.K. along the coast of the North Sea in 2008. During the test, a jamming unit was positioned 25 m above ground level with a maximum power of 1.58 W. These tests demonstrated that relatively small jamming units can

---

3   "Nulling is a technique to reduce unwanted interference by selecting specifically against some characteristic of the interference" (LeComte, Henion and Schultz 1994).

affect GNSS reception over great distances (Grant et al. 2008).

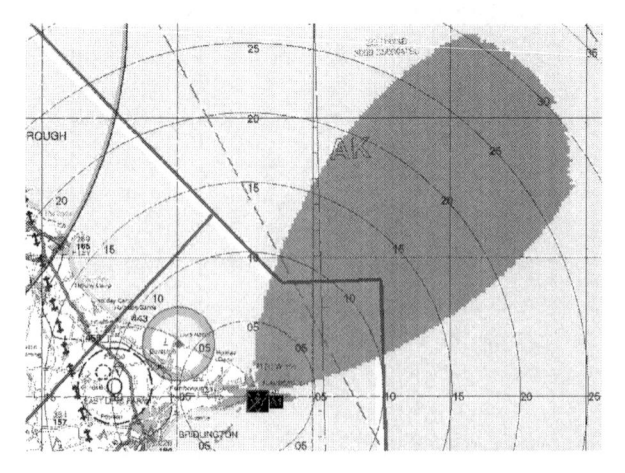

Figure 7: Coverage area of the GPS jamming unit at 25 m above ground level on maximum power of 1.58 W ERP. (Grant et al. 2008—Image courtesy of DSTL), ranges in km

THREAT SCENARIOS

At this time, the most likely GPS maritime threat scenarios to consider include:

Jamming of a port or other congested waterway by an individual or small group of non-state actors using small, portable jammers. Rapid movement of these individuals, coupled with intermittent use of the jammer(s) would make it very difficult for local law enforcement officials to track and arrest the perpetrators quickly. Attacks of this type can lead to significant economic losses as well as loss of confidence by system users.

State-sponsored GNSS Jamming. The most well-documented examples of state sponsored jamming attacks occur in the Republic of Korea (Seo and Kim 2013). The findings of Seo and Kim by attack date, jammer location, affected area and disruption are summarized in Figure 8–Table 1. On three different occasions, the Republic of Korea was subjected to intentional, high-power jamming by North Korea over a wide area. The sources of these attacks appear to have been large truck-mounted jamming units placed at strategic geographic locations. Amongst the many attacks that have been conducted, the 2012 attack affected over 1,000 aircraft and 250 ships (Seo and Kim 2013).

By June of 2016, United Press International (UPI.com) reported that North Korea had sent over 2100 jamming signals to the south since January resulting in widespread interference and disruptions to South Korea (Shim 2016).

Intentional High-Power Jamming of Republic of Korea			
Dates	August 23–26, 2010	March 4–14, 2011	August 28—May 13, 2012
Jammer Locations	Kaesong	Kaesong and Mt. Kumgang	Kaesong
Affected Areas	Gimpo, Paju, Gangwon	Gimpo, Paju, Gangwon	Gimpo, Paju, Gangwon
GPS Disruptions	181 cell towers, 15 aircraft, 1 military vessel	145 cell towers, 106 aircraft, 10 vessels	1,016 aircraft, 254 vessels

Table 1: Source: (Seo and Kim 2013)

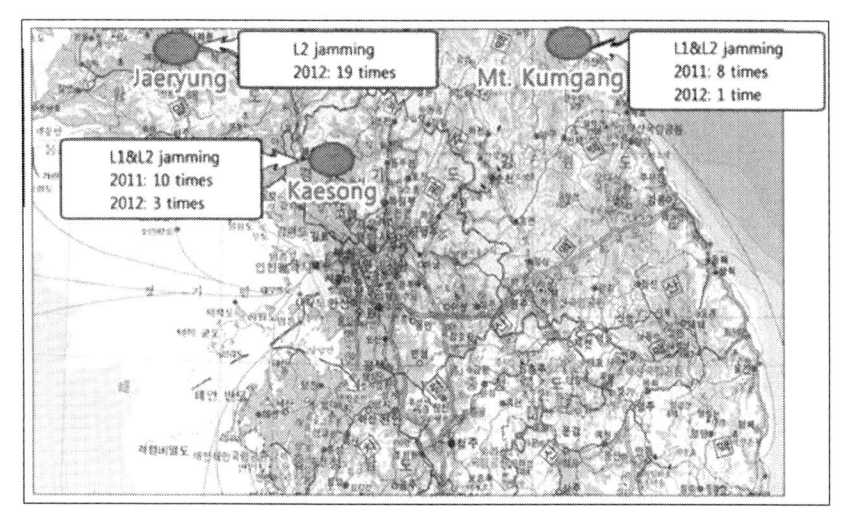

Figure 8: Location of North Korean Jammers. (Seo and Kim 2013)

State-sponsored Spoofing. Eventually, spoofing may pose a significant maritime threat to GNSS as it has the potential to lead vessels astray into dangerous waters, resulting in significant loss of life (cruise liners and ferries) or environmental damage. Presently, spoofing requires a level of technical sophistication that is normally presented through nation states. However, small groups have conducted successful spoofing tests, most notably students at the University of Texas under Professor Todd Humphreys.[4]

Primary Defenses Against Jamming

Improved Maritime Training and Education. One of the most cost-effective counter measures to defend against the intentional or unintentional jamming of GNSS signals is crew training. If trained and educated properly, commercial

---

4 University of Texas spoofing tests are viewable on YouTube, "Spoofing on the High Seas." http://www.youtube.com/watch?v=ctw9ECgJ8L0#t=38.

ship crews should be capable of operating their vessels in GNSS compromised environments. Ship crews should be taught how GNSS systems interact with ship systems and how to recognize when GNSS signals may have been compromised. The maritime industry should also be encouraged to maintain basic seamanship skills, such as dead reckoning and the ability to use piloting instruments. Routine ship drills should include signal loss and spoofing of the signal.

Improved Equipment. Development continues on new GPS receivers that can identify non-GPS signals by their relative location (jamming and spoofing signals come from the terrestrial locations not satellites) and their strength (jamming and spoofing signals must by necessity be stronger than GPS satellite-generated signals). In addition to receiver signal strength alarms and specialized antennas, the effects of intentional jamming could be mitigated through the use of inertial navigation systems (INS) and radio frequency (RF) jamming detectors. However, at this point in time it is unclear when such equipment would be available to and employed by the commercial industry, or how much it will cost.

Installation of Powerful Alternate Ground Based PNT Systems. Coastal nations most at risk should consider the installation of alternate (back-up) or complementary, land-based PNT systems, such as enhanced LORAN (known as eLoran). Both the United Kingdom and the Republic of Korea are in the process of installing eLoran systems. Seo and Kim reported that the Republic of Korea initially proposed five locations for eLoran transmitters as shown in Figure 9. The benefit of such systems is to provide PNT users with a more resilient PNT signal—one that is too powerful to be effectively jammed or spoofed. Also the low frequency of the powerful terrestrial eLoran signals permits PNT reception in GNSS denied environments. Examples include indoor structures (especially heavy buildings), underground in parking garages and basements, underwater, urban canyons, and dense foliage.

CONCLUSIONS

Worldwide dependence on Global Navigation Satellite Systems (GNSS) continues to grow. Ongoing advancements in jamming technology and the availability of small, portable jammers constitute a significant threat to maritime commerce and safety. In the face of a GNSS jamming attack, most commercial ports could be forced to suspend operations until the source of the interference is located and suppressed. It is very possible that a group of individuals operating small, portable jammers could force the closure of a major seaport or international maritime chokepoint. The economic consequences of such an attack could run into the billions of dollars.

In the long-term we also anticipate that more powerful jamming technolo-

CHAPTER 1

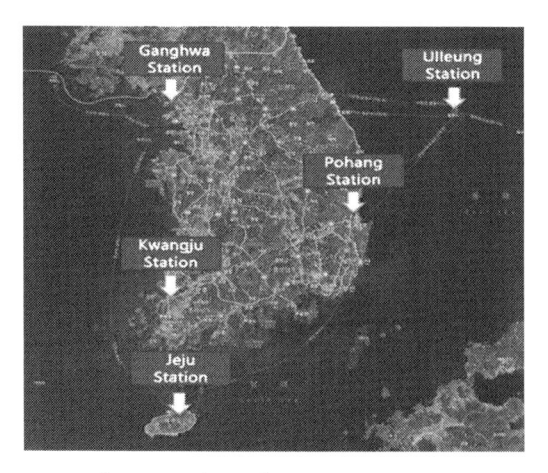

Figure 9: Location of proposed South Korean eLoran transmitters (Seo and Kim 2013)

gy and delivery systems (such as broadband jammers and drones) will become widely available and constitute two of the greatest threats to GNSS. The maritime community needs to become more vigilant, actively train to recognize and respond to both jamming and spoofing attacks, and encourage the immediate installation of complementary PNT systems such as eLoran for strategic maritime locations.

Research Tips

1. For over half a century, Loran A and Loran C provided an alternate means of positioning independent of GNSS fixing, celestial, radar or terrestrial piloting when in or near U.S. waters. In 2010 when the USCG halted its Loran C transmissions, GNSS/GPS became the de facto sole source for nearly all position fixing. For several decades, in U.S. offshore environments, Loran C had provided a confirmation of GNSS positioning that was considered critical by the USCG, NTSB and other entities concerned with vessel safety. Consider researching investigation results for groundings like the M/V ROYAL MAJESTY as far back as 1995 (Grounding of the Panamanian Passenger Ship Royal Majesty). Keywords include Loran C, Chayka, sole-source positioning, Terrestrial PNT, GPS and GNSS Vulnerabilities.

2. NAVSTAR (U.S. GPS) information including signals, infrastructure and future is provided in great detail at http://www.gps.gov/. Improvements to GPS security through upgrades in satellites, ground networks and receivers are presented. Another important U.S. site for GPS/NAVSTAR information

is the U.S. Coast Guard Navigation Center at http://www.navcen.uscg.gov/?pageName=gpsmain.

3. GNSS jamming and spoofing incidents are seldom reported as users are often unaware their receiver is experiencing interference receiving the satellite signals. Whether the interference is unintentional or purposeful, the impacts to an unwary navigator may be quite serious. Keywords to consider here are GPS/GNSS jamming, GPS/GNSS spoofing, GPS/GNSS meaconing, ECDIS, eCharts, precise positioning. Search out jamming reports such as the incident referred to in the USCG Marine Safety Alert at http://www.uscg.mil/hq/cg5/cg545/alerts/0116.pdf. The U.S. DoD's Purposeful Interference Response Team (PIRT) has been instrumental in promoting recognition and reporting of interference incidents worldwide.

4. Both GPS and GLONASS GNSS have suffered serious failures in the recent past where position fixing and/or precision timing became compromised for a large segment of users. Consider reviewing the two most recent events: GPS on January 26, 2016 and GLONASS on April 1, 2014 to gain perspective on the serious consequences that can develop.

5. eLoran Systems and eLoran Studies are searchable in, for example, private trade publications and government sites covering GNSS PNT. For instance, the DHS S&T website is reporting on precision timing by eLoran in releases like DHS S&T Demonstrates Precision Timing ... —Homeland Security which may also be sited on corporate websites like DHS and UrsaNav Successfully Demonstrate Timing Inside NYSE ... Another governmental (United Kingdom and Ireland) site that has much information on eLoran is at the General Lighthouse Authority: http://www.gla-rrnav.org/

6. Emerging technologies pose constant security challenges for all sectors including Maritime Transportation. GNSS sole-source issues continue to be a concern for other reasons as, for instance, drone technology expands. Drones can easily carry jammers and increase their efficiency by raising the jamming signal altitude and other reasons. Another looming technology, autonomous ships, will need to rely as never before on electronic positioning and specifically GNSS fixing. As the sole-source available with no navigator onboard, GNSS signals and positioning will need to leave no room for error. Consider: "The prudent navigator will not rely solely on any single aid to navigation." (USCG Notices to Mariners)

CHAPTER 1

REFERENCES

Aristova, Victoria. 2016. "GLONASS." 10th Annual Baška GNSS Conference, Croatia. Sponsored by Royal Institute of Navigation, University of Zagreb and University of Rijeka.

Grant, Allen, Paul Williams, Nick Ward, and Sally Basker. 2008. "GPS Jamming and the Impact on Maritime Navigation." http://www.navnin.nl/NIN/Downloads/GLAs%20-%20GPS%20Jamming%20and%20the%20Impact%20on%20Maritime%20Navigation.pdf.

International Maritime Organization. 2016. "Integrated Bridge System (IBS)." http://www.imo.org/en/OurWork/Safety/SafetyTopics/Pages/IntegratedBridgeSystems.aspx.

Jones, Michael. 2011. "The Civilian Battlefield: Protecting GNSS Receivers from Interference and Jamming." *Inside GNSS*. http://www.insidegnss.com/auto/marapr11-Jones.pdf.

Last, David. 2016. "GNSS/PNT/eLoran." 10th Annual Baška GNSS Conference, Croatia. Sponsored by Royal Institute of Navigation, University of Zagreb and University of Rijeka.

LeComte, Wiliam, Scott R. Henion, and Peter A. Schultz. 1994. "Adaptive Wideband Optical Nulling for an Antenna System." Proc. SPIR 2155, Optoelectronic Signal Processing for Phased-Array Antennas IV, 256.

Moskoff, David. 2014. "GPS Jammers a Top Concern in Maritime Cyber Readiness." *Professional Mariner*. http://www.professionalmariner.com/June-July-2014/GPS-jammers/.

"National Coordination Office for Space-Based Positioning, Navigation, and Timing." 2014. www.gps.gov.

Ruddock, Alan. 2013. "Sports Equipment: Is GPS the Best Route to Performance Analysis?" *Peak Performance Lite*. www.pponline.co.uk.

Seo, Jiwon, and Mincheol Kim. 2013. "eLoran in Korea—Current Status and Future Plans." http://www.govexec.com/media/gbc/docs/pdfs_edit/061813bb2.pdf.

Shim, Elizabeth. 2016. "North Korea sent 2100 GPS Jamming signals to South."UPI.com.http://www.upi.com/Top_News/World-News/2016/06/29/North-Korea-sent-2100-GPS-jamming-signals-to-South/8211467212439/.

"Ship Technology Global." 2014. www.ship-technology.com.

The Mitre Corporation. 2014. www.mitre.org.

University of Texas at Austin. 2014. "Professor Todd Humphreys' Research Team Demonstrates First Successful GPS Spoofing of UAV." *Aerospace Engineering and Engineering Mechanics.* http://www.ae.utexas.edu/news/504-todd-humphreys-research-team-demonstrates-first-successful-uav-spoofing.

USCG Marine Safety Alert 01-16. 2016. "Global Navigation Satellite Systems—Trust, But Verify, Report Disruptions Immediately." USCG Inspections & Compliance Directorate. http://www.uscg.mil/hq/cg5/cg545/alerts/0116.pdf.

U.S. Department of Homeland Security. 2016. "DHS S&T Demonstrates Precision Timing Technology at New York Stock Exchange." DHS Science & Technology Press Release. https://www.dhs.gov/science-and-technology/news/2016/04/20/st-demonstrates-precision-timing-technology-ny-stock-exchange.

CHAPTER 2:
CYBER SECURITY CONSIDERATIONS FOR THE MARITIME ENVIRONMENT

Lillian Ablon
RAND Corporation

Our world—and in particular the maritime environment—is increasingly digital, connected, on-demand, and smart. More devices are connected to networks and the Internet, and more devices have computing components—we call this the Internet of Things (IoT). The number of connected devices is growing: Gartner estimates that in 2014, there were 4.9 billion connected devices (Gartner 2014). In 2020, that number is expected to jump to 25 billion connected devices (Evans 2011). In the maritime environment, connected devices consist of everything from computerized cargo and terminal management systems, GPS and dynamic navigation systems, networked gantry cranes, RFID tracking capabilities, fiber optic security cameras, fuel farms, logistics management systems, communications, monitoring, and situational awareness systems, and many more. This Internet of Things allows for greater connectivity, accessibility, and functionality: operations can be managed more efficiently, information can be shared in real-time, and situational awareness can be maintained. For example, GPS can use the Internet to get faster acquisition time. Security cameras can provide digital "eyes" on a property. And RFID tagging and networked logistics systems provide better and more efficient management of systems and cargo.

The massive growth of networked and connected devices also increases the attack surface, enables cyber physical effects, and impacts overall security costs. This is due in large part to an increased number of connected devices, a lack of emphasis on security, and little awareness of what actually is connected.[1]

---

1 For example, a 2013 Brookings report on cyber vulnerabilities of US Port Facilities found that "landlord ports often have little awareness of what networked systems are being run by their lessees and almost no awareness of what, if any, cyber security measures are being taken to protect these systems" (Kramek 2013).

Software Vulnerabilities and the Human Element Enable a Vulnerable Environment

The pervasiveness of software vulnerabilities and the weakness of the human element are big enablers in making the cyber security environment so vulnerable.

Modern day equipment that we rely on can contain millions of lines of code, creating a large attack surface with plenty of opportunities for attackers. While code certainly exists in the digital realm (e.g., browsers and operating systems have anywhere from 8 to 40 million lines of code), code also is responsible for running components in the physical realm (e.g., civilian and military aircraft contain 14–25 million lines of code, commercial and military vehicles consist of over 100 million lines of code), and in particular the systems important to the maritime environment: for example, a 40-m vessel has more than 5 million lines of executable code; Raytheon's Total Ship's Computing Environment contains almost 7 million lines of code; and navigation systems have more than 20 million lines of code.

Software vulnerabilities are used to gain access and implement effects. Software bugs (which can turn into vulnerabilities to be exploited) remain in code: even after thorough review, approximately 1 bug exists per every 2,000 lines of code (Libicki, Ablon, and Webb 2015), and zero-day software vulnerabilities-vulnerabilities unknown to the public and for which no patch or fix exists-exist an average of 6.9 years before detection (Ablon and Bogart 2017). We can expect vulnerabilities to continue to exist and persist, especially as code reuse among vendors is common, there is no accountability or liability for bad code, and secure coding is not typically part of computer science curriculum. Furthermore, the future of security research is unknown: policy that limits or restricts the research puts everyone at risk—if no WhiteHats can explore products for vulnerabilities (to fix them), then only the BlackHats are looking for vulnerabilities (to exploit them).

Another weak aspect is the human element, and the desire for functionality and ease of use, which often leads to security holes. More people are connected to and interact with technology: whether or not they want to be and whether or not they are security aware. Attacks exploiting the human element are growing: in 2014, five out of every six large companies were targeted with spear-phishing attacks (Symantec 2014). And functionality almost always trumps security. As one example in the maritime environment, an unwitting user innocuously plugged his smartphone into a ship network to charge, unknowingly sending malware to jump to the ship's navigation systems causing all electronic charts to go dark; another example is of someone using operational wifi networks to check personal social media accounts.

CHAPTER 2

Different Cyber Threat Actors, Each with Their own Motivation and Capabilities, Can Wreak Havoc in the Maritime Environment

Several types of malicious cyber actors may want to take advantage of both software and human vulnerabilities (Robinson et al. 2013). These actors, from state-sponsored, to cyber criminals, hacktivists, and even cyber terrorists, range in motivation, but can each wreak havoc on connected technology in the maritime environment. Malicious cyber actors might attack GPS capabilities: spoofing location or jamming usage which could lead to shutting down operations at major U.S. ports. They might carry out low-level hoaxes or Distributed Denial of Service (DDoS) attacks—attacks that are cheap to carry out and remediate, but expensive when limited resources get tied up and bigger attacks can go unobserved. Data can be manipulated or corrupted: in 2013, drug smugglers hacked into container terminals in Rotterdam, and changed the location and delivery times of containers that had drugs in them to avoid detection. Cyber threat actors could manipulate data or information to allow illicit activity to go unnoticed, or to alter weather data causing rerouting of vessels into bad weather. This is especially concerning, because weather is an important factor for those in the maritime environment. And threat actors could potentially destroy SCADA or ICS components leading to effects that are physically damaging.

Addressing Maritime Cyber Threats is a Critically Important, Complex, and Difficult Problem

Cyber security—keeping systems up and running, and protected against cyber threats, is important to the maritime environment. A huge part of the nation's commerce and economy relies on the maritime environment: maritime trade is more than 30% of the U.S. global domestic product, and the 360 ports that are part of the Maritime Transportation System account for 95% of U.S. trade, and $1.3 trillion in economic activity.

The maritime environment is increasingly dependent on critical infrastructure made up of computer networks, connected devices, and millions of lines of code—and will continue to be more and more reliant. Events that happen on cyber infrastructure can have real-world impact: manipulating weather data, spoofing GPS coordinates, denying or degrading information, or destroying SCADA or industrial control system components, etc. which could all lead to damaging effects. No longer do cyber security effects live in just the digital realm: it is not just crashing code, but crashing vehicles and vessels.

One aspect that makes securing the maritime environment especially difficult is that it involves both public and private stakeholders. This raises challenges of securing infrastructure that the Coast Guard does not own; operating on open networks (i.e., walled off networks like SIPRNet are not always options); convincing private sector stakeholders to adhere to cyber norms (e.g., conducting vulnerability assessments and developing remediation and resilience plans); and persuading third parties to invest in cyber security tools, training, and defenses—largely because cyber security is often seen as a cost, rather than an investment. Furthermore, current security focuses on physical security, instead of cyber security. This may be because it is difficult to see the gains of cyber security—no attacker will tell a target when they were unsuccessful due to good defenses and protection, and because cyber security gear, tools, and devices can be costly—estimated worldwide spending on cyber security is approaching $70 billion per year, growing at roughly 10%–15% annually. Lastly, tools to hack are cheap (e.g., $300 buys you a radio tool to emulate GPS and falsify GPS locations), discovery of a breach or attack can be slow (FireEye (2015) estimates it takes 205 days to discover a compromise, and Verizon (2014) found that 70%–80% of companies discovered from a third party that they had been breached), and attribution is difficult (especially with the increased use of anonymizing networks like Tor and I2P, encryption capabilities, and virtual private networks), making cyber attacks easy and accessible for cyber threat actors.

What Can Be Done?

Securing the maritime environment requires solutions in both science and technology (S&T) and policy. The U.S. Coast Guard's cyber strategy (2015), was a positive first step. Next steps should include implementing it, having more awareness, and starting to take cyber security seriously across all areas. There are several areas of focus in order to develop these solutions.

First, developing ideal and appropriate cyber security investment and acquisition processes. The ideal is to incorporate security early on—that is, ensure that measures like secure infrastructure and encryption systems are "baked in" from the start, rather than "bolted on" as an afterthought, as well as keep up with advances in technology—especially those related to the Internet of Things. Investments may be in non-physical areas, such as setting up appropriate ways to share information about threats and security practices with others in the maritime and cyber security community, and building and retaining a capable cyber work force. Determining and keeping up with the ideal investment level and acquisition process is a challenge, and will continue to be an even bigger challenge as we become more dependent on networks and technology, and because there is no-one-size-fits-all solution. Evaluation is needed on what the right balance of investment is, and in what areas.

Second, ensuring infrastructure is protected and stays up and running—especially infrastructure owned by third parties. This infrastructure will likely become more vulnerable as more devices become connected, increasing the avenues for possible attack. A joint government–industry–academia approach may be helpful here. There is likely duplication of effort, and thus efficiencies to be gained. A possible first place to evaluate could be a comprehensive update to the Maritime Transportation Security Act. Other solutions could involve developing cyber response capabilities, given a compromise has occurred.

Third, standardizing and prioritizing risk assessments, to include recovery and resilience plans after an incident has occurred—that is, understanding the consequences of an attack, and how to recover after an incident has occurred. This includes systematically identifying (and keeping track of) systems, networks, and data that are critical to Coast Guard missions and the maritime environment. Risk assessments, and risk mitigation plans should become a priority, as should the understanding that deception and manipulation by adversaries is possible.[2] Kramek (2013) found that of the top six ports in the United States, only one of them (the Port of Long Beach—had performed a vulnerability assessment, built their own cyber infrastructure, and invested $1 million in COTS cyber tools, to include backups and redundant measures; but they did not have a cyber response plan, so maintaining continuation of operations and resilience after an attack may be ad hoc. Risk consists of understanding threats (and those threat actors that may want to do harm to the maritime environment), vulnerabilities, and possible impact or consequences of an attack. Understanding the potential impact means exploring all possibilities for damage, beyond just the technical: disruptions could have ripple effects on the economy or could affect safety and security with adversaries seeing us as an easier target. It is important to run through a variety of scenarios in order to understand what resources one should ask for after a cyber attack occurs. Thus, information sharing efforts among both public and private stakeholders should include all three of these elements. And once risk assessments have been conducted, it may be crucial to ensure that private sector players adhere to the standards set.

Fourth, reducing software vulnerabilities; and developing appropriate liability policies, for code libraries used by traditional computing components, as well as newly connected IoT devices. This could be done through encouraging security research as well as bug bounty programs; creating a culture to bake in security from the start with secure design and architecture, and considerations

---

[2] The National Institute of Standards and Technology (NIST 2012) provides an outline for conducting risk assessments.

for liability legislation—holding companies responsible for bad code. There should also be a culture of cyber security: all employees should have regular security awareness training, and companies should consider hiring red teams to test the vulnerability of software as well as the human element. Limited resources may mean constraints on being able to hire, train, and retain capable workforce, as well as have a dedicated IT and security staff. As such, considerations may be desired for outsourcing "as a service" options, isolating parts of one's network, and restricting BYOD and device usage.

Research into these areas of cyber security investment and acquisition, securing infrastructure, standardizing, and prioritizing risk—to include resilience measures, and reducing software vulnerabilities could significantly contribute to filling the S&T needs and crafting better policy solutions to secure the maritime environment.

References

Ablon, Lillian and Andy Bogart, *Zero Days, Thousands of Nights: The Life and Times of Zero-Day Vulnerabilities and Their Exploits*, Santa Monica, Calif.: RAND Corporation, RR-1751-RC, 2017. As of March 09, 2017: http://www.rand.org/pubs/research_reports/RR1751.html.

Evans, Dave. 2011. *The Internet of Things: How the Next Evolution of the Internet is Changing Everything*. CISCO White Paper 1: 1–11. http://www.cisco.com/c/dam/en_us/about/ac79/docs/innov/IoT_IBSG_0411FINAL.pdf.

FireEye. 2015. "M-Trends 2014: A View from the Front Lines." https://www2.fireeye.com/rs/fireye/images/rpt-m-trends-2015.pdf.

Gartner. 2014. "Gartner Says 4.9 Billion Connected 'Things' Will Be In Use in 2015." *Gartner*. http://www.gartner.com/newsroom/id/2905717.

Kramek, Commander Joseph. 2013. *The Critical Infrastructure Gap: US Port Facilities and Cyber Vulnerabilities*. The Brookings Institution. http://www.brookings.edu/research/papers/2013/07/03-cyber-ports-security-kramek.

Libicki, Martin C., Lillian Ablon, and Tim Webb. 2015. *The Defender's Dilemma: Charting a Course Toward Cybersecurity*. Santa Monica, CA: RAND Corporation. http://www.rand.org/pubs/research_reports/RR1024.html.

CHAPTER 2

NIST. 2012. *Guide for Conducting Risk Assessments.* NIST Special Publication: 800-30-Rev1. http://csrc.nist.gov/publications/nistpubs/800-30-rev1/sp800_30_r1.pdf.

Robinson, Neil, Luke Gribbon, Veronika Horvath, and Kate Robertson. 2013. *Cyber-security Threat Characterisation: A Rapid Comparative Analysis.* Santa Monica, CA: RAND Corporation. http://www.rand.org/pubs/research_reports/RR235.html.

Symantec. 2014. "Internet Security Threat Report." Symantec.com. http://www.symantec.com/content/en/us/enterprise/other_resources/b-istr_main_report_v19_21291018.en-us.pdf.

U.S. Coast Guard. 2015. "United States Coast Guard Cyber Strategy." https://www.uscg.mil/seniorleadership/DOCS/cyber.pdf.

Verizon. 2014. "2014 Data Breach Investigations Report." http://www.verizonenterprise.com/DBIR/2014/reports/rp_dbir-2014-executive-summary_en_xg.pdf.

CHAPTER 3:
U.S. COAST GUARD CYBER THREAT INFORMATION SHARING

Jennifer M. Konon
U.S. Coast Guard

Abstract

This study sought to identify the conditions associated with cases of effective cyber threat information sharing from the Coast Guard to marine transportation system partners, to aid in the development of a framework for cyber information sharing. Using a phenomenological approach, interviews of Coast Guard intelligence personnel provided data about their use of information sharing mechanisms and the conditions that might affect their sharing of cyber threat information with marine transportation system partners. Findings showed that stronger relationships encouraged information sharing, as did strong leadership support and the establishment of Area Maritime Security Committee cyber subcommittees. Perception of risk, including concern regarding sharing sensitive information, discouraged information sharing. Subjects who did not share cyber threat information most frequently referred to a lack of policy or training guiding them in this sharing. This study offered recommendations for improving Coast Guard cyber threat information sharing with marine transportation system partners.

Introduction

Timely dissemination of actionable information is critical to enable mitigation of vulnerabilities and threats. With our nation's infrastructure increasingly reliant on functioning and secure cyber networks and cyber-based systems, the sharing of cyber threat information[1] between government agencies and private sector partners has become paramount (McCarthy et al. 2009). This has led to the call for improvements in cyber threat information sharing between the federal government and the private sector. One focus area is the U.S. Coast Guard's performance in sharing cyber threat information with its partners in the marine transportation system, including owners and operators of ports, waterways, vessels, bridges, facilities, and their intermodal connections (US Government Accountability Office [GAO] 2014).

While the general topic of information sharing has been studied fairly extensively, very few of these studies focus on government agencies sharing information with the private sector. Roughly half of the studies reviewed for this research focused on information sharing within a single company or organization (see Liebowitz and Chen 2003; Huysman and de Wit 2004; Liao et al. 2004; Barua, Ravindran, and Whinston 2007; Hatala and Lutta 2009). The other half looked at information sharing across organizations within the same sector, commonly referred to as cross-boundary information sharing. The majority of these studies focused on single-level cross-boundary information sharing at either the local, state, or federal government level (see Dawes 1996; Liu and Chetal 2005; Luna, Gil, and Betiny 2006; Pardo et al. 2006; Pardo, Gil-Garcia, and Burke 2008). Among the few studies on information sharing between government and private entities, the University of Washington's 2014 report entitled "Maritime Operational Information Sharing Analysis" is most relevant to this research, which is focused on Coast Guard information sharing with private sector partners. The University of Washington studied safety and security information sharing across the entire maritime sector in the Puget Sound region, including federal, state, local, and private partners. As of the date of this publication, however, there have been no studies specifically focused on cyber threat information sharing between the Coast Guard and its marine transportation system partners. Using a phenomenological approach, this research sought to identify the conditions associated with cases of effective cyber threat information sharing from the Coast Guard to marine transportation system partners, with the goal of assisting the Coast Guard in improving its overall cyber threat information sharing efforts.

The following section briefly outlines the Coast Guard's role in maritime infrastructure cyber security, the mechanisms the Coast Guard uses to share cyber threat information with private sector partners, and the need for the Coast Guard to improve this sharing. The next section describes this study's

---

1   For the purposes of this research, the term *cyber threat information* refers to raw information or finished intelligence about techniques or exploits made through cyberspace "that may adversely affect computers, software, a network, an organization's operation, an industry, or the Internet itself" (Government Accountability Office 2012).

overall research design and methodological approach. Following this, a discussion of relevant literature and the theoretical justification for each of conditions considered in this study, as well as the outcome. The findings of this research then are provided, along with the conclusions drawn therefrom. The final section offers recommendations for further research.

BACKGROUND

The nation's marine transportation system relies significantly on cyber-based systems for both control and management functions. Networked logistics management systems track cargo from its origin to destination, allowing opportunistic criminals to track their smuggled cargo virtually undetected, or even permitting pirates to identify the precise location of valuable cargo onboard ships to expedite their illegal operations (Kramek 2013). Highly automated industrial control systems are becoming more common throughout maritime ports, ranging from systems that control pumps and valves governing fuel transfers to gantry crane systems that offload shipping containers. Degradation of these systems could result in physical effects leading to injury or death, destruction of the marine environment, and significant disruptions to commerce (Department of Homeland Security [DHS] 2016).

The Coast Guard's role in protecting the marine transportation system is firmly established in the organization's existing authorities. The 2002 *Maritime Transportation Security Act* outlined the Coast Guard's authority and responsibility to ensure the security of the marine transportation system against all hazards and threats. While focus has traditionally been on physical security, the growth in cyber vulnerabilities and threats has led to a natural extension into the realm of cyber security. The *National Infrastructure Protection Plan* designated the Coast Guard as the Sector-Specific Agency for Maritime Transportation, outlining the Coast Guard's obligation to protect the marine transportation system from both physical and cyber security threats (DHS 2009). Protecting the marine transportation system against cyber threats is one of the primary goals of the Coast Guard's Cyber Strategy (US Coast Guard [USCG] 2015).

Information sharing between the government and private sector is vital to enabling the protection and resilience of our nation's critical infrastructure. This fact is highlighted in the 2012 National Infrastructure Advisory Council's report *Intelligence Information Sharing*:

> Information on threats to infrastructure and their likely impact underlies nearly every security decision made by owners and operators, including which assets to protect, how to make

operations more resilient, how to plan for potential disasters, when to ramp up to higher levels of security, and how to respond in the immediate aftermath of a disaster (1).

To further establish the role of federal agencies in cyber security of critical infrastructure, Executive Order 13636 and Presidential Policy Directive 21, both issued in February 2013, directed Sector-Specific Agencies to assess critical infrastructure vulnerabilities, assist infrastructure owners and operators in strengthening their infrastructure against cyber threats, and aid in incident response. These documents emphasized the role of federal agencies and private industry partners in forming strong collaborative relationships and sharing information to protect infrastructure against cyber threats. In December 2015, the Cyber Security Information Sharing Act (Public Law 113-114) directed the Secretary of Homeland Security, in consultation with the heads of appropriate federal entities (e.g., the Coast Guard), to develop and enact procedures to promote timely sharing of cyber threat indicators, cyber security best practices, and defensive measures with non-federal entities such as private and public critical infrastructure owners and operators.

The Coast Guard uses several mechanisms to facilitate sharing cyber threat information with maritime private sector partners. At the Sector level, Area Maritime Security Committees (AMSCs) provide a forum for timely sharing of threat information with marine transportation system owners and operators, law enforcement and intelligence agencies, emergency responders, and port managers (GAO 2006). At the national level, the Coast Guard representation at the DHS National Cyber Security and Communications Integration Center enables continuous engagement with private sector entities sharing information related to cyber threats and incidents (The White House 2015). Additionally, the Coast Guard maintains the secure Internet portal "Homeport" to share sensitive information with marine transportation system partners. Despite the existence of these lines of communication between the Coast Guard and its private sector partners, a 2014 Government Accountability Office report indicated that these mechanisms have proved insufficient for effective sharing of cyber threat information. The report found a lack of consistent and active use of the mechanisms in place to share such information. An October 2015 update to this report re-iterated the need for the Coast Guard to improve its cyber threat information sharing, noting that private sector maritime stakeholders still lacked a national-level mechanism for such information exchange (GAO 2015). The Coast Guard recognizes the challenge of creating a framework for effective cyber threat information sharing. The 2015 *Coast Guard Cyber Strategy* outlined this goal:

CHAPTER 3

> The Coast Guard must use intelligence and investigative capabilities, and leverage the Intelligence Community and law enforcement partners, to enhance mission performance. To this end, the Coast Guard will ... develop policies and guidelines for ensuring the rapid dissemination of cyber threat indicators to maritime partners and stakeholders to enhance the safety, security, and resiliency of the Marine Transportation System (28).

By studying conditions associated with cases of effective cyber threat information sharing from the Coast Guard to marine transportation system partners, this research sought to assist the Coast Guard in improving its cyber threat information sharing efforts.

Research Design and Methodological Approach

This study used a phenomenological approach to gain insight into and draw patterns describing how interview subjects experienced the process of cyber threat information with their private sector partners. The interview subjects consisted of full-time intelligence personnel working at the "Sector level" of the Coast Guard, either at Coast Guard Sectors or Marine Safety Units. The units at the Sector level of the Coast Guard are operational units that fall under Coast Guard Districts and oversee subordinate units such as Small Boat Stations. They may cover one to ten U.S. states depending on the operational tempo and the geographic area. As part of their duties, intelligence personnel at these units share threat information with private sector partners, including cyber threat information with marine transportation system partners. Of the potential 38 subjects, 34 subjects completed interviews (an 89.4% participation rate). Interview questions asked about conditions potentially affecting subjects' cyber threat information sharing with marine transportation system partners, as well as the extent and nature of such sharing (i.e., the frequency and specific types of mechanisms used). Questions were designed to collect data about the conditions affecting information sharing and determine the presence or absence of the outcome of effective information sharing for each case. Existing theory informed the choice of conditions and outcome, which are described in detail in section 4. Interviews were semi-structured and consisted of structured questions intermixed with unstructured, follow-on questions used to gain clarification or further insights. In the event that a subject was not the sole intelligence member working at a unit (occurring in 29 cases), they were instructed to answer some questions in a way that represented the intelligence staff's activities as a whole. For example, when asked about sharing information using formal mechanisms, the subject was asked to consider not only his actions but also the actions of other full-time intelligence members of the staff.

The first interview question asked whether the subject received cyber threat information from any source. The goal of this questioning was to establish a baseline and determine whether it would be appropriate to proceed with the remainder of the interview. After all, in order to share cyber threat information, the subject would have to possess such information in the first place. However, even if subjects replied that they did not receive any cyber threat information, this would not mean that they had no information to share. Coast Guard intelligence personnel are also expected to collect information through their own research or observations, and therefore could also obtain cyber threat information even if they did not receive this information from "push" sources. Hence, a negative response to this question did not obviate the rest of the interview; rather, simply called for additional follow-on questions to clarify whether the subject had any cyber threat information to share. In all cases, subjects responded that they either received cyber threat information from outside sources or obtained such information through their own research. Thus, all 34 interviews proceeded in full.

This study used a phenomenological approach to gain insight into and draw patterns describing how subjects experienced the process of cyber threat information with their private sector partners. Prior to data collection, the conditions and the outcome to be considered were identified based on the current information sharing theory. The size and nature of this study, consisting of in-depth interviews with 34 Coast Guard individuals possessing direct experience in information sharing with private sector partners, was well-suited for the phenomenological approach. This approach generally seeks to provide a researcher with better understanding of others' experiences with a social phenomenon by studying multiple perspectives on the topic (Leedy and Ormrod 2010).

Limitations in Scope

Certain limitations of this research design are important to note. The scope of this research was limited to one direction of information flow from Sector-level Coast Guard intelligence personnel to their marine transportation system partners, rather than from Coast Guard personnel in other specialties or working at other levels of the Coast Guard. Any extrapolation of the findings and conclusions of this research to other areas of the Coast Guard must therefore keep this in mind. Additionally, since the interview subjects consisted solely of Coast Guard personnel, the responses regarding the relevance and timeliness of the information shared did not represent whether the recipients felt the information was timely or relevant. The necessarily limited scope of the research design made this bias unavoidable.

Limitations in Methodology

There are also limitations related to the methodology used in this research. First, any conclusion from this analysis depends on the selection of cases and conditions. Ample research into existing information sharing theory validated the selection of conditions, but there were certainly additional relevant conditions that may have been overlooked. Similarly, any conclusions drawn are limited to the cases actually studied. Finally, one must be careful to avoid the impression that this research proves any "cause and effect" relationship; rather, the analysis merely approximates causes based on the conditions and cases studied.

INFORMATION SHARING: CONDITIONS IDENTIFIED BY THEORY

The concept of information sharing is characterized in various ways throughout the literature. At times, "knowledge" is used interchangeably with "information;" however, a more precise definition helps to maintain clarity. According to Wilson (2010), knowledge often implies "a set of mental processes involving understanding and learning" (3). What someone actually shares is the information about what she knows, and not knowledge itself, which is formed by each recipient through his individual learning process. Studies on information sharing usually assume an element of reciprocity, where two or more parties each receive information from and give information to each other (Wilson 2010; Pilerot 2012). However, this study focused solely on one direction of information sharing between the Coast Guard and private sector partners; namely, Coast Guard personnel giving information to private sector recipients. Research about the other direction of this information sharing, i.e., the Coast Guard receiving cyber threat information from its private sector partners, is currently being researched by others (including Egan et al., to appear).

Previous research has shown that conditions or factors affecting information sharing are highly contextual in nature, thus supporting a qualitative methodological approach that takes this into consideration. According to Pardo et al. (2006), "achieving an understanding about how these factors influence knowledge sharing is difficult because knowledge in practice is intimately linked to context. Examining knowledge sharing processes requires attention to how particular factors present in different contexts" (296). As Fedorowicz, Gogan, and Williams (2007) elaborate further, information sharing is influenced both by the external environment and the internal environments of the entities sharing information. This contextual nature of information sharing makes it a topic well-suited for a phenomenological approach that assesses each case in a qualitative fashion.

The list of conditions that may affect information sharing, either by facilitating or posing barriers to sharing, is long and varied depending on the organizational context of who is sharing and the mode of sharing. The remainder of this section outlines the conditions and outcome chosen for this research.

Relationships

According to a University of Washington (2014) study, "when it comes down to it, it is all about relationships" (2). This conclusion was drawn by researchers who conducted an ethnographic exploration of maritime operational information sharing between the Coast Guard and other partners, including those in the private sector. When asked about receiving information from their Coast Guard partners, marine transportation system operators claimed that strong relationships, built through face-to-face interactions at meetings and other routine operations, helped ensure information sharing during emergency incidents.

The literature abounds with evidence that information sharing relies largely on relationships, with stronger relationships facilitating increased sharing (Carlin and Womack 1999; Liao et al. 2004; Li and Lin 2006). Hatala and Lutta (2009) pointed out that relationships are used not only to share information, but also to identify those who have information to share. Other studies have found that information sharing helps strengthen inter-organizational relationships, which in turn encourages further sharing (Carlin and Womack 1999; Yang and Maxwell 2011).

One of the key barriers to building strong relationships between Coast Guard personnel and their private sector partners lies in the high turnover rate of Coast Guard members. The average length of time that a Coast Guard intelligence member stays at a Sector or Marine Safety Unit is 3 years, after which time he or she will be re-assigned to a different location and replaced by a member coming from a different unit. In contrast, marine transportation system owners and operators typically remain in in their position for considerably longer, sometimes decades. As noted by the aforementioned University of Washington study, this personnel turnover often creates gaps in relationship-based information sharing, as it can be quite difficult to maintain relationships through such transitions. On the other hand, in their study of information sharing within an organization, Hatala and Lutta (2009) found that turnover of individuals had little impact on overall information sharing once initial social network structures had been established.

Four interview questions were designed to ascertain whether the subject or a member of their staff had established strong relationships with marine transportation partners. These questions asked Coast Guard members how long they had been at their current unit, whether anyone on their staff had been working there for more than 5 years, and if so, whether that person had established relationships with local marine transportation system partners. Subjects were also asked whether the quality of their relationships with marine transportation system partners affected their sharing of cyber threat information with these partners.

Trust

The literature reveals trust as one of the most important factors affecting information sharing. Within the context of information sharing, trust is characterized by emotional bonds, shared values or goals, goodwill, and integrity among entities sharing information (Luna, Gil, and Betiny 2006). According to Pardo, Gil-Garcia, and Burke (2008), it is important in any type of collaboration. An abundance of studies have found that partners who trust each other are more likely to share information (Nahapiet and Ghoshal 1998; Dyer and Chu 2003; Li and Lin 2006; Barua, Ravindran, and Whinston 2007; Willem and Buelens 2007; Wu, Hsu, and Yeh 2007; Choi, Kang, and Lee 2008; Wilson 2010; University of Washington 2014). Specifically, Pardo et al. (2006) found that trust served "as the foundation for forming shared mental sets and establishing an unobstructed channel for communication" (305). University of Washington researchers (2014) found that trust between partners promoted positive views about the validity of the information, thus encouraging more sharing. Trust was also a frequent prerequisite to formal inter-agency information sharing agreements.

Research has found trust to be vital to information sharing during critical or emergency incidents, although different studies found that trust and information sharing affected each other in varying ways. The 2014 University of Washington study found that information sharing during incidents relied heavily on the trust maritime partners had previously established during routine operations. On the other hand, Ibrahim and Allen (2012) discovered that during emergency response incidents, proactive information sharing actually helped rebuild degraded trust in a relatively quick fashion.

Several studies on trust and information sharing are particularly relevant to the research on hand. For example, Liu and Chetal (2005) showed trust to be a necessary precursor to sharing sensitive information. This observation is particularly pertinent to this study, as cyber threat information is often quite sensitive, due to its potential revelations about vulnerabilities or its derivation from classified information that could damage national security interests if revealed. Additionally, Dodgson (1993) reported that inter-organizational information sharing can survive personnel turnover (a common issue with Coast Guard members) if broad-based trust is established within the cultures of the organizations.

Copious evidence throughout the literature showed that trust affects information sharing; thus, trust was included as a condition in this study. Coast Guard subjects were asked if they trusted their marine transportation system partners as a whole.

Understanding of Role

While not as commonly studied as other factors affecting information sharing, such as relationships, leadership, and trust, the extent to which an individual understand their role in information sharing has nevertheless been shown to be relevant. Several researchers showed information sharing hindered due to lack of clearly defined roles and responsibilities for sharing. Direction for whom to share with, what to share, and how to prioritize information sharing was found necessary for effective information sharing (Small and Sage 2005/2006). Others found that clear definitions of information sharing roles fostered trust between partners, which in turn promoted information sharing (Pardo, Gil-Garcia, and Burke 2008; University of Washington 2014).

To establish the effect of understanding one's role in information sharing, interviews inquired whether the Coast Guard subjects understood their role in sharing cyber threat information with marine transportation system partners.

Understanding of Information

Another less-commonly studied condition of information sharing is the extent to which an individual understands the information to be shared. A lack of understanding may delay or inhibit a person from sharing that information. As Yang and Maxwell (2011) describe, an individual may feel that they need to dedicate great time and effort to learn more about the information before they can share it effectively. Additionally, they may worry that sharing the information will lead the recipients to request more information, or ask questions that they are not prepared to answer; in either case, they may decide to simply keep the information to themselves to avoid extra work or embarrassment.

Previous studies revealed that individuals more confident in their understanding feel greater ownership of the information, and therefore more readily share it (Constant, Kiesler, and Sproull 1994; Hatala and Lutta 2009). Additionally, even if an individual does not fully understand the information, they may know whom to contact if faced with difficult questions about the information. Knowing who has the needed information is referred to as *transactive knowledge*, and can help encourage non-experts to share information that they do not fully understand themselves (Brauner and Becker 2006).

Coast Guard subjects were asked whether they thought they had sufficient knowledge or understanding of cyber threat information to be able to provide the "so what and what does this mean to me" context for their partners.

Risk

A person's perception of risk regarding information sharing has been shown to negatively impact sharing. Information sharing may be considered risky due to the information sharer's lack of control over how the recipients use or further disseminate the information (Dawes 1996; Pardo, Gil-Garcia, and Burke 2008; Yang and Wu 2014). Yang and Wu (2014) point out that individuals may also hesitate to share information if they are concerned about its quality; i.e., they may only share information that they believe is accurate and complete. According to Pardo et al. (2006), unless such risks are mitigated effectively, information sharing may be prevented or limited.

A subject's perception of risk was included in this study as a condition of the outcome. Coast Guard subjects were asked whether there were any risks or if they had any concerns regarding sharing cyber threat information with their marine transportation system partners.

Leadership Support

Several studies have highlighted the role of leadership in providing vision, support, and guidance toward information sharing (see Small and Sage 2005/2006; Li and Lin 2006; Zhang, Faerman, and Cresswell 2006; Pardo, Gil-Garcia, and Burke 2008; Yang and Wu 2014). Carlin and Womack (1999) stressed a leader's part in promoting information sharing by serving as a positive model in this regard. Yang and Maxwell (2011) described how leadership may affect information sharing in a variety of ways, by providing resources, encouraging employee participation, or clarifying employee information sharing roles and responsibilities.

This study asked subjects whether their chain of command supported their sharing of cyber threat information with marine transportation system partners.

Policies

Many organizations have policies that regulate information sharing activities. The effect of such policies receives a fairly dichotomous review in the literature, with most researchers acknowledging both positive and negative effects. On the positive side, policies can help clarify how to share information and to whom it should be shared, empowering personnel to engage in information sharing practices. Policies can also impose an obligation to share information (Yang and Wu 2014). Organizations reluctant to share information may benefit from policies that explicitly authorize such activity. However, as Dawes (1996) points out, employees may want policies that allow but do not mandate information sharing.

On the other hand, several studies have shown policies to constrain information sharing (Zhang and Dawes, 2006; 6 et al. 2007; University of Washington 2014; Yang and Wu 2014). Kim and Lee (2006) proposed that policies may prevent the formation of information-sharing communities or restrict the sharing of sensitive information. Policies may indirectly discourage information sharing by neither explicitly allowing nor disallowing it, leading employees to "play it safe" by keeping information to themselves (Yang and Wu 2014). As one marine transportation system representative put it, "there are good reasons to do work informally ... If you make certain things policy, your hands are tied to work and share information in a certain way" (University of Washington 2014, 49). In essence, an increase in formal policies does not always lead to an increase in information sharing.

Interviews inquired whether subjects were aware of any currently existing Coast Guard policies or doctrine that support cyber threat information with marine transportation system partners. In fact, it was already known that the Coast Guard is still working on developing policies relating to cyber threat information sharing, and no such policies have yet been implemented. The intent of such questioning was to ascertain the subjects' perceptions of the existence of such policies, rather than to identify the actual existence of policies.

Outcome: Effective Information Sharing

One of the aims of this research was to identify the combinations of conditions associated with "effective" sharing of cyber threat information from the Coast Guard to marine transportation system partners. The outcome of *effective information sharing* was defined as routine sharing of timely and relevant information. This characterization was based on the literature's portrayal of effective information sharing. In several studies, the term *effective* was seamlessly interchanged with *successful, quality,* or *productive* (e.g., see Drake, Steckler, and Koch 2004; Yang and Maxwell 2011). Effective information sharing was often paired with *timeliness* and *relevance* (Mohr and Sohi 1995; Drake, Steckler, and Koch 2004; Li and Lin 2006; Fawcett et al. 2007; Hsu et al. 2008; Carminati, Ferrari, and Guglielmi 2011). Several researchers also referenced the value and accuracy of the information (Mohr and Sohi 1995; Fawcett et al. 2007; Yang and Maxwell 2011). The assessment of these latter two qualities was beyond the scope of the present research, and therefore not included.

TIMELINESS AND RELEVANCE

The final two interview questions asked the Coast Guard members if they considered the cyber threat information they shared with marine transportation system partners to be timely and/or relevant. For information sharing to be considered timely, it had to be received or acquired by the Coast Guard sharer in a timely fashion, and then likewise shared with marine transportation system partners. Timely sharing of information that was already outdated when the Coast Guard member received it was not considered timely overall. Subjects were also asked if they considered the information they shared to be relevant. For the qualities of timeliness and relevance, no definition of the terms was provided to the subjects. Rather, each subject was free to interpret the terms based on his own.

FREQUENCY OF SHARING

Some researchers also evaluated information sharing based on *information flow* or the frequent updating of shared information. These researchers did not, however, further define this concept (Zhou and Benton 2007; Yang and Maxwell 2011). In this study, the outcome was caveated with a consideration for how routinely the information sharing occurred. To characterize whether information sharing in each case was routine, subjects were asked about the mechanisms they used to share information; specifically, formal, informal, and web-based Information Technology (IT) systems. Interview questions asked the Coast Guard members if they used formal mechanisms, such as established groups or committees, to share cyber threat information with their marine transportation system partners; informal mechanisms, such as ad hoc phone calls, emails, or visits; and web-based IT systems, such as the Coast Guard Homeport Internet Portal (Homeport) or the Homeland Security Information Network (HSIN).

As Pardo, Gil-Garcia, and Burke (2008) noted, formal mechanisms often contribute toward trust between partners, which in turn encourages information sharing. In contrast, Willem and Buelens (2007) found that formal mechanisms resulted in a small but negative impact on information sharing. University of Washington researchers (2014) showed that less formal methods of communication such as emails, phone calls, and face-to-face interactions during routine operations more positively impacted information sharing. Finally, the literature extensively discussed the use of IT systems to share information, albeit the discussion is somewhat divided on how IT systems affect information sharing. Some researchers found a notable positive association between IT applications and effective information sharing (Kim and Lee 2006). The majority of researchers, however, observed that IT systems had a minimally positive to negative impact on information sharing, often due to IT incompatibility between partners or the effort required to learn how to use such systems (Jarvenpaa and Staples 2000; Huysman and de Wit 2004; Zhang and Dawes 2006; Barua, Ravindran, and Whinston 2007; Choi, Kang, and Lee 2008; Hatala and Lutta 2009; University of Washington 2014).

It was originally intended to consider the use of formal, informal, and IT system mechanisms as conditions affecting the outcome, as supported by the treatment of such mechanisms in the literature. However, it was later deemed more appropriate to use these conditions to help determine whether the outcome of effective information sharing was present for each case. The aforementioned literature all treated the presence or absence of such mechanisms as conditions affecting information sharing. The questions for this study, however, did not ask about the availability of these mechanisms; in fact, it was already known that *all* subjects had formal, informal, and IT mechanisms available for use in information sharing. Every subject was stationed at a unit with an established Area Maritime Security Committee, the prime example of a formal mechanism used by the Coast Guard to share information with private sector partners. All subjects also had informal mechanisms, such as email systems and phones, at their disposal. Likewise, they all had access to the web-based IT systems Homeport and HSIN. Thus, if the questions had asked about the availability of such mechanisms, the condition the same for every case, regardless of the outcome.

Instead, subjects were asked if they used each of the three types of mechanisms to share cyber threat information with their marine transportation system partners. While it was already known these mechanisms were available, it was unknown whether the subjects actually used them. Follow-on questions to positive responses asked which specific mechanisms were used and how frequently. These responses provided valuable information toward deciding whether each case portrayed information sharing that was routine, sporadic, or simply non-existent. Theory did not define a standard for routine. Fortunately, for this study cases either fell toward one end of the spectrum or the other, making assessment relatively easy. The Coast Guard members either shared cyber threat information at least every couple of months (and in some cases on weekly basis), barely at all (at most once a year), or never. The cases in which information sharing occurred at least once every couple of months were considered to have routine sharing; anything less than that was not considered routine.

CHAPTER 3

OVERALL OUTCOME

The outcome of effective information sharing was considered to be present only if all three requirements were fulfilled; i.e., there was routine sharing of timely and relevant information. If one or more of these qualities were not satisfied, then the outcome was determined to be absent. For example, routine sharing of irrelevant or untimely information was not deemed effective; neither was sharing of timely and relevant information that only occurred on a very limited basis.

Inter-relatedness of Conditions

The literature supports an analytical perspective that considers the inter-relatedness of the aforementioned conditions. Several researchers have noted the general complexity of the relationships between the various conditions that may affect information sharing, describing how each condition can influence others (Small and Sage 2005/2006; Fedorowicz, Gogan, and Williams 2007; Yang and Maxwell 2011). Others focused on the interplay between specific factors. For example, several researchers identified a strong role that trust played in mitigating perceptions of risk in information sharing, because trust in one's partners helped decrease the concern over the lack of control over what the recipient would do with the information (Hart and Saunders 1997; Pardo et al. 2006; Pardo, Gil-Garcia, and Burke 2008; Wilson 2010). Researchers also found a close connection between trust and relationships, noting that strong relationships depend on trust (Pardo et al. 2004; University of Washington 2014).

Expectations

Based on the current information theory throughout the literature, as well as the experience of this author as a former Sector-level intelligence officer, it was expected that the Coast Guard process for sharing cyber threat information with marine transportation system partners can be improved through strengthening trusting relationships with these partners, increasing leadership support toward this sharing, reducing any perceived risk in sharing such information, and through increasing intelligence members' understandings of their role in sharing this information in addition to their understanding of this cyber threat information itself.

FINDINGS

Analysis of the data addressed the overall research question regarding what conditions contribute to effective cyber threat information sharing from Sector-level Coast Guard members to marine transportation system partners. This section presents the findings of the phenomenological study of 34 interviews of Sector-level Coast Guard intelligence members. This discussion describes overall patterns that emerged from the analysis.

Cases of Effective Information Sharing

The overall outcome of effective information sharing, defined as the combination of routine sharing of timely and relevant information, occurred in 5 out of 34 cases (14.7%). Interestingly, these cases did not consistently align with the conditions as expected according to current information sharing theory. Based on theory, it was expected that effective information sharing would correlated with strong relationships, trust, understanding of one's role in sharing information, understanding of the information, leadership support toward sharing, and the absence of perception of risk in sharing. As shown by Table 1, the only condition that was consistent with theory was the condition of leadership support, which was always present in the cases of effective sharing.

RELATIONSHIP	TRUST	ROLE	INFO	RISK	LEADERSHIP	N
0	0	1	1	1	1	2
0	1	0	1	1	1	1
1	1	0	0	1	1	1
1	1	0	0	0	1	1

Key:
0 = absence of condition
1 = presence of condition
N = number of cases with each configuration of conditions

Table 1: Cases of Effective Information Sharing

To add to this finding, it is also informative to know how many cases each had sharing of either timely or relevant information, regardless of whether the information sharing consisted of all three factors of timeliness, relevance, and routine. Nine Coast Guard members stated that they shared cyber threat information with their marine transportation system (MTS) partners in a timely manner. These subjects also stated, on the other hand, that they rarely to never *received* the information in a timely manner; i.e., it was usually outdated by the time they received it. As several subjects explained, the information was often already in the news or had been circulated by other government agencies, so it was not new by the time they shared it with MTS partners. These subjects stated that they still shared this information, however, because they recognized that some cyber threats were persistent and therefore may be relevant to MTS partners who were still be vulnerable to certain threats.

CHAPTER 3

Seventeen Coast Guard intelligence members stated that, in their opinions, the cyber threat information they shared with MTS partners was relevant to their partners. They all explained that the information they shared was relevant, but they also received a lot of cyber threat information that they deemed not relevant to their MTS partners. As an anecdotal example, one subject stated that he might consider one out of every 20 cyber threat products he received to be relevant enough to share. Two subjects responded that they thought the information they shared was relevant to some of their MTS partners, but they suspected that they were not reaching the partners who would get the most value from the information. For example, the MTS partners with whom they interacted may work in security, and not deal with IT systems, and therefore would have little direct use for information about cyber threats.

Figure 1 summarizes subjects' responses pertaining to whether the cyber threat information they shared with MTS partners was timely, relevant, or routine. As shown, 17 subjects shared information they considered relevant, 9 subjects shared information they deemed timely, and 6 subjects shared cyber threat information routinely. The intersection of all three characteristics of timeliness, relevance, and routine sharing, considered as overall effective information sharing, is shown as occurring in five cases.

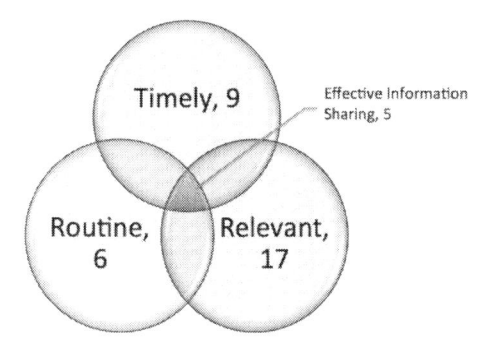

Figure 1: Cases of Timely, Routine, or Relevant Information Sharing

Relationships

To detect the potential effect of relationships on effective information sharing, subjects were asked whether they thought the quality of their relationships with MTS partners affected their cyber threat information sharing with them. Nineteen subjects responded yes, and elaborated that they shared more with partners with whom they had stronger relationships. This response coincides with information sharing theory, which contends that stronger relationships lead to more effective information sharing. The remaining 15 subjects stated that that the quality of their relationships did not affect their information sharing, but it is important to note that their responses were not all based on the same reasons. Eight subjects responded in the negative because, as they stated, they either did not share information with their MTS partners at all, or they did not have relationships with their MTS partners. Three said that they have good relationships with all of their MTS partners, so they shared with all of them equally, and three said that they shared equally with their MTS partners regardless of the nature of these relationships. Thus, the majority of the responses aligned with the expectation that a stronger relationship encourages effective information sharing.

Trust

When asked if they trusted their MTS partners, the overwhelming majority of subjects (30) responded clearly in the affirmative and indicated strong relationships with these partners, supporting the inter-relatedness of the conditions of trust and relationships. A handful of Coast Guard members elaborated that private sector partners are typically driven by profits and business sustainability, and are not expected to share the same goals as their public servant Coast Guard partners. The subjects also stated, however, that these differences in goals did not preclude trust from existing between the two groups.

Understanding of Role

Responses to the question of whether the subjects understood their role in sharing cyber threat information with their MTS partners provided valuable data toward addressing to what degree the Coast Guard's process for sharing cyber threat information with MTS partners may be improved. The numeric total of how many subjects responded yes (13) versus no (21) does not tell the whole story. Fifteen subjects, some who responded yes and some who responded no, stated explicitly that the extent of their understanding of their role was based solely on their own self-interpretation of what this role should be. Twelve specifically expressed the desire or need for more guidance regarding this role, saying that they either did not know how to share, what to share, when to share, or with whom they should share cyber threat information.

Two Coast Guard intelligence members volunteered that they would benefit most from policies or guidance that not only addressed their role in MTS cyber security as intelligence personnel, but also addressed the role of other personnel at the Sector level. Four other subjects indicated they would benefit from education and training on cyber security in general, as they felt this would aid them in understanding their role in sharing this information. Thus, not all subjects saw a finite distinction between the matters of understanding their role in sharing the information and understanding the information itself.

Understanding of Information

Coast Guard subjects were asked whether they thought they had sufficient knowledge or understanding of cyber threat information to be able to provide the "so what and what does this mean to me" context for their MTS partners. Ten subjects replied that they might be able to answer some questions about the information they shared, depending on how technical the information was or from where the information originated. These same 10 subjects also expressed confidence in knowing whom they could ask if they encountered questions they could not answer themselves. This transactive knowledge is important to recognize, as it is unrealistic to expect that all Coast Guard intelligence members would know enough about the cyber realm to be able to fully understand cyber threat information. A separate group of 10 subjects responded that they did feel they had sufficient knowledge or understanding of cyber threat information, while the remaining 14 responded simply in the negative. In summary, a total of 20 out of 34 subjects either thought they understood cyber threat information well enough to provide sufficient context for their MTS partners, or knew whom they could ask for assistance if needed.

Risk

This research sought to discover whether Coast Guard intelligence personnel perceive risk in sharing cyber threat information with their MTS partners. Identifying these risks or concerns would be a first step in mitigating these potential obstacles to sharing. A total of 19 subjects stated that they perceived risk in sharing cyber threat information with their MTS partners. Fifteen subjects stated they did not know what the policies were regarding sharing SBU information, including information marked "For Official Use Only" (FOUO) or "Law Enforcement Sensitive" (LES), with private sector MTS partners. While two subjects thought that it might be appropriate to share cyber threat information marked FOUO with their MTS partners because they were doing so in an official capacity and had a legitimate reason for sharing the information, the remaining 13 stated that they would not feel comfortable sharing this information without receiving explicit permission from the originator of the information. Three subjects said that it often took too much time obtain this permission, at times as long as 6 months, rendering the information moot before they could share it. Thus, they often decided to simply not share information marked FOUO. Only four subjects stated that they had no reservations sharing FOUO information with their MTS partners. Overall, a Coast Guard intelligence member's lack of knowledge of policies governing sharing of sensitive information with private sector partners was shown to negatively affect information sharing.

Leadership Support

In the overwhelming majority of cases, subjects responded that their leadership strongly supported such sharing. In one case, the interviewee responded that leadership would support sharing of cyber threat information as long as it was approved by legal staff; in another case, the subject responded that such support was probable but unknown because no attempts had been made to share such information.

Significantly, the five cases demonstrating effective information sharing only had the presence of the condition of leadership support in common. Leadership support did not always result in effective sharing, as the remaining 29 cases without effective information sharing all had some level of leadership support as well. However, it is informative to look more closely at the nature of this leadership support in the cases of effective sharing, as this can help explain why these cases all had positive outcomes despite the manifestations of their other conditions being inconsistent with theory.

Data regarding the existence of AMSC cyber subcommittees was collected during follow-on questions to the inquiry about the use of formal mechanisms to share cyber threat information. It was expected that the presence of this formal mechanism would promote effective sharing, since cyber subcommittees are established with the express purpose of sharing this information. To cast a wider net for detection of the potential effect of this condition, Coast Guard subjects were also asked if their units were in the process of establishing a cyber subcommittee, if one did not yet exist. It was supposed that a unit attempting to form a cyber subcommittee demonstrated a notable level of commitment to cyber threat information sharing, and therefore may correlate with effective sharing.

Of the five cases of effective information sharing, two of these cases had established cyber subcommittees, and two were in the process of establishing these committees. This is indicative of particularly strong Sector-level leadership support for cyber threat information sharing with MTS partners. In fact, even in the fifth case, where no attempt had been made to form a cyber subcommittee, the Coast Guard subject indicated that his leadership very strongly promoted communication with MTS partners, and often directed him to share cyber threat information without waiting for him to do so on his own initiative. The number of MTS partners in this case was much smaller than usual, which could help explain the lack of formation of an AMSC cyber subcommittee; such a small number of MTS partners may render a formally established group unnecessary to ensure effective information sharing.

The nature of leadership support for cyber threat information sharing also provides insight into the single case where the five main conditions all aligned with effective information sharing according to theory, but still corresponded to a negative outcome. In this case, the Coast Guard intelligence member had strong relationships with and trust in his MTS partners, understood cyber threat information, and did not perceive any risk in sharing such information. He stated that he understood what his role would be in sharing cyber threat information with MTS partners, but it was not considered to be his responsibility *at his unit*. Although his leadership would not object to his sharing of cyber threat information, a non-intelligence member at his unit fulfilled this role instead. A lack of strong direction from his leadership encouraging him to share cyber threat information with MTS partners, therefore, may have resulted in this intelligence member's lack of participation in such activities. Thus, findings show that particularly strong leadership support corresponded consistently with effective information cyber threat information sharing. This finding is in agreement with expectations based on theory and experience.

Policies

The condition of a subject's perception of policies regarding cyber threat information sharing did not add relevant data for this research and therefore was not included in the final analysis. It was originally thought that whether the interviewees perceived such policies to exist would influence their information sharing. However, a thorough review of the literature revealed no theoretical support for this line of thinking. All previous research about the effect of policies was based on actual policies that regulated information sharing, not the subjects' perceptions about whether policies existed. Moreover, the majority of subjects replied that they were not aware of any Coast Guard policies that support cyber threat information sharing. Eight subjects responded that they believed such policies must exist, but they could not cite any relevant cyber threat information sharing policies. Thus, subjects' perceptions of policies appeared to have no direct impact on their information sharing.

Conclusion

While the theory emphasizes relationships, trust, and absence of risk as key factors promoting effective information sharing, this study shows that these conditions are not determinative in Coast Guard information sharing with MTS partners. Effective cyber threat information sharing occurred even when Coast Guard members had not yet developed strong, trusting relationships with their MTS partners. Coast Guard members shared despite not fully understanding their role in sharing cyber threat information, and despite not fully understanding the information itself. They shared despite perceiving risks or having concerns about doing so. While strong leadership support did not always result in effective sharing, all cases of effective sharing had this condition in common.

Outside of focusing only on the five cases of effective sharing, further analysis of overall patterns from interview responses provide additional insight into the current state of Sector-level Coast Guard cyber threat information sharing. These findings showed that stronger relationships generally encouraged information sharing, while the perception of risk, including concern regarding sharing sensitive information, discouraged information sharing, as did a lack of understanding of one's role in sharing and a lack of understanding of the information itself. When their unit leaders strongly encouraged them to share cyber threat information with MTS partners, Coast Guard intelligence members did so despite the absence of strong relationships, absence of trust, and presence of risk. On the other hand, in the case where all other conditions were favorable toward effective sharing, the Coast Guard member did not share information because his leadership did not consider that role to be the intelligence member's responsibility. Overall, therefore, findings showed that the condition of strong leadership support, above all other conditions, was the most important factor toward effective sharing.

Recommendations

In summary, this research has revealed that there is room for improvement in Sector-level Coast Guard cyber threat information sharing with private sector marine transportation system partners. There are several ways to improve Sector-level Coast Guard intelligence member cyber threat information sharing with MTS partners. This section offers recommendations for practical measures to improve this sharing. Suggestions for potential areas of further research are also provided.

Recommendations for Improvement of Cyber Threat Information Sharing

Recommendation 1: At the Sector level, Coast Guard leaders must unequivocally support their intelligence staff's role in sharing cyber threat information with MTS partners, while also fostering a unity of effort among all Sector-level personnel who share in this role. Where appropriate, Sector Commanders should also lead the establishment of AMSC cyber subcommittees and ensure active participation from their intelligence staff.

Recommendation 2: At the Headquarters level, Coast Guard leaders must develop policies and training that clearly guide Sector-level intelligence members in their role of sharing cyber threat information with private sector partners. This training should also include an introduction to cyber threats, vulnerabilities, and cyber security measures pertaining to the marine transportation system.

Recommendations for Future Research

This study was limited in scope and leaves areas for future research that could significantly contribute to the body of work on this topic. This section offers suggestions for any individuals interested in further researching Coast Guard cyber threat information sharing.

Recommendation 1: The perspective of this research was solely from that of the Coast Guard side of the partnership between that organization and private sector marine transportation partners. Only Coast Guard members were interviewed, and the data includes only their viewpoints about matters such as the timeliness and the relevance of the information they shared. It would have also been informative to collect similar data from marine transportation partners, especially regarding such matters as the relevance and timeliness of the information they received from their Coast Guard partners. It is recommended, therefore, that future researchers collect data from this group of subjects to gain vital insights that could inform how the Coast Guard may improve in sharing cyber threat information.

Recommendation 2: The focus of this research was also limited to how Coast Guard intelligence members share information. Other Sector-level Coast Guard members play a part in sharing cyber threat information with MTS partners. Along the same vein, this research looked at information sharing only at the Sector level of the Coast Guard. To obtain a more holistic view of how the Coast Guard shares cyber threat information, this research could be expanded to include non-intelligence members as well as cyber threat information sharing occurring at other levels of the Coast Guard.

Recommendation 3: Coast Guard cyber threat information sharing with private sector partners continues to evolve as new sharing mechanisms emerge. Future research could consider the role of the newly established Maritime and Port Security Information Sharing and Analysis Center, a public–private partnership designed to advance maritime cyber resiliency. Researchers could study what impact this Center may have on cyber threat information sharing at the Sector level of the Coast Guard.

Recommendation 4: Information that is shared must also be stored, disseminated, and applied appropriately. Future research could focus on Coast Guard and private sector practices of *knowledge management* to identify best practices and areas for improvement, and inform how these practices may in turn foster further information sharing.

Recommendation 5: Finally, future research could seek to identify to what extent sharing of cyber threat information differs from sharing of other types of information. Research could examine whether the factors that affect information sharing are universal across all types of information sharing, or whether some of the factors affecting cyber threat information sharing are unique to this subset of information sharing.

REFERENCES

6, Perri, Christine Bellamy, Charles Raab, Adam Warren, and Cate Heeney. 2007. "Institutional Shaping of Inter-Agency Working: Managing Tensions Between Collaborative Working and Client Confidentiality." Journal of Public Administration and Research 17 (3): 379–404. https://dspace.lboro.ac.uk/dspacespui/bitstream/2134/4482/3/J_PART_0906.pdf (accessed April 16, 2016).

Barua, Anitesh, Suryanarayanan Ravindran, and Andrew B. Whinston. 2007. "Enabling Information Sharing Within Organizations." Information Technology Management 8 (1): 31–45. http://search.proquest.com/docview/194455479/fulltextPDF/8E8407B00A3949CBPQ/1?accountid=10504 (accessed April 16, 2016).

Brauner, Elisabeth, and Albrecht Becker. 2006. "Beyond Knowledge Sharing: The Management of Transactive Knowledge Systems." *Knowledge and Process Management* 13 (1): 62–71. doi:10.1002/kpm.240 (accessed April 16, 2016).

Carlin, Stephanie, and Alexandria Womack. 1999. *Creating a Knowledge-Sharing Culture.* eds. Susan Elliot. Houston, TX: American Productivity & Quality Center.

Carminati, Barbara, Elena Ferrari, and Michele Guglielmi. 2011. "Secure Information Sharing on Support of Emergency Management." *IEEE International Conference on Privacy, Security, and Trust,* 988–995. https://www.opensource.gov/providers/ieee/ielx5/6112285/6113084/06113250.

pdf?tp=&arnumber=6113250&isnumber=6113084 (accessed April 16, 2016).

Choi, Young, Young Sik Kang, and Heeseok Lee. 2008. "The Effects of Socio-Technical Enablers on Knowledge Sharing: An Exploratory Examination." *Journal of Information Science* 34 (5): 742–754. https://www.researchgate.net/profile/Young_Sik_Kang/publication/220195812_The_effects_of_socio-technical_enablers_on_knowledge_sharing_An_exploratory_examination/links/00b7d52f9b2fe6ed8c000000.pdf (accessed April 16, 2016).

Constant, David, Sara Kiesler, and Lee Sproull. 1994. "What's Mine is Ours, or Is It? A Study of Attitudes about Information Sharing." *Information Systems Research* 5 (4): 400–421. https://www.cs.cmu.edu/~kiesler/publications/PDFs/Constant1994WhatsMine%20.pdf (accessed April 16, 2016).

Dawes, Sharon S. 1996. "Interagency Information Sharing: Expected Benefits, Manageable Risks." *Journal of Policy Analysis and Management* 15 (3): 377–396. http://www.docfoc.com/dawes-1996-interagency-information-sharing-yVbBh (accessed April 16, 2016).

Dodgson, Mark. 1993. "Learning, Trust, and Technological Collaboration." *Human Relations* 46 (1): 77–95. http://search.proquest.com/docview/231489517?accountid=10504 (accessed April 17, 2016).

Drake, David B., Nicole A. Steckler, and Marianne J. Koch. 2004. "Information Sharing in and Across Government Agencies: The Role and Influence of Scientist, Politician, and Bureaucrat Subcultures." *Social Science Computer Research* 22 (1): 67–84. http://ssc.sagepub.com/content/22/1/67 (accessed November 10, 2015). abstract.

Dyer, Jeffrey H., and Wujin Chu. 2003. "The Role of Trustworthiness in Reducing Transaction Costs and Improving Performance: Empirical Evidence from the United States, Japan, and Korea." *Organization Science* 14 (1): 57–68. http://www.jstor.org/stable/3086033 (accessed April 16, 2016).

Egan, Dennis, Darby Hering, Paul Kantor, Christie Nelson, and Fred S. Roberts. 2017. "Information Sharing for Maritime Cyber Risk Management." In *Issues In Maritime Cyber Security*, eds. Joseph DiRenzo, Nicole K. Drumhiller, and Fred S. Roberts. Westphalia Press: to appear.

Fawcett, Stanley E., Paul Osterhaus, Gregory M. Magnan, James C. Brau, and Matthew W. McCarter. 2007. "Information Sharing and Supply Chain Performance: The Role of Connectivity and Willingness." *Supply Chain Management: An International Journal* 12 (5): 358–368. http://business. utsa. edu/faculty/mmccarter/files/Fawcett%20et%20al%202007%20 SCMIJ.pdf (accessed April 16, 2016).

Fedorowicz, Jane, Janis L. Gogan, and Christine B. Williams. 2007. "A Collaborative Network for First Responders: Lessons from the CapWIN Case." *Government Information Quarterly* 24 (October): 785–807. doi:10.1016/j.giq.2007.06.001 (accessed April 16, 2016).

Hart, Paul, and Carol Saunders. 1997. "Power and Trust: Critical Factors in the Adoption and Use of Electronic Data Interchange." *Organization Science* 8 (1): 23–42. http://www.jstor. org/stable/2635226 (accessed January 27, 2016).

Hatala, John-Paul, and Joseph George Lutta. 2009. "Managing Information Sharing Within an Organizational Setting: A Social Network Perspective." *Performance Improvement Quarterly* 21 (4): 5–33. http://www. performancexpress.org/wp-content/uploads/2011/11/Managing-Information-Sharing.pdf (accessed November 17, 2015).

Hsu, Chin-Chun, Vijay R. Kannan, Keah-Choon Tan, and G. Keong Leong. 2008. "Information Sharing: Buyer-Supplier Relationships, and Firm Performance." *International Journal of Physical Distribution & Logistics Management* 38 (4): 296–310. http://search.proquest.com/docview/ 232593952?accountid=10504 (accessed April 16, 2016).

Huysman, Marleen, and Dirk de Wit. 2004. "Practices of Managing Knowledge Sharing: Towards a Second Wave of Knowledge Management." *Knowledge and Process Management* 11 (2): 81–92. http://dx.doi.org/ 10.1002/ kpm.192 (accessed April 16, 2016).

Ibrahim, Nurain Hassan, and David Allen. 2012. "Information Sharing and Trust During Major Incidents: Findings from the Oil Industry." *Journal of the American Society for Information Science and Technology* 63 (10): 1916–1928. doi:10.1002/asi.22676 (accessed April 17, 2016).

Jarvenpaa, S.L., and D.S. Staples. 2000. "The Use of Collaborative Electronic Media for Information Sharing: An Exploratory Study of Determinants." *Journal of Strategic Information Systems* 9: 129–154. http://scholar.cci. utk.edu/beyond-downloads/publications/use-collaborative-electronic-

CHAPTER 3

media-information-sharing-exploratory (accessed April 16, 2016).

Kim, Soonhee, and Hyangsoo Lee. 2006. "The Impact of Organizational Context and Information Technology on Employee Knowledge-Sharing Capabilities." *Public Administration Review* 66 (3): 370–385. http://search.proquest.com/docview/197172507/fulltextPDF/DC19AFA79AB74010PQ/1?accountid=10504 (accessed April 16, 2016).

Kramek, Jospeh. 2013. The Critical Infrastructure Gap: U.S. Port Facilities and Cyber Vulnerabilities. Brookings Institution/Center for 21st Century Security and Intelligence. July. http://www.brookings.edu/~/media/research/files/papers/2013/07/02%20cyber%20port%20security%20kramek/03%20cyber%20port%20security%20kramek.pdf (accessed October 29, 2015).

Leedy, Paul D., and Jeanne Ellis Ormrod. 2010. *Practical Research: Planning and Design*. 9th ed. Upper Saddle River, NJ: Merrill.

Li, Suhong, and Binshan Lin. 2006. "Accessing Information Sharing and Information Quality in Supply Chain Management." *Decision Support Systems* 42 (March): 1641–1656. doi:10.1016/j.dss.2006.02.011 (accessed April 16, 2016).

Liao, Shu-hsien, Juo-chiang Chang, Shih-chieh Cheng, and Chia-mei Kuo. 2004. "Employee Relationship and Knowledge Sharing: A Case Study of a Taiwanese Finance and Securities Firm." *Knowledge Management Research & Practice* 2 (April): 24–34. http://www.palgrave-ournals.com/ kmrp/journal/v2 /n1/full/8500016a.html (accessed April 16, 2016).

Liebowitz, Jay, and Yan Chen. 2003. "Knowledge Sharing Proficiencies: The Key to Knowledge Management." In *Handbook on Knowledge Management 1: Knowledge Matters*, eds. C.W. Holsapple. Berlin: Springer-Verlag, 409–424.

Liu, Peng, and Amit Chetal. 2005. "Trust-Based Secure Information Sharing Between Federal Government Agencies." *Journal of the American Society for Information Science & Technology* 56: 283–298. doi:10.1002/asi.20117/abstract (accessed April 16, 2016).

Luna, Luis, Ramon Gil, and Cinthia Betiny. 2006. "Collaborative Digital Government in Mexico: Some Lessons from Federal Web-Based Inter-Organizational Information Integration Initiatives." *Proceedings of* the *Twelfth Americas Conference on Information Systems* (December), 2375–2384. http://aisel.aisnet.org/cgi/viewcontent.cgi?article=1841& context=amcis2006 (accessed April 16, 2016).

McCarthy, John A., Chris Burrow, Maeve Dion, and Olivia Pacheco. 2009. "Cyberpower and Critical Infrastructure Protection: A Critical Assessment of Federal Efforts." In *Cyberpower and National Security*, eds. Franklin D. Kramer, Stuart H. Starr, and Larry K. Wentz. Washington, DC: Potomac Books, Inc., 543–556.

Mohr, Jakki J., and Ravipreet Sohi. 1995. "Communication Flows in Distribution Channels: Impact on Assessments of Communication Quality and Satisfaction." *Journal of Retailing* 71 (4): 393–416. http://digitalcommons.unl.edu/cgi/viewcontent.cgi?article=1038&context=marketingfacpub (accessed April 16, 2016).

Nahapiet, Janine, and Sumantra Ghoshal. 1998. "Social Capital, Intellectual Capital, and the Organizational Advantage." *The Academy of Management Review* 23 (2): 242–266. http://www.jstor.org/stable/259373 (accessed April 16, 2016).

Pardo, Theresa A., Anthony M. Cresswell, Sharon S. Dawes, and G. Brian Burke. 2004. "Modeling the Social & Technical Processes of Interorganizational Information Integration." *Proceedings of the 37th Hawaii International Conference on System Sciences*, 1–8. https://www.computer.org/csdl/proceedings/hicss/2004/2056/05/205650120a.pdf (accessed April 16, 2016).

Pardo, Theresa A., Anthony M. Cresswell, Fiona Thompson, and Jing Zhang. 2006. "Knowledge Sharing in Cross-Boundary Information Sharing Development in the Public Sector." *Information Technology Management* 7 (4): 293–313. http://link.springer.com/article/10.1007%2Fs10799-006-0278-6 (accessed April 16, 2016).

Pardo, Theresa A., J. Ramon Gil-Garcia, and G. Brian Burke. 2008. "Building Response Capacity Through Cross-boundary Information Sharing: The Critical Role of Trust." *Center for Technology in Government Working Paper*, no. 6. https://www.ctg.albany.edu/publications/working/building_response /building_ response.pdf (accessed November 25, 2015).

Pilerot, Ola. 2012. "LIS Research on Information Sharing Activities—People, Places, or Information." *Journal of Documentation* 68 (4): 559–581.

doi:10.1108/00220411211239110 (accessed April 16, 2016).

Presidential Policy Directive 21 (PPD-21). "Critical Infrastructure Security and Resilience." February 12, 2013. https://www.whitehouse.gov/the-press-office/2013/02/12/presidential-policy-directive-critical-infrastructure-security-and-resil (accessed October 28, 2015).

Small, Cynthia T., and Andrew P. Sage. 2005/2006. "Knowledge Management and Knowledge Sharing: A Review." *Information Knowledge Systems Management* 5: 153–169. http://web.b.ebscohost.com/ehost/pdfviewer/pdfviewer?vid=5&sid=828329ad-6532-4962-a0d8-39571480d9c0%40 sessionmgr113&hid=118 (accessed November 10, 2015).

The White House. Executive Order 13636. "Improving Critical Infrastructure Cybersecurity." *Federal Register* 78, no. 33 (February 19, 2013): 11737–11744. https://federalregister.gov/a/2013-03915 (accessed May 12, 2016).

University of Washington Department of Human Centered Design & Engineering. 2014. *Maritime Operational Information Sharing Analysis.* June. http://www.hcde.washington.edu/files/MOISA1-Final-Report_10-April-2015.pdf (accessed March 28, 2016).

US Coast Guard. 2015. *United States Coast Guard Cyber Strategy.* Washington, D.C. June. https://www.uscg.mil/seniorleadership/DOCS/cyber.pdf (accessed October 28, 2015).

US Department of Homeland Security. 2009. *National Infrastructure Protection Plan: Partnering to Enhance Protection and Resiliency.* January. https://www.dhs.gov/xlibrary/assets/NIPP_Plan.pdf (accessed October 28, 2015).

US Department of Homeland Security. 2016. *Consequences to Seaport Operations from Malicious Cyber Activity.* March 3. http://www.maritimedelriv.com/Port_Security/DHS/DHS_Files/OCIA_Consequences_to_Seaport_Operations_from_Malicious_Cyber_Activity.pdf (accessed June 25, 2016).

US Government Accountability Office. 2006. *Maritime Security: Information-Sharing Efforts are Improving.* July. http://www.gao.gov/assets/120/114342.pdf (accessed October 28, 2015).

US Government Accountability Office. 2012. *Cybersecurity: Threats Impacting the Nation*. April. http://www.gao.gov/products/GAO-12-666T (accessed October 25, 2015).

US Government Accountability Office. 2014. *Maritime Critical Infrastructure Protection: DHS Needs to Better Address Port Cybersecurity*. June. http://www.gao.gov/assets/670/663828.pdf (accessed October 28, 2015).

US Government Accountability Office. 2015. *Maritime Critical Infrastructure Protection: DHS Needs to Enhance Efforts to Address Port Security*. October. http://www.gao.gov/assets/680/672973.pdf (accessed October 29, 2015).

Willem, Annick, and Marc Buelens. 2007. "Knowledge Sharing in Public Sector Organizations: The Effect of Organizational Characteristics on Interdepartmental Knowledge Sharing." *Journal of Public Administration Research and Theory: J-Part* 17 (4): 581–606. http://www.jstor.org/stable/25096342 (accessed April 16, 2016).

Wilson, T.D. 2010. "Information Sharing: An Exploration of the Literature and Some Propositions." *Information Research* 15 (4). http://files.eric.ed.gov/fulltext/EJ912770.pdf (accessed November 16, 2015).

Wu, Wei-Li, Bi-Fen Hsu, and Ryh-Song Yeh. 2007. "Fostering the Determinants of Knowledge Transfer: A Team-Level Analysis." *Journal of Information Science* 33 (3): 326–339. http://jis.sagepub.com/cgi/content/abstract/33/3/326 (accessed April 16, 2016).

Yang, Tung-Mou, and Terrence A. Maxwell. 2011. "Information-Sharing in Public Organizations: A Literature Review of Interpersonal, Intra-Organizational and Inter-Organizational Success Factors." *Government Information Quarterly* 28: 164–175. doi:10.1016/j.giq.2010.06.008 (accessed April 16, 2016.).

Yang, Tung-Mou, and Yi-Jung Wu. 2014. "Exploring the Determinants of Cross-Boundary Information Sharing in the Public Sector: An e-Government Case Study in Taiwan." *Journal of Information Science* 40 (5): 649–668. http://jis.sagepub.com/content/40/5/649.full.pdf+html (accessed November 10, 2015).

Zhang, Jing, and Sharon S. Dawes. 2006. "Expectations and Perceptions of Benefits, Barriers, and Success in Public Sector Knowledge Networks." *Public Performance & Management Review* 29 (4): 433–466. http://www.

jstor.org/stable/20447606 (accessed April 17, 2016).

Zhang, Jing, Sue R. Faerman, and Anthony M. Cresswell. 2006. "The Effect of Organizational/Technical Factors and the Nature of Knowledge on Knowledge Sharing." *Proceedings of the 39th Hawaii International Conference on System Sciences*, 1–10. https://www.computer.org/csdl/proceedings/hicss/2006/2507/04/250740074a.pdf (accessed April 17, 2016).

Zhou, Honggeng, and W.C. Benton Jr. 2007. "Supply Chain Practice and Information Sharing." *Journal of Operations Management* 25: 1348–1365. ftp://mail.im.tku.edu.tw/Prof_Shyur/Enterprise%20Information%20System/11080614521405974.pdf (Accessed April 17, 2016).

CHAPTER 4:
NAVIGATING THE CYBER THREATS TO THE U.S. MARITIME TRANSPORTATION SYSTEM

Jim Twist, Blake Rhoades, and Ernest Wong
Army Cyber Institute

Abstract

Our ports are critically important to our economic strength and national security. However, our increasing dependence on technology has exposed considerable vulnerabilities in our reliance upon the cyber domain. Our maritime transportation system today gains considerable efficiencies from computer applications, technologies, and systems that help drive us toward improved safety, higher operational throughput, and advanced analytics that yield greater profits and revenues. Unfortunately, cyber adversaries present us with a whole host of cyber security challenges that have the potential to become catastrophic events. Furthermore, we also have lacked a holistic understanding of the entire spectrum of cyber threats that can impact the maritime domain. In this chapter, not only do we address these shortcomings by characterizing the various threats to the cyber domain, we also identify a number of capabilities that we have at our disposal to reduce the overall threats to our nation's critical maritime infrastructure.

Background—A Need for Increased Cyber Threat Awareness Throughout Our MTS

Over the past several years, our nation has become increasingly dependent on technology. The internet of things has turned our globally connected society into one that must cope with the insecurity of things, and our cyber vulnerabilities make not just our individual citizens and corporate entities more unsafe but also jeopardize our national security and economic prosperity. While cyber security professionals have focused considerable attention and resources on the protection of the U.S. power grid and our globally connected financial systems, the concern placed on vulnerabilities that the cyber domain presents to our port facilities and maritime transportation system (MTS) has been fairly muted. In 2013, Joseph Kramek exposed a number of deficiencies with our maritime cyber security posture, and his monograph has done an outstanding job educating our nation on the urgent need to increase the amount of attention paid on the security of the networked systems that undergird our port operations (Kramek 2013). That same year, the Defense Science Board developed a useful model and taxonomy for characterizing adversaries in the cyber domain (U.S. Defense Science Board 2013). While these studies have done a good job informing the U.S. public about real-world cyber threats, what continues to be lacking is a holistic look at the full spectrum of cyber threats to our critical maritime infrastructure. This chapter aims to remedy this shortcoming by providing a more comprehensive understanding of the cyber vulnerabilities within our MTS. Moreover, this chapter attempts to educate and equip our U.S. Coast Guard (USCG) with the tools necessary to reduce the likelihood of a broad-scale complex cyber attack that has the potential to cripple our shipping and port operations.

Throughout the history of our nation, U.S. foreign trade has been predominately sea-based. Furthermore, U.S. maritime operations have become increasingly reliant upon computer systems and networks to function. Currently, these networked facilities process more than $1.3 trillion in cargo each year; however, our port facilities are highly vulnerable to many forms of cyber attack (U.S. Department of Commerce 2016). At present, based on our nation's nascent and lackluster efforts to defend this underemphasized area of the cyber domain, the overall risk of our MTS to cyber attacks from the full spectrum of cyber adversaries remain a serious concern.

Fortunately, our maritime transportation systems and port cyber security have recently received increased attention from our political leaders and policy-makers. On December 18, 2015, the House of Representatives approved the "Strengthening Cybersecurity Information Sharing and Coordination in our Ports Act of 2015." This bill names the USCG as the lead government agency charged with managing port-wide cyber security, and it establishes requirements for integrating cyber security with port security activities currently in place in accordance with the Maritime Transportation Security Act (MTSA) and International Ship and Port Security code (Marine Log 2016). This bill also expands upon Presidential Executive Order 13636, "Improving

Critical Infrastructure Cybersecurity," which focuses on three main issues: information sharing, privacy, and the adoption of cyber security practices. The directive instructs the U.S. Department of Homeland Security to incentivize participation in a voluntary framework and to share best practices across the various industries considered critical national infrastructure. Unfortunately, this voluntary program has not been widely adopted, and as a result, the USCG has focused more of its limited resources and priorities on closing physical security gaps in the MTS. There continues to exist a large gulf between the vulnerabilities inherent in the technologies we rely upon within the MTS to function and that of the security posture for our MTS cyber domain needed to deter, prevent, respond, and recover from a broad-scale cyber attack. Today, even unsophisticated cyber threats could cause serious damage given the lack of standards, poor security practices, and fragility of our commercial shipping processes which are not fundamentally resilient to disruptions, be they accidental, deliberate, man-made, or acts of nature. We must all be more aware of the wide array of cyber actors capable of conducting offensive cyber activities against our MTS. With cyber threats becoming more sophisticated every day, we must begin the hard work of preparing and protecting ourselves against adversaries who intend to do our nation great harm, are backed by considerable financial resources, and maintain global cyber space platforms that help to nullify the benefits of our oceans barriers which once kept our nation relatively safe against potential adversaries. Without basic cyber security in place, attacks from advanced threat actors taking advantage of the numerous vulnerabilities in the network architecture of the MTS can wreak havoc on our trade, economy, and critical lines of communication. Our defensive posture, capability, and capacity must improve in order to reduce the attack surfaces that a cyber adversary can exploit to compromise our maritime infrastructure, ports, and ships.

In spite of the increased use of technology for our port operations, very few components within the sector have appropriately assessed risk to their infrastructure. A Brookings Institute study in 2013 found that only 1 in 6 U.S. port authorities had methodically assessed the cyber risk to its organizational assets (Kramek 2013). Although responsible for maritime security in the U.S., the USCG has also experienced difficulties in being able to appropriately prioritize its assessment of cyber risks to maritime environments. Fortunately, the USCG in 2015 released a comprehensive strategy for dealing with cyber threats to our port systems, which aimed to place greater emphasis on this critical issue impacting our national security and economic prosperity (Young 2015).

Gaining a Better Understanding of Cyber Threats: A Taxonomy of Cyber Adversaries

To effectively combat the cyber threats posed against our maritime critical infrastructure, it is important to first gain an understanding of the full spectrum of cyber adversaries that present a danger to our ships and ports. The U.S. Defense Science Board has developed a notional taxonomy to classify cyber threats (as shown in Figure 1 below). Tier 1 and Tier 2 threat actors present a nuisance to operations by exploiting known vulnerabilities. Tier 3 and Tier 4 threats are capable of conducting cyber intelligence, surveillance, and reconnaissance operations to discover previously unknown vulnerabilities within a network to include undisclosed flaws in commercial hardware and software. A Tier 5 or Tier 6 threat actor is able to use the full spectrum of cyber operations to create vulnerabilities where none existed before. A Tier 5 and 6 actor has extensive resources, incorporates operational design and planning, and institutes long-term campaigns against targeted resources, systems, and infrastructure, which constitutes an existential threat to our national security.

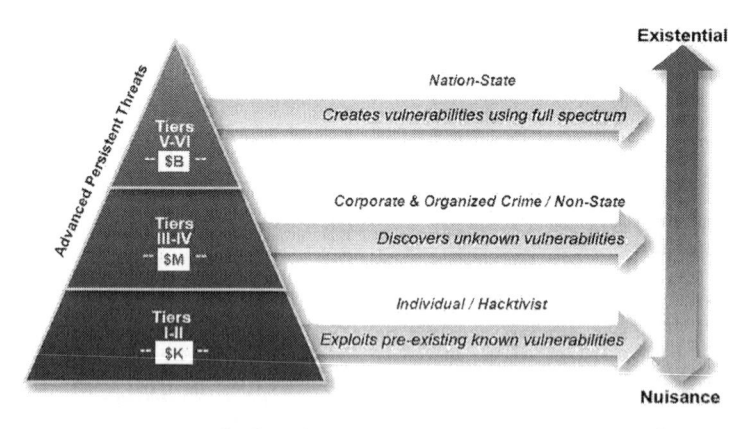

Figure 1: Taxonomy of cyber threats and adversaries to our ports (U.S. Defense Science Board 2013)

A sharpened discernment of vulnerabilities also improves our capability to protect and defend our cyber infrastructure. A cyber vulnerability is a weakness that allows an attacker to penetrate a system for unintended usage. A system vulnerability is the intersection of a system susceptibility or flaw, access to the flaw, and the capability to exploit the flaw (U.S. Air Force 2009). Unfortunately, vulnerabilities prevalent at U.S. port facilities can endanger both our national security and economic well-being. At the macro level, weaknesses in our shipping sector create a vulnerability to the U.S. economy; at the micro level, our shipping sector is composed of a number of network vulnerabilities that permits adversaries to engage in nefarious activities that elevate our risk to cyber attacks. In order to better understand these vulnerabilities, this chapter will examine both the macro and micro perspectives that create this heightened potential for systemic, complex, and catastrophic disruption.

ONE OF AMERICA'S BIGGEST ECONOMIC VULNERABILITIES: OUR PORTS

E-commerce has revolutionized the world economy, allowing users to quickly find and purchase goods from across national borders. While the methods for shopping have changed, the ways in which these goods are transported have not varied much: the global maritime economy continues to thrive, propelled by the growing demand created through the internet. The majority of U.S. trade is transported by ship, making the shipping industry one of the most critical components to our national transportation system and the U.S. economy. By weight, U.S. seaports currently process 99.4% of the nation's overseas cargo (American Association of Port Authorities).

While the shipping sector is one of the biggest contributors to our nation's economic clout, it is also one of the most vulnerable to disruption. To sustain the commerce system that supplies our national economy and promote greater efficiency in our national supply chain, U.S. ports have increasingly operated toward a "zero inventory, just-in-time delivery" system. This system represents a fragile balance of integrated processes, which reveals an inherently elevated level of risk even outside of the cyber security context. Slight disruptions in the system would generate ripple effects throughout our economy, similar to the way blood flow is vital to the human heart: a blockage of just a few second can cause immense damage to this vital organ. Stoppage at one of our major ports would cause extensive impairment due to the critical roles our largest ports have in business, trade, and even support to military operations. As port facilities and processes become more automated, they will inevitably become increasingly vulnerable to cyber attacks—an unfortunate circumstance that can only be partially mitigated through risk management assessments and appropriate response and recovery procedures.

Such a broad-scale disruption became evident in February of 2015, when 29 West Coast ports shut down over a labor dispute. A Washington Council on International trade later estimated that the disruption immediately cost the state of Washington over $770 million, and this estimate only accounts for delayed shipping costs and idle truck and transportation fees (The Columbian 2016). Kevin O'Marah, Head of Research for SCM World, a cross-industry learning community aiming to advance worldwide supply chain management, estimates that the labor standoff cost the U.S. economy $2 billion per day during the 10 days of complete shutdown. As a result of the loss of throughput in the west coast ports, the U.S. economy posted a 15% year-over-year decline in exports for the month of January (Elementum News Desk 2015). Therefore, it should be unmistakable that disruptions in our MTS—whether or not they are the result of a cyber attack—can severely and negatively impact the U.S. economy.

All of our ports are vulnerable, but each individual port also exhibits unique characteristics and has distinctive qualities that can increase their exposure to cyber attacks. The individual roles that our largest ports play in the overall macro economy also exposes them to being unique targeting opportunities from an adversary's perspective. The Port of Los Angles, for instance, imports over $130 billion in commodities from China each year and together with the Port of Long Beach, processes state foreign fuel imports that account for over 50% of California's fuel supply and imports. The Port of Houston Authority, similarly, accounts for 25% of the nation's oil imports and is vital to the health of the American economy (Kramek 2013). The Port of Beaumont is home to one of the Military Surface Deployment and Distribution Command's (SDDC) port-handling battalions, which is responsible for packaging and shipping war materials to combat zones across the globe. Beaumont is an example of concentrated infrastructure within a limited geographic area which increases overall risk to national security from disruption. U.S. Representative Ted Poe described the port's importance to national security in the following way: "Our refineries along the channel here produce the majority of the jet fuel that is used by our aviators ... because of the petrochemical plants, because of the refineries, because of the ship channel, and it is a national security issue that we take care of the port of Beaumont" (Myers 2008). The Port of Beaumont is an attractive target for foreign cyber powers, and an adversary could potentially disrupt up to 50% of the war materials being transported overseas to the combat theater of operations by disrupting operations there (Myers 2008). The specialized function of each of these ports varies, and each port impacts our economy in uniquely different ways. It is the particular characteristics of each individual port that can present different objectives for an adversary's campaign plan or strategic goals.

Gaining Entry into MTS Networks: Systemic Flaws in Cyber Security

As the lead government agency for the protection of our ports, the USCG is uniquely postured to defend the cyber terrain, particularly those belonging to port facilities. While much of the U.S. military is limited to the protecting of Department of Defense and government infrastructure due to *posse comitatus* restrictions, the USCG has military, intelligence, and law enforcement authorities that allow it considerably more leverage and resources in addressing cyber threats on the .mil, .gov, and .com domains (U.S. Coast Guard 2016a). Despite these broad authorities, the USCG has struggled to gain momentum with this complex mission. While it has initiated the program, it has not yet completed the requirements from a cyber security perspective. The 2013 Brookings Institute report details the poor state of cyber security at many of our ports, and it adds that at the time, nearly all of the USCG's focus had been directed exclusively toward physical security (Kramek 2013). Despite this, the USCG does not bear sole responsibility for the vulnerabilities in our ports' cyber preparedness; private sector facility owners had also appropriated much of their budget and FEMA grants toward physical security, but have not yet placed an emphasis toward cyber security. This is due, in large part, to cyber being a relatively new risk element, and the emphasis of the Port Security Grant Program and the Marine Transportation Security Act of 2002 geared toward physical security and preparedness. Ultimately, this report found that "lack of standards and enforcement authorities" coupled with the lack of prudent actions by the private owners has generated an ideal environment for would-be cyber attackers (Kramek 2013).

Fortunately, the USCG has taken a series of positive steps to more appropriately prioritize cyber space protection. In his 2016 budgetary remarks to congress, USCG Commandant Admiral Paul Zukunft identified cyber security of ports as one of the top priorities for the service.[1] In the last year, the USCG has also made headway in enumerating cyber risks and has outlined its mitigation strategy by publishing the "USCG Cyber Strategy" in June of 2015. Additionally, it increased its cyber security remediation budget by $5.2 million in order to secure its own networks, and it is actively engaging the private sector to promote security at the ports (U.S. Coast Guard 2016b).

Unfortunately, these measures fall short of fully addressing the scope of the USCG's complex cyber challenges. Its budget requirements do not reflect the appropriate amount of resources needed to remedy the vast gulf in its present vulnerabilities with what must be achieved to prevent a potentially calamitous attack from a cyber threat, even against an unsophisticated Tier 1 and 2 adversary. Furthermore, a review of its appropriations by mission breakdown reveals that it is not clear what funds, if any, are being committed to improving port security. Of the roughly $7 billion budget request submitted, $59 million was requested to upgrade Command, Control, Communications, Computers, Intelligence Surveillance, and Reconnaissance (C4ISR). While useful in enhanc-

---

1   As described by the 2017 USCG budget in brief, priorities are (1) The Rise of Transnational Crime, (2) Southern Maritime Border Security, (3) Increasing Maritime Commerce, and (4) Emerging Cyber Risks.

ing USCG capacity for control, there is no specific mention of cyber security initiatives. Looking at how the goals of reducing cyber security risks are to be executed, its stated plan is to "coordinate cyber regulatory and technical assistance activities across Federal, state and local maritime industry stakeholders" (U.S. Coast Guard 2016b). From the proposal itself, it is not clear who will be executing this mission, or what assets, resources, or financial support will be committed once a broad-scale cyber attack occurs. Other than "coordination," there are no performance measurements to indicate and assess whether or not a cyber adversary has been stopped and whether or not response and recovery efforts have been successful. This could be an important consideration in the outcome of a cyber-related event which impacts the USCG. Its credibility would be at stake should the USCG declare victory against a cyber threat without knowing for sure if the network is free of an adversary's malware. This budget also reflects an overwhelming emphasis with traditional USCG missions that are tied chiefly to physical security.

To begin to address our complex cyber security challenges, we must commit not only financial resources to technological solutions, but more importantly, invest in human capital, develop sound processes, and practice, train, and rehearse against the full range of cyber threats that have the means and motive to exploit vulnerabilities and infiltrate our cyber infrastructure, no matter how seemingly secure. Good cyber hygiene and the employment of best practices in private businesses cyber security will not be enough to defend against the advanced persistent threats in the Tier 5 and Tier 6 threat categories. The importance and sizable impact of our nation's ports and MTS should compel our government to be prepared for enduring cyber disruption resulting from conflict with another nation-state or their proxies.

Despite the USCG's positive steps toward addressing and confronting cyber challenges to our MTS, adversaries today are looking to take advantage of the inherent vulnerabilities throughout our network of connected systems. Our maritime sector has become increasingly vulnerable to cyber attacks as it continues to rely more heavily on computers for key functions, to include navigating the waters, managing the vast quantities of cargo transported into and out of our ports, controlling storage environments, allocating power and energy, providing weather forecasts, and steering ships and operating vessels. Ironically, even the automated security systems that have been designed to help make our ports more secure have become yet another attack vector for the cyber adversary. As a result, to ensure that our MTS can continue to perform its critically important functions, it is imperative that we be fully cognizant of how inherent vulnerabilities within our automated and connected systems can endanger the MTS.

CHAPTER 4

GPS and AIS: A Critical Dependency for Ports and Ships

The global positioning system (GPS) Communication System is a key component of the maritime economy that has become a key driver of efficiency for U.S. commerce. Thousands of ships around the globe utilize GPS at any given moment to navigate the seas. A disruption to this system would disable the vast majority of these vessels; without it, many ships today no longer have the capability to revert back to analog skills such as navigating by the stars (Prudente 2015). Recognizing that these perishable skills are necessary to providing sufficient resiliency should these electronic tools fail, the USCG Academy, United States Merchant Marine Academy, and six state maritime academies continue to train the basics of celestial navigation and basic paper chart navigation. However, the U.S. Naval Academy only recently reinstated celestial navigation into its curriculum after recognizing that this foundational skill adds resiliency should automated systems fail.

Not only is GPS vulnerable to cyber attacks, data from GPS can be extremely valuable to foreign intelligence services. Perhaps even more disconcerting is the potential for data manipulation within GPS. With spoofing of the network developing into a common technique that enables attackers to misdirect and exploit unwitting targets, this placing our ships and port facilities at increased risk. Our dependence on GPS has created a vulnerability that our cyber adversaries can use to their advantage, and the disruption of this global network would, no doubt, have significant repercussions for our economy.

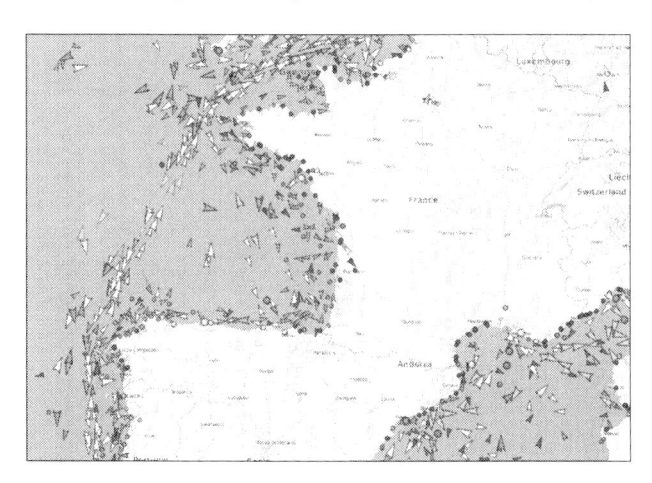

Figure 2: Vessel finder real-time tracking (Fleetmon n.d.)

The automatic identification system (AIS) is another essential maritime network and prtocol that can be attacked and spoofed (an AIS screenshot is shown in Figure 2). Designed to supplement marine radar systems and serve as a redundant method for preventing ship collisions at sea, the AIS connects near real-time information between nearby ships through AIS base stations and satellites. However, because the AIS has permitted open-source tracking of most major vessels navigating the seas, adversaries can use the system for illicit activities by gaining access to key data such as a ship's AIS serial number, name, and cargo weight (Fleetmon n.d.). More sophisticated adversaries can exploit the AIS to more easily triangulate, track, and target our ships at sea. And because such a threat can turn off access to GPS and AIS while vessels are operating, prudence dictates that ship operators be able to fall back on alternate navigation methods.

THE RELATIVE EASE OF OBTAINING DATA THAT EXPOSE VULNERABILITIES IN OUR MTS

To illustrate how relatively easy it is for adversaries to gain information on vulnerabilities to our maritime systems, including Supervisory Control and Data Acquisition/Internal Control Systems (SCADA/ICS), and to highlight just how much information is readily available through the open source, search engines such as those available at Censys.io and Shodan.io can quickly provide adversaries with a cyber scan of a port facility's network architecture. By initiating a simple free-text search on "SCADA" or "Modbus 53," a serial communications protocol used in SCADA/ICS systems (such as one depicted in Figure 3), adversaries can easily locate SCADA systems that are connected online, even though they were originally designed to be part of a more secure closed network. More sophisticated queries can reveal all SCADA/ICS systems that are connected within a port facility and show their flaws.

Port facilities and ship operators are using computer systems to also manage the vast amounts of cargo that are transported in and out of ports. With over 2 billion tons of materials processed through U.S. ports each year, as of 2014, computer systems play a critical role in tracking and accurately processing these goods (U.S. Army Corps of Engineers, 2016). These systems are important because they are interconnected with other systems that serve the same functions, and they enable the tracking of cargo as it travels around the globe. Whether it be through RFID technologies, GPS trackers, or manual scanning and logging, cargo is tracked and placed into databases that log their locations, containers, origins, and destinations. Such information is valuable to illicit actors who either wish to intercept and steal the cargo while in transit or to adversaries who want to manipulate the information in those databases to meet their own illicit intent and purposes. A recent instance of such an incident was reported in March of 2016 when maritime pirates exploited a vul-

CHAPTER 4

nerability in a shipping company's content management system; they used it to access records of shipping routes, schedules, and container contents, and as a result, these pirates were able to target specific ships and even specific containers and steal what was most valuable to them in a matter of hours (Zorabedian 2016).

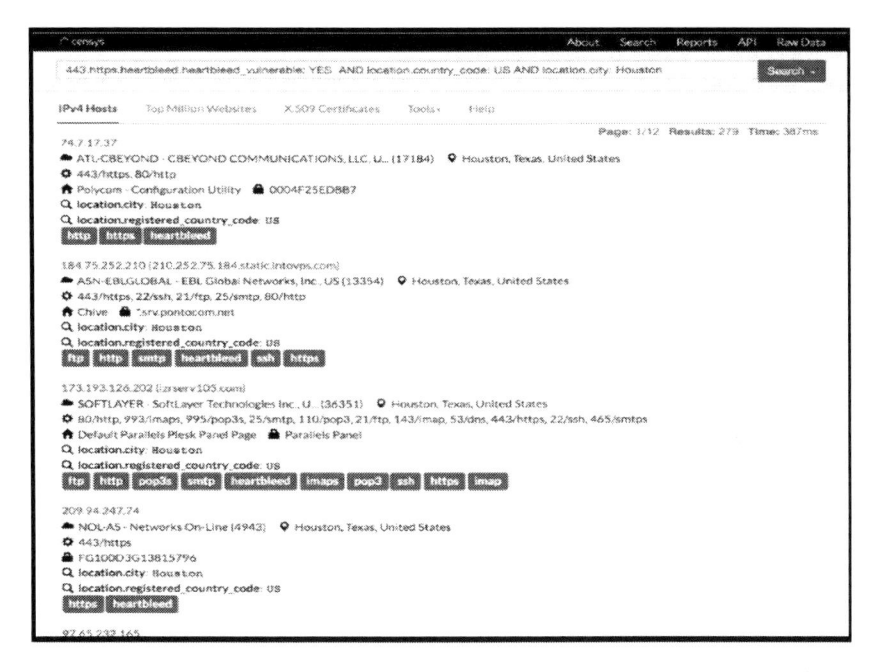

Figure 3: Screenshot of a Censys.io Query Revealing Computer System Vulnerabilities at Houston, TX[2]

COMPREHENDING THE FULL SPECTRUM OF CYBER SECURITY THREATS

There exists a wide range of threat actors who have the capability and intent necessary to conduct an effective cyber attack against U.S. ports. From hacktivist groups conducting nuisance computer-related activities at the Tier 1 and Tier 2 levels to rival nation states such as China, Russia, North Korea, and Iran capable of protracted campaigns at the Tier 5 and Tier 6 levels against our ports, the capability varies widely and presents a complex threat picture that we must be cognizant of if we are to successfully protect our ports and mari-

---

2   A Censys.io query with the parameters "heartbleed_vulnerable: YES AND location.city: Houston" will return IP addresses for those systems that are vulnerable to Heartbleed exploits within the Houston area.

time infrastructure. We consider criminal enterprises at the Tier 3 and 4 levels to be one of the more likely threat actors to conduct a limited attack against an individual U.S. port. For these illicit organizations, any physical damage to our port facilities may simply be a by-product of their true goals. Criminal organizations may seek to hamper the activities of rival groups, and shutting down port security and operations for a limited time during opportune periods may provide these illicit groups with advantages against one another as well as over law enforcement. Creating a widespread disturbance in a major port, especially one that threatens ecological disaster from a ruptured pipeline or refinery explosion, would be a way to divert law enforcement personnel and security away from the criminal enterprise's primary objective.

What is also important to realize is the huge financial resources that many Tier 3 and 4 level criminal enterprises already have at their disposal through cyber attacks. Identity theft and ransomware exploits have helped to provide a steady income stream for many of these actors, which in turn, support even more complex cyber attacks. Furthermore, because computer-enabled operations have relatively low cost of barriers to entry, criminal organizations that employ cyber attacks are able to create a disproportionate relationship between the effects they want to have on the target environment in comparison with the necessary capital required to adequately mount a cyber attack. In an environment where cyber security is weak or non-existent, the attacker's job is usually a trivial matter. To make things even worse, these criminal organizations do not even need to have the expertise or inherent capability for conducting the attack themselves. They can simply outsource their cyber requirements through the dark web, where any number of hackers-for-hire will sell their computer hacking skills for the right price.

A screenshot from a Tor-enabled website (shown on the next page in Figure 4) reveals there are any number of individuals who are willing to sell their expertise to anyone willing to pay them for their services. These individuals are known as black-hat hackers, crackers, and dark-side hackers. They leverage their expertise within the cyber environment to suit their own needs: they are able to obscure their true identities through the anonymizing feature of the Tor network; they tend to be more sophisticated and experienced users who purposefully make it extremely problematic for law enforcement agencies to attribute cyber attacks back to them; and they oftentimes can execute their cyber attacks from anywhere in the world, where they are out of reach of more stringent law enforcement and authorities. The one saving grace to the extraordinary challenges that these types of cyber criminals pose is that they are mutually entwined with the relative well-being of existing business models that help to define consumer spending and corporate profitability. If consumers no

CHAPTER 4

longer can afford to buy goods and services, and companies can no longer stay in business and sell to consumers due to systems disruption, these cyber criminals reduce the likelihood that they can be rewarded for their illicit efforts and expertise. Therefore, the threat of a large disruption to our cyber infrastructure from these types of cyber adversaries remain relatively low, in spite of the real and significant impact that they have on the security and economic welfare of our citizens, companies, and government.

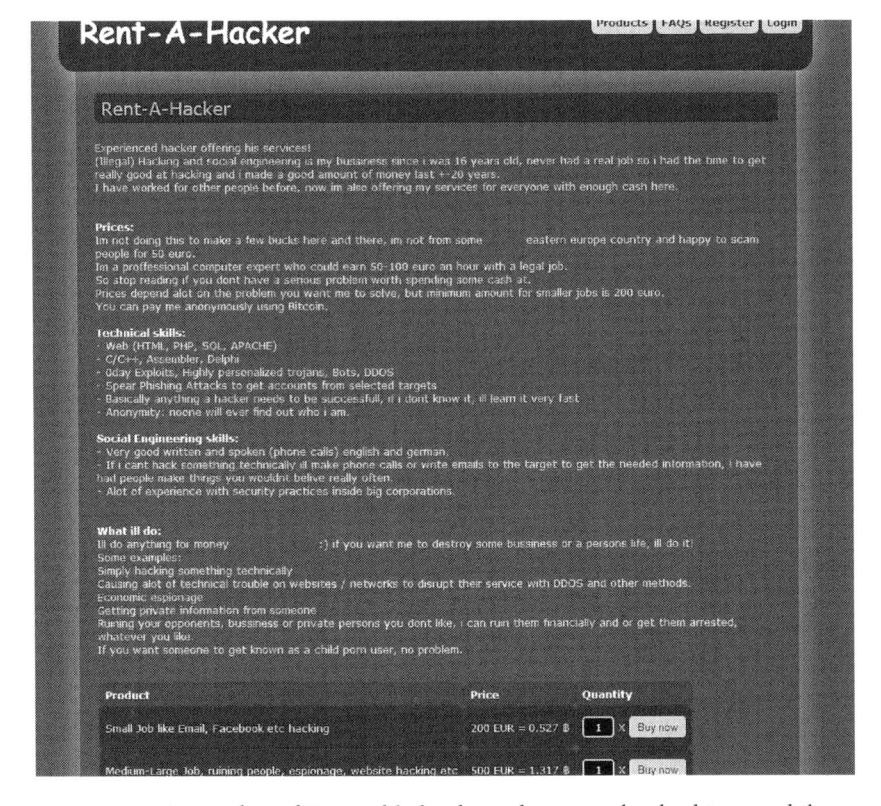

Figure 4: Screenshot of Tor-enabled website showing cyber hacking capabilities for a fee (Anonymous 2016)

Among the Tier 5 and 6 levels of predominately state sponsored adversaries, arguably the most capable and advanced persistent threat is Russia which, according to a number of cyber security experts, has already executed a number of sophisticated cyber attacks against critical infrastructure as part of their military strategy. Russia's suspected computer network attack against the Ukrainian electric power infrastructure in December 2015 demonstrates advanced planning, surveillance, reconnaissance, infiltration, and coordination of cyber capabilities with traditional military operations. While the power was only out for just 1–6 h for all the areas hit in Ukraine, a U.S. investigative report has found that it has taken months for a number of the utility control centers to become fully operational. Furthermore, these hackers could have done much more damage, if only they had decided to physically destroy substation equipment as well. But as a first-of-its-kind attack, the power disruption in the Ukraine has set an ominous precedent for the safety and security of power grids around the world (Zetter 2016). One can now anticipate how a cyber attack that escalates to the next progressive step in which power is disrupted to major cities, key facilities, airports, and seaports is no longer just science fiction—what happened in Ukraine is a harbinger of what is to come from cyber warfare.

Terrorist groups that want to cause harm to our nation may be considered merely a Tier 1 or 2 level threat in terms of their cyber sophistication. However, as discussed earlier in this chapter, these groups can offset their lack of skills and expertise by hiring computer hackers who have the capability and capacity to execute what they do not know how to perform, thereby elevating terrorist groups to more existential cyber threats to our national security. It may surprise some people to learn that Islamic State has already actively employed the dark web and has leveraged this area of cyber terrain to establish a propaganda site.[3] Furthermore, their sheer presence on the dark web places them in close digital proximity to more sophisticated cyber actors and criminal enterprises who do possess more advanced cyber skills which they currently lack. Over time, Islamic State and other terrorists will benefit from the proliferation of knowledge on the dark web and improve their technical understanding of the employment of cyber capabilities.

Other concerns include the numerous malware kits and exploits that are for sale on the dark web. As discussed earlier in this chapter, unsecure and vulnerable components of a network architecture's topology can be easily discovered. Knowledge of how to take advantage of these vulnerabilities is freely accessible on the internet. A terrorist is able to buy a malware kit to take over a vulnerable router for as little as $500. Being able to purchase cyber capabilities and tools with digital currency is another advantage that terrorists can employ. Additionally, recent reports over the past year suggest that the Islamic State has made it a priority to rapidly improve their cyber capabilities. Consequently, we see terrorists currently hovering at the Tier 3 and 4 level threat, but their motivation and determination to strike directly at the U.S. make them the most likely and pre-eminent cyber threat against our nation's critical infrastructure.

What makes these terrorist threats even more problematic is that these

---

[3] The Islamic State's dark web site has since been taken down.

groups have shown that they have the capacity to create a potential nexus for other cyber threat actors. For example, the Islamic State might successfully radicalize a "lone wolf" to conduct an attack from the inside. This would be particularly harmful for individuals with network administrator rights, privileges, and access where they work. The Islamic State already possesses the financial means to outsource a cyber attack to a criminal enterprise or independent black-hat hacker. Terrorist organizations could also act as a proxy to Tier 5 and 6 level nation-state actors who wish to employ cyber attacks without attribution. For example, Russia is reputed to have used proxies in their 2008 cyber campaign against Georgia in the form of the cyber criminal organizations, such as the Russian Business Network (Cilluffo 2016). Proxies provide a means of plausible deniability and can be especially attractive to those nation-states that have not sufficiently addressed policies and measures to establish retaliatory responses, attributional requirements, or international agreements on the prosecution of reciprocity.

Insider threats present extraordinarily large amounts of risk in the cyber environment. Anyone with access into their company's private databases has the potential to cause considerable harm. There are a myriad of reasons why someone on the inside might want to inflict harm on their own organization, including mental illness, extreme duress, workplace dissatisfaction, or enticement to become an agent of a foreign power. Furthermore, potentially harmful acts are oftentimes committed without ill will or malicious intent, resulting rather from mistakes or procedural errors. Insiders can oftentimes be enticed to perform seemingly harmless acts, such as granting entry, revealing group passwords, and providing organizational procedures and details to others who mask their affiliation as nation-state, criminal, or terrorist actors. Insider threats can also reveal more critical information to outsiders, such as virtual or close access to network components, data with intellectual property value, or knowledge of the defensive posture within their networks, and may even serve as conduits or hosts for the execution of a computer network attack. Accordingly, the cyber vulnerabilities exposed by organizational insiders are immense and can provide cyber adversaries with initial footholds into and reconnaissance of an organization.

Factors that Increase Difficulty in Attaining Competent Cyber Security

Many organizations tend to not invest too heavily in cyber security as this impedes ease of use and functionality. Executives and decision-making authorities tend to keep obsolete technology in place because it remains compatible with other systems in the architecture and investment in IT upgrades is costly. The overwhelming bias in security is favored toward protection and physical security rather than information and cyber security. In order to improve organizational efficiency, many organizations have chosen to link third-party services and providers into their business networks. In an effort to enhance their customer service and reliability, these organizations enter into mutual agreements that grant access to their network topologies. The more access that is given to a supposedly closed network, the greater the attack surface a cyber adversary can leverage. This usually occurs through social engineering and phishing attempts to gain entry into the systems. Unfortunately, once cyber adversaries have gained initial entry they try to elevate their level of access, take further control of the system, and hold data, processes, and functionality at risk or possibly for ransom.[4] In SCADA/ICS systems that were originally designed to be on closed networks, vulnerabilities are introduced each time new connections and IP-addressed devices are added to the network without cyber security considerations being fully scrutinized. Even at organizations that are fortunate enough to have network security specialists, the process of conducting traffic analysis and defense-in-depth security planning is time intensive. This burdensome task is made even more difficult when considering that many engineering components within network architectures communicate using industrial protocols that tend to be obsolete and antiquated. Furthermore, there is a scarcity of knowledgeable and trained personnel who are capable of managing the whole host of IT requirements for all our networks. Due to these many factors that weaken cyber security within our networks, perhaps the most prudent action that each organization that is networked into the cyber environment needs to take is to plan for contingencies and emergency responses when a cyber failure occurs.

Proactive Measures for Increasing a Cyber Adversary's Cost to Attack

While total prevention against a cyber attack is not possible, healthy cyber hygiene practices and relatively inexpensive cyber protection measures can provide sufficient baseline prevention such that the incidence of successful attacks becomes manageable. Furthermore, it is important to realize that it takes not only technical means to help monitor and detect a cyber threat, but, perhaps more importantly, a network of people with sound processes to quickly, effectively, and efficiently respond to the determined actions of an active and

---

4 An example of this third party form of cyber-attack occurred when the U.S. Office of Personnel Management had its sensitive data breached giving attackers security clearance records of over twenty million Americans. "OPM Breach Analysis: Update," June 9, 2015, http://www.threatconnect.com/opm-breach-analysis-update/ (accessed August 22, 2015).

CHAPTER 4

intelligent cyber adversary. Finally, it is important to establish a robust intelligence sharing architecture that can immediately alert key nodes in our cyber environment of an attack. This is necessary so that it can be treated at the localized level rather than allowing the event to escalate to a regional or national problem. Containing the damage will allow for much faster response measures and lead to much faster recovery from the incident. Rapid response and recovery will help to restore the people's trust and confidence in the security of our cyber infrastructure, and in the federal government's abilities to meet these challenges. These best practices are consistent with the prevailing construct that the U.S. Department of Homeland Security has provided on cyber security (as shown in Figure 5).

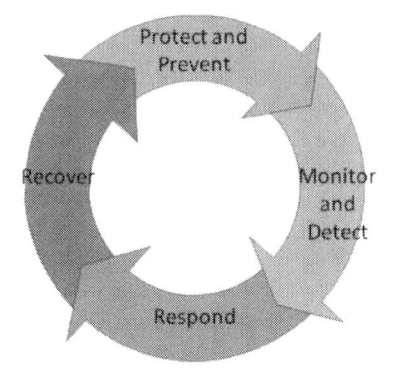

Figure 5: U.S. Department of Homeland Security's prevailing cyber security construct (U.S. DHS 2011)

The USCG's Roles and Responsibilities in Cyber Defense

The USCG is uniquely positioned to confront the challenges with building broad-based cyber security in our nation's ports and critical maritime infrastructure. To institute change across the key stakeholders that include domestic private industry, various echelons of the U.S. government, and foreign owned personnel and property, the USCG will have to make use of the various authorities, roles, and relationships at its disposal. As a law enforcement agency, a key component of the Department of Homeland Security (DHS), an arm of the military, and a member of the intelligence community, the USCG has the reach to engage the various partners with a stake in maritime cyber security. More importantly, it can hold these disparate organizations accountable in their respective responsibilities toward collective security.

As the coordinating body for maritime cyber security, the USCG can draw together elements across the Federal Government to establish a Joint Task Force to push forward a systematic plan that establishes baseline cyber security practices in the maritime industries, promotes sensible policies to enhance mutual support and response, links cyber security to ongoing physical security practices, and enforces punitive action against legal transgressions that expose national security to unnecessary risks against all levels of cyber threat actors. While it clearly does not have the personnel to take on these additional responsibilities on its own given the wide range of missions already assigned to the service, the USCG already coordinates the actions between these key stakeholders to ensure that port operations function on a daily basis.

Uniquely positioned within each of the three primary departments with cyber security responsibilities for the U.S. federal government (as shown in Figure 6 which outlines the various responsibilities within the three primary departments responsible for overseeing and coordinating cyber security issues), the USCG has the authority to institute and regulate baseline security standards as established by the National Institute of Standards and Technology. While these cyber security standards and best practices have been in place for a long time, they are not systematically enforced. In their role as a protection service within DHS with law enforcement capacity, the USCG will be able to wield the necessary authority to enforce compliance. Through guidance or regulation, the USCG will also be able to promote systematic cyber hygiene standards across the various industries, businesses, and multiechelon government activities in our port system. The implementation of modern firewalls, software patches, elevated defensive posture for high priority components and programs, use of encryption, and protective measures to ensure data integrity will go a long way toward reducing the odds of success that cyber adversaries can pose to our ports.

To further enhance the security of the MTS, the USCG can leverage its responsibilities as a member of the intelligence community and as an arm of law enforcement. As the lead for the Maritime and Ports Cyber Security Task Force, the USCG can bring together all the necessary partners to begin the hard work of preparing to respond to more sophisticated cyber attacks. This will require data and intelligence sharing on cyber threats, prioritization of key assets to protect and defend, joint planning, exercises, and coordinated rehearsals with all relevant government agencies and private industry partners to cyber attacks.

In response to a complex broad-scale cyber attack that jeopardizes the operation of our ports for an extended period of time, the USCG can exercise its authorities as the Maritime Cyber Security and Ports Joint Task Force and leverage the resources of Cyber Protection Teams (CPTs) which can help to protect and defend critical assets necessary to US TRANSCOM, Naval Facilities, the Merchant Marine, and all of the various sub-components of critical infrastructure that support continued economic activity in peace, and support

CHAPTER 4

Figure 6: U.S. federal cybersecurity operations team (U.S. Department of Homeland Security 2016)

to projection of military power in time of conflict. Oil refineries, pipelines, telecommunications, sensors and navigations systems, and the heavy machinery supporting logistics and sustainment are some of the key assets we want to deny an adversary from disrupting.

CONCLUSION

To instill comprehensive cyber security in our critical maritime infrastructure and ports, it is imperative that the U.S. government be able to protect our cyber terrain from the full range of threats. Practicing good cyber hygiene on our cyber networks is essential. By implementing relatively low-cost solutions such as the use of modern firewalls and de-militarized zones for networks, applying updated security patches to platforms, and employing multifactor authentication on our networked systems, we can reduce the attack surface that adversaries are actively seeking to initiate a cyber attack that disrupts the MTS. By taking these measures, we increase the overall costs to our adversaries in conducting such attacks, and thereby, help to create a more defensible cyber security environment (as illustrated in Figure 7).

The Attack Math

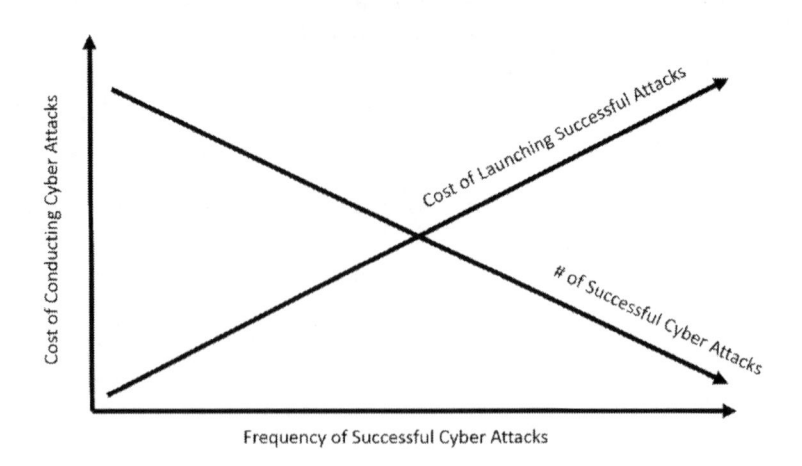

Figure 7: As the cost of launching a cyber attack increases the probability of having a successful attack decreases (McLaughlin 2015)

Because cyber security will never yield a perfectly safe cyber environment, the USCG will need to lead planning, practice, and support exercise response to a broad-scale maritime cyber attack. While the USCG can regulate and require MTS stakeholders to attain baseline cyber hygiene and protective standards, it is the responsibility of the entire maritime community to develop and implement protocols and practices that fit their business operations. Furthermore, the more that the entire maritime community can do to cooperate on standards and practices that increase the costs to a potential cyber adversary for carrying out a successful cyber attack on our MTS, the better prepared the USCG will be to respond to such an event. To bolster the resiliency of our MTS, the USCG will have to know how to request assets at the national level that can help in its response, and it will have to coordinate effectively with various governmental organizations and private industry partners. Based on its current authorities, missions, and roles, the USCG is well-positioned to lead these efforts. Furthermore, by partnering with other law enforcement agencies and intelligence communities, the USCG can proactively not only share and inform its key stakeholders of likely threats and modes of attack that can impact the MTS, it can better prepare its own forces for trends that our adversaries are taking to make our nation less secure.

Throughout this chapter, we have addressed the wide array of cyber related threats and vulnerabilities throughout our MTS. Most Tier 1 and 2 adversaries can be prevented from disrupting our maritime networks with good cyber security and hygiene. Tier 3 and 4 threats that can disrupt our MTS with cascading effects will require a more advanced level of preparedness, coordination, training, and defensive posture. This takes a commitment of time and resources to implement. Complex cyber attacks that originate from Tier 5 and 6 level actors are those that will seriously degrade our MTS which requires the most amount of coordination, preparation, and planning to mitigate against. Advanced persistent threats gain entry into our networks through multiple attack surfaces and will continue to maintain access as long as possible. They have the requisite skills to cover their tracks by scrubbing security logs and deleting electronic files that help to obscure their intentions. They are able to penetrate the network topology through multiple layers and regain entry into targeted systems using boutique malware signatures undetectable to existing commercial security vendors. These are the kinds of complex attacks capable of keeping large systems and infrastructures off-line for extended periods of time.

Accordingly, against a determined Tier 5 and 6 level threat that has compromised the integrity of our networks, we also must seriously contemplate under what circumstances we shut down our systems so that we not only contain the problem but begin the task of repairing, cleaning, and removing the infection within our systems—a daunting task with enormous ramifications which have been highlighted throughout this chapter. Despite the negative repercussions with taking such a drastic step, our entire MTS will be in a much better position to recover from a complex cyber attack if we begin the hard work of analyzing, coordinating, table-topping, rehearsing, and exercising these cyber incidences now while we have the time to properly prepare, rather than reacting to them after they occur. In order to prepare for a major incident response and implement a recovery strategy on the scale of a major port or multiple ports, it will be necessary to develop a comprehensive strategy for each USCG sector of responsibility. These sector strategies must address data storage and back-up, recovery sites, processing agreements to manage shared risk, and emergency management teams. It will fall to the USCG to ensure that each private enterprise at the port in question has a business continuity plan that will meet government requirements for protecting critical maritime infrastructure necessary for national security. Detailed and deliberate planning is needed to assess and specify requirements such as attainable recovery controls, recovery time objectives (RTO), and recovery point objectives (RPO). Ultimately, the goal is to maximize the operations of the MTS while minimizing the disruptions.

The task of securing our maritime transportation system and ports from cyber adversaries is indeed daunting. However, with appropriate tools, procedures, policies, and understanding of these cyber threats, the USCG as the lead collaborator among a team of MTS stakeholders is the ideal organization with the capability and capacity to reduce the likelihood of cyber attacks as well as prepare to defend against the existential broad-scale cyber threat. To succeed in this endeavor, it will take a committed long-term joint effort across many government agencies to make our ports and shipping industry more secure. Even in a greatly improved cyber security environment, we will not be able to protect against the most complex threats. It is therefore critical that we plan to be flexible, redundant, and cyber resilient in our approach to this challenge. Our responses must be prepared, coordinated, and rehearsed in advance so that we can quickly and systematically recover from a disruptive cyber incident before it turns into a crisis.

REFERENCES

American Association of Port Authorities. "U.S. Port Industry." http://www.aapa-ports.org/Industry/content.cfm?ItemNumber=1022 (accessed April 16, 2016).

Anonymous. 2016. "Rent-A-Hacker." March 15. 2016.2ogmrlfzdthnwkez.onion/index.php (accessed March 15).

Cilluffo, Frank. 2016. "Emerging Cyber Threats to the United States." Center for Cyber & Homeland Security, (February 25): 1–14. (accessed May 4, 2016).

Elementum News Desk. 2015. "The Real Cost of the West Coast Port Strike, Part 1." April 10. http://news.elementum.com/the-real-cost-of-the-west-coast-port-strike-pt.-1 (accessed July 7, 2016).

Fleetmon. "Satellite AIS—Space-borne Vessel Tracking." Germany. https://www.fleetmon.com/services/satellite-ais/ (accessed May 5, 2016).

Kramek, Joseph. 2013. *The Critical Infrastructure Gap: U.S. Port Facilities and Cyber Vulnerabilities*. Washington: Brookings Institution. http://www.brookings.edu/research/papers/2013/07/03-cyber-ports-security-kramek (accessed April 12, 2016).

Marine Log. 2016. "From Conversation to Implementation: Maritime and Port Cybersecurity." March 11. http://www.marinelog.com/index.

CHAPTER 4

php?option=com_k2&view=item&id=10688:from-conversation-to-implementation-maritime-and-port-cybersecurity&Itemid=230 (accessed April 18, 2016).

McLaughlin, Mark. 2015. "Prevention: Can it be Done?" In *Navigating the Digital Age,* 3-8. Chicago. Caxton Business & Legal, Inc. https://www.securityroundtable.org/wp-content/uploads/2015/09/Cybersecurity-9780996498203-no_marks.pdf.

Myers, Ryan. 2008 "Port Dedicated New Military Headquarters." *Beaumont Enterprise,* November 13. http://www.beaumontenterprise.com/news/article/Port-dedicated-new-military-headquarters-776096.php (accessed July 6, 2016).

Prudente, Tim. 2015. "Seeing Stars, Again: Naval Academy Reinstates Celestial Navigation." *Annapolis Capital Gazette,* October 12. http://www.capitalgazette.com/news/ph-ac-cn-celestial-navigation-1014-20151009-story.html (accessed May 5, 2016).

The Columbian. 2016. "In Our View: Act to Keep Trade Moving." March 3. http://www.columbian.com/news/2016/mar/03/in-our-view-act-to-keep-trade-moving/ (accessed April 14, 2016).

U.S. Air Force. 2009. "U.S. Air Force Software Protection Initiative." December 1. http://www.spi.dod.mil/tenets.htm (accessed April 12, 2016).

U.S. Army Corps of Engineers. 2016. *Tonnage for Selected U.S. Ports in 2014.* Navigation Data Center. http://www.navigationdatacenter.us/wcsc/porttons14.html (accessed May 5, 2016).

U.S. Coast Guard. 2016a. "United States Coast Guard Cyber Strategy." April. Washington, D.C. https://www.uscg.mil/seniorleadership/DOCS/cyber.pdf (accessed April 19, 2016).

-U.S. Coast Guard. 2016b. "United States Coast Guard Posture Statement, 2016 Budget in Brief, and 2014 Performance Highlights." https://www.uscg.mil/budget/docs/2016_Budget_in_Brief.pdf (accessed April 29, 2016).

U.S. Defense Science Board. 2013. *Cyber Security and Reliability in a Digital Cloud.* Task Force. Washington: U.S. Government. http://www.acq.osd.mil/dsb/reports/CyberCloud.pdf (accessed April 14, 2016).

U.S. Department of Commerce. 2016. *The Logistics and Transportation Industry in the United States*. Washington: U.S. Government. http://selectusa.commerce.gov/industry-snapshots/logistics-and-transportation-industry-united-states (accessed April 14, 2016).

U.S. Department of Homeland Security. 2011. "Enabling Distributed Security in Cyberspace: Building a Healthy and Resilient Cyber Ecosystem with Automated Collective Action." March 23. https://www.dhs.gov/enabling-distributed-security-cyberspace (accessed April 22, 2016).

Young, Stephanie. 2015. "Release of U.S. Coast Guard Cyber Strategy." Coast Guard Compass: Official Blog of the U.S. Coast Guard. June 16. http://coastguard.dodlive.mil/2015/06/release-of-u-s-coast-guard-cyber-strategy/ (accessed April 25, 2016).

Zetter, Kim. 2016. "Inside the Cunning, Unprecedented Hack of Ukraine's Power Grid." *Wired*, March 3. https://www.wired.com/2016/03/inside-cunning-unprecedented-hack-ukraines-power-grid/ (accessed April 15, 2016).

Zorabedian, John. 2016. "Pirates Hacked Shipping Company to Steal Info for Efficient Hijackings." *Naked Security*, March 7. https://nakedsecurity.sophos.com/2016/03/07/pirates-hacked-shipping-company-to-steal-info-for-efficient-hijackings/ (accessed April 19, 2016).

CHAPTER 5:
CYBER SEAWORTHINESS: A CALL TO ACTION

Kevin S. Cook and David L. Nichols
U.S. Coast Guard (Ret)

> *The Coast Guard Vessel Traffic Service (VTS) receives a call from the Master of the M/V TANKSHIP ONE stating that his ship is five miles from the sea buoy and is having a problem. The VTS watchstander requests to know the nature of the problem, to which the Master replies that there are numerous alarms sounding without cause, and the radar continues to show AIS targets in the ship's path although it is clear. The Master requests to transit to the cargo berth and says he will fix his casualty at the pier. After discussion with the Captain of the Port, the watchstander conveys the message,* **"Sir, your vessel is not considered seaworthy at this time and is denied entry."**

This is not a future scenario. Ships have had navigation systems and computer controls rendered useless causing operational shut downs. Like many land-based operations, ships have become increasing dependent on automation making them more and more vulnerable to a cyber incident, both intentional and unintentional. But unlike traditional shipboard casualties such as engine or steering failures, the ability to detect and repair a cyber casualty can be far more challenging. The question is what can and should be done about shipboard cyber security now.

If history has taught us nothing else, reactionary planning in the light of known hazards is not the optimal way to mitigate risk. There is too much at stake. Both the Oil Pollution Act of 1990 and the Maritime Transportation Security Act were products of reactionary legislation, precipitated by events that cost an estimated $7 billion and $3.3 trillion respectively. Cyber attacks are no different in regards to the need for deliberate planning and preparation to mitigate vulnerabilities and consequences, and they have the compounding risk of potentially global simultaneous impact. According to the 2015 Ponemon Institute Cost of Cyber Crime Study, the average cost of a single cyber attack globally is $7.7 million, and $15.4 million in the United States, up 19% from the previous year. Arguably, so much more is at stake with global shipping compared with other industries. With 95% of goods being shipped by sea globally, maritime commerce is the life blood of the world economy. The ability to disrupt the flow of goods by sea, potentially on a large scale, cannot be overlooked. While much thought has been given on how to mitigate the threat of maritime cyber incidents, we must view cyber stability and security on board ships just as critical to the vessel's seaworthiness as steering and firefighting systems.

The good news is that the maritime industry has a long history of tackling such issues on an international scale. Given the global impact of the cyber threat, the IMO must be involved in any regime designed to combat the cyber threat. More so than with other large issues such as environmental protection and safety of life at sea for which IMO has developed a robust set of rules, a single cyber event has the unique potential to occur at various localities around the world simultaneously. Therefore, a full understanding of cyber threats that might be generated in one country must be considered when creating rules to protect vessels transiting the territorial seas of other countries from that cyber threat. Through engagement with IMO, a consensus can be reached on how best to deal with cyber threats and requirements can be created to mitigate those threats. As with other requirements developed at IMO, cyber rules can lead to the development and exercising of shipboard plans designed to handle cyber threats. Third-party audits, and flag and port state enforcement, can ensure that operators are acting responsibly in carrying out their plans.

The real challenge lies in determining when a cyber event has rendered a vessel unseaworthy and therefore subject to flag or port state control or sanctions. For example, is it appropriate to direct a vessel to anchorage simply because they had a brief, but explained, malfunction of their AIS? The answer to these types of questions has been challenging. As the Coast Guard Cyber Strategy recognized: "In the U.S., the rapid growth and evolution of cyber technology has continually challenged our Executive, Legislative, and Judicial branches of government" (United States Coast Guard 2015a, 16). Partly for this reason, governments have been slow to develop prescriptive rules regarding cyber security, though such rules are not unprecedented. In 2009, the U.S. Nuclear Regulatory Commission issued prescriptive rules requiring licensees of nuclear power plants to "provide high assurance that digital computer and communications systems and networks are adequately protected against cyber attacks" (United States Nuclear Regulatory Commission 2009). That regulation goes on to describe specific systems that must be protected and the types of attacks from which they must be protected. Likewise, as recently as December 2015, the European Union adopted the Network and Information Security Directive which is expected to result in binding legislation by 2017 and would include cyber security rules for vessels and maritime interests. In the U.S., the Coast Guard relies on existing International Maritime Organization (IMO) documents such as the International Safety Management Code (ISM) and International Ship and Port Facility Security Code (ISPS) to provide guidance on handling cyber threats as well as the National Institute of Standards and Technology *Framework for Improving Critical Infrastructure Security*.

So what constitutes cyber unseaworthiness and when should it result in an intervention by a flag or port state? The answer could matter a lot. In the United States, for example, a vessel bound for or departing from ports or places in the United States is required to notify the Coast Guard of any "hazardous condition either aboard a vessel or caused by a vessel or its operation...." (United States Coast Guard 2015b). This, in turn, could result in the Captain of the Port directing the vessel to anchorage or to cease operations pending resolution of the cyber event (United States Coast Guard 2015c). As previously mentioned, these rules are much more easily applied to traditional mechanical or structural failure than a cyber event. That being said, we must find a consistent, standard, international approach to prevent and mitigate cyber threats, as well as define when a cyber event has happened. Similarly, it is feasible that a vessel could be deemed unseaworthy for purposes of a charter or carriage of goods contract at the start of a voyage due to a cyber incident or lack of cyber readiness thus resulting in contractual damages as well. If the United Nations Convention on Contracts for the International Carriage of Goods Wholly or Partly by Sea (the "Rotterdam Rules") adopted by the General Assembly in 2008 becomes international law, then it is also arguable that a vessel could be made unseaworthy at any point in her voyage due to a cyber incident also resulting in contractual damages.

Whether a prescriptive means of preventing and mitigating cyber threats is the right answer is up for debate. On the one hand, a non-prescriptive, guidance approach allows for maximum flexibility, yet can create inconsistency and lacks full accountability. Whatever the answer is, we must take immediate action in light of the yearly increase in cyber incidents over the last decade. Vessel seaworthiness is and will increasingly continue to be affected by cyber events.

Yes, this is a difficult problem to tackle; but the building blocks of an effective regulatory regime are well known. It takes the whole of strong international leadership IMO consensus on new requirements, classification society surveys, flag state enforcement, port state enforcement, contingency plans for vessels and port facilities, exercises, and professional mariners to solve such a complex international issue. While the physical world is clearly more amenable to a straight-forward regulatory regime, cyber security has emerged as today's greatest challenge and must be taken seriously.

Returning to the introductory scenario, there are two possible outcomes.

> **Either,** *the Master takes his ship to anchor and calls the owner. Neither of them knows what to do. The Flag State contacts the Port State, and they conference in the classification society; none of them are certain what standards to apply. After a lengthy delay, Flag State, Port State, and Class are finally willing to "sign-off" on the ship's cyber seaworthiness. Unfortunately, the ad hoc nature of the casualty resolution has shaken the confidence of intended cargo facility representatives and they do not want the perceived risk of this ship at their dock. The delay and vessel rejection will likely cost the owner millions of dollars.*

> **Or,** *the Master takes his ship to anchorage and takes out his approved Cyber Response Plan to identify and call the pre-contracted cyber response contractor. The classification society cyber surveyor attends the vessel to re-validate the Cyber Certificate of Seaworthiness on behalf of Flag State. On behalf of the Port State, the local Coast Guard inspectors, augmented by inspectors from the Coast Guard Cyber National Center of Expertise, confirm the ship's cyber seaworthiness. The Coast Guard Captain of the Port releases the ship to enter port and conduct cargo operations.*

The preference would clearly be for the latter outcome, but unfortunately it is not a current reality. If fairly assessed, the current state of cyber preparedness in the maritime world could only be deemed "unseaworthy." The time to move toward an international cyber standard is now as the risk of a serious cyber incident at sea only increases with the passage of time.

References

United States Coast Guard. 2015a. *Cyber Strategy.* Washington, D.C.

United States Coast Guard. 2015b. "Force Majeure." *Code of Federal Regulations*, title 33, section 160.215 (2015): 610. (January 30). https://www.gpo.gov/fdsys/pkg/CFR-2015-title33-vol2/pdf/CFR-2015-title33-vol2-sec160-215.pdf

United States Coast Guard. 2015c. "Possession and Control of Vessels." *Code of Federal Regulations,* title 3 CFR Section 6.04-8 (2015): 63–64. https://www.gpo.gov/fdsys/pkg/CFR-2015-title33-vol1/pdf/CFR-2015-title33-vol1-sec6-04-8.pdf.

United States Nuclear Regulatory Commission. 2009. "Protection of Digital Computer and Communication Systems and Networks." *Code of Federal Regulations*, title 10, section 73.54 (2009): 518–519. https://www.gpo.gov/fdsys/pkg/CFR-2016-title10-vol2/pdf/CFR-2016-title10-vol2-sec73-54.pdf.

PART II:

Policy

CHAPTER 6:
CYBER RISKS IN THE MARITIME TRANSPORTATION SYSTEM

Charles D. Michel, Paul F. Thomas, and Andrew E. Tucci
U.S. Coast Guard

Historic Background and Coast Guard Mission

The U.S. Coast Guard has a long history of protecting our nation from all manner of threats and hazards. When Alexander Hamilton founded what was then called the Revenue Marine, he charged those early sailors with patrolling our coasts and protecting our ports with vigilance.

Piracy and smuggling were the main threats of the day, but soon enough other risks appeared. Boiler explosions, navigation hazards, and fires on merchant vessels all threatened the safety of the nation's marine transportation system. The Coast Guard, including our various predecessor agencies, developed the capabilities needed to protect the nation from those and other risks, including oil spills, the dominance of foreign flag ships for our overseas trade, and terrorism. Stemming from the sabotage at Black Tom's Island in New York in 1916, the Coast Guard established Captains of the Port[1] whose duties center on port wide risks and maritime critical infrastructure protection.

Today, cyber-related risks are unquestionably a large and rapidly growing portion of all the risks our ports, facilities, and vessels face. The Coast Guard must address this threat in continuing to achieve our mission of protecting the safety, security, and stewardship of America's waters.

Cyber Risks and the Marine Transportation System

The U.S. Coast Guard is proud of our service to the country. We are also grateful for the professionalism and cooperation of the marine industry in helping to

---

1 See http://www.uscg.mil/hq/cg5/cg544/docs/Captain%20of%20the%20Port.pdf for details.

ISSUES IN MARITIME CYBER SECURITY

> *The Coast Guard's mission is to reduce the risk of deaths, injuries, property damage, environmental impacts, and disruptions to the MTS itself. Accordingly, our focus is on industrial control and other systems that could lead to these types of events. The integrity of IT systems that handle, for example, financial transactions is not, per se, a Coast Guard concern. Sound cyber risk programs will look at all types of risk, and operators need to be alert for the possibility that low risk or administrative IT systems may provide a network connection or backdoor to higher risk systems.*

build and operate the safest, most secure Marine Transportation System (MTS) in the world. The ports, terminals, vessels, related infrastructure and, most importantly the people that operate it drive the American economy and are vital to the nation's strength and prosperity.

Vessel and facility operators use computers and cyber-dependent technologies for navigation, communications, engineering, cargo, ballast, safety, environmental control, and many other purposes. Emergency systems such as security monitoring, fire detection, and alarms increasingly rely on cyber technology. Collectively these technologies enable the MTS to operate with an impressive record of efficiency and reliability.

While these cyber systems create benefits, they also introduce risk. Exploitation, misuse, or simple failure of cyber systems can cause injury or death, harm the marine environment, or disrupt vital trade activity. For example, vessels rely almost exclusively on networked GPS-based systems for navigation, while facilities often use the same technologies for cargo tracking and control. Each provides multiple sources of failure, either through a disruption to the GPS signal, or malware that impacts the way the signal is interpreted, displayed, and used on the vessel or facility.

Cyber vulnerabilities are in no way limited to GPS. Engineering and other systems are equally vulnerable. The Coast Guard and other authorities have documented cyber-related impacts on technologies ranging from container terminal operations ashore to offshore platform stability and dynamic positioning systems for offshore supply vessels. While in some cases modern day pirates and smugglers have been the source of these events, others have been the result of non-targeted malware or relatively unsophisticated insider threats. Even legitimate functions, such as remotely driven software updates, could disable vital systems if done at the wrong time or under the wrong conditions.

Commercial pressure and the ever increasing demand for speed, efficiency, centralized control, and convenience creates incentives to make greater and more integrated use of these systems. This in turn increases vulnerability and the "attack surface" available to hackers and criminals, as well as to simple misuse.

CHAPTER 6

The engine control room on a modern cruise ship. Photo credit: LCDR Eric Allen, USCG

Vessel and facility operators must be able to recognize cyber risks alongside more conventional threats and vulnerabilities. Once recognized, operators should address them via established safety and security regimens, such as security plans, safety management systems, and company policies.

Coast Guard Strategic Approach

The Coast Guard's operating model for all types of risk is to prevent incidents, accidents, and attacks whenever possible, and to be prepared to respond to those events when they do occur. Both have a role in the Coast Guard's recently released cyber strategy. Appendix 1, the cyber risk "bowtie model," illustrates some of the prevention and response related aspects of this approach.

The Prevention side of this equation is to identify and establish broadly accepted industry standards that reduce the likelihood of an incident occurring. In developing Prevention standards and programs for cyber and other vulnerabilities, the following principles apply:

Principles of the Coast Guard's Prevention Program

The Coast Guard's prevention standards are *risk based.* That is, they correlate the degree of protection with the potential consequences. For example, vessels and facilities that handle liquefied natural gas are subject to greater requirements than those that handle most other products. For any individual vessel or facility, vital systems such as firefighting, lifesaving, and communications are generally given more scrutiny than those with only a secondary influence on safety or security.

In addressing potential cyber vulnerabilities, the Coast Guard will follow a similar risk-based approach. While a vessel or facility may have any number of

Coast Guard personnel observing the security and safety control systems at a marine terminal. USCG photo

cyber dependent systems, our concern is with those few where failure or exploitation of the system might result in significant safety, security, or environmental consequences.

A second principle is that the Coast Guard uses **performance standards** wherever possible. That is, the purpose of our standards is to achieve a high degree of safety and security performance—to protect the mariners, facility workers and vessel passengers from harm, to protect the marine environment, and to avoid damage to property and equipment. There are many ways to accomplish that goal, and the Coast Guard strives to allow industry the greatest flexibility. In some cases, such as with our Maritime Transportation Security Act requirements, our regulations are almost entirely performance based. Even in cases where more prescriptive requirements are appropriate, such as engineering standards, the Coast Guard allows and encourages industry to propose alternative methods that achieve an equivalent level of safety or security.

Despite the technical nature of cyber systems, the Coast Guard believes that the principle of performance standards can and should be part of any vessel or facility's approach to reducing cyber risks. In some cases, an operator may choose to mitigate a cyber vulnerability through an established technical protocol. In other cases, training programs, physical access controls, or a simple manual backup may be a better option. The business needs of the organization should serve to identify the best method of reducing the risk.

A third aspect of the Coast Guard's Prevention model is that our standards reflect the **unique risks of the marine environment**. Heat, vibration, salt water, weather, and other factors require standards suitable for this environment. Coast Guard approval of items such as fire extinguishers and marine wiring reflect this reality.

The marine environment includes unique risks that any cyber risk management effort must address. These include serious consequences to people, the environment, prop-

> *The Coast Guard's Prevention Program:*
>
> *Risk Based,*
>
> *Performance Oriented,*
>
> *Customized to the unique marine environment*

erty, and the marine transportation system as a whole. The Coast Guard's cyber risk management program is concerned with these special maritime risks. Businesses certainly face other cyber risks, such as the loss of proprietary or financial data. These risks, while very real, are not unique to the maritime environment and are outside the Coast Guard's mission. The technical aspects of cyber security are also not uniquely maritime. Computers onboard a vessel or on a marine facility are no different from those in other environments, and the threats they face come in one and zeros wherever the computer is located and without regard to its ultimate function. Technical protocols need to be appropriate for the system and threat in question. They need no modification for vessel or marine facility use.

> *Cyber risks are an international threat. The Coast Guard is working with the International Maritime Organization to improve cyber risk management for vessels and ports subject to SOLAS and the ISPS Code.*
>
> **IMO**

Response, Investigation, and Recovery

Because we cannot expect to prevent all incidents (cyber related or otherwise), preparedness on the consequence management side is equally important to reducing the overall risk to the public and MTS. In many cases, addressing the consequences of a cyber event—such as an oil spill caused by computer controlled pump—is no different than if the incident had no cyber aspect. In such an incident, the responsible party would activate their spill response plan under the direction of the Coast Guard and other agency officials.

> *Appendix 2 describes cyber notification requirements. Notification triggers any needed immediate response actions and alerts the COTP to a possible port-wide threat. The Coast Guard will also support the Federal Bureau of Investigation and others in the investigation of cyber related crimes.*

The Coast Guard investigates pollution incidents, marine casualties and certain other incidents to determine the factors that led to the incident and prevent reoccurrences. If the investigation reveals a cyber nexus, the Coast Guard will work with law enforcement and other appropriate agencies to gather evidence and support criminal prosecution. In all cases, the Coast Guard will typically require the operator to conduct tests or inspections to ensure a system is safe before resuming

> **The NIST Framework identifies the following core functions:**
>
> Identify
>
> Protect
>
> Detect
>
> Respond
>
> Recover

normal operations. For cyber incidents, that process might include measures to ensure a system is free of malware or known vulnerabilities.

How Can Vessel and Facility Operators Manage Cyber Risks?

The marine industry has a long history of success in risk management. Mariners and port workers identify and evaluate risks on every watch and shift. Vessel and facility operators should view cyber along with the physical, human factor, and other risks they already face. The NIST Framework provides guidance on how to accomplish this. The first step is to identify and evaluate the sources of risk.

While physical and personnel risks are relatively easy to identify, cyber risks pose a unique challenge. Cyber vulnerabilities are invisible to the casual observer and cyber attacks can originate from anywhere in the world. Information technology specialists can help, but their focus is often with routine business applications. IT specialists may not fully recognize the various operational systems on a vessel or waterfront, the potential consequences should they fail, or have an operator's perspective on potential non-technical (and lower cost) solutions.

Risk Assessment

To assess cyber risk, assemble a team that includes operators, emergency managers, safety, security, and information technology specialists. Very briefly, their risk assessment process would proceed as follows:

- Inventory cyber dependent systems that perform or support vital operational, safety, security, or environmental protection functions.

- Map any connections between these systems and other networks. Note which systems are accessible via routine internet connection and for portable media such as USB and CD drives. This step in the process helps to identify potential **vulnerabilities**. Note that even systems with no connection to the internet whatsoever are still subject to insider threats and simple technical failures.

- For each system, discuss the potential **consequences** if the system was exploited, malfunctioned, was unavailable, or simply failed under "worst case scenario" situations. Remember, Murphy's Law always applies, and adversaries may combine a cyber attack with a physical attack.

- Considering both the vulnerability and the potential consequences, evaluate the relative risk for each system. Systems with multiple vulnerabilities and high potential consequences have higher risk than those with few vulnerabilities and low potential consequences.

Risk Mitigation

Once the team recognizes their cyber risks, the organization can select mitigation strategies to reduce that risk. Prevention/protection strategies reduce vulnerabilities and the frequency of successful attacks or adverse events. While high-risk systems should naturally have more robust protection strategies, this does not necessarily equate to sophisticated technical solutions. For example, physical access control and training may be sufficient for systems where the primary vulnerability is an insider threat. Where risk managers choose technical solutions, they must also recognize their limitations. Many systems are only capable of recognizing and blocking known threats. Unfortunately, the pace of innovation in the malware world is increasing, zero day exploits are common, and a strategy that relies exclusively on a perimeter defense designed to filter out known threats will not be successful.

> The term **Defense in Depth** refers to a multi-faceted and multi-layered approach to cyber defense. Defense in depth considers the various people, technology, and operating policies an organization might adopt. It includes protection, detection, response, and recovery activities. Defense in depth recognizes that no single strategy can ensure security.

Operators can also reduce risk at the consequence end. For example, manual backups may be appropriate for situations where the cyber failure is disruptive, but does not include immediate life, safety, or environmental impacts. Manual backups can be an excellent way of building cyber resilience—provided the manual system is reliable and personnel still know how to use it!

Exercises can help identify the procedures an organization may need to take to isolate a suspect system, purge it of malware, and safely resume operations. Including a cyber aspect into an existing security, natural disaster, or environmental response plan can help an organization prepare for a cyber incident with an "all hazards" approach.

The teamwork approach among operators, IT specialists, and other risk

> There are many private and public resources available to help companies address cyber risks, including ICS-CERT. Identifying these resources in advance and designating specific personnel with the responsibility to contact them will improve preparedness.
>
> **ICS-CERT**
> INDUSTRIAL CONTROL SYSTEMS CYBER EMERGENCY RESPONSE TEAM

managers is vital. Only a multitalented team can develop multitalented solutions. Regardless of the strategy chosen, operators need to see risk assessment and risk mitigation as continuous processes, not one-time events. While this is true for any risk an organization may face, the rapid change in technology and its ever increasing use in society make this especially important.

Risk Management

Once an organization has identified, evaluated, and mitigated cyber-related risks to an acceptable level, it must still do two things to maintain that condition. First, organizations need to incorporate their cyber procedures into appropriate internal policy and operating requirements. These will vary from organization to organization, but may include the following:

- Safety Management System/ISO procedures
- MTSA required security plans
- Operations manuals
- Continuity of Operations/Continuity of Business plans
- Company training programs and policies

Second, because no risk is static, organizations must view cyber security as a *process*, and establish a regular schedule to review cyber risks, re-evaluate the need for mitigation measures, and ensure personnel understand and can follow good cyber practices. Rapid changes in technology and ubiquitous cyber threats make this concept especially important. Ultimately, an organization should strive to incorporate cyber into an existing culture of safety, security, and risk management.

A final point is that an organization's leadership must recognize and visibly support a strong cyber culture as an "all hands" responsibility, not just simple IT function. With the backing of senior leadership, an organization can develop this strong cyber culture needed to keep the operations safe, secure, and efficient.

CHAPTER 6

Conclusion

Despite the apparent complexity and scale of cyber threats, we can and are adding cyber to a long list of risks the maritime industry and the Coast Guard have overcome. More senior members of the Coast Guard, and of industry can look back on their careers and see great advances in environmental stewardship, safety, and conventional security. Those accomplishments reflect a cooperative approach that establishes meaningful standards to address real risks, devises flexible strategies to meet those standards, and shares responsibilities to maintain those systems over time. We have strengthened our nation and ensured that our ports and waterways are a safe place to live, conduct business, and link our economy to the world.

While cyber risk management certainly requires some technical skills from the current and next generation of leaders, it will succeed on the foundation of those of us (this authors included) that still think an A-60 bulkhead is the best firewall for any situation.

Appendix 1: Cyber Risk Bowtie Model

The model below depicts cyber risk management activities. On the left, the model notes several types of attack or threat vectors. These range from sophisticated, targeted attacks from "Advanced Persistent Threats" (including, but not limited to nation-states), down to a simple technical error, such as improper software update. The term "insider threats" also represents a broad range of actors—from those with special access and a desire to inflict deliberate harm on an organization to those who unknowingly introduce malware by clicking on the wrong link or plugging a personal smart phone or other device into a USB drive or other port.

Cyber Risk Bowtie Model

All activities must take place against a backdrop of the training, education, and policies needed to promote a culture of cyber security.

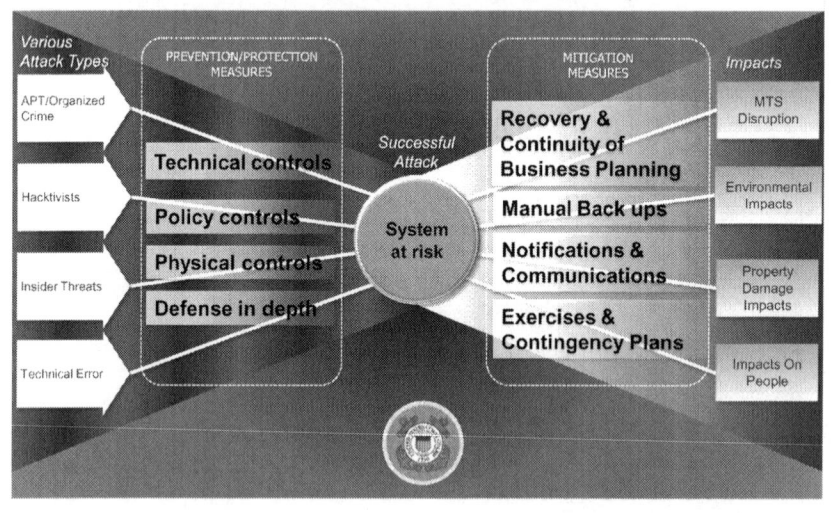

Various Attack Types	Prevention/Protection Measures		Mitigation Measures	Impacts
APT/Organized Crime	Technical controls	Successful Attack	Recovery & Continuity of Business Planning	MTS Disruption
Hacktivists	Policy controls	System at risk	Manual Back ups	Environmental Impacts
Insider Threats	Physical controls		Notifications & Communications	Property Damage Impacts
Technical Error	Defense in depth		Exercises & Contingency Plans	Impacts On People

Prevention/Protection measures reduce the likelihood of an incident by creating barriers to the malware or other measures that can compromise a system. These include technical measures, policy and training, and physical access controls. Once an incident has occurred, communications, response, and contingency plans reduce the impact of the event and promote rapid recovery. An organization with strong cyber resilience will consider all types of threats, institute both protection and response procedures to reduce risk, and promote a strong culture of cyber security through training, education, and leadership.

CHAPTER 6

APPENDIX 2: CYBER INCIDENT NOTIFICATIONS AND INVESTIGATIONS

Coast Guard regulations[2] require MTSA regulated vessel and facility operators to report suspicious activity, breaches of security, and Transportation Security Incidents to the U.S. Coast Guard. This includes incidents and activities with a cyber nexus. In cases of a TSI or other emergent incident, notification enables the Coast Guard and other security partners to take immediate action to protect the port and respond to the threat. Suspicious activity reports provide the Captain of the Port with information that, in combination with other sources, may indicate a port-wide threat.

In practice, cyber incident reporting has some unique challenges. In many cases computer security monitoring, such as intrusion detection, is done remotely rather than at the vessel or facility operator level. Detecting a cyber incident, recognizing the potential for it to impact systems related to Coast Guard requirements, and relaying that information to the Coast Guard as well as the vessel or facility operator in a timely manner is not as straightforward as it might be for a physical security incident.

The definition of "suspicious activity" in a cyber context is also problematic. Larger organizations may experience near-constant attacks on their firewalls or routinely find malware on various networked systems. Reporting every such incident is neither practical nor desired.

The Coast Guard and industry have a shared goal of keeping our nation safe, secure and protecting our marine transportation system. Organizations must report cyber incidents that threaten that goal, affect vital systems, or impair functions described in Coast Guard security plans. Our purpose is to promote mutual security, never to punish those who make a judgment call in good faith.

The Coast Guard also recognizes that cyber incident reporting requires diligent attention to confidentiality. As of this writing, several federal government organizations accept or require cyber incident reports. Agencies are working to streamline these systems in a way that minimizes the impact on industry, maximizes security, and ensures that agencies have access to the information they need to carry out their responsibilities. While the nuances of that effort are beyond the scope of this paper, suffice to say that this is a complex task, and that the Coast Guard and other agencies ask for patience, cooperation, and suggestions on accomplishing this goal.

The National Response Center (NRC) is the designated reporting point for Coast Guard regulated vessels and facilities. The NRC is staffed by trained professionals who treat all security reports as Protected Critical Infrastructure

---

2   33 CFR 101.305.

Information. Distribution of these reports is limited to law enforcement agencies on a need to know basis. In cases where extreme discretion is appropriate, vessel and facility operators have the option of reporting an incident directly to the local Captain of the Port, with a follow up call to the NRC providing only generic information for documentation purposes. Regardless of how a report is made, the Coast Guard will share the information with the FBI, and with other agencies with cyber security responsibilities. With the help of those agencies, we will facilitate efforts to help the impacted vessel or facility operator recover from the incident, resume operations, and support prosecution efforts.

CHAPTER 6

Appendix 3: Cyber Security Roles and Responsibilities

A full discussion of the various cyber security related authorities and responsibilities within the federal government is beyond the scope of this paper. Broadly speaking, the Department of Homeland Security is primarily responsible for critical infrastructure protection, the Department of Justice is primarily responsible for criminal investigations, while the Department of Defense is responsible for national defense.

	DHS	DOJ	DOD
Lead role	Protection, Information Sharing	Investigation and Prosecution	National Defense
Responsibilities	Coordinate national response to significant cyber incidents Disseminate domestic cyber threat and vulnerability analysis Protect critical infrastructure Secure federal civilian systems Investigate cyber crimes under DHS jurisdiction Coordinate cyber threat investigations	Prosecute cyber crimes Investigate cyber crimes Lead domestic national security operations Conduct domestic collection and analysis of cyber threat intelligence Coordinate cyber threat investigations	Defend the nation from attack Gather foreign cyber threat intelligence Secure national security and military systems Support the national protection, prevention, mitigation of, and recovery from cyber incidents Investigate cyber crimes under military jurisdiction

These descriptions are best understood as generalizations. Individual agencies often have their own, unique authorities. For example, within DHS, the U.S. Secret Service has authority to investigate and prosecute certain types of computer fraud and other cyber crimes.

The U.S. Coast Guard, as a member of the Department of Homeland Security, has responsibility to help protect the nation's maritime critical infrastructure, and to promote safety and security in the Marine Transportation System. As a member of the U.S. Armed Forces, the Coast Guard works closely with the Department of Defense, including U.S. Cyber Command, in defending the nation. As a law enforcement agency, the Coast Guard has authority to investigate violations of all federal crimes with a maritime nexus (14 U.S.C.). Finally, the Coast Guard is a member of the intelligence community, providing us access to many sources of information that can help us with our mission to protect the American people.

CHAPTER 7:
BARRIERS AGAINST CYBER SECURITY THREATS: USING BARRIER MANAGEMENT TO VISUALIZE AND ADMINISTER CYBER SECURITY COUNTERMEASURES

Pål B. Kristoffersen
DNV GL, Norway

Abstract

The recommended approach to reduce the risk of cyber security threats is to establish effective countermeasures. Moreover, an asset owner needs confidence that countermeasures are sufficient and correctly performed. In traditional safety systems, barrier management and bow-tie method are established methodologies to both reduce probability and consequence of hazards. In this paper, DNV GL explains how the bow-tie method and barrier management can be used to visualize and manage cyber security threats and countermeasures. Barrier management adds value by both focusing on reduced probability of cyber security incidents and by reducing the consequences of such incidents. Barrier management is effective to make sure multiple independent barriers are in place, and it is effective to make sure the performance of the barriers is maintained over time. Samples are given from studies of cyber security vulnerabilities in the Norwegian maritime and oil and gas sectors. Barriers for malware-related incidents and denial of service attacks are shown.

Introduction

Most organizations within the maritime and offshore oil and gas sectors are aware that they may be a target for cyber security threats. Indeed, they may also have started to establish a defense against such incidents. However, only a few organizations have established a systematic approach and very few are aware of their actual risk for such incidents. The objective of this paper is to suggest a systematic approach to manage cyber security threats. The main idea is to learn from safety management practices in industry, where the barrier concept has been used for years. The industry's cumulative safety knowledge is highly relevant to implement a risk-based barrier strategy for cyber security threats.

Background

The maritime and offshore oil and gas sectors have achieved significant cost saving by using digital systems. Such systems are used in all parts of the industry, including central infrastructure, communication, navigation, and control systems onboard ships and platforms. Digital vulnerabilities in general information systems are well known, and incidents based on such vulnerabilities have been well publicized. For example, Somali pirates have exploited online navigational data to choose which vessels to target for hijacking. Additionally, criminal organizations have infiltrated computers in the Port of Antwerp to smuggle drugs and delete the falsified cargo records (Wagstaff 2014).

Focus on vulnerabilities in automation, control and safety systems is a more recent concern. The Stuxnet worm in 2010 (Kushner 2013) showed how a worm could gain access to process control systems and launch a complex attack to manipulate and destroy industrial components. In 2013, researchers from the University of Texas demonstrated the possibility to change a vessel's direction by interfering with its GPS signal (UT Austin 2013). If terrorist organizations are able to control vital components in navigation or ship control systems, the consequences may be a significant loss of lives and/or an environmental disaster. Given that the maritime transportation system carries approximately 90% of international commerce, any significant disruption will also have substantial economic consequences. Cyber risks must therefore be given the same attention and priority as traditional safety issues.

The Generic Approach to Cyber Threats

The generic approach to reduce the risk of cyber security incidents is to establish countermeasures. If the countermeasures are sufficient and correctly deployed, the risk may be acceptable. In Figure 1, this approach is illustrated based on the common criteria model (ISO/IEC 15408 2009). To administer such cyber security countermeasures is a challenge. The process must be organized and documented in conjunction with proven risk management methods. Since the challenge for cyber security is similar to many safety issues, the well-known risk methodology for safety and, in particular, barrier management philosophy with the bow-tie method can be used.

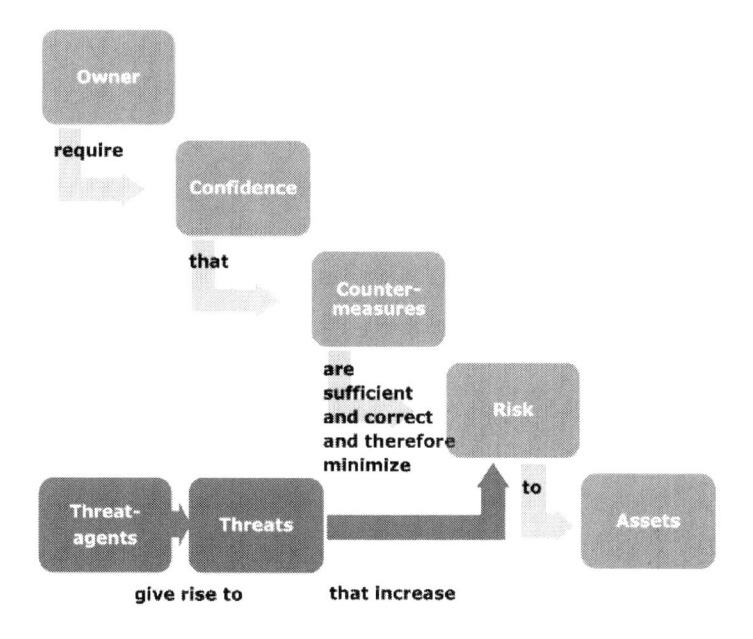

Figure 1. Generic approach to cyber threats

BARRIER MANAGEMENT

The barrier management philosophy has been used for years to manage risks for fire and explosion. Indeed, the rationale for using barriers dates back to the energy barrier principle that was introduced in 1961 (Gibson 1961). In simple terms, the basic idea of barrier management is to establish a set of barriers to hinder an incident from occurring and also to establish a set of barriers to mitigate the consequences of an event if one does occur. Such barriers may consist of technical elements (e.g., a fire extinguisher, an interlock or automatic shutdown), operational elements (e.g., the task to inspect fire extinguishers, test interlocks or automatic shutdowns), and also organizational elements (e.g., training the firefighter or procedures to require limited operations or even a shutdown if certain barriers are not available or have been degraded).

Effective barrier management calls for "coordinated activities to establish and maintain barriers so that they maintain their function at all times" (PSA 2013). Conversely, the barrier management concept also recognizes that an incident may have many causes and many consequences and the model also considers that none of the barriers are either 100% reliable or 100% effective. To address less than 100% reliability or effectiveness and multiple causes and consequences, sound barrier management typically requires implementing a number of redundant and independent barriers. For the first example, the barrier management model and the bow-tie method for malware and denial of service (DOS) attacks are explained below. Some cyber security countermeasures in these samples, such as testing and training, are not usually defined as barriers, but for simplicity they are handled as such in this chapter.

The Bow-Tie Method

The bow-tie method is a well-accepted method to visualize barriers. In fact, a bow tie figure looks similar to a bow-tie, in which the knot represents the "top event", the undesirable initiating event. Next, the "left wing" of the bow-tie represents the different threats and the barriers employed to prevent the incident from happening. Then, the "right wing" of the bow-tie shows the different consequences and the barriers to reduce them. As bow-ties clearly show risks as well as preventive and corrective measures, the bow-ties are often linked or incorporated into risk management systems. Additionally, bow-ties may facilitate more rigorous internal and external audits.

The advantages of using the bow-tie method for cyber security threats are:

- Both barriers to prevent the incident from happening and barriers to reduce the consequence. -In maritime cyber security defense there has been too little focus on barriers to reduce the consequences.

- Multiple independent barriers. -There is too much trust in a single barrier as e.g. a firewall.

- Focus on maintaining barrier quality over time. -With no maintenance, many cyber security barriers will be outdated in short time.

Barriers Against Malware

Malware, short for malicious software, is any evil software that may harm an organization. The term includes computer viruses, worms, trojan horses, ransomware, spyware, adware, scareware, root-kit, and similar programs. In Norwegian studies of both the maritime and the oil and gas sectors, malware spread during system upgrades and maintenance is among the top-10 vulnerabilities (Lysneutvalget 2015). Memory sticks with infected files may be the most important source. A simplified bow-tie showing barriers against malware is shown in Figure 2.

CHAPTER 7

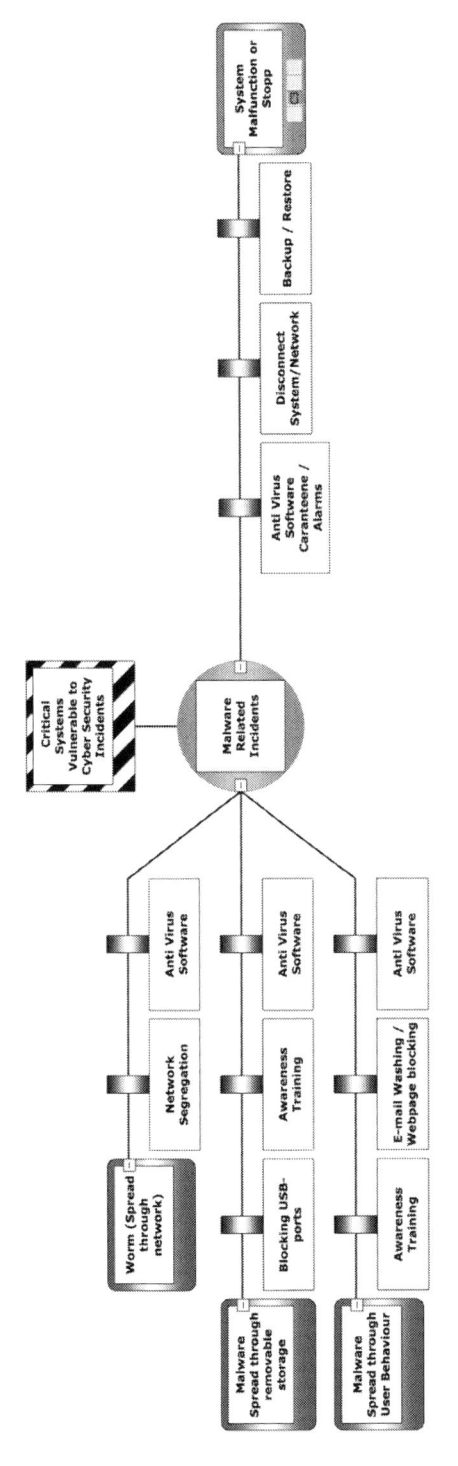

Figure 2. Barriers against malware

Figure 2 shows the threats categorized by how the malware can get into the system. Normally, this happens through the network (worms), through removable storage media or based on user behavior. Social engineering is increasingly used to exploit weaknesses in user behavior.

In this example, the barriers to prevent a worm are network segregation and limiting the traffic allowed to flow between network segments. Thus, only defined traffic (protocols) and nodes (addresses) are allowed. Special intrusion prevention systems (IPS) that recognize and block known malicious patterns are available, but these are mainly targeted at larger information systems and they may not be suitable for industrial control systems. However, anti-virus and anti-spyware software are mandatory for all malware threats as well, including the regime to keep the signatures updated. Updating anti-virus and spyware software for ships may require manual procedures since data traffic to ships may be limited due to segmentation and limited bandwidth. System hardening and patch management are needed barriers for malware given that changes are tested, particularly for critical systems. Lastly, a software patch inventory system is essential to know what patches have been installed in all components.

Next, barriers to malware that can be introduced via removable storage include blocking or disabling the USB ports on all devices where the risk of such incidents is unacceptable. Undeniably, guidelines for the accepted use of removable storage-devices and awareness training are important for both employees and contractors.

Barriers to hinder unwanted user behavior or prevent social engineering are primarily guidelines and awareness training. For example, opening e-mail attachments is the most common source of malware, so procedures for handling e-mail attachments are essential, however, this must be reinforced with random tests and audits. Other procedures and common practices include segregating devices where users handle e-mail or web browsing from critical equipment. Similarly, both incoming and outgoing e-mails should be washed, and web-traffic may either be washed or limited, however, whitelists are preferred.

If malware or spyware is introduced, the barriers to reduce the consequence of vital control systems stopping or malfunctioning is primary handled by anti-virus software. These packages work by attempting to quarantine the malware and generating alarms. Even so, the antivirus system may not effectively quarantine all malware. Therefore, procedures to detach critical networks from the internet and to detach the production network from the general IT network must be established. So, procedures to restore infected systems are required as well as training personnel in identifying and responding to malware attacks.

CHAPTER 7

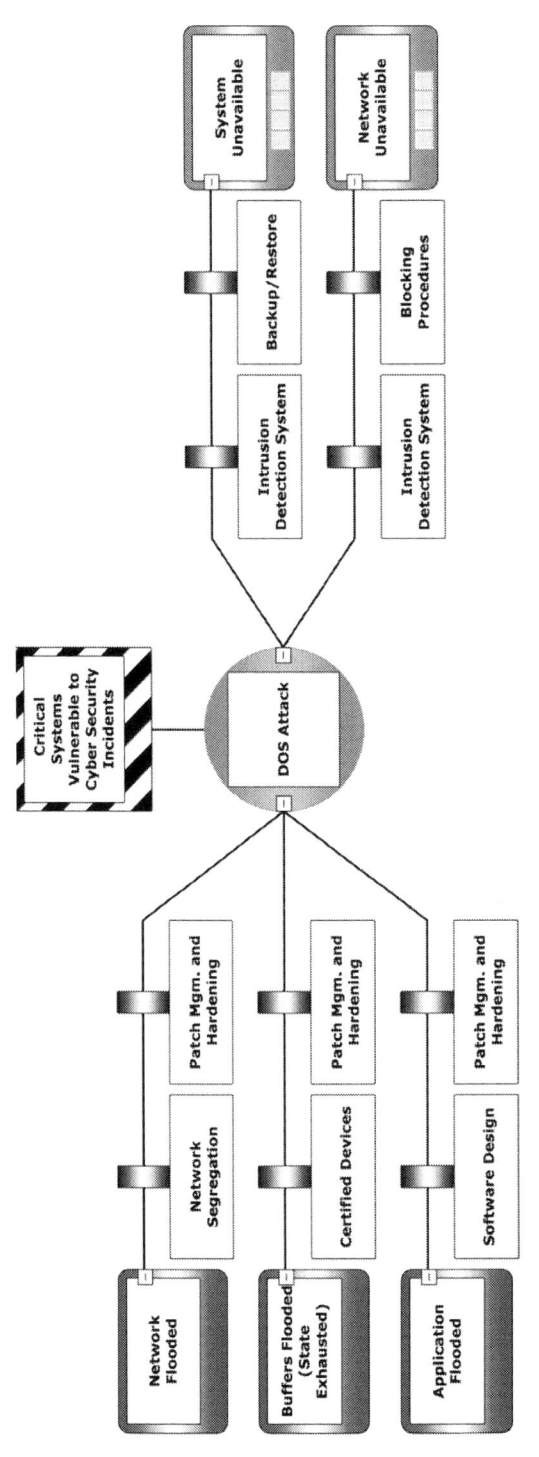

Figure 3. Barriers against DOS attack

Barriers Against Denial of Service Attacks

A denial of service attack makes a system or network resource unavailable to its intended users (deny the service). A coordinated DOS attack from multiple sources is called a distributed denial of service attack (DDOS). Either of these may also be executed to camouflage other attacks, such as espionage or hacking. Testing performed by a maritime test-laboratory (Marine Cybernetics) of DOS attacks on critical maritime systems has shown dramatic consequences, and has led to bug-fixing on these devices. Figure 3 shows a simplified bow-tie for a DOS attack against an industrial control system.

In this scenario, the attacker floods the network, floods buffers in devices (e.g., state buffers in the firewall) or floods the application (e.g., by doing frequent resource intensive operations).

Barriers to prevent a network flooding threat are to segregate the network and limit the traffic allowed to flow between the segments. Only a minimum of defined traffic (protocols) and nodes (addresses) should be allowed. Special devices to block DOS attacks ("cleaning centers" or "scrubbing centers") are available, but they are mainly targeted at larger information systems and may not be suitable for industrial control systems onboard a ship, offshore rig, or terminal. As with malware barriers, software updates and patch management are important barriers as well as hardening of systems for all types of attacks. Typical barriers to prevent buffer flooding are the use of security certified devices and to configure the devices according to certificate requirements. Application flooding is primarily addressed by building resistance into the application design so that it is resistant against "legal input with evil intent." Then, coupled with strong user authentication and authorization, an effective barrier to application flooding can be established.

In order to reduce the consequences of a DOS attack, the unwanted traffic must be identified; there are special intrusion detection systems (IDS) or DOS detection systems available, but these may not be suitable for the smaller shipboard and offshore industrial control systems. Another approach for a DOS barrier is to use simpler devices to monitor the network traffic and establish procedures to monitor these logs. Then, if abnormalities are detected, the source network should then be blocked and systems may be restored.

Sufficient and Correctly Implemented Barriers

As previously stated, the owner of the asset needs confidence that the countermeasures are sufficient and correctly performed. Ideally, this confidence is based on a risk assessment where the risks are found to be acceptable or "as low as reasonable practical" (ALARP). In the ISO/IEC 15408 standard (ISO/IEC

15408 2009), a special assurance technique to give the required confidence is described. This technique can then be used to confirm that specific requirements are fulfilled. Sufficiency and correctness may also relate to what is "best practice" and, preferably, best practice is based on standards. For the maritime sector, there are a number of standards relevant for cyber security, such as ISO/IEC 27001 (ISO/IEC 27001 2013a), ISO/IEC 27002 (ISO/IEC 27002 2013b), ISA/IEC 62443-3-3 (ISA/IEC 62443-3-3 2013), and NIST 800-82 (NIST 800-82 2011), which are particularly relevant.

To verify that a barrier is correctly established, a trusted vendor, a trusted independent party, or the company's own resources may be used for verification. Generally, a combination of these three approaches is recommended. For general information systems and components, such as operating systems, routers, firewalls, etc., the common criteria model is often used in which a third party (the evaluator) has verified that the system or component is secured according to a security target or a profile. For maritime industrial control systems, however, the new regime based on the ISA/IEC 62443-3-3 standard (ISA/IEC 62443-3-3 2013), in which the product or system shall comply with certain "security levels," may be more relevant.

If the owner of the assets verifies the barrier quality, audits, and testing are typical methods. An audit may be performed on, e.g., the filter definitions in a firewall and testing will usually be a penetration test. There are a large number of tools and partners for penetration testing of variable quality. In any case, certified resources are preferred. Organizations such as SANS (CWE/SANS 2011) and OWASP (OWASP 2013) have defined "top vulnerabilities," and testing should verify the barriers against these vulnerabilities, at a minimum.

Barrier Management

Effective barrier management requires maintaining barriers so that they maintain their function at all times. Therefore, developing and establishing a "barrier strategy" and "performance standards" are vital activities for successful barrier management. A "barrier strategy" should require that all barriers must have sufficient capacity, availability and response times to fulfill their function, while performance standards shall ensure that the barriers are suitable and fully effective for the type of hazards identified. Many cyber security barriers are ineffective because there are no specified performance requirements and no means to follow-up on these requirements. For example, devices are logging security events, but no one is analyzing the log-files or initiating corrective actions. Similarly, the patch system is not fully implemented for all software programs and components.

Conclusion

Bow-tie methodology and barrier management is a highly effective way to visualize and manage cyber security threats. Used together, these allow identification of the threats, development, and establishment of barriers to prevent incidents and mitigate the consequences of a cyber event and to monitor the cyber defense for weak or insufficient protective measures. By implementing a barrier strategy, and by integrating this strategy in the risk management system, cyber risks may be reduced to an acceptable level. Nevertheless, for further risk mitigation the "risk avoidance" option should still be remembered: "Not all critical systems need an internet access."

References

CWE/SANS. 2011. *TOP 25 Most Dangerous Software Errors.* SANS Institute. https://www.sans.org/top25-software-errors/.

Gibson, James. 1961. *The Contribution of Experimental Psychology to the Formulation of the Problem of Safety—A Brief for Basic Research.* New York: Association for the Aid of Crippled Children.

IEC 62443-3-3. 2013. *Security for Industrial Automation and Control Systems, System Security Requirements and Security Levels.* International Electrotechnical Commission. http://isa99.isa.org/Public/Documents/ISA-62443-3-3-EX.pdf.

ISO/IEC 15408. 2009. *Information Technology—Security Techniques—Evaluation Criteria for IT Security.* International Organization for Standardization. https://www.iso.org/standard/50341.html.

ISO/IEC 27001. 2013a. *Information Technology—Security Techniques—Information Security Management Systems—Requirements.* International Organization for Standardization. https://www.iso.org/standard/54534.html.

ISO/IEC 27002. 2013b. *Information Technology—Security Techniques—Code of Practice for Information Security Management.* International Organization for Standardization. http://www.iso27001security.com/html/27002.html.

Kushner, David. 2013. *The Real Story of Stuxnet*. IEEE Spectrum. http://spectrum.ieee.org/telecom/security/the-real-story-of-stuxnet.

Lysneutvalget. 2015. *Digital Vulnerabilities in Maritime Sector*. Ministry of Justice and Public Security. https://www.regjeringen.no/contentassets/fe88e9ea8a354bd1b63bc0022469f644/no/sved/7.pdf (accessed June 4, 2016).

"Marine Cybernetics: World-Leading Verification and Testing Services for Control Systems." 2016. DNV-GL.com. https://www.marinecybernetics.com.

NIST 800-82. 2011. *Guide to Industrial Control Systems (ICS) Security*. National Institute of Standards and Technology.

OWASP. 2013. *OWASP Top Ten Project*. Open Web Application Security Project. https://www.owasp.org/index.php/Category:OWASP_Top_Ten_Project.

PSA. 2013. *Principles for Barrier Management in the Petroleum Industry*. Petroleum Safety Authority Norway. http://www.ptil.no/getfile.php/1319891/PDF/Barrierenotatet%202013%20engelsk%20april.pdf.

UT Austin. 2013. "UT Austin Researchers Successfully Spoof an $80 million Yacht at Sea." *UT News*, July 29. http://news.utexas.edu/2013/07/29/ut-austin-researchers-successfully-spoof-an-80-million-yacht-at-sea.

Wagstaff, Jeremy. 2014. "All at Sea: Global Shipping Fleet Exposed to Hacking Threat." *Reuters*, April 23. http://www.reuters.com/assets/print?aid=USBREA3M20820140423.

CHAPTER 8:
CYBER WARFARE AND MARITIME SECURITY: A CALL FOR INTERNATIONAL REGULATION

Laura Sturdevant
American Military University

Abstract

Cyber warfare has quickly evolved into the fifth battlespace. Current international policies such as the Law of Armed Conflict and the Tallinn Manual do not properly address the unique aspects of it, particularly in the maritime domain. This lack of regulation could allow cyber weapons to be developed into Weapons of Mass Destruction greatly impacting the maritime sector. Current research identifies the necessity of a new international treaty, but fails to address how to proceed if an attack were to occur. This paper proposes the need for an international treaty that defines, protects, and criminalization of cyber warfare in order to enhance the potential for maritime cyber security.

Introduction

As the United States involvement in Afghanistan is coming to a close, and the American military starts to prepare for whatever the next conflict will be, it finds itself at a crossroads. Previously, its wars have been mostly fought in the conventional sense—military weapons against other military weapons—with a set of internationally agreed upon rules to guide its conduct. Even in the ever-evolving asymmetric warfare against terrorism the U.S. military attempted to adhere to the Law of Armed Conflict (LOAC) as much as possible. However, as it steps back and assesses where it needs to develop next, it becomes apparent that cyber warfare is a relatively new field where the military does not quite know how to proceed. The development of newer and better ships, tanks, and airplanes is simple to apply to existing laws. In the cyber world, however, there is a hesitation because there are no clear cut lines of what is allowed and what is not. As Dombrowski and Demchak explain, "Cybered [sic] conflict is here to stay and must be taken seriously even if cyber war in the conventional sense—that is, combat deaths—is not likely" (2014, 87). In the realm of maritime cyber security, can the adversary develop a virus that can infect the commercial shipping manifest tracking systems of another state it is at war against in order to inhibit that state's resupply its troops? Would this be considered an attack on noncombatants and thus illegal? Is hacking into the network of corporations in order to gain intelligence on weapons systems they are developing for other nations legal or not? Is sending out an electromagnetic pulse attack against a port that also impacts a local hospital adhering to LOAC?

The lack of cyber warfare regulation is concerning and dangerous. Anyone involved in the cyber realm needs to call for an agreed upon definition of cyber warfare. What different attacks constitute an individual attack or an attack against a nation before it becomes the new "wild, wild west," also needs to be further specified. This paper will discuss how the current construct of LOAC does not properly address cyber warfare; and how this lack of regulation allows nation states to infringe upon another with relative ease or the threat of international repercussions. As such, it will argue for the formulation of an international treaty specific to cyber conduct that allows for individuals, corporations, and states to legitimately claim it was maliciously attacked by some other force, thus calling for protection and action.

Additionally, this paper will argue the need to pursue international criminalization of cyber attacks in order to make the treaty resilient. This will give the signatories of the international cyber treaty appropriate legal avenues to pursue a response to cyber attacks rather than cyber retaliation. Without this twofold approach, the world will have no ability to prevent cyber-weapons from possibly being developed into Weapons of Mass Destruction (WMDs). In turn, those concerned with maritime cyber security will see that tactical threats such as GPS jamming, coupled with the lack of international regulation, poses a growing risk to maritime cyber security. Any attempt to secure the maritime cyber domain will be ultimately futile without international legal protection.

Background

The majority of literature available on cyber warfare, cyber attacks, and cyber weapons fall into four loose categories: attempting to define the cyber realm, attempting to define cyber security threats, discussions of what an international treaty on cyber warfare should look like, and nation-state policies on cyberspace. In regards to defining the battlespace, there are articles like Johan Eriksson et al.'s discussion on Internet ownership provides a forum among the authors with the goal of "discussing what actors are controlling what aspects of Internet usage and under what conditions" (Eriksson et al. 2009, 205). They analyze the theories of state-centrism and public–private partnerships, as well as "[advocating for] more critical perspectives emphasizing complexity, interactivity, and discourse" (Eriksson et al. 2009, 205). The forum provided a variety of perspectives that informs the reader of the ever-increasing interconnectedness of the Internet. Namely, the evolution of the Internet has molded it in a way that easy delineation among nation-states is near impossible. In the maritime domain, this difficult discrimination further blurs the lines between commercial and government assets.

Countering this work, Hansen and Nissenbaum create a theoretical framework allowing them to "define and theorize the cyber sector of security ..." (Hansen and Nissenbaum 2009, 1171). Here they define cyber security as "three distinct forms of securitizations: hypersecuritizations, everyday security practices, and technifications" (Hansen and Nissenbaum 2009, 1155). With hypersecuritization, the "what if" scenario comes into play and the inclination is the attempt to secure the cyber infrastructure against any attacks. Everyday security practices are the actions taken by individuals, private organizations, and businesses to protect their day-to-day operations (Hansen and Nissenbaum 2009). Finally, technification is essentially the actions and discourse provided by scientists with the expert knowledge of the inner, technical workings of the cyber world, as the "knowledge required to master the field of computer security is daunting and often not available to the broader public" (Hansen and Nissenbaum 2009, 1166–1167).

While the Internet was created in the 1960s, it did not come into prominence until the 1990s and early 2000s. As such, some researchers during that time, such as James Lewis of the Center for Strategic and International Studies, suggest "infrastructures in large industrial countries are resistant to cyber-attack" (2002, 10). However, as the cyber realm has become more robust and integrated, so has the cyber threat. Whole research facilities have been implemented to analyze cyber threats, such as Georgia Tech's Information Security Center (GTISC). In their 2014 report, the GTISC notes prevalent risks, such as cloud data, insecure but connected devices, the cost of defending against cyber attacks, and more (Lee and Rotoloni 2014). Separately, Bill Gertz highlights how cyber warfare can be utilized to level the playing field against an opponent stronger in traditional warfare, such as the United States (Gertz 2012). This creates the potential for a cyber arms race; particularly if no international agreement is reached, at least between the key players like the United States,

Russia, and China (Markoff and Kramer 2009).

However, the threat is not just related to companies and countries. A prime example of this is Rollins and Wilson, who delve into cyber threats presented by terrorism. The authors explain terrorists' reliance on telecommunication platforms such as the Internet and cell phones has risen sharply "for supporting organizational activities and for gaining expertise to achieve operational goals" (Rollins and Wilson 2007, 1). Their report "examines possible terrorists' objectives and computer vulnerabilities that might lead to an attempted cyber-attack" and possible policies that could prevent such an event (Rollins and Wilson 2007, 1). This is important as groups such as the Islamic State of Iraq and the Levant continue to exploit cyber hackers to support their cause, such as the proliferation of propaganda.

However, much of the literature on cyber warfare and cyber-attacks encompass a dialogue discussing the international regulation of cyber warfare, or rather the lack thereof. As the European Security Review states, "it is a generally acknowledged truth that warfare is not what it used to be" (Levarska 2013, 2). It is proposed that an international treaty on cyber warfare needs to "have universal applicability" and "consist of relatively ambiguous norms which would permit the regulation of future technological developments in the cyberspace and cultural differences in perceptions of cyberspace" (Levarska 2013, 1). Academic papers like this one often highlight current products, such as the "Tallinn Manual on the International Law Applicable to Cyber Warfare" and prove how it is not enough for regulation (Levarska 2013). They also discuss how more complications to these products exist because "the international community might not be prepared for creating a treaty with global outreach because the use of Internet is linked to sensitive issues such as the right to privacy or the right to free speech ..." (Levarska 2013, 15).

Similarly, Rex Hughes is a proponent of an international treaty on cyber warfare. In his "A Treaty for Cyberspace", he outlines the stances of the major cyber players in 2010: the United States, UK, South Korea, India, China, Russia, and Israel (Hughes 2010). Hughes assesses how the principles of the Law of Armed Conflict (LOAC), the internationally recognized parameter for guiding conventional warfare, apply to cyber warfare. He postulates "the primary objective of a multilateral cyber warfare treaty should be to regulate this method of warfare and its consequences" and, as such, cyberspace has become "the fifth battlespace" (Hughes 2010, 536, 540).

Every battlespace, however, will have weapons. Louise Arimatsu from the Chatham House takes a different route and centers her focus on cyber weapons, the malware actually creating the havoc. She argues "before considering whether an arms control treaty is a feasible option" the question must be asked: "What is a cyber-weapon?" (Arimatsu 2012, 91). This is an intriguing angle to approach the problem set, as "most of the malicious codes or malware that would fall within the parameters of a cyber weapon are designed to have an *indirect* kinetic outcome which may, or may not, result in the listed outcomes" of being designed to "kill, injury, or disable people, or to damage or destroy property" (Arimatsu 2012, 97). Thus far, most cyber attacks had no kinetic outcomes, such as the 2011–2013 attack on the Antwerp port IT system by organized criminals to facilitate the smuggling of drugs (Caponi and Belmont 2014).

Additional insight on cyber warfare can be found in the national strategies of key international players. While official information on Russia and China is limited, discussion on their strategies and positions can be found in statements made by leadership and actions taken to propose new bills and treaties at the UN or within other international communities. A clear intelligence gap emerges as reporters may improperly convey the stances of the other nations and/or add their own personal bias when writing their articles. Furthermore, U.S. defense entities, such as the US Coast Guard, have created their own cyber strategies that compartmentalize the national strategy into more concise, targeted lines of efforts.

Thus, there are generally four categories of research discussing cyber warfare: how to define the cyber realm, how to define cyber security threats, discussions of what an international treaty on cyber warfare should look like, and nation-state policies on cyberspace. However, there is no discussion on how to give an international treaty on cyber space the theoretical legs to stand on. Given how much global trade is handled by the maritime domain, around 90% of the European Union's (EU) external trade and more than 43% of the internal trade, in addition to an international treaty, there is a need to pursue international criminalization of cyber attacks, cyber weapons, and cyber crime (European Network and Information Security Agency 2011). First, however, it must be asked: how do the current international regulations such as LOAC apply to cyber warfare and maritime security?

Lack of International Regulation

Cyber Warfare and LOAC

While cyber warfare has been distinguished as other than conventional warfare, the general inclination is to still apply the core LOAC principles to cyber conduct for ethical purposes. The first principle, distinction, stipulates that "combatants are required to distinguish themselves as such" and "combatants must target only military objectives ..." (Solis 2010, 251). As Rex Hughes then explains, "persons [sic] who commit combatant acts without authorization are subject to criminal prosecution" (Hughes 2010, 537). The difficulty with applying this principle to cyber warfare is "if a cyber-attack is launched from thousands of miles away by an anonymous or covert force, there is no reliable way to apply [distinction]" (Hughes 2010, 537). In most cases, the goal of a cyber attack is to remain anonymous or covert, so asking those responsible to add a mark of distinction on their attack would be foolish (Arimatsu 2012). Thus, given that "large sections of what matters to the maritime services now overlap with traditional military, intelligence, and even commercial operations across the nation and globe," if a maritime port servicing both naval and commercial ships comes under attack from an unknown actor, the distinction principle would be violated (Dombrowski and Demchak 2014, 88). Furthermore, more trouble lies in the fact that, once the attack is launched, it takes near-impossible surgical precision in its execution to ensure the attack is isolated to the intended targets, and is not spread to non-military objectives (Hughes 2010; Arimatsu 2012).

The next principle, military necessity, holds that "a state may do anything that is not unlawful to defeat the enemy" (Solis 2010, 259). In regards to cyber warfare thus far, there remains very little litigation on what an illegal attack may be, thus a belligerent is, in essence, given free rein to commit whichever cyber attacks it deems necessary to bring the war to an end. The danger in this can be likened to the US Civil War in which cotton crops were attacked in order to restrict the financial income for the Confederate Army (Solis 2010). Similarly, an aggressor may insert malicious code to jam the automated cranes loading and unloading naval cargo ships, which would then impact the military's ability to resupply their troops overseas, but would also cause a potentially huge clog in the commercial shipping industry. Fitton et al. notes, "we are living in an age of open source cyber weapons where it is possible to download programmes [sic] which target industrial control systems, weapons, ships, and more for free from the Internet ... [which] can have an adverse effect on the security of your own system" (2015, 9). Without legal guidelines, an adversary can attack the financial institutions, manufacturing facilities, and other such targets that normally would be restricted via cyber attack.

However, another principle, perfidy, "is designed to regulate the targeting of certain facilities that are historically considered legal sanctuaries during a time of war" such as hospitals and prisons (Hughes 2010, 539). This would prevent adversaries from attacking those facilities deemed as legal under the military necessity principle. Nevertheless, the inability to ensure the containment of a cyber attack puts the network of hospitals, prisons, and other sanctuaries at risk. An attack on naval communications nodes that intends to limit warfighting ships ability to coordinate attacks with one another could unintentionally block the distress beacon of a ship in trouble, impacting the US Coast Guard's ability to conduct necessary search and rescue operations (Knox 2015). The debate is then whether attacking an adversary's network counts as an attack on the facility or not.

A protection against this could be the "ban [of using] certain weapons with potential to cause grave damage beyond their original targets" (Hughes 2010, 538). This is known as the principle of unnecessary suffering (Hughes 2010). For example, an update laced with viruses or Trojans targeting the navigation systems of naval vessels mistakenly transmitted to non-military ships would be considered unnecessary suffering. Another instance Hughes uses would be hardware that is altered and distributed by a country with the intent to infect military computers, but the proliferation extends into the civilian spectrum (Hughes 2010). "Errors in coding lead to system vulnerabilities which can be exploited by adversaries for malevolent ends," such as a flaw in highly proliferated operating systems like Windows XP (Fitton et al. 2015, 7–8). An attack on the military's Information Communication and Technology (ICT) systems could also impact the same vulnerabilities in civilian ICT systems. However, without an international treaty delineating what counts as an indiscriminate cyber attack, cyber actors are free to do as they please.

Additionally, according to LOAC, all attacks must be proportional to the intended end (Solis 2010). "It weighs the military advantage obtained against the corresponding harm inflicted ... As with military necessity, the fundamental challenge in applying proportionality to a cyber-attack in how to differentiate between civilian and military targets" (Hughes 2010, 538). First off, an attack may be launched against main computers controlling a dual-use critical infrastructure, such as an oil field or a naval port, in order to inhibit the enemy's ability to fight the war, but consideration must be given to the impact on civilian life. Also, it may not be immediately obvious what the military advantage would be if the cyber attack was launched from across the world (Hughes 2010). Finally, the military has increased its usage of civilian infrastructure, particularly in the cyber world (Hughes 2010). This makes the distinction between the two even more difficult. "ICT is increasingly used to enable essential maritime operations, from navigation to propulsion, from freight management to traffic control communications, etc." (European Network and Information Security Agency 2011, 1). An attack on a dual-use port may cause disproportional impacts, as well, if the military infrequently uses the port.

As a final principle, Hughes discusses the standard of neutrality and how, if declared, those involved in a war agree to not attack the declarer if they, in turn, do not support either side (Hughes 2010). Respecting this principle is problematic due to the integrated nature of the cyber world today. In order to deliver a virus to the intended target, it may have to travel through many different routers and terminals, which could include ones located or belonging to the neutral state, something a state could argue violates the declared state's neutrality (Hughes 2010). Thus, in the global village, would these computers be defined as borderless or owned by the neutral state? Furthermore, in the maritime cyber domain, do companies registered under multiple countries constitute as part of the country at war or the one that remains neutral? While part of that company facilitates the shipment of goods for the aggressor's troops, does that mean every part of that company is thus open for cyber attack?

These are just a few of the dilemmas presented through applying the LOAC to cyber warfare. It thus becomes apparent that an international discussion must be had to further flesh out the legal parameters. There is no precedent to utilize, but a foundation must be set. This begs the question: just how is the international community supposed to achieve this?

Solutions and Recommendations

A Call for an International Treaty including Criminalization

In his article *The Everywhere War*, Derek Gregory uses the escalating situation between Georgia and Russia in 2008 as an example of cyber attacks being used to coincide with real world conflict where "Georgian government servers were subjected to coordinated barrages of millions of requests that overloaded and eventually shut them down" (Gregory 2011, 245). Georgia accused Russia of the attack. When Russian troops did enter South Ossetia, a posting to a hacker forum invited the hackers to a cyber-militia forum where they would be "given target lists and instructions" (Gregory 2011, 245). The danger in this is that "it suggests an emerging model of cyber warfare that involves both the outsourcing of cyber-attacks and the militarisation [sic] of cybercrime" (Gregory 2011, 245). This creates ambiguity on whether the hackers' actions would be legal, as they could claim to be conducting them in the name of a national entity, who could then refuse this claim due to the covert nature of the operation.

Earlier, in 2007, "Estonian government networks were harassed by a denial of service attack by unknown foreign intruders" after Estonia moved a Russian war memorial (NATO Review 2015). While "the attacks were more like cyber riots than crippling attacks," it prompted Estonia to become NATO's Cooperative Cyber Defense (CCD) Center of Excellence (NATO Review 2015). In 2008, the CCD Center of Excellence tasked 20 top legal academics and practitioners to analyze how "the current framework of international law applied to cyber warfare" (Levarska 2013, 3). The result was the "Tallinn Manual on the International Law Applicable to Cyber Warfare," which discussed many issues involving LOAC and cyber warfare, but disagreement among the group of experts remained, such as "whether cyber acts which do not involve intervention and do not cause physical damage can still violate sovereignty of a state" (Levarska 2013, 5). Therefore, the Tallinn Manual is a great foundation for an international treaty regulating cyber conduct, but there is much room for improvement.

The maritime domain also has treaties and conventions such as the Safety of Life at Sea (SOLAS), monitored by the International Maritime Organization (IMO). SOLAS is "generally regarded as the most important of all the international treaties concerning the safety of merchant ships" with the first version adopted in 1914 and the fourth, and most recent, in 1960 (International Maritime Organization 2017). While amendments are relatively easy, and take much less time than other international agreements (minimum length of time from amendment circulation to entry into force is 24 months), there are still no provisions specifically addressing cyber warfare or cyber attacks.

As the *European Security Review* highlights, computing technology is increasing and reinventing itself at an exponential rate, thus an international treaty drafted might become obsolete by the time the legislative process completes itself (Levarska 2013). Therefore, it argues a better approach would be to "create a new treaty laying down the most fundamental principles while leaving a considerable room for their interpretation in accordance with the level of technological development," much like how the Geneva Conventions and LOAC have been adapted for any technological improvement or invention in conventional warfare (Levarska 2013, 13). As the European Network and Information Security Agency (ENISA) recommends, "A platform for further consultation and coordination on maritime cyber security ... is desirable at this level" (2011, 12). The IMO and SOLAS would be an appropriate entity to spearhead this effort in order to align international efforts and add cyber security to the "minimum standards for the construction, equipment, and operation of ships, compatible with their safety" (US Coast Guard 2015, 2).

Others have argued the focus should not be on cyberattacks, but rather cyber weapons or the method the attack conducted. One such proponent is Louise Arimatsu from the Chatham House, who argues that "since malicious codes are designed to have different—and sometimes multiple—intended objectives, distinguishing between exploitation, intelligence-gathering, disruptions, and conduct that is the prelude to something more serious will be challenging at best" (Arimatsu 2012, 107). Additionally, "cyber weapons are relatively inexpensive and widely accessible to non-state actors," as seen when the Islamic State of Iraq and the Levant hacked the US Central Command's YouTube and Twitter accounts in January 2015 (Arimatsu 2012, 108). Moreover, "the identity of the originating party behind a significant cyber-attack can be concealed with relative ease ... [and] it would be impossible to destroy all the copies of the malicious code which may be stored in countless digital devices across the globe ..." (Arimatsu 2012, 108). Therefore, it becomes imperative if an international treaty is drafted on cyber warfare, the language needs to include a discussion on cyber weapons.

Research by Fitton et al. (2015) and ENISA both also identify the human threat to cyber security, whether intentional or unintentional. Fitton et al. (2015) use the Nigerian Lottery spear phishing email (the victims received an email saying they have won the Nigerian Lottery and need to reply with their bank account details to receive their prize) to emphasize how a low-tech cyber attack had widespread impact. Furthermore, hacktivists, cyber-criminal gangs, and state actors can utilize "social engineering, deception, identity theft, bribery, and blackmail" against employees and workers of their intended targets (Fitton et al. 2015, 15). For example, a ship may be tracked via the social networks utilized by the employees, such as geo-tags on their Facebook posts or analyzing their Instagram posts for identifying features (Fitton et al. 2015). Awareness of the cyber threats is another prevalent issue, which increases the risks that vulnerabilities in their ICT systems or their own cyber usage could be exploited by adversaries without notice (European Network and Information Security Agency 2011). Both advocate "preventing, spotting and defending against cyber-attacks requires educating, training, and drilling staff, so they can efficiently respond to attacks, spot errors, and continue to operate under cyber-attack conditions" (Fitton et al. 2015, 5). Increased awareness could also help motivate governments and international entities to create cyber regulations as citizens become more aware of the risks not being addressed.

Another component with the act of a cyber attack is the notion of cybercrime, such as theft and fraud (Adoption of Convention on Cybercrime 2001). A cyber attack may be launched in order to steal manufacturing secrets or to delete any remnants of a government entity's involvement in an event. Furthermore, criminals could exploit the cyber realm to hack into a bank's infrastructure to steal their assets. As such, "deterring and punishing computer criminals requires a legal structure that will support detection and successful prosecution of offenders" (Adoption of Convention on Cybercrime 2001, 890). The Convention on Cyber-Crime was adopted by the Council of Europe, the United States, Canada, Japan, and South Africa, which became the first international treaty addressing cybercrimes such as "fraud, theft, and the distribution and sale of child pornography" (Adoption of Convention on

Cybercrime 2001, 889). However, this treaty does not address greater cybercrimes and cyber attacks, such as those perpetuated against Georgia and Estonia. Therefore, greater discussion must be had in regards to criminalization of those acts and how to properly prosecute them. This is imperative to the maritime domain, as it is often used for the transportation of illegal goods, such as illicit narcotics. As highlighted above, cyber criminals were able to attack the Antwerp port from 2011 to 2015 in order to facilitate cocaine and heroin trade (Caponi and Belmont 2014). Furthermore, "data theft, for criminal purposes, may increase as a direct result of insufficient cyber security measures—or measures not sufficiently matching the complexity of the ICT environment involved" (European Network and Information Security Agency 2011, 9).

A caveat must also be added that an international treaty on cyber warfare assumes the aggressors will follow the laws created. Non-state actors, like that of terrorist groups, have no obligation to conduct themselves per international law, as has been seen in the everyday actions of most of these groups. As noted above, "terrorist's use of the Internet and other telecommunications devices is growing both in terms of reliance for supporting organizational activities and for gaining expertise to achieve operational goals" (Rollins and Wilson 2007, 1). However, these laws will protect the defenders so they may legitimately claim when an attack has been made, similar to how the United States used the indiscriminate nature of attacking a civilian infrastructure on 9/11 as justification to take action against al Qaeda in Afghanistan in 2011.

Cyber Warfare and WMDs

The danger of unregulated cyber warfare likens itself to the development and progression of nuclear warfare in that, without international attention to regulation, cyber weapons could eventually develop to the size and scale that could produce WMD-level effects. This paper utilizes a broad over-arching definition, in which WMDs are "weapons that cause massive destruction or kill large numbers of people, which does not necessarily include or exclude [chemical, biological, radiological and nuclear] weapons" (Enemark 2011, 383). The additional hazard of WMD-level cyber-attacks is that, due to the interconnectedness of the Internet, the catastrophic effects will not be isolated to just the target, but will inevitably impact innocent bystanders as well. "Proponents of including cyber-attacks under the WMD label ... discuss the ability of cyber-attacks to destroy critical governance or commercial infrastructure, or argue that cyber-attacks will be used to produce secondary effects resembling [nuclear, biological and chemical] weapons" (Kaminski 2012, 13). While this may be true, without internationally recognized legal parameters defining a cyber weapon, a cyber attack, and an expansion of LOAC as applicable to cyber warfare coupled with the criminalization of said cyber attacks, the victims of WMD-level cyber-attacks will have no legal justification to pursue retribution for their attacks. Therefore, a cyber attack cannot be classified as a WMD, even if it may produce that level of effect, until an international treaty and criminalization of cyber warfare is achieved.

Conclusion

The development of the cyber world in the last few decades has created a myriad of legal issues stemming "from both the rapid spread of cyber warfare and the lack of precedent to guide international regulation of cyberspace intrusions" (Hughes 2010, 533). In light of the global village of the Internet, steps must be taken to protect the innocent. Without such, there is a danger of hypersecuritization, which would infringe upon the lives of everyday citizens. Furthermore, the legal ambiguity essentially creates cyber anarchy where it is "anyone's game" in the virtual world. Thus, with the fact that cyber warfare is "both relatively cheap and readily available, making it all the more alluring for small states and non-state actors to deploy their weapons of technological knowledge, information, and skill," belligerents can attack their targets with relative ease and without threat of international incursion (Hughes 2010, 541). The principles of LOAC, the Tallinn Manual, SOLAS, and the Convention on Cyber-Crime stand as a good starting points for the discussion, but decisions must be made in order to guide state actions. While this will only ensure law-abiding actors will act accordingly, it will provide justification against perpetrators who violate the established parameters. Furthermore, it can help stem the evolution of cyber weapons and attacks into ones that could cause mass destruction.

Global cyber security is a critical node for the maritime logistical network and must be protected in order to secure it from adversaries. Dombrowski and Demchak surmise it perfectly for the maritime domain: "the more the Navy [and the rest of the maritime sector] is able to answer the systemic cyber challenges and reduce the scale, proximity, and precision advantages attackers enjoy today, the better prepared [they] will be for the bordered, encrypted, and technologically diverse future international system" (2014, 90). Most of those concerned with maritime cyber security tend to focus on tactical, immediate cyber threats such as the hacking of their systems or GPS jamming. However, in an analysis of the lack of international regulation of cyber warfare and the potential direction it could spiral into, it becomes evident that true maritime cyber security cannot be obtained and maintained without the development of an international treaty that defines, protects, and criminalizes against cyber warfare.

CHAPTER 8

REFERENCES

"Adoption of Convention on Cybercrime." 2001. *The American Journal of International Law* 95 (4): 889–891.

Arimatsu, Louise. 2012. "A Treaty for Governing Cyber-Weapons: Potential Benefits and Practical Limitations." *4th International Conference on Cyber Conflict* 91–109.

Caponi, Steven L., and Kate B. Belmont. 2014. "Maritime Cybersecurity: A Growing, Unanswered Threat." *The Maritime Executive* (October 24): 1–4.

Dombrowski, Peter, and Chris C. Demchak. 2014. "Cyber War, Cybered Conflict, and the Maritime Domain." *Naval War College Review* 67 (2): 71–97.

Enemark, Christian. 2011. "Farewell to WMD: The Language and Science of Mass Destruction." *Contemporary Security Policy* (August 26): 382–400.

Eriksson, Johan, Giampiero Giacomello, Hamoud Salhi, Myriam Dunn Cavelty, J.P. Singh, and M.I. Franklin. 2009. "Who Controls the Internet? Beyond the Obstinacy or Obsolescence of the State." *International Studies Review* 11 (1): 205–230.

European Network and Information Security Agency. 2011. "Analysis of Cyber Security Aspects in the Maritime Sector." *European Union* (November): 1–25.

Fitton, Oliver, Daniel Prince, Basil Germond, and Mark Lacy. 2015. "The Future of Maritime Cyber Security." *Security Lancaster* (2015): 1–34.

Gertz, Bill. 2012. "Inside the Ring: New WMD Threats." *The Washington Post*, October 10.

Gregory, Derek. 2011. "The Everywhere War." *The Geographical Journal* 177 (3): 238–250.

Hansen, Lene, and Helen Nissenbaum. 2009. "Digital Disaster, Cyber Security, and the Copenhagen School." *International Studies Quarterly* 53 (4): 1155–1175.

Hughes, Rex. 2010. "A Treaty for Cyberspace." *International Affairs (Royal Institute of International Affairs 1944-)* 86 (2): 523–541.

International Maritime Organization. 2017. International Convention for the Safety of Life at Sea (SOLAS), 1974. http://www.imo.org/en/About/Conventions/ListOfConventions/Pages/International-Convention-for-the-Safety-of-Life-at-Sea-(SOLAS),-1974.aspx.

Kaminski, Ryan. 2012. "Clash of Interpretations: Cyberattacks as 'Weapons of Mass Destruction.'" *Council on Foreign Relations* (2012): 1–13.

Knox, Jodie. 2015. "Coast Guard Commandant on Cyber in the Maritime Domain." *COAST GUARD Maritime Commons* (June 15): 1–6.

Lee, Wenke, and Bo Rotoloni. 2014. "Emerging Cyber Threats Report 2014." *Georgia Tech Cyber Security Summit* 1–13.

Levarska, Nina. 2013. "European Security Review: Regulation of Cyber-warfare: Interpretation versus Creation." *International Security Information Service, Europe*, 70 (December): 1–15.

Lewis, James A. 2002. "Assessing the Risks of Cyber Terrorism, Cyber War and Other Cyber Threats." *Center for Strategic and International Studies* (December): 1–12.

Markoff, John, and Andrew E. Kramer. 2009. "US and Russia Differ on a Treaty for Cyberspace." *The New York Times*, June 27. http://www.nytimes.com/2009/06/28/world/28cyber.html?pagewanted=all&_r=0.

NATO Review. 2015. "The History of Cyber Attacks- a Timeline." *NATO Review Magazine*. www.nato.int/docu/review/2013/cyber/timeline/EN/index.htm.

Rollins, John, and Clay Wilson. 2007. "Terrorist Capabilities for Cyberattack: Overview and Policy Issues." *CRS Report for Congress* RL33123 (January 22): 1–25.

Solis, Gary D. 2010. *The Law of Armed Conflict: International Humanitarian Law in War.* Cambridge: Cambridge University Press.

US Coast Guard. 2015. *US Coast Guard Cyber Strategy: The US Coast Guard's Vision for Operating in the Cyber Domain.* GPO (June): 1–43.

CHAPTER 9:
COMMERCIAL MARITIME CYBER SECURITY RISK AND VULNERABILITY: WHAT WE DID NOT KNOW, WE DID NOT KNOW

Bruce G. Clark
Chair, California Marine and Intermodal Transportation System Advisory Council (CALMITSAC)

Abstract

Government and commercial sector operators and policy makers in the maritime domain are beginning to recognize the potentials for cyber attacks on both land and sea. Recent experience has demonstrated significant, although to date largely isolated intrusion and disruption of ship and facility computerized management and control systems—and most targeted at criminal rather than terror-driven motivations. These documented motivations have included: stealing money, clandestinely moving cargo, stealing sensitive information, and causing operational disruption and loss. Evidence of clear motivation tied to terrorism-related goals has been elusive to date, but as the capability and capacity of cyber intrusion is repeatedly demonstrated and documented, it is reasonable to expect that malefactors embracing terrorist agendas will begin to exploit weaknesses in the cyber realm. Clearly—there is a demonstrable threat, but the level of risk and the cumulative vulnerability resultant from that risk is yet to be fully researched or completely understood. Until this critical assessment is undertaken and researched at all levels and applied to all operational conditions, the maritime community will remain unsure as to the potential ultimate worst case negative results and the quantifiable impacts necessary to access risk and to develop protective and mitigation strategies based on those risks. Intuitively, all operators within the maritime domain understand that the march of technology—so necessary to improve efficiencies and reduce costs—also sometimes create unintended weaknesses or breaches in the cyber security "curtain wall" surrounding the critical command, control and data management systems that are becoming more and more commonplace aboard ships—as well as between ship and shore—for all of the necessary communication of instructions, changes in plans, etc. Older vessels with older generation, more mechanically based control systems are less vulnerable to cyber-attacks simply because they are less "integrated", while newer vessels with fully integrated bridge, engine and communications systems—all designed to "talk" to each other, share common data and are linked in real time to shore stations via satellites—are more vulnerable to attack by bad actors. Just how vulnerable these systems are—or will be as the automation trend continues—must be the intense subject of study by all who operate upon the seas and the greater transportation system as a whole. To attain a modicum of systems resiliency and redundancy, new control systems need to incorporate solutions that balance the necessary continued engagement of the human element, continuing systems technological efficiencies, retention of mechanical redundancy, and design of integral systems protections.

Keywords: Maritime Cyber Domain, Maritime Cyber Risk Assessment, Integrated Bridge and Ship Management Systems, Maritime Cyber Threat, Maritime Cyber Vulnerability, Cyber Redundancy, Cyber Resiliency

Introduction

"The complexity of (computer system) hardware and software creates great capability, but this complexity spawns vulnerabilities and lowers the visibility of intrusions. Cyber systems' responsiveness to instruction makes them invaluably flexible; but it also permits small changes in a component's design or direction to degrade or subvert system behavior. These systems' empowerment of users to retrieve and manipulate data democratizes capabilities, but this great benefit removes safeguards present in systems that require hierarchies of human approvals. In sum, cyber systems nourish us, but at the same time they weaken and poison us."

—*The Honorable Richard Danzig, Former Secretary of the Navy, RAND Corp Fellow and Board Member at the Center for New American Security—July 2014*

CHAPTER 9

Potential cyber threats to the maritime transportation sysem are not limited to possible impacts upon big ships and blue water operations. The MTS comprises all modes of transportation with a nexus at ports: vessels, terminals, petrochemical refineries, railroads, pipelines, and highway operations—and in so far as they are all expanding use of computer-based technologies to manage commodity flows and efficiencies, they are all at risk for cyber intrusion, cyber hijacking or worse. According to the Center for New American Security, over 120 countries and an unknown number of criminals and terrorist groups are developing cyber attack and spying capabilities that arguably have or could disrupt all modes of the MTS (Danzig 2014). Even so, it is not necessary to look to vast international conspiracies and malicious state actors to identify areas of major cyber risk and associated vulnerabilities. The vast majority of cyber intrusions still arise from simple phishing attacks within an email message triggered by unwary employees or via an infected USB drive. Many of these intrusions are specifically targeted to a particular organization or company for criminal intent.

Presently, worldwide between 70% and 90% of malware attacks are unique to and targeted at specific, single organizations. The most difficult purposeful and deliberate attacks to defend against are so-called Zero Day attacks—attacks that are hidden and/or previously unknown and that, as a consequence, require a proactive and ongoing defense as a critical systems maintenance function. Systems risk is not limited to in house computer management and communications components.

The more interconnection occurs via smart equipment technology and increased dependence upon third-party providers, the greater the potential for cascading systems disruptions and failure. One major service provider—BAE Systems, tasked with protecting the digital assets of more than 5,500 organizations worldwide—has recently developed cyber protections specifically addressing the increased exposure of Cloud-Based communications and e-mail systems.

These security measures are based on techniques which analyze e-mail communications in the cloud for malicious content and intent, before it reaches the recipient (BAE Systems 2015). Even with the advent of emerging protective tools and strategies, the first—and ultimately the last—line of defense remains with the system administrators and users.

The increasing frequency of identified cyber intrusions and attacks across the full spectrum of world government and commercial operations—including a growing rate of occurrence within the MTS—clearly reveals the risk, but does not identify or quantify the overall vulnerability nor provide as yet a clear assessment or profile of the perpetrators useful in countering their intentions.

Who Then in General are These "Bad Actors" and What Do They Want?

As a good starting point, John Weathington (2015) describes six categories of "Cyber Villains"—summarized here—who may choose to exploit cyber system vulnerabilities for the basic motivations described:

Cybervillain #1: The Revenger

The insider attack precipitated by the disgruntled current or prior employee poses the largest threat for most organizations. Imagine what would happen if one of your system administrators decided to go postal on one of your mission-critical systems? "Profile what normal behavior looks like and anticipate what attack behavior looks like. The combination of both detecting unusual behavior and recognizing attack behavior will protect your fortress from those within the walls" (Weathington 2015).

Cybervillain #2: The Martyr

Anyone passionate about a cause including terrorist organizations might revel at the opportunity to use a large American icon for the purposes of trumpeting their cause.

To defend against martyrs, it's important for your data scientists to constantly know what is trending. "A cyber attack does a martyr no good if they can't get good publicity" or levy economic havoc, physical destruction, and even associated deaths (Weathington 2015).

Cybervillain #3: The Spy

"Espionage is an insidious cyber attack that can cause monumental damage; this makes the Cyber Spy one your most formidable villains. Information that's private, classified, or otherwise confidential can fetch a huge price tag from the right buyer ... In contrast to the Martyr, the Spy wants to stay in your systems undetected, for as long as possible. Dormant cyber spies are very difficult to detect; but, when they 'wake up,' you must identify them quickly" (Weathington 2015).

Cybervillain #4: The Thief

"As the means of exchanging of money has rapidly evolved from paper to electronic so have the methods of stealing it. Trillions of dollars are exchanged electronically every day, and there's no sense for someone to break into a bank and crack a safe, when they have even a small chance of tapping into this river of electronic money that's gushing by them every day. The Cyber Thief wants to go undetected like the Cyber Spy, but they won't want to stay around very long if they're smart" (Weathington 2015).

Cybervillain #5: The Washer

"Money laundering is another crime that's gone cyber. The Cyber Washer is a specialist at turning dirty money into clean money using electronic means. Drug dealers, terrorists, and other bad guys turn to the Cyber Washer to make their illicit funds look legitimate; Cyber Washers do this by moving money around and covering up tracks" (Weathington 2015).

Cybervillain #6: The Bragger

"Some cyber attackers just want bragging rights ... Cyber Braggers love the challenge of hacking into something that's not supposed to be hacked. Cyber Braggers are the least threatening from the perspective of real damages but can be extremely difficult to defend against because they're often the best hackers in the world" (Weathington 2015).

What are some Things to Look for in the Internal Workforce?

Vectors for cyber attacks can come from many different sources including the internal workforce. A recent analysis by Kroll Associates of client cyber cases across all industries found that 51% of historic, documented breaches were tied to insiders. A comparison of these cases shows that (Bancroft 2014):

- Some perpetrators—but not all—were malicious people motivated by employment dissatisfaction;
- Some were unwary facilitators with no intent to cause disruptions but through carelessness mishandled data or disposed of information improperly providing access to malefactors;
- Some were tricked into divulging confidential information or authorizing fraudulent transactions or disbursements.

Some warning signs that an employee might be committing or facilitating a cyber crime are simple to remember and can provide a ready preventative to monitor activity. These warning signs might include:

- Working odd hours without authorization;

- Disregarding company policies about installing personal software or hardware;
- Taking short trips to foreign countries for unexplained reasons;
- Buying things they cannot afford;
- Taking proprietary or other information home in hard copy form and/or on thumb drives, computer disks, or email (Bancroft 2014).

Considering that e-mail remains the most common method of cyber intrusion, an active follow-up program for confirmation of suspicious or questionable requests for access or action remains a simple and effective tool against possible fraud or other malicious activity. When in doubt—contact the purported requestor or sender and verify the information by direct contact via telephone or Skype or similar means.

Worldwide Cyber Risk is Growing

CISCO Systems is on record predicting that by the year 2020, 50 billion different types of devices will be connected or connectable as a part of the "Internet of Things" (Macaulay, Buckalew, and Chung 2015). As systems continue to become more sophisticated and integrated—sharing data and information between what used to be totally isolated systems—the chances of either intentional or accidental introduction of malware or other stealth programs into increasingly critical control systems rises dramatically. The methods of introduction of cyber attacks need not be anywhere as complex as the systems themselves: by far, the introduction of bad actors into control systems still commonly employs the use of simple USB data sticks or via malicious attachments to e-mails that are activated by unwary employees. The STUXNET virus that caused so much damage to Iran's nuclear program centrifuges is believed to have been introduced by a human hand via a USB drive. In conjunction with the advance of fully integrated ships control systems is the ongoing development of "Ghost Ships" that will operate without the use of human crews. Major companies like Rolls Royce and governments such as the European Union are investing heavily in these technologies that will allow full control and operation of vessels remotely from the shore.

In the light speed environment of increasingly sophisticated interconnected controls systems that fully embraces Moore's Law and the thesis that technology doubles at a rate of every 18 months, maritime stakeholders have no alternative but to evaluate both the benefits and the potential vulnerabilities created by such dynamic change. Therefore, any protective strategy must include expanded, enhanced and repetitive education and training in addition to practical hardware and software hardening. Before any of this can occur however, users and operators within the maritime community must be able to identify and quantify the risk—and risk is currently not well defined or understood in the cyber realm.

Security Background for the M.T.S.: How Did We Get Here?

Prior to the 9/11 terrorism event, the United States was not focused on possible transportation related vulnerabilities in any really tangible way. In particular, security planning, prevention, and mitigation in the maritime mode were akin to *The 3 Monkeys Approach: A See No Evil, Hear No Evil, Speak No Evil* posture that did virtually nothing to enhance transportation security beyond extremely basic physical barriers and requirements carried over from WWII.

The maritime security threat was generally seen through a frosted lens of historic political dynamics between large state actors, within a context of conventional and "Cold" warfare, without recognizing the growing vulnerability of actions from small but motivated groups of disaffected persons—both internal and external—developing at least since the 1970s. Risk mitigation tools were almost always physical solutions involving fences, gates, and guards and the largest vulnerability was tied to criminal pilferage and organized theft—not large scale disruptions for socio-economic, religious, or political means. In the midst of the Cold War and post-Soviet Era, maritime domain security vulnerabilities were not evaluated as realistic risks requiring anything more than minimal attention. As a result, our security policies and practices—particularly in the maritime mode—were essentially set at the lowest priority and dramatically underemphasized when competing for scarce resources

Following WWII, within the USCG—where mission areas historically have been prioritized by the influence and pressures of Congress and the reactionary impacts of the "disaster du jour," maritime and port security missions were not well funded or staffed. Resources were directed at other mission areas such as maritime safety and environmental protection for example.

U.S. port and maritime security—and the means and methods to adjudge threat, risk, and vulnerability associated with it—has—particularly for the last 100 years—been driven by a cyclical process more dependent upon public and political perceptions of "need" and by the allocation of scarce resource and budget assets, than by a reasonable risk determination based on real circumstances or a need for preparatory prophylaxis.

As a practical result, a reasonable and sustainable homeland defense and port security capability have not been constituted and maintained in the United States—a necessary action needed to establish a routine day-to-day operational "habit" for all levels of public and private activity.

A typical example of this respnse cycle—illustrated as Figure 1 (Clark, Nincic, and Fidler 2007)—shows the historical and cyclical process of security capability in response to a credible threat and/or event from a "relaxed peace-time posture" where significant force assets either were not constituted as a part of the standing capability or were extremely limited in form and function; followed by re-creation of resources and capability over a period of months; deployment of assets; sustained operations for the duration of the conflict or crisis; resolution of the conflict or crisis; and ultimate demobilization and de-commissioning of the now extraneous force package no longer deemed necessary by the political decision makers who hold the budgetary checkbook.

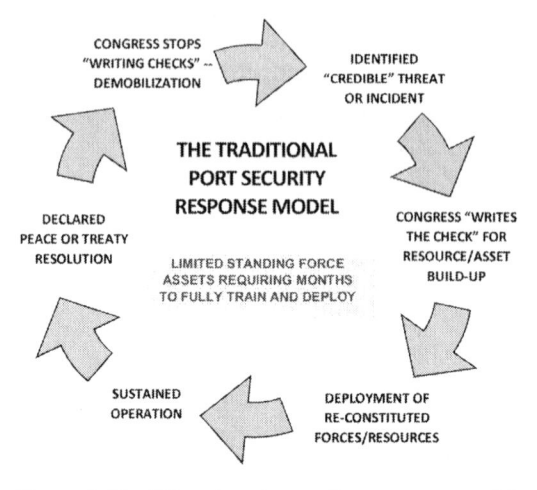

Figure 1: Traditional port security response model

From the earliest days of the Republic, the concept of maintaining a standing Army or Navy of any significance was routinely defeated by Congress as an *unnecessary expenditure* of public treasure. This routine extended to other branches and agencies of government—so much so that absent a clear threat (what we would call "credible" today) or a direct attack on U.S. interests, even non-military agency capabilities were reduced to a bare minimum between the periods of hostile activities. Part of this was tied to a sense that eminent threat would arise via some overseas series of events, requiring a significant period of time to coalesce into a "real threat"—and giving the nation the necessary interval to re-constitute the required defensive and offensive capabilities. Another part of this routine was based on the concept that the United States—surrounded by vast seas and relatively friendly neighbors on all sides—could not be effectively attacked by even the most determined aggressor. The final element was the practical restriction of available tax revenues with which to fund ongoing—and expensive—defense and security infrastructure and force asset components. Indeed, until the advent of the personal income tax during WWI (specifically designed as a *temporary* measure to fund the war effort)—funding to operate the federal government derived only from taxes and tariff revenues on international and domestic interstate trade as well as program specific federal government bonds. As a purely practical matter, there were strict limitations on the likely amount of predictable revenue flowing to the federal government in any particular fiscal year for ongoing support of required programs for the benefit of the commonweal. As a result, in the absence of a clear and present danger to the contrary, limited funds were expended elsewhere as the public demand and political response dictated.

By 1992—which similarly compares to pre-1983 prior to the Reagan defense build-up, the Coast Guard was expending less than 4.2% of its budget on *both* its National Defense (joint DOD missions) and Port Security (domestic) missions ($124.9 MM). These funds were dedicated to the combined use for cutter activities in the Persian Gulf in support of the First Iraq Wars—Desert Shield and Desert Storm—in addition to any other uses for domestic port and harbor security—primarily at those installations directly supporting the war logistics effort—typically termed strategic out load ports and facilities. Therefore, what "port security" activities were underway were tied directly to national defense, direct support of the war efforts and the immediate aftermath. By CY 2000 (prior to 9/11)—although retained as a formal USCG mission—the Coast Guard was expending *no more than 1% of its overall budget allocation* ($250 MM in FY 2000) directly on port and maritime security related missions in *both* domestic and foreign operations (The Interagency Commission on Crime and Security in U.S. Seaports Report—Fall 2000). In no small measure due to the assignment and allocation of enforcement resources and associated priorities elsewhere—as well as a general lack of focused interest on the part of the commercial business community and the general public—an outside assessment of the *state of "security"* at America's ports was generally described in the fall of 2000 as only *poor to fair*.

The Coast Guard itself estimates that the port security budget for all of FY 2001—*including supplemental funding for the 9/11 attacks response*—was about $250 MM or approximately 5% of its overall departmental budget for that year. Nevertheless—in the immediate aftermath of 9/11, the USCG's operational responsibilities—and the re-assignment of personnel and assets—expanded to *encompass nearly 50% of its total operations* during this period (Seapower Almanac 2006, U.S. Navy League). This budget increased dramatically to about $1.5 BB for FY 2005 (Source: USCG Budget in Brief, FY 2005) or a percentage total budget for Port Security missions of *approximately 25% on average for every year since 2001* in the aftermath of the September 11, 2001 attacks. As a personal example illustrating the dynamics of this decision process, the author participated in a USCG Pacific Area Flag and Senior Staff conference on Sept 8/9, 2001 where port and harbor security issues were raised as a topic of debate specific to planned resource allocations. Senior members of staff discounted the importance or necessity of enhanced allocation of resources to support and improve the port security mission posture. Proponents of increased support cited intelligence reports and activities of foreign terrorist groups—most prominently Al Qaeda, the Tamil Tigers, Abu Sayef and others—who were demonstrating interest, intent, and the capability to execute attacks in the maritime mode. Opponents opined that such transportation focused terrorist activity could never occur in the United States, and therefore, scarce resources should be allocated to more critical USCG mission requirements.

The opponents carried the day. Two days later, as the author was beginning a temporary duty assignment in Washington DC, hijacked commercial aircraft crashed into the twin towers of the New York World Trade Center and the Pentagon, while a fourth aircraft was sacrificed by brave Americans who crashed that airliner into a Pennsylvania field ... in that instant, terrorism and homeland security was suddenly the top priority. The United States, however, was ill prepared to address the emergent threat with little in the way of existing tangible port security resources, plans or policies. Nevertheless, in relatively short order the United States lead the international community at the International Maritime Organization (IMO) to develop worldwide baseline maritime security requirements as amendments to the Convention of Safety of Life at Sea (SOLAS Convention) that became the International Ship and Port Facility Security Code (ISPS Code). The ISPS Code was later implemented by regulation in the United States as the Maritime Transportation Security Act (MTSA). The MTSA provided the first consistent and comprehensive outline of regulations and requirements for ship and facility security in the United States, but largely still focused on physical security enhancements and implementation of specific security plans, policies and procedures.

The USCG over time has developed the Maritime Security Risk Assessment Model (MSRAM) to evaluate risk applied to national, regional, port and facility/vessel specific vulnerabilities and priorities. This tool does not yet include risk assessment algorithms that address cyber security. Therefore, identification of cyber security vulnerabilities and implementation of mitigation factors are not currently addressed or included under MTSA requirements. The biggest reason for this is a deficiency in understanding the complexity and scope of maritime systems vulnerabilities. In the common risk assessment modality where:

$$RISK = Threat \times Vulnerability \times Consequence$$

CHAPTER 9

The MTS is at a major disadvantage where virtually none of cyber security risk variables are adequately quantified. Intuitively, it is known there is a cyber risk problem, but the lack of definitive knowledge, information and data necessary to effectively define the risk limits the ability to craft effective defenses and solutions. Until that process is complete up to the present interval, no "best practices" can be fully developed or solutions fielded to best mitigate cyber security risk. Following that initial cyber risk assessment, the evaluation process must continue apace with development and deployment of new technologies to keep up with the evolving risk and threat variables.

But in a real world where about 90% of all U.S. Commerce—valued at $1.4 trillion USD in goods value per year (AAPA 2008)—travels by water at some point through 360 commercial (public and private) U.S. ports supporting over 13 million direct and indirect maritime related jobs, it is clear that all risk elements effecting the maritime transportation system—including cyber security challenges—requires better evaluation and inclusion as a key factor in developing an overall, comprehensive maritime security strategy.

Complicating this organizational paradigm is the many layers of operational interdependency of shippers, ship operators, terminal operators, port authorities, and land-based transportation networks (rail, pipeline, and highway) that contribute to potential security risks both physical and cyber related.

The Government Accountability Office (GAO) has highlighted the following "short list" of general maritime cyber security deficiencies in a June 2014 report:

> While the Coast Guard initiated a number of activities

> *Many of our ports are landlords, meaning they lease space to terminal operators who in turn conduct the business that moves through the ports. We have ports that are operated by state authorities, operating ports that are a part of larger state and regional government infrastructure; we have small, large and medium ports as well as ports that specialize in container, energy, break bulk and Roll on/Roll off cargo. At our core, ports are facilitators of partnerships that further the global, regional and national economies. At any one time our ports will be accessed by vessel, truck and rail lines, creating a complex, yet efficient center of economic activity. Ports are living and evolving entities that directly reflect the economic health of our nation. It is vital that our ports continue to have the flexibility and fluidity to meet the changing global trading trends and dynamics, while ensuring that our communities and commerce remain safe and secure.*
>
> *AAPA Advisory Letter to USCG*
> *April 15, 2015*

and coordinating strategies to improve physical security in specific ports, it has not conducted a risk assessment that fully addresses cyber-related threats, vulnerabilities, and consequences. Coast Guard officials stated that they intend to conduct such an assessment in the future, but did not provide details to show how it would address cybersecurity. Until the Coast Guard completes a thorough assessment of cyber risks in the maritime environment, the ability of stakeholders to appropriately plan and allocate resources to protect ports and other maritime facilities will be limited.

• Maritime security plans required by law and regulation generally did not identify or address potential cyber-related threats or vulnerabilities. This was because the guidance issued by Coast Guard for developing these plans did not require cyber elements to be addressed. Officials stated that guidance for the next set of updated plans, due for update in 2014, will include cybersecurity requirements. However, in the absence of a comprehensive risk assessment, the revised guidance may not adequately address cyber-related risks to the maritime environment.

• The degree to which information-sharing mechanisms (e.g., councils) were active and shared cybersecurity-related information varied. Specifically, the Coast Guard established a government coordinating council to share information among government entities, but it is unclear to what extent this body has shared information related to cybersecurity. In addition, a sector coordinating council for sharing information among nonfederal stakeholders is no longer active, and the Coast Guard has not convinced stakeholders to reestablish it. Until the Coast Guard improves these mechanisms, maritime stakeholders in different locations are at greater risk of not being aware of, and thus not mitigating, cyber-based threats.

• Under a program to provide security-related grants to ports, FEMA identified enhancing cybersecurity capabilities as a funding priority for the first time in fiscal year 2013 and has provided guidance for cybersecurity-related proposals. However, the agency has not consulted cybersecurity-related subject matter experts to inform the multi-level review of cyber-related proposals—partly because FEMA has downsized the expert panel that reviews grants. Also, because the Coast Guard has not assessed cyber-related risks in the maritime risk assessment, grant applicants and FEMA have not been able to use this information to inform funding proposals and decisions. As a result, FEMA is limited in its ability to ensure that the program is effectively addressing cyber-related risks in the maritime environment (U.S. Government Accountability Office 2014).

CHAPTER 9

Prompted by this GAO report, DHS and the United States Coast Guard (USCG) have commenced a multidisciplinary, multistakeholder analysis and discussion comprising the issue of the USCG Cyber Strategy (2015) and so the national debate has begun. The CCICADA DHS Center of Excellence Maritime Risk Forum in March 2015 (Rutgers University), the CSU California Maritime Academy hosted Maritime Cyber Risk Summit in June 2015, and the USC CREATE Center hosted Maritime Cyber event in March 2016 all have developed a baseline discussion dialog for maritime cyber defense and cyber security "next steps". The findings from the CCICADA forum are posted online and the report from the Summit has been publicly released by the USCG. Results of the USC forum target prioritization of cyber research and development projects.[1] Other events are planned in a sustained program encouraged by the DHS and USCG and all activities will inform the process for the USCG resulting in a final Best Practices policy decision and implementation action. The intent and scope of this action will balance identified vulnerabilities against mitigation recommendations and requirements.

The good news in the midst of this process are multiple efforts by other entities looking toward development of best practices and standards that can/should be applied in the maritime cyber domain to assure security goals are attained.

The bad news is that most of these efforts are uncoordinated responses to the emergent and largely unquantified threat. At the operational level, the majority of U.S. public ports are landlords that lease facilities to terminal operators and responsibility for adequate cyber security—as well as MTSA compliant physical security—is layered across multiple organizational and operational entities.

Complicating the overall task—there is as of yet no common maritime cyber security standard of practice to provide a baseline for development of competent protective and mitigation measures. Organizations engaging in related but different aspects of cyber security include at the national level the National Academy of Sciences Transportation Research Board Maritime Committee (TRB-MC); the National Institute of Standards and Technology (NIST) and on the international stage by the International Standards Organization (ISO), the International Maritime Organizations Maritime Safety Committee (IMO MSC), and the International Electrotechnical Commission (IEC) where two standards are in active development (IEC 62940—Integrated Communications Standard and IEC 61162-460—Security Levels for Integrated Vessel Bridge Networks).

These initiatives are laudable and necessary, but ultimately more effective and efficient if collaborative in nature and implementation.

---

1 For more information, please see USC CREATE website: www.create.usc.edu

One of many questions to be answered during this process will be: *In the United States, is maritime cyber security regulation required (for example under revision and amendment of the MTSA) or is directive recommendation such as those contained via the NVIC process—and voluntary compliance—all that is necessary—or perhaps is a combination of both elements a better solution following the ISPS Code model where Part A of the code is mandatory and Part B is optional implementation guidance?*

While this process is far from complete and remains an open question, agencies and organizations around the world are actively pursuing the guidance option. A major consortium comprised of the Baltic and International Maritime Council (BIMCO), Intertanko, Intercargo and the International Chamber of Commerce have already suggested development and implementation of voluntary international standards is the way forward. These organizations and others are advising the International Maritime Organization (IMO) to pursue this course. The USCG is also evaluating this option among others for implementation in the United States. Inclusion and encouragement of international ship classification societies such as ABS, Det Norske Veritas/GL, and others should also be a part of overall cyber security strategy and planning.

Some of these activities are already underway and more are expected. For example, RINA (The Italian Maritime Registry—Registro Italiano Navale) is "working with major cruise ship operators including Carnival, MSC Cruises and Moby Lines to introduce a preventative, risk assessment and management" approach to operations and maintenance that includes cyber hygiene practices (Laursen 2015a). RINA is also about to launch a full package of ferry safety initiatives, which will cover "specific training for crews including behavioral training to avoid and manage crises," fire risk mitigation, "enhanced planned maintenance, condition-based monitoring, and food and bacteriological risk management"—all motivated by the M/V Costa Concordia vessel casualty and other similar events (Laursen 2015a). The author, working with the Golden Gate Ferry Division of the Golden Gate Bridge and Transit District in San Francisco, CA has developed similar practical training courses customized to local ferry operators—some using state of the art maritime training simulators. A cyber security training element could easily be added to this program once standards and requirements are realized.

Bureau Veritas, DNV/GL, ABS, Class NK, and others associated under the International Association of Classification Societies (IACS) are similarly engaged in development of voluntary Cyber Security guidelines and Best Practice Standards—however none of these initiatives will be enforceable without implementing legislation and regulations. The new Chairman of IACS—ABS Chairman, President and CEO Christopher J. Wiernicki sees the organization as an important tool "in supporting the industry at a time when more stringent regulatory requirements are being formulated and implemented" worldwide (International Association of Classification Societies LTD 2015). IACS will also continue development "of a cyber-system safety framework that addresses control systems, software quality assurance, data integrity and cyber security enhancing an initiative that was commenced in 2014" (Fonseca 2015).

In the midst of development of cyber protective measures, the debate as to whether voluntary or mandated compliance by regulation of emergent cyber security procedures is raging. Most commercial entities favor adoption of voluntary standards—while governments tend to favor standardization of practice via regulation. History and experience shows that voluntary best practices will be embraced by some and ignored by others, largely dependent on the following factors:

- Calculation and perception of the risk/benefit analysis,
- Economic health of the enterprise,
- Influence of the risk management industry (maritime P&I clubs),
- The frequency of individual corporate negative experiences.

Arguably, a more balanced approach to provide options for maximum flexibility of solution design, dependent upon the specific needs and requirements of the operator under a proscribed set of baseline regulatory requirements, is the best proven way forward.

The Current Challenge

It can be argued that the maritime industry is particularly susceptible to cyber attacks due to the expanding interconnection of systems aboard ships. Subject to all of same potential vulnerabilities as land-based enterprises, the maritime domain literally includes many more "moving parts" ranging from ships at sea, to containers moving ashore via automated terminals and onward to destinations serviced by truck and rail. Further, the frequent rotation of multinational ships' crews also magnifies the risk of an insider providing malicious access to the network by accident, specific intent, or for criminal financial reward. In situations where ships can increasingly connect their networks via the Internet of Things (IoT)—whether from land or at sea, cyber criminals, and others can gain access remotely without an insider physically on board.

Ships—like most everything in the world today—"depend more and more on complex, [interconnected] computer systems. When each system is isolated and serving only one function ... a software failure ... [has] limited consequences. But when the systems become integrated the risk of a software failure bringing down the entire ship" rises due to possible cascading effects (Laursen

2015). Once cyber criminals or terrorists have access to interconnected systems controlling navigation, propulsion, fire suppression, fresh water, grey water, and ballast water systems, communications and electrical power—anything is possible. To highlight the cyber security challenges in the maritime operational environment, it may help to visualize the following scenario modified and updated from the opening pages of "Maritime Security" by Kenneth Gale Hawkes published in October 1989—and still a valuable resource in describing maritime security challenges:

> The super large 18,500 TEU container ship is operating normally as she makes her way toward San Francisco's Golden Gate. She is the pride of the fleet and only a year old, configured with the latest in fully integrated control systems. The approach has been largely uneventful and routine and since bringing the Pilot aboard at the sea buoy, she has been steaming at a fast 25 knots as she endeavors to make her time slot assignment at the Port of Oakland.
>
> Concurrently, a large partially laden tanker is outbound for Los Angeles about a half a mile east of the bridge.
>
> As she transits the traffic separation for the final approach into San Francisco Bay, the container ship suddenly loses control of main propulsion and electronic navigation systems simultaneously. Now within 100 yards of the bridge, the ship CPA closes on the southern bridge tower and will not respond to helm or to engine control. Helpless to avoid the inevitable, the Master sounds the general alarm for collision alert and makes an emergency broadcast to USCG VTS San Francisco.
>
> The outbound tanker observes the change in course and attempts to avoid the area, but does not have sufficient sea room to stop or reverse. As the 1,200-foot-long container ship allides with the bridge tower, the flood tide carries her sideways blocking the main channel. With insufficient sea room remaining, the tanker collides with the container ship, breeching the forward peak, center and wing tanks, releasing a load of refined gasoline into the water and— finding an ignition source-, bursting into flame. Both the container ship and the tanker are engulfed in fire, while entangled in the main channel and the container ship is fast aground to the bridge tower. Both ships are holed and taking on water. Crew members caught on both ships begin to self-evacuate into the chilling waters of SF Bay.
>
> Faced with a fire, threat to life and property as well as a large environmental spill, the USCG Sector Commander (and Captain of

CHAPTER 9

the Port) orders the waterway closed and the Golden Gate Bridge District also closes the bridge in both directions until the fire is extinguished and a damage assessment for the structure can be performed and evaluated. Fire rages, gasoline continues to spill, six SF Bay area ports are now effectively closed, one of the main transit hubs into and out of the City.

Possible? ... Unfortunately, yes ... Critically vulnerable and at risk? ... The consensus again is "Yes," but as of yet it is unknown how catastrophic the effects would be either short or long term. In any case, recent events in the maritime sector have government, military, and commercial operators all scrambling to determine vulnerabilities and define the risk within their own operational areas.

Efficiency and economics are driving systems integration in the maritime shipping business. The more sophisticated control systems become, the more vulnerability is created as these systems are fully integrated so that their "decisions" are increasingly linked together. These systems are becoming very common on newer vessels and are contributing to the lessening of manning requirements along with a concurrent reduction of operating cost. But there are always direct, indirect—and sometimes unrecognized and unseen—costs to be paid for progress. One such cost might well be increased vulnerability to cyber attacks with potentially millions of dollars at risk. According to Steven Caponi and Kate Belmont writing for the major U.S. law firm Blank Rome LLP:

> Given the interconnectivity of the maritime industry and paramount need to keep ports moving with speed and efficiency, a cyber attack on just one of the major EU or U.S. ports would send a significant negative ripple throughout the entire industry. With the ability to impact so many nations and peoples at once, the maritime industry presents a fruitful target for both private and political actors. Both the GAO and ENISA agree that the soft underbelly of the maritime industry is its reliance on Information and Communication Technology ("ICT") in order to optimize its operations. ICT systems used by ships, ports, and other facilities are frequently controlled remotely from locations both inside and outside of the U.S. Responsibility to actively defend against the risks of a cyber attack and be in a position to effectively respond to an incident rest squarely on the shoulders of individual ship owners, shipping companies, port operators, and others involved in the maritime industry. The failure to assume this responsibility will undoubtedly lead to serious and potentially devastating consequences, including government fines, direct losses, third-party liability, lost customers, and reputational damage that cannot be repaired (Caponi and Belmont 2014).

Recent documented experience has revealed vulnerabilities in automated control systems in MTS cargo management and ship navigation systems tied to GPS. There appears to be solid evidence of successful cyber attacks on equipment Supervisory Control and Data Acquisition (SCADA) and other Process Logic Controller (PLC) operated systems aboard floating oil platforms. Siemens, one of the world's largest manufacturers of electronic control systems—recently issued a warning to customers about not one, but two, remote execution faults discovered embedded in the WinCC system operating software. It was Siemen's system software that was targeted by the Stuxnet cyber virus in Iran in 2010, but these vulnerabilities are not confined to Siemens. All integrated operating systems follow the same basic architecture and the more data sharing and transmission that occurs from ship to shore, the more opportunity for unintentional failures and intentional hostile acts arises (Tung 2014). Brian Lord, managing director of PGI Cyber—a UK based cyber security firm—has highlighted that the trend for centralized electronic controls aboard ships, with the IMO pushing for ECDIS to become the focal point, brings an inherent weakness. "Every system must have a backup so that it can be operated on its own; otherwise attackers will find it easier to take control" cautions Lord (Hidden in Full View 2015). Mark Gazit of Israeli based security firm ThetaRay goes further saying:

> Ships, boats, patrol vessels, rigs, and naval resources can communicate using radio and IP-based communications of different capacities, relying heavily on satellite communications. While these enable them better situational awareness and beyond-line-of-sight operations, they are considered insecure comparing to other types of encrypted communications used by different critical infrastructure. For example, flaws in AIS can allow attackers to hijack communications of existing vessels, create fake vessels, trigger false SOS or collision alerts, and even permanently disable AIS tracking on any vessel. Another main issue in this sector is the human factor. The maritime organization is designed to serve the vessel's most pertinent needs, manning it with an operational crew that rarely includes a cyber security professional, or an information security officer. This is another factor that presents a challenge to the security of communications and cyber safety of vessels and their port/ashore personnel (Hidden in Full View 2015).

The human element, however, is a sword that can cut both directions depending on how it is defined and applied. Technology advocates promoting fully automated systems see human engagement as a weak link contributing to enhanced vulnerability of computerized systems by way of human error or ac-

CHAPTER 9

tive interference. The contrary view sees humans as providing the last line of defense should a cascading failure within interconnected systems occur. The author's view is that a balance between these positions is best to assure maximum flexibility and resilience.

There is already a growing list of nefarious examples of cyber intrusion and hijacking of control and business systems directly affecting the MTS. Some of these specific examples include:

- *Cyber Crime—Port of Antwerp 2013 Shipping Container Misappropriation:* In late 2013, Belgian government and Port of Antwerp officials revealed a cyber intrusion ongoing at least since 2011 that allowed criminal elements to gain access to terminal container tracking systems. The malefactors used this access to divert and release targeted containers containing smuggled narcotics to their own trucks and then to delete any record of the container transit—all accomplished without the knowledge of the shipping line or terminal operator.

- *Facilitating Piracy:* Somalian pirates have hired hackers to break into the computer network of a shipping company to locate ships carrying precious cargo or fewer security guards. They also use AIS tracking as a "shopping list" to identify and follow the most valuable vessel targets for possible seizure. These techniques have resulted in at least one verified case of the successful capture of a ship by pirates.

- *Cyber Fraud:* One major example of cyber fraud within the maritime community is the "cyber-attack that cost World Fuel Services (WFS) an estimated $18 million" USD (Caponi and Belmont 2014/2015). "Impersonating the United States Defense Logistics Agency, cyber criminals used fake credentials to send an email seeking to participate in a tender for a large amount of fuel. WFS received the offer to participate in the tender, took the email at face value and purchased 17,000 MT of marine gas oil from" the Danish fuel bunkering company "Monjasa that was then delivered to a tanker known as the *Ocean Pearl* while it was off the Ivory Coast. Upon submission of the invoice, the government agency responded that it had no record of the fuel tender" and refused to pay (Caponi and Belmont 2014/2015).

- *AIS, GPS and ECDIS Spoofing, and Jamming:* All three systems are critical tools used for safe navigation and operation of ships. The first critical point is that the automated information system (AIS) can be switched off at the vessel source whether intentionally or otherwise—creating a "ghost ship" that cannot be "seen" by other ships and ports. Another key challenge with AIS is that it currently has no integral security or encryption protocols. Since AIS was designed to be a safety system freely available to everyone, information broadcast over the system is automatically assumed to be legitimate and accurate.

AIS: Trend Micro Inc.—the third largest systems security firm in the world—in October 2013 demonstrated how the AIS system could be easily penetrated and spoofed into providing the attacker with a range of nefarious possibilities. Using equipment having a cost averaging about $200.00 USD, the staff at Trend Micro was able to accomplish the following:

- Modification of all ship details, including position, course, cargo, speed, and name. Creation of "ghost" vessels at any global location, which would be recognized by receivers as genuine vessels.

- Send false weather information to a vessel to have them divert around a non-existent storm.

- Trigger a false collision warning alert, for some vessels resulting in a course adjustment.

- The ability to impersonate marine authorities to trick the vessel crew into, e.g., disabling their AIS transmitter rendering them invisible to anyone but the attackers themselves.

- Create "ghost" search and rescue helicopters

- Create a fake man-over-board distress beacon, triggering the alarm on nearby vessels.

- Cause vessels to increase the frequency with which they transmit AIS data, resulting in all vessels and marine authorities being flooded by data in essentially a denial-of-service attack (Cyberkeel 2014).

GPS: Maritime navigation is now heavily dependent on Global Positioning System technology to pinpoint vessel positions to a high degree of accuracy and is preferred by many over more cumbersome paper chart methods. Some state actors are seeking to disrupt and manipulate this technology for nefarious purposes as described here: South Korea has been subject to annual GPS jamming attacks by its North Korean neighbor since 2010. Over that period, jamming has

CHAPTER 9

extended over longer periods, with the longest" attack documented as "being a continuous 16-day attack, employing various frequencies, techniques and signal strengths. As the jamming periods increased each year, they affected more and more GPS users. In 2013, South Korean officials estimated that 1,016 aircraft lost GPS signals, as did 254 ships and a large number of cellphone towers" (Sheridan 2013).

Further—"The jamming seriously affected South Korea's substantial merchant fleet and its key trans-Pacific export trade to the U.S., for which GPS has become a standard navaid," increasing the time required to authenticate signals and/or revert to other techniques to verify positions. The "North Korean GPS jamming attacks underscore the vulnerability of the very low-powered satellite-based GPS signals, and South Korea recently announced a nationwide e-LORAN navigation project under which the entire country will be covered by the new system in 2016, with a 20-meter position accuracy" (Sheridan 2013).

In another example of GPS vulnerability, "in July 2013 a research team from the University of Texas managed to take control of the navigational systems of an $80 million dollar 210-foot yacht—the" *White Rose of Drachs* (shown at right, source charteryacht.com)—in the Mediterranean (Cyberkeel 2014). They accomplished this using customized equipment, which cost only $3,000.00 USD to build. "Essentially they injected their own radio signals into the vessel's GPS antennas, which enabled them to steer the vessel as they saw fit. Whilst they were doing this, the vessel's GPS systems reported that the vessel was moving steadily in a straight line, with no indications of changes. The captain of the vessel, who had given permission to perform the test, stated that: '[They] did a number of attacks and basically we on the bridge were absolutely unaware of any difference'" (Cyberkeel 2014).

Even though not legal in every jurisdiction, "powerful GPS jammers are readily available on the commercial market [and] are easy to obtain. Units with jamming ranges of up to [400] meters" (up to 1,200 feet) are available (Cyberkeel 2014). Disabling or disrupting "GPS can present a significant challenge" to safe navigation (Cyberkeel 2014).

The UK and Irish General Lighthouse Authority performed a similar test on the vessel NLV Pole Star (Cyberkeel 2014; Photo on left: commons. wikimedia.org). "Powerful GPS jamming equipment was directed [at] a specific patch of ocean [and] the vessel was sailed into the zone to record developments. As the vessel entered the jamming zone a range of services failed: the vessel's DGPS receivers, the AIS transponder, the dynamic positioning system, the ship's gyro calibration system and the digital selective calling system" (Cyberkeel 2014).

"The crew was able to cope with multiple alarms as they had been expecting this to happen. However, on a modern vessel the bridge might on some cases be single-manned at night, causing significant problems should such a situation occur. Although the Pole Star [8's] crew was expecting GPS failure, material unexpected problems were seen. The vessel's Electronic Chart Display & Information System (ECDIS) was not updated due to the failure of the GPS input, resulting in a static screen. ECDIS is the normal mode of positioning on board Pole Star 8 (with paper chart backup,) and during the periods of jamming some crew members became frustrated when trying to [use] the ECDIS. This resulted in the monitor being switched off" (Cyberkeel 2014).

"In addition to the vessels themselves, some automated container terminal systems use GPS to facilitate the [automatic] placement and movement of containers and can similarly be jammed, which would cause significant congestion problems" since many of these new operations have greatly reduced the use of human beings on the container apron (Cyberkeel 2014).

ECDIS: The Electronic Chart Display and Information System (ECDIS)—"is the computer system usually installed on the bridge of the ship and used by navigation officers as an aid to traditional paper chart navigation—often supplanting traditional navigation" (Cyberkeel 2014). While it can be argued that this may not be a best practice, the International Maritime Organization is calling "for ECDIS to completely replace the use of paper-based navigation" (Cyberkeel 2014).

"ECDIS is interconnected with a wide range of other systems and sensors such as radar, Navigational Telex (NAVTEX), AIS, Sailing Directions, Position Fixing, Speed Log, Echo Sounder, anemometer, and fathometer. These sensor feeds are often connected to the shipboard network, which in turn has a gateway to the Internet" (Cyberkeel 2014). Electronic digital "navigational charts are either downloaded into ECDIS directly via the Internet or loaded from CD/DVD or USB memory disk manually by the personnel" (Cyberkeel 2014).

When the U.S. Navy warship USS Guardian struck a protected reef and ran aground off the Philippines in 2013, the Navy in part blamed incorrect electronic digital charts. (Command Investigation, May 2013). The naval digital chart in use aboard the Guardian placed the reef 8 nautical miles in error of its actual position and the crew failed to validate the information with other means. (Maritime Accident Casebook, January 19, 2013) A NATO-accredited think-tank said the case illustrated "the dangers of exclusive reliance upon electronic systems, particularly if they are found vulnerable to cyber-attack" (Wagstaff 2014).

In January 2014, the U.S.-based security firm The NCC Group demonstrated that ECDIS can be penetrated and manipulated (Cyberkeel 2014). NCC tested an ECDIS product from a major manufacturer to see whether system penetration was possible. Several system security weaknesses were identified including the clandestine ability to read, download, replace, or delete any file stored on the server hosting the ECDIS software (Cyberkeel 2014). NCC demonstrated intrusions "could be achieved by various means, such as the introduction of a virus via [a] portable USB disk by a crew member or any other visitor to the vessel; or using an unpatched vulnerability via the Internet—either directly or via one of the multiple systems linked into ECDIS. Once such unauthorized access is obtained, attackers [would] be able to interact with the shipboard network and everything to which it is connected" (Cyberkeel 2014).

- *CyberKeel Container Carrier Management System Penetration Testing:* From February to September 2014, Denmark-based CyberKeel investigated cyber vulnerability for possible intrusion into 50 of the world's largest major container shipping carriers. The research consisted of two relatively simple tests: one was the use of online applications to perform shipment track and trace, schedule inquires and e-mail query forms to validate they were safeguarded against insertion of malicious code. The other test was the use of a search tool to identify carrier hardware then cross-match those systems to known hacker resources to determine the probability of whether those systems had already likely been compromised. The results of this research revealed that 37 of the 50 carriers appeared vulnerable to simple penetration attacks (Cyberkeel 2014).

- *Dynamic Positioning Systems (DPS):* DPS is the prevailing technology that mobile drilling platforms (ships) to maintain the precision necessary to perform drilling operations in the open ocean for gas and oil exploration. It involves the application of an intricate ballet of computer-assisted station keeping using engines, rudders and propulsion units tied to GPS and other means to maintain an exact location and reduce the risk of failure of the drill string (Wagstaff 2014). "According to security company ThetaRay, a cyber-attack on a floating oil rig off the coast of Africa" resulted in tilting the vessel sufficiently to force a shutdown of drilling operations, requiring a team of experts a week to identify, isolate and remedy (Cyberkeel 2014).

Technology is Our Friend, Except When it is Not

When it works, sophisticated technology is close too miraculous; when it does not work the results can run the gamut from inconvenience too catastrophic. Most or all of us have experienced standing in the check-out line at the local market when the power fails, and the scanner and the registers stop functioning. The helpless cashier has no method of manual operation of the register, so the business is effectively closed until power is restored and hopefully the computerized system reboots without difficulty. Translate this simple situation to a large, heavy ship moving at high speed where nearly everything is connected to make operations more efficient, and it is not a hard to visualize the worst case result.

Since 100% systems reliability or perfection can never be attained, the full impact from short term and long-term practical effects—and the associated business costs—of routine systems failures associated with fully integrated "smart" control systems cannot be fully estimated, let alone the effects of a concerted maritime terrorism attack such as conceptualized above.

Many new systems are not configured with manual "work arounds" that will allow human control to be easily and effectively exercised. Training of mariners is also beginning to change with more emphasis on technical systems management and electronic operation of navigation and propulsion systems—and less time focused on the perishable skills of practical seamanship.

"Over the past five decades, computer controls have been integrated into innumerable operational and business processes across diverse industries, including the shipping industry, resulting in considerable improvements in safety, accuracy and profitability," but also creating heretofore unseen vulnerabilities and weaknesses (Bancroft 2014). While the benefits have been clearly documented, the increased reliance on technology for even the most basic operations can leave any operation or enterprise exposed to complete "business interruption or, in a worst-case scenario, absolute continuity failure" (Bancroft 2014). The "number of potential risk scenarios is significant and keeps growing" (Bancroft 2014). System intruders "employ whatever hacking technology works, often tailored to specific" industries and "targets of opportunity" (Bancroft 2014).

> *The growing reliance on automated systems makes the domestic and global supply chains vulnerable to potential criminal and terrorist cyber-attacks. The flow of goods through ports is dependent in large part on networked computer systems, controlled by human operators, but that also work independently, and programmed to move containers and goods in a very precise, orchestrated manner. Should these processes be compromised, a cascading effect might be created, disrupting the goods movement supply chain through ports and across the entire country.*
>
> AAPA
> Cybersecurity Government Policy Paper
> March 2014

It is likely the maritime community has (or soon will) experience similar concerns as those surfacing in the airline industry, where pilots used to "flying by wire" using technology that essentially flies the airplane without human intervention are left underprepared to fly the aircraft manually in an emergency. "There is widespread concern among pilots and air carriers that as the presence of automation increases in the airline cockpit, pilots are losing the skills they still need to fly the airplane the 'old-fashioned way' when the computers crash," said Steve Casner, coauthor of "The Retention of Manual Flying Skills in the Automated Cockpit" and research psychologist at NASA's Ames Research Center (Homeland Security News Wire 2014a). Professor Richard Clegg, managing director at the Lloyds Register Foundation, has described this weakness as approaching a "... blind reliance on technology (where) new technology can, ironically, diminish the role of the seafarer. That includes over-confidence and over-reliance on, and lack of understanding, of the technology. Higher operator skills are needed but the operator roles become more mundane. Traditional hands-on experience is lost as the operator roles focus more on monitoring" (World Maritime News 2015d).

On the other hand, some positive benefits of integrated technology can function to remove human personnel from hazardous operational environments, exposure to work-related chemicals as well as reduce the overall costs and increase the efficiencies of performing these tasks. History shows that modern jet aircraft will fly on autopilot until they run out of fuel and crash into land or sea.—Similar incidents occur aboard ships where reliance on systems technology providing flawed or misinterpreted data has resulted in serious incidents including groundings, collisions, and allisions. A fundamental question arises: *Should the industry follow that path with increasingly automated ships and no human control element aboard?* Such alternatives are currently under consideration within the European Union with the development of drone ship concepts operated remotely from shore stations.

The more we connect to a system of systems—the Internet of Things, the greater the overall vulnerability and the increased risk. Older ships with simpler, mechanical, and unlinked analog control systems are actually less vulnerable to attack, and are designed with "work arounds" that allow the crew to manage an untoward event by literally turning knobs, pushing buttons, and pulling leavers. Modern vessels however are being designed with extensive labor saving devices and interconnected systems that create the potential for critical single points of failure that can immobilize multiple ships systems simultaneously. The US Navy—engaged in major redesign and reconfiguration of ships systems to drive down human crew requirements and improve efficiency—has already experienced multiple major catastrophic failures of integrated ships control systems that resulted in complete loss of control of the vessel for a period of hours and continuing effects for days.

Commencing in 1998 with the USS YORKTOWN, an Aegis class cruiser and extending to the USS Freedom, the first Littoral Combat Ship, the latest class of LPD amphibious ships and even to the new aircraft carrier USS STENNIS in the last few years, the Navy has experienced ongoing problems and challenges in implementing increasingly sophisticated integrated systems technology —in at least one case resulting in a combat ship DIW—"Dead In The Water"—for hours while a crashed computer system was troubleshot and rebooted—with significant tech support required from the land-based civilian systems operator. The YORKTOWN was so disabled she was required to be towed into port and repaired fully over a two-day period (Slabodkin 1998).

Considering the U.S. Navy has had the most experience with integrated maritime control systems and lessons learned from decades of problems should inform transition of similar systems to commercial ships. Yet the Navy continues to drive to a single integrated system solution. The latest example

CHAPTER 9

of this is CANES, Consolidated Afloat Networks and Enterprise Services, a fully integrated intelligent controls system including internal and external ship to shore information and data sharing. Commenced in 2010, surviving numerous contract bid challenges and valued at $2.53 billion USD, Northrup Grumman is the lead design and implementation vendor, set to compete with six other approved vendors for installation of the entire program. The CANES program is expected to be completed by August 2022 (Chips Magazine 2013).

Eventually CANES will be installed into 192 ships and shore facilities with the first ships being the destroyers USS MILIUS (DDG-69) and USS Mc-CAMPBELL (DDG-89) and the carrier USS STENNIS (CVN-74) intended to consolidate the workings of up to 41 ship and ship to shore control and communications system networks into only three: one open source network for unclassified materials, one dedicated to classified materials up to the secret level (SIPRNet), and one network dedicated solely to management of top-secret information and documents (JWICS). Surprisingly, although clearly designed to increase efficiency, security and reduce overall system vulnerability, CANES classified and unclassified data streams still "crossover" to make the best use of limited bandwidth using a common Automated Digital Network System (ADNS). ADNS collects all digital traffic through a single router, sends the information via a single encrypted waveform packet relayed through a defense satellite, and is then split apart by a second ADNS router when received at the destination.

The obvious question here then is: *Does this equipment integration make the overall system of systems more or less vulnerable?* Even as the Navy remains committed to CANES it is wary and troubled by this uncertainty. In a dialogue with CHIPS Magazine, Rear Adm. William Leighter stated,

> While it is true the U.S. Navy is the most technologically advanced Navy in the world; we are subject to the same end of life and end of support issues that face industry and home computer users. Operating systems and software are continually being upgraded and advanced, and older software eventually, is no longer maintained by the manufacturer. This means that they stop providing security patches and the software becomes increasingly vulnerable.
>
> The Navy is the ultimate wireless customer—we don't trail fiber cables behind our ships which connect to the shore. CANES, along with our new satellite gear, the Navy Multiband Terminal (NMT), and upgrades to our Automated Digital Network System (ADNS), provide the network and connectivity allowing the Navy to accomplish those missions and to assure our command and control.

Every "aperture" that can receive electromagnetic signals, whether it's a sensor or a wireless network, is a potential gateway for hackers to connect to your systems. Navy "red teams" that emulate a potential adversary's techniques regularly scan the Navy's ships, aircraft, and bases for just such vulnerabilities.

We probe for those things and we correct them as we find them. If the bad guys do get into your network, though, they can follow the digital links to all sorts of unpleasant places. Connectivity can become a security problem (Chips Magazine 2013).

THE COMMERCIAL SECTOR LEVERAGES AND FOLLOWS DEFENSE INNOVATION

But—why should the public and commercial industry care what the defense establishment does? After all—the basic mission goals and objectives differ dramatically between commercial and defense/security driven organizations. Weapons systems and strategies bear little resemblance to a stake of cargo containers and multinational goods movement and international trade activities.

The answer is that modern commercial businesses nearly always look to innovation and technology developed and field tested by the government and military sources for eventual adaption to merchant ships. Indeed many, if not all, of the technology providers to the military also field company divisions pushing these products out to the commercial sector. Is it reasonable to assume that the commercial products will be any more secure and reliable than those designed and installed on military combatants? Prudent analysis would indicate a negative conclusion, but the trend once started continues to gain a seemingly unstoppable momentum.

One driving force behind this rush to everything technologically new, bright and shiny is a concept sometimes defined as the "demand for micro-perfection" where advances in process and technology conspire to create false demands for "maximum improvements" in operational efficiency and mission accomplishments. Since attainment of 100% of anything is never possible, any goal established with this objective as a primary outcome will fail. Viewed in this way, rather than advance attainment of positive results, micro-perfection typically generates macro-risk because we assume we are making the process perfect and fool-proof. Yet, this is the operational paradigm pervasive within an environment where an often misunderstood concept of "make it better" is always perceived to be the "best" solution (Welsh and Harvey 2014). Like the author, the reader might enquire about what ever happened to the concept of *if it works, don't fix it* as a more practical, effective and manageable operational paradigm.

In one example of maritime super systems integration—and a good example of the truly global nature of emergent technology—the major German shipping operator Hamburg-Sud has contracted for three 10,600 TEU container ships to be configured with a new controls package by L-3 Systems South Korea engineering affiliate—using the NACOS Platinum navigation, automation, and control systems (illustration, right, from Wärtsilä SAM Electronics) developed and sold by the German based Wärtsilä SAM Electronics; proudly announcing:

> ... navigation systems [that] each comprise X- and S-band radars linked to a series of five multifunction Multi-pilot workstations for centralized control of all main radar, ECDIS and conning operations, in addition to those for automatic steering, track control and voyage planning. The scope of supply includes AIS, VDR, differential GPS, Doppler log, echo sounder and wind/weather navaids, as well as a bridge navigational watch alarm system (BNWAS). Associated communications facilities cover a full range of GMDSS A3 equipment, inclusive of a KVH TracPhone FB500... The design architecture also provides the integration of a fuel-efficient vessel performance system to reduce emissions from shipping vessels and associated running costs. [The system] supports a full range of remote access and diagnostic functions (Howard 2014; Wärtsilä 2016).

The major European big box ship carrier MAERSK Lines announced in March 2016 that 27 of its new build ship and 47 existing ships in its fleet will also be fitted with the same NACOS controls system (ShipTechnology 2016). These advances in systems integration are clearly beneficial to the improvement of operational efficiencies, but also potentially vulnerable to intrusion and attack as system capabilities and capacities expand to facilitate "remote access" and increase real time data to data transfer and connectivity between what used to be completely separate ship control systems.

The NACOS Platinum system is only one example of a multitude of integrated control packages either already in production or on the verge of entering the marketplace. Other major equipment engineering, ship building, and design firms such as Lockheed Martin, Northrop Grumman, Raytheon, Hyundai Heavy Industries, Siemens and Rolls Royce are equally engaged in integrated control systems research that will transfer control decisions to computer-based decision-making and away from human interaction and engagement. Wow—perhaps it really is time to put our feet up on the bridge rail, think happy thoughts and let the ship drive herself while we order cocktails from the Lido Deck.

GHOST SHIPS WITHOUT HUMAN CREWS

Or perhaps the ultimate future vision is the operation of "Ghost Ships" (conceptualized below by Rolls-Royce; www.images.dailytech.com) that are fully automated and controlled remotely from shore stations. Another dimension of micro-perfection assumes that removal of the human element aboard ships at sea is the desirable ultimate future outcome and this drives the ghost ship concept that industry has been exploring for over a decade. Industry estimates of freighter costs estimate that 44% of the average total is associated with supporting vessel crews. "Rising fuel costs in recent years have [also] meant slower and less fuel-consuming voyages are more economical, but long journeys are also less attractive" to men and women at sea, so finding willing crews from around the world is becoming an increasingly difficult challenge (Carroll 2014). As a result, active research in the development of completely autonomous vessels—operated without any humans aboard—is actively underway worldwide.

U.S.-based Sea Machines CEO Michael Gordon Johnson foresees a number of applications for autonomous vessels including oil spill response where an unmanned system as more efficient, able to operate for longer periods and safer than crewed vessels (Laursen 2015b). Hybrid systems are also envisioned where a manned vessel would form the centerpiece "of the operation with one or more unmanned boats working in formation with it. The master of the manned boat would have control of the other vessels, or alternatively, they could be controlled from a shore base. "Our whole concept is to develop autonomous control systems and unmanned work boats," says Johnson. 'This is the next natural evolution of the maritime space—to go from the current, manned vessels that are out there to adding additional (increased) levels of automation." (Laursen 2015b).

Oskar Levander, Vice President for innovation, engineering and technology at Rolls-Royce Marine based in Finland suggests that "the first unmanned commercial ships are likely to be locally operated vessels since single flag states can permit their operation" within their territorial waters before international regulations are adopted (World Maritime News 2015e). Mr. Levander perceives that passenger and roll on/roll off (ro/ro) "ferries would be a prime candidate for early autonomous adoption because they operate within a confined area and in addition "there is a clear desire to address the crew cost." Studies indicate most essential technology building blocks already exist, but practical marine solutions will still require some development efforts"—and the public will still

CHAPTER 9

require some convincing that this approach is as safe—or safer—than currently crewed operations (Carroll 2014). Movement toward this level of autonomous control has been an incremental but continuous process covering every aspect of vessel operations. For example, as a first step Rolls-Royce is initially planning for improvements in engine efficiency of up to 20% directly tied to autonomous control systems and a corresponding reduction fuel usage and emissions, while exploring other aspects moving toward 100% autonomous operations.

The march to fully integrated and automated systems seems inexorable, even as some experts are making the case for a "Back to the Future" vision of returning to analog back-up capability as a redundant failsafe measure— and the retention of human crews. One major proponent of this approach is former US Navy Secretary Richard Danzig—now a fellow at the RAND Corporation—who argues that system operators should "merge your system with something that is analog, physical, or human so that if the system is subverted digitally it has a second barrier to go through," he said. "If I really care about something then I want something that is not just a digital input but a human or secondary consideration" (Danzig 2014). Even with the calls for caution, the push to fully automated systems continues to accelerate.

Joining private sector efforts like Rolls-Royce over the last several years, the European Union (EU) has committed the equivalent of $3.85 million USD to the development of Project MUNIN (Maritime Unmanned Navigation through Intelligence in Networks) that is intended to provide for 100% control of shipping from remote shore stations. Rolls Royce's Blue Ocean Team is one commercial entity actively working on this concept and was just awarded another EU grant (July 2015) underwritten by Finland's TEKES (Funding Agency for Technology and Innovation) valued at $7.3 million USD in support of The Advanced Autonomous Waterborne Applications Initiative. The AAWAI is focused on development of concepts "to explore the economic, social, legal, regulatory and technological factors which need to be addressed to make autonomous ships a reality" (Haun 2015b). The EU has also funded the EfficienSea2 initiative to a level of $11million USD, which is being spearheaded by the Danish Maritime Authority and 32 public and private partner organizations from 12 countries. This initiative has a primary goal of establishing a "maritime cloud" communication tool to manage the exchange of vessel to vessel and ship to shore information transfer and facilitate further integration of navigation, maneuvering, and other ship control systems. Maritime cyber security protocols and strategies are a major part of this effort (Haun 2015a).

Just added to this mix of researchers from around the world is Plymouth University in the United Kingdom who are leading a "pioneering project to design, build and sail the world's first full-sized, fully autonomous unmanned ship across the Atlantic Ocean. The Mayflower Autonomous Research Ship, codenamed MARS, will be powered by renewable energy technology, and will carry a variety of drones through which it will conduct experiments during the crossing. MARS is expected to be constructed within the next two-and-a-half years" and is scheduled for its maiden trans-Atlantic voyage in 2020 (World Maritime News 2015c).

Also recently announced (Aug 14, 2015) is a renewed investment and continued push from China's Maritime Safety Administration (MSA) and Wuhan University for expansion of a project commenced in 2012 that has already resulted in four national systems patents for "non-stop networked sea supervision, intelligent cruise and rescue, motorized multipoint coverage by shore-vessel based detection and control platform managing a certain amount of unmanned cruise and rescue vessels on a low-cost basis, which would effectively improve the ability and efficiency in maritime cruise and rescue" (World Maritime News 2015a).

Shoreside Operations are not Exempt or Immune from Cyber Disruptions

The same pressures and factors influencing systems and operational design aboard ships are also changing the operational environment at marine terminals ashore. In the EU and now in at least two locations in the United States, fully automated container terminals are being constructed and tested. On the U.S. west coast, the Port of Long Beach is the site of a major terminal experiment being advanced by Hong Kong based Orient Overseas Container Line (OOCL) that involves the total automation of container management from ship to truck or rail distribution.

Artist Rendering, Long Beach Container Terminal (OOCL), courtesy Port of Long Beach

The new facility, expected to be in partial operation by 2016 and full oper-

CHAPTER 9

ation by 2020, utilizes the largest and most technologically advanced gantry cranes in the United States, capable of a "maximum outreach of 226 feet," allowing "them to reach over container ships carrying containers up" to 24 units "across (side by side)—large enough to support the 18,000 TEU Maersk Triple-E class containership which carries 23 rows of containers" across the beam from port and starboard (Almeida 2014).

OOCL Terminal, Courtesy Port of Long Beach

Although each crane retains a human operator in the cab, the vast majority of standard operations typically involved with the on-loading and offloading of containers is a well-choreographed and highly computerized activity overseen by the human "operator." Longshoremen and ships' crew still work the containerized cargo aboard the vessel, but once lifted off the ship and transferred ashore, every step of the container management process is fully automated. The container is set onto a 100% electric battery operated automated guided vehicle (AGV) that is programed to carry the container across the apron and into the container stacks to a pre-designated position. From there, dependent upon distribution mode and destination, the fully integrated and GPS guided apron crane system retrieves and transports the selected container and places it onto the waiting, pre-staged drayage truck or rail car. Throughout this process, there is no requirement for any human activity or intervention on the terminal apron or in the container stacks—therefore the risk of human caused error, accident and injury is eliminated, labor costs are reduced, and efficiency in cargo management enhanced (Almeida 2014).

The author has seen the LBCT terminal in operation and it is indeed quite impressive. But—there is also a documented downside: The terminal operator revealed that during system testing, the global positioning system (GPS)-dependent container tracking and transportation system suddenly and mysteriously failed. As a result of an investigation, it appears that a person parked on the nearby public access road was using a GPS jammer to override the tracking system on his cell phone so his wife and others would not know his location.

The jammer was sufficiently powerful to disrupt GPS signaling in the immediate area—including the container yard—disrupting operations for hours. Although not an intentional act intended to disrupt the terminal, it is a cautionary tale that illustrates that any system with a single critical point of failure is vulnerable even if not the intended target.

As integrated systems technologies continue to advance, so too will the integrated operational relationships between the transportation modes comprising the MTS, especially the connective interface between truck, rail, and ship terminal. Automated systems management seen in the maritime mode is and will continue to be similarly implemented as we look to the ongoing development of drone trucks and smart rail. In every case, GPS is increasingly a major element in assuring control, location plotting, positioning, and safe operations. The wide spread employment of GPS—a government provided, maintained and operated satellite based system—is a double edged sword. While as an extremely precise tool provided at no cost to users makes it extremely attractive to design engineers who have incorporated it into everything from warehouse robots, self-driving vehicles, cell phones, air planes, and ship navigation systems, the government can switch off GPS during a national emergency and—as shown—can be jammed and/or spoofed with a relatively low level of sophistication—or simply by accident.

Currently—there are no alternatives or options to the government controlled, satellite operated GPS. President George W. Bush directed the US DOT in 2004 to develop and implement a viable back-up alternative:

> In coordination with the Secretary of Homeland Security, develop, acquire, operate, and maintain backup position, navigation, and timing capabilities that can support critical transportation, homeland security, and other critical civil and commercial infrastructure applications within the United States, in the event of a disruption of the Global Positioning System or other space-based positioning, navigation, and timing services, consistent with Homeland Security Presidential Directive-7, Critical Infrastructure Identification, Prioritization, and Protection, dated December 17, 2003 (U.S. Space-Based Positioning, Navigation, and Timing Policy 2004).

President Barack Obama restated and reinforced this directive in 2010 by ordering all departments of government to cooperate and attain: "Invest(-ment) in domestic capabilities and support international activities to detect, mitigate, and increase resiliency to harmful interference to GPS, and identify and implement, as necessary and appropriate, redundant, and back-up systems or approaches for critical infrastructure, key resources, and mission-essential functions" (Executive Order 13636). President Obama again directed, by Executive Order 13636 in February 2013, expanded cooperation among agencies and operators where the national cyber policy was again tied to expanding threat and vulnerability and described as: "The cyber threat to critical infrastructure continues to grow and represents one of the most serious national security challenges we must confront. The national and economic security of the United States depends on the reliable functioning of the Nation's critical infrastructure in the face of such threats" (Executive Order 13636).

Although the US Department of Transportation (DOT) in collaboration with the US Department of Defense (DOD) and the US Department of Homeland Security (DHS)—have been ordered to make cyber security "happen," no real progress has been realized to date—more than a decade following the first Presidential Directive. Five prominent members of Congress (as late as August 2015) continue to push for alternative augmentation to GPS via systems such as enhanced LORAN (e-LORAN) that will provide for backup to a potential widespread GPS failure. eLORAN is a very low frequency, land based positioning system that follows onto the LORAN-C system used prior to the global deployment of the satellite based GPS and similar systems now in primary use. Broadcasting at 200 kHz, eLORAN will be harder to jam and therefore more resilient than GPS, while providing a good augmentation redundancy for critical transportation choke points such as port and harbor approaches. While the political will and the support of the commercial sector appear solid and growing, the direction of Congress and the allocation of the capital resources required have yet to materialize (Garamendi et al. 2015).

In typical form, the military is not waiting for Congress to act. Defense Advanced Research Projects Agency (DARPA) has initiated its own GPS hardening/augmentation program that may provide a model for eventual commercial adoption and implementation. Rockwell Collins has been selected by "DARPA to develop technologies that could serve as a backup to GPS. The research, being conducted as part of DARPA's Spatial, Temporal and Orientation Information in Contested Environments (STOIC) program, aims to reduce warfighter dependence on GPS for modern military operations" (Rockwell Collins 2015).

> The goal of the STOIC program is to develop positioning, navigation, and timing (PNT) systems that provide GPS-independent PNT, achieving timing that far surpasses GPS levels of performance. The program is comprised of three primary elements that, when integrated, have the potential to provide global PNT independent of GPS, including long-range robust reference signals, ultra-stable tactical clocks, and multifunctional systems that provide PNT information between cooperative users in contested environments (Rockwell Collins 2015).

"STOIC technology could augment GPS, or it may act as a substitute for GPS in contested environments where GPS is degraded or denied," said John Borghese, vice president of the Rockwell Collins Advanced Technology Center. "The time-transfer and ranging capabilities we are developing seek to enable distributed platforms to cooperatively locate targets, employ jamming in a surgical fashion, and serve as a backup to GPS for relative navigation" (Rockwell Collins 2015).

Lack of Management Understanding is Complicating the Cyber Security Challenge

Overarching all of these challenges is a lack of knowledge and appropriate emphasis on cyber security issues by chief executives and senior managers. Proper operation of communications and computerized control systems are still largely perceived as the responsibility of IT departments by senior managers—a consequence of the typical division of labor within organizations of all types and recognition that high technology is the domain of an extremely smart sub-set of technical wizards who speak a language most of us are still struggling to comprehend. Yet the insidious dependence on ever-expanding integrated systems masks the potential downside risks all the more difficult to quantify as connectivity and information exchange continuous to advance, thus often obscuring the required necessary decision-making and capital allocations.

Managers are often so insulated from the possible consequences of a cyber attack or intrusion as to make them actively resist systems assessments and audits, believing there is nothing to fix. Michael Van Gemert, former manager of systems and controls for Lloyd's Register Drilling Integrity Services—one of few international classification societies that actually require offshore drilling rigs include mobile vessel computer system certification audits as a certification requirement—says "I get people fighting me all the time on it…they tell me it's a closed system (that is not connected to the Internet and therefore not vulnerable)" (Shauk 2013). In one such instance, a system audit for an operator in the Gulf of Mexico revealed such an extensive infestation of malware that it required about 10 days to restore operations. In another instance, a vessel based drilling unit enroute from South Korea to South America was infected with malware that affected all ship systems for a full 19 days before normal operations resumed (Wagstaff 2014). Even more challenging, those "C-Level" managers (CEO, CFO, CIO, etc.) have become subject to personal liability and risk, losing their jobs from the consequences of corporate cyber lapses and insurance coverage for these events is often excluded from standard policies.

CHAPTER 9

The most publicized example of public accountability for a cyber intrusion is the forced resignation of Target CEO Gregg Steinhafel for not taking adequate steps to protect customers' data electronic data. Specialized insurance is available but expensive and often still does not cover liabilities accruing from negligence, malfeasance, or omission (Homeland Security News Wire 2015a).

First Steps to Maritime Cyber Security

To address all of these emergent issues, the United States Coast Guard has recently published (USCG, June 2016; summary highlight on the following page) the national Cyber Strategy to provide a plan of action for comprehensive, collaborative analysis and operations to address cyber vulnerabilities within the service and applicable to the MTS. The first of several linked multi-disciplinary, public/private symposiums and summits convened to engage discussions in this process occurred in March 2015 (hosted by Rutgers University/AMU in New Jersey), in June 2015 (hosted by The California Maritime Academy in the San Francisco Bay Region) and in March 2016 (hosted by the USC CREATE DHS Center of Excellence).

The primary discussions at these conferences have not encompassed whether a reasonable and practical set of objectives and uniform standards are necessary—they are—but rather how these standards are to be universally developed and employed across an international operational framework with a layered regulatory environment.

Key questions to be asked and reliably answered immediately include:

Who should take the lead on developing Maritime Cyber Standards of Practice?

Since all operations share many common operational mechanisms, yet differ significantly in areas of emphasis and scope of operation—What categories and levels of risk mitigation are required?

How should they be applied both within and without international borders?

Should these standards be voluntary or compulsory implemented by way of international and national laws and regulations?

What mechanism for compliance oversight and remedial penalties are necessary or required, if any?

USCG Cyber Strategy
(Highlight Summary) – June 2015

To fully ensure the Coast Guard is able to perform its essential missions in the 21st Century, it must fully embrace cyberspace as an operational domain. To this end, the Coast Guard will focus on three specific strategic priorities in the cyber domain over the next ten years:

- Defending Cyberspace
- Enabling Operations
- Protecting Infrastructure

Defending Cyberspace: Secure and resilient Coast Guard IT systems and networks are essential for overall mission success. To ensure the full scope of Coast Guard capabilities are as effective and efficient as possible, the Coast Guard must serve as a model agency in protecting information infrastructure and building a more resilient Coast Guard network.

Enabling Operations: To operate effectively within the cyber domain, the Coast Guard must develop and leverage a diverse set of cyber capabilities and authorities. Cyberspace operations, inside and outside Coast Guard information and communications networks and systems, can help detect, deter, disable, and defeat adversaries. Robust intelligence, law enforcement, and maritime and military cyber programs are essential to enhancing the effectiveness of Coast Guard operations, and deterring, preventing, and responding to malicious activity targeting critical maritime infrastructure. Coast Guard leaders must recognize that cyber capabilities are a critical enabler of success across all missions, and ensure that these capabilities are leveraged by commanders and decision-makers at all levels.

Protecting Infrastructure: maritime critical infrastructure and the MTS are vital to our economy, national security, and national defense. The MTS includes ocean carriers, coastwise shipping along our shores, the Western rivers and Great Lakes, and the nation's ports and terminals. Cyber systems enable the MTS to operate with unprecedented speed and efficiency. Those same cyber systems also create potential vulnerabilities. as the maritime transportation Sector Specific agency (as defined by the national infrastructure protection plan), the Coast Guard must lead the unity of effort required to protect maritime critical infrastructure from attacks, accidents, and disasters.

CHAPTER 9

> ***Ensuring Long-term Success:*** in support of the three strategic priorities, this Strategy identifies a number of cross-cutting support factors that will ensure the Coast Guard's long-term success in meeting the Service's strategic goals in the cyber domain. These include:
>
> - *recognition of cyberspace as an operational domain,*
> - *developing cyber guidance and defining mission space,*
> - *leveraging partnerships to build knowledge, resource capacity, and an understanding of **MTS** cyber vulnerabilities,*
> - *sharing of real-time information,*
> - *organizing for success,*
> - *building a well-trained cyber workforce, and **making thoughtful future cyber investments***

Debating these issues and resolving a common cyber security outline strategy are the first steps toward building the necessary uniform Cyber Security Standards of Practice. Ultimately to be fully effective these efforts must be coordinated with all of the other ongoing international efforts—commercial, public and public.

Ultimately, the International Maritime Organization (IMO) must draft SOLAS driven, treaty obligated standards—similar to the ISPS Code model—that will require national regulatory implementation by all SOLAS treaty signatories worldwide as the most comprehensive and universally consistent cyber security and cyber defense template.

Solutions and Recommendations:

Successful and effective maritime cyber security requires cooperation and collaboration across the full spectrum of MTS stakeholders: Government, Commercial, Public, and Private. The best solutions will be arrived at via consensus as much as practicable, with the participation, acceptance and endorsement of the commercial and operational elements of the MTS. Agencies can—and perhaps need to—provide baseline regulations to standardize and guide the process, but there is also a requirement for flexibility that allows commercial enterprises to fit best practices and lessons learned to each particular operational environment.

In the interim, public and private entities are continuing to explore pos-

> Like physical security, which continually adapts to changes in buildings and new threat vectors, cybersecurity also requires an ongoing commitment to responding to the rapidly changing cyber threat environment. Just as annual physical security exercises are conducted to ensure good working processes, annual cybersecurity exercises are recommended and should include a port's law enforcement partners to ensure appropriate notifications, forensics preservation, and investigation processes meet the port's needs.
>
> AAPA
> Cybersecurity Government Policy Paper
> March 2014

sible cyber security strategies and protective measures. The US Navy—currently leading active, practical research in cyber protection and resilience in the maritime mode—just announced (September 21, 2015) the development of the Resilient Hull, Mechanical, and Electrical Security (RHIMES) system "designed to prevent an attacker from disabling or taking control of programmable logic controllers" (PLC's)—"the hardware components that interface with physical systems on the ship" (Freeman 2015). US Navy Chief of Naval Research Rear Adm. Mat Winter says "this technology will help the Navy protect its shipboard physical systems, but it may also have important applications to protecting our nation's physical infrastructure" (Freeman 2015).

As "Dr. Ryan Craven, a program officer of the Cyber Security and Complex Software Systems Program in the Mathematics Computer and Information Sciences Division of the Office of Naval Research" explains:

> Traditionally, computer security systems protect against previously identified malicious code. When new threats appear, security firms have to update their databases and issue new signatures. Because security companies react to the appearance of new threats, they are always one step behind. Plus, a hacker can make small changes to their virus to avoid being detected by a signature. RHIMES relies on advanced cyber resiliency techniques to introduce diversity and stop entire classes of attacks at once. Most physical controllers have redundant backups in place that have the same core programming, he explained. These backups allow the system to remain operational in the event of a physical controller failure. But without diversity in their programming, if one gets hacked, they all get hacked. "Functionally, all of the controllers do the same thing, but RHIMES introduces diversity via a slightly different implementation for each controller's program," Dr. Craven explained. "In the event of a cyber-attack, RHIMES makes it so that a different hack is required to

exploit each controller. The same exact exploit can't be used against more than one controller" (Freeman 2015).

MITIGATION STRATEGIES

So where to commence and where to focus? Below are some key elements that must be addressed to move the process forward in the required uniform and standardized way.

Standardized Best Practices, Requirements, and Expectations: The USCG Cyber Strategy is a good first start in providing a criticality outline of goals and objectives for attainment of functional maritime cyber security. Specific directive requirements and expectations for implementation of the strategy are required. To achieve a comprehensive Standard of Practice and an associated set of best practices will require an exhaustive analysis of existing best practices from industry and government agencies, and engagement of academic and research entities who are already focused on the challenges such as NIST, the NAS Transportation Research Board, DoD, DHS Academic Centers of Excellence, IMO, ISO, BIMCO, the USCG, and maritime operators and representative organizations such as AAPA and others.

The National Institute of Standards and Technology (NIST) estimates that worldwide there are over 200 agencies and organizations actively engaged in developing security related IT and ICS standards (Hogan and Newton 2015). Each of these requires evaluation and analysis for commonalities, differences and applicability to operations in the maritime operational environment—and the best of these should form the basis of an international standard of practice. It is reasonable to assume the IMO (internationally) and the USCG (nationally) will coordinate and lead this effort for the MTS.

Systems Redundancy and Resiliency is Key: Redundancy is not the same thing as Resiliency. Resiliency is the ability of a system or system element to recover capability after a failure or deliberate attack. Redundancy is a design factor that allows minimum capabilities to be retained even if a total systems failure occurs. Both factors should be standard requirements for modern ship design and enhanced by training of the human crew to respond and mitigate. Although certainly possible in the coming years due to advances in technology, the maritime community, and international regulations should stop short of eliminating the human factor aboard ship and retain and enhance the ability to exert manual control over critical systems aboard vessels at sea when necessity arises.

Comprehensive Protection of Systems must extend beyond front end security protocols to include the use of internal system programs that will identify, isolate and render safe malicious attacker software. A beginning for this process would be the requirement to perform specific assessment of cyber systems for all maritime transportation system stakeholders. Assessments should map the connectivity and operation of all cyber systems to identify systems vulnerabilities. The assessment will inform determination of relative risk that will define the security planning process required for each specific entity. As all operations vary dependent upon the operational environment, there will be no single solution that will maximize cyber security—just as there is no way to attain 100% security short of ceasing all activities.

Knowing what to do if a cyber security breach occurs or is suspected is a critical preparation planning and mitigation factor.

Whether it stems from a disgruntled employee, a mole planted by an organized crime gang or a sophisticated hacker, when an information security issue is discovered, the proper response depends on first ascertaining:

1) When did the security breach occur? Is it still happening?

2) Where did it originate—internally or externally?

3) How and why did the incident occur? For example, did a malicious intruder exploit network access privileges to steal your data for financial gain, or did an employee accidentally disclose sensitive information via email?

4) What was compromised—intellectual property, personal data, network operations, etc.? (Bancroft 2014).

It is strongly recommended that "to avoid spreading malware throughout [a] network or destroying the trail of evidence, the hacked organization and its IT department" is advised to "not try to "fix" a suspected problem on their own without the assistance of" third-party experts (Bancroft 2014). Whenever possible, affected systems should be designed and programed to achieve isolation so that forensic investigation can occur.

Cyber Risk Insurance is another option that cannot prevent cyber-related attacks but can assist a stricken maritime operator with recovery and restoration activities. The challenge remains that many insurance underwriters and P&L Clubs do not yet well understand the nature of the risk, and therefore availability of adequate coverage remains limited, possibly deficient in scope and improperly priced. "As insurers look to offer their cyber coverage, they are finding it difficult to assess risks. (The industry) is seeking cybersecurity professionals to build a centralized cyber team, but there is a shortage of qualified talent. It is hard for insurers and brokers to find people able to handle the product," insurance broker Munich Re's Andreas Schlayer has said (Homeland Security News Wire 2014b).

The absence of internal cyber experts means many insurers well able to calculate traditional physical risk are unable to appropriately identify clients' vulnerability to cyber security intrusions. In some cases, without really even knowing what questions to ask, some P&I Clubs simply require clients to submit descriptions of "security procedures in place, rather than conducting thorough security audits" to determine whether the assumptions underpinning these procedures are valid and correct (Homeland Security News Wire 2014b). "'There's a real risk that insurance companies are not appropriately pricing the risk,' said Bryan Rose, managing director with Stroz Friedberg, a firm that investigates cyberattacks", and therefore the protective strategy and insurance coverage is inadequate (Homeland Security News Wire 2014b). Appropriate and accurate tools for the assessment of risk and determination of costs that support and enhance application of "Best Practices" are a fundamental aspect of overall cyber security and protection of the MTS.

Mariner Education and Professional Development Training in practical seamanship skills must remain a critical component of maritime education. When systems fail—and they will—the mariners aboard must be able to perform basic functions necessary to safely operate the ship. Concurrently, a new hierarchical curriculum to train mariners and managers on cyber risk and cyber security must be developed and standardized across the nation. Further, a baseline of best practices and guidelines is necessary to develop competent educational pedagogy and to assure adequate procedures are implemented across the full spectrum of the integrated elements of the modern MTS.

MTS Management Training in cyber security vulnerabilities is a critical element in prioritizing business decisions that affect availability of company resources required to harden company IT and control systems and implement proactive solutions. Cyber vulnerability has largely been the province of computer room specialists and not board-room decision-makers. Until the threat and vulnerability is clearly defined, the risk will remain ill-defined and the resources withheld and/or diverted to other needs with a greater perceived return on investment. Training need not be so sophisticated that only IT specialists can comprehend the totality of the threat. On the contrary, continuous reminders to managers and employees of uniform best policies and practices are a simple, but effective first step for enhancement of systems security. For example, the US Navy issued a very simple five-point advisory to all personnel, including contractors, for the use and management of e-mail—still the most common method of cyber intrusion via phishing attacks:

- Never share anything online you would not tell directly to the enemy

- Never post private or personnel information.

- Assume any information you share electronically will be made public.

- Phishing scams tend to have common characteristics that make them easy to identify:
    o Spelling and punctuation errors
    o Scare tactics to entice a target to provide personnel information or follow imbedded links
    o Sensational subject lines to entice targets to click on attached links or provide personal information
    o Include a redirect to malicious URL's which require you to input usernames and passwords to access.
    o Try to appear genuine by using legitimate operational terms, key words and accurate personal information.
    o Fake or unknown sender (U.S. Fleet Cyber Command 2014).

- "When in doubt about a suspicious email from a supposed bank," Other Similar Entity or Any Other E-Mail in Question, *independently call* the purported source to verify the message and notify your IT Department if fraud or cyber intrusion is suspected (U.S. Fleet Cyber Command 2014).

These simple steps, and strictly regulating the use of USB thumb drive data sticks that can easily transfer malware and other viruses—are major initial steps in reducing the window of opportunity for cyber intrusion.

Red Team Testing: Ultimately to clearly define vulnerabilities, there remains a critical need for expanded practical shipboard and land-based terminal research and testing to evaluate the redundancy of control systems on real ships and in real freight yards in an operational environment. "Red Teams" are comprised of dedicated technology savvy subject matter experts whose mission is to attack, subvert, disrupt and hijack computer based control systems to identify weaknesses and vulnerabilities. These findings are shared with company management. The active support and cooperation of global ship and terminal operators is the key to this endeavor and they will need to be willing to expose and share identified vulnerabilities as they are identified so that universal protection strategies may be developed, enhanced and implemented across the MTS.

Build Internal Cyber Defenses with People—and not just Systems—Safeguards: Consider the Willie Sutton or Frank Abagnale solution. Sutton was a renowned bank robber and Abagnale a master check forger and impersonator—both of whom ended up working for society to deter and catch other criminals. The logic here supports the premise of incorporating system hackers into your protective strategy. "To beat hackers, you have to think like them ... True computer security pros are always hacking systems, all the time, at least mentally. They have the mind-set to automatically think of ways to break into almost any system they come across. By looking at systems through the eyes of a hacker, you can better identify weaknesses and create defenses. The best anti-hackers are hackers themselves" (Grimes 2011). Beyond that, consider implementing internal rules that limit—or revoke—access to systems by those employees who have repeatedly shown vulnerability to phishing or other common cyber intrusion techniques. DHS Chief Information Security Officer Paul Beckman has said "that it is astonishing how often even senior managers and other high-ranking officials click on" a phishing link during Red Team systems testing, saying that "... if it was a true attack, rather than a compliance test, such carelessness could result in serious damage (having given the attacker 'ownership' of the system." (Homeland Security News Wire 2015b).

Conclusion

> Most senior policymakers have grown up in a world when cyber security was not a prominent issue ... Most of the technologies are evolving faster than our comprehension of them.
>
> *Hon. Richard Danzig, Former Secretary of the Navy, RAND Fellow*
> *July 2014*

The advance of technology in the maritime domain—both land and sea operations—can be a double edged sword that should be carefully analyzed prior to full implementation as this may have unintended critical consequences. Cyber policy and protection are often described as comprising a 3P Security System (People, Process, and Product). To attain maximum protection requires continuous vigilance and adaption to a rapidly changing technology environment and a reasonable balance between all three of these key elements. To achieve this requires one to dispassionately analyze the prospective way forward to decide upon best practices and influence the approach for decades to come. Therefore, there are several existential "facts" to consider now:

Just because a thing can be done, does not mean it should be done. In the final analysis the wisdom of completely removing the human element from physical command and control over vessels that continue to increase in size and tonnage must be scrutinized. On the brink of attaining Ghost Ship technology where ships of all sizes and types can be operated without human crews and controlled completely from the shore, it remains an open question as to whether or not it should be done.

Quicker and faster is not always cheaper and better overall. For every advance, there are those that will look to exploit progress for criminal, ideological, or overtly political gain. The smartest approach is to consider holistic solutions that balance efficiency goals against safety and security precepts—essentially a broad and comprehensive cost/benefit analysis to best define the calculus of wise decision-making.

Systems Redundancy—the Process of Design for Failure—must be a key component of current vessel upgrades and future vessel design and construction. This is a different but complimentary concept than resiliency—the ability of integrated systems to respond and adapt to outside intrusion, interference, or internal failure. Arguably attainment of both are necessary, yet perhaps it is fair to state that of these redundancy is the more "certain" remedy as a counter-cyber security protective strategy as technology continues to advance at light speed. Possible redundancy augments might include the re-installation of analog control systems (switches, levers, knobs, and buttons) as backups as well as the continued requirements for training and education of mariners in basic pencil and paper navigation techniques and skills. There is an absolute need for these redundant elements when the power goes out and/or the systems fail.

Experience demonstrates that systems will fail whether by accident or intent. To survive the next negative event there is a need to know with certainty what would happen in the aftermath of a major cyber incident and the resultant dynamic world-wide ripple effects. The historical record shows that systems disruptions, equipment failures, and labor disputes have caused significant and costly world-wide impacts to global trade, transportation and international economies. These examples demonstrate how fragile the interconnected global trade system has become—and also illuminate the potential target value these systems and technologies represent to criminals, activists, and terrorists. To plan and prepare protective and mitigation strategies effectively requires a reasonably accurate estimate of vulnerability to best define the risk and develop optimized remedial means and methods. To date, this risk has not yet been well and fully understood and remains the single greatest critical research area where resources and funding is required to support a comprehensive, collaborative national effort.

References

Almeida, Rob. 2014. "New Technology Installed at OOCL's Long Beach Container Terminal." May 28. http://gcaptain.com/new-technology-installed-oocl-long-beach-container-terminal/.

American Association of Port Authorities (AAPA). 2008. "U.S. Public Port Facts." July. http://www.aapa-ports.org/files/PDFs/facts.pdf.

American Association of Port Authorities (AAPA). 2014. "Cybersecurity." March. http://aapa.files.cms-plus.com/CybersecurityFinal_139542061 1523_3.pdf.

American Association of Port Authorities (AAPA). 2015. "Re: Docket Number USCG-2014-1020." Advisory Letter. April 15. http://aapa.files.cms-plus.com/AAPA%20Cybersecurity%20Comments%20-%20 Docket%20Number%20USCG-2014-1020%20April%2015.pdf.

BAE Systems. 2015. "BAE Launches Cloud Based Cyber Security Tools." BAE Corporate Press Release. September 3. Published on ASDNews.com.

Bancroft, Collum. 2014. "Cyber Crime and the Shipping Industry." December 9. http://www.kroll.com/en-us/intelligence-center/blog/cyber-crime-shipping-industry.

Caponi, Steven L., and Kate B. Belmont. 2014. "Maritime Cybersecurity: A Growing Threat Goes Unanswered." Blank Rome LLP. October 27.

https://www.blankrome.com/index.cfm?contentID=37&itemID=3420.

Caponi, Steven L., and Kate B. Belmont. 2014/2015. "Old Dog New Tricks." Blank Rome LLP. December/January. http://www.blankrome.com/index.cfm/publicationid/43ee2fc4-6f38-4f34-9ffd-c6a8a90c25ec/siteFiles/News/pdfPage.cfm?contentID=37&itemID=3469.

Carroll, Michael. 2014. "Are Unmanned Vessels the Future of Global Shipping?" *Newsweek*. June 13. http://europe.newsweek.com/are-unmanned-vessels-future-global-shipping-262376.

Chips Magazine. 2013. "Rear Adm. William E. Leigher Talks About the Importance of CANES." *Chips*. July/September. http://www.doncio.navy.mil/CHIPS/ArticleDetails.aspx?ID=4722.

Clark, Bruce G., Donna Nincic, and Nevin Fidler. 2007. "Protecting America's Ports: Are We There Yet?" Prepared for US MARAD under Grant #DTM A1H050041. October.

Cyberkeel. 2014. "Maritime Cyber-Risks." October 15. Copenhagen, Denmark. http://www.cyberkeel.com/images/pdf-files/Whitepaper.pdf.

Danzig, Richard J. 2014. "Surviving on a Diet of Poisoned Fruit: Reducing the National Security Risks of America's Cyber Dependencies." Center for New American Security. July 21. http://www.cnas.org/surviving-diet-poisoned-fruit#.V-qyWK21U2Y.

Executive Order 13636. 2013. Improving Critical Infrastructure Cybersecurity. 78 C.F.R. 11737 (February 12, 2013).

Fonseca, Joseph. 2015. "IACS Chairman Meets Industry Stakeholders." *Marine-Link*. September 26. http://www.marinelink.com/news/stakeholders-chairman398476.aspx.

Freeman, Bob. 2015. "A New Defense for Navy Ships: Protection from Cyber Attacks." Story Number: NNS150918-05. America's Navy. September 18. http://www.navy.mil/submit/display.asp?story_id=91131.

Garamendi, John, Frank A. LoBiondo, Peter A. DeFazio, Bill Shuster, and Walter B. Jones. 2015. "Congressional Letter Addressed to Robert O. Work and Victor Mendez Requesting Support of Enhanced LORAN and National P.N.T Resiliency." U.S. Congress. August 31.

CHAPTER 9

Grimes, Roger A. 2011. "To Beat Hackers, You Have To Think Like Them." *InfoWorld*. June 7. http://www.infoworld.com/article/2622041/hacking/to-beat-hackers--you-have-to-think-like-them.html.

Haun, Eric. 2015a. "Danelec to Lead Working Group in EU EfficienSea2 Project." *MarineLink*. May 20. http://www.marinelink.com/news/efficiensea-working391618.

Haun, Eric. 2015b. "Rolls-Royce to Lead Autonomous Ship Research." *MarineLink*. July 2. http://www.marinelink.com/news/rollsroyce-autonomous394020.

Hawkes, Kenneth Gale. 1989. *Maritime Security*. Cornell Maritime Press, Centreville MD USA

"Hidden in Full View." 2015. PortStrategy.com. February 10. http://www.portstrategy.com/news101/port-operations/planning-and-design/hidden-in-full-view.

Hogan, Michael, and Elaine Newton, eds. 2015. "Report on Strategic U.S. Government Engagement in International Standardization to Achieve U.S. Objectives for Cybersecurity." NIST Interagency Report (NISTIR) 8074 Volume 1 (Draft). (August). National Institute of Standards and Technology. http://csrc.nist.gov/publications/drafts/nistir-8074/nistir_8074_vol1_draft_report.pdf.

Homeland Security News Wire. 2014a. "Cockpit Automation Causes Pilots to Lose Critical Thinking Skills." December 2. http://www.homelandsecuritynewswire.com/dr20141202-cockpit-automation-causes-pilots-to-lose-critical-thinking-skills.

Homeland Security News Wire. 2014b. "Demand for Cyber Attack Insurance Grows but Challenges Remain." July 16. http://www.homelandsecuritynewswire.com/dr20140716-demand-for-cyberattack-insurance-grows-but-challenges-remain.

Homeland Security News Wire. 2015a. "CEO Responsibilities for Data Breach." February 13. http://www.homelandsecuritynewswire.com/dr20150213-ceo-responsibilities-for-data-breach.

Homeland Security News Wire. 2015b. "Clearance of Employees Who Repeatedly Fall for Phishing Scams Should Be Revoked Says Experts." September 22. http://www.homelandsecuritynewswire.com/dr201509

22-clearance-of-employees-who-repeatedly-fall-for-phishing-scams-should-be-revoked-experts.

Howard, Michelle. 2014. "New Hamburg Süd Ships to Feature NACOS Platinum Systems." *MarineLink*. November 25. http://www.marinelink.com/news/platinum-hamburg-feature381397.aspx.

International Association of Classification Societies LTD. 2015. "New IACS Chairman to Focus on Goal-Based Standards, Cyber System Safety and Quality." September 14. http://www.iacs.org.uk/news/article.aspx?newsid=192.

Laursen, Wendy. 2015a. "Confronting the Safety Challenge." *The Maritime Executive*. September 10. http://www.maritime-executive.com/magazine/confronting-the-safety-challenge.

Laursen, Wendy. 2015b. "Taking Autonomy One Step at a Time." *The Maritime Executive*. September 13. http://www.maritime-executive.com/article/taking-autonomy-one-step-at-a-time.

Macaulay, James, Lauren Buckalew, and Gina Chung. 2015. "Internet of Things in Logistics: A Collaborative Report by DHL and Cisco on Implication and Use Cases for the Logistics Industry." DHL Trend Research and Cisco Consulting Services. http://www.dhl.com/content/dam/Local_Images/g0/New_aboutus/innovation/DHLTrendReport_Internet_of_things.pdf.

National Space Policy of the United States of America. 2010. Whitehouse.com June 28. https://www.whitehouse.gov/sites/default/files/national_space_policy_6-28-10.pdf.

Rockwell Collins. 2015. "Rockwell Collins Wins DARPA Award to Develop GPS Backup Technologies for Contested Environments." September 17. https://www.rockwellcollins.com/Data/News/2015_Cal_Yr/GS/FY15GSNR54-STOIC.aspx.

Shauk, Zain. 2013. "Malware Offshore: Danger Lurks Where the Chips Fail." Fuel Fix. April 29. http://fuelfix.com/blog/2013/04/29/malware-offshore-danger-lurks-where-the-chips-fail/.

Sheridan, John. 2013. "South Korea to Install eLORAN to Counter North Korean GPS Jamming." *Aviation International News*. August 2. http://www.ainonline.com/aviation-news/aviation-international-news/2013-08-02/south-korea-install-eloran-counter-north-korean-gps-jamming.

ShipTechnology. 2016. "Wärtsilä to Deliver NACOS Platinum System to Maersk's New-Builds." ShipTechnology.com. March 23. http://www.ship-technology.com/news/newswrtsil-to-deliver-nacos-platinum-system-to-maersks-new-builds-4846183.

Slabodkin, Gregory. 1998. "Software Glitches Leave Navy Smart Ship Dead in the Water." *GCN Magazine*. July 13. https://gcn.com/Articles/1998/07/13/Software-glitches-leave-Navy-Smart-Ship-dead-in-the-water.aspx.

Tung, Liam. 2014. "Siemens SCADA Flaw Likely Hit in Recent Attacks." *ZDNet*. November 28. http://www.zdnet.com/article/siemens-scada-flaw-likely-hit-in-recent-attacks/.

U.S. Coast Guard. 2015. "United States Coast Guard Cyber Strategy." June. Washington, D.C. https://www.uscg.mil/seniorleadership/DOCS/cyber.pdf.

U.S. Fleet Cyber Command. 2014. "5 Things Sailors Need to Know About Social Media, Phishing, Security." *Navy Dispatch Newspaper*. October 8. http://navynews.com/News2014/news1014/100914-Sailors-Should-Know.html.

U.S. Government Accountability Office. 2014. "Maritime Critical Infrastructure Protection: DHS Needs to Better Address Port Cybersecurity." June 5. GAO-14-459.

U.S. Navy League. *SEAPOWER Almanac for 2006*. Arlington, VA.

"U.S. Space-Based Positioning, Navigation, and Timing Policy: Fact Sheet." 2004. Whitehouse. December 15. https://www.whitehouse.gov/files/documents/ostp/Issues/FactSheetSPACE-BASEDPOSITIONINGNAVIGATIONTIMING.pdf.

Wagstaff, Jeremy. 2014. "All at Sea: Global Shipping Fleet Exposed to Hacking Threat." *Reuters*. April 23. http://www.reuters.com/article/us-cybersecurity-shipping-idUSBREA3M20820140424.

Wärtsilä. 2016. "Wärtsilä NACOS Platinum Navigation Automation Control System." June. http://www.wartsila.com/docs/default-source/product-files/ivcs/ivc/brochure-o-ea-nacos-platinum-toplevel.pdf.

Weathington, John. 2015. "6 Types of Cybervillains That Are No Match for Your Data Scientists." *TechRepublic*. July 18. http://www.techrepublic.com/article/6-types-of-cybervillains-that-are-no-match-for-your-data-scientists/.

Welsh, Larry D., and John D. Harvey Jr. 2014. "Independent Review of the Department of Defense Nuclear Enterprise." U.S. Department of Defense. June 2. https://www.hsdl.org/?view&did=759459.

World Maritime News. 2015a. "China Presses On with Unmanned Ship Development Project." August 14. http://worldmaritimenews.com/archives/168917/china-presses-on-with-unmanned-ship-development-project/.

World Maritime News. 2015b. "New IACS Chairman Pinpoints Top Priorities." September 16. http://worldmaritimenews.com/archives/171626/new-iacs-chairman-pinpoints-top-priorities/.

World Maritime News. 2015c. "Plymouth University to Build First Ever Unmanned Ship." August 7. http://worldmaritimenews.com/archives/168517/plymouth-university-to-build-first-ever-unmanned-ship/.

World Maritime News. 2015d. "Reliance on Technology a Double Edged Sword." September 11. http://worldmaritimenews.com/archives/171462/reliance-on-technology-a-double-edged-sword/.

World Maritime News. 2015e. "Rolls Royce: First Unmanned Ferry Could Be Just Five Years Away." September 14. http://worldmaritimenews.com/archives/171582/rolls-royce-first-unmanned-ferry-could-be-just-five-years-away/.

CHAPTER 10:
MARITIME CYBER SECURITY:
THE UNAVOIDABLE WAVE OF CHANGE

Kate B. Belmont, Esq.
Blank Rome LLP

INTRODUCTION

The maritime industry has been one of the greatest agents of change throughout history, but is also one that holds fast to tradition and is necessarily rooted in regularity and dependability. It is this paradox that has allowed the maritime industry to be exposed to one of the most critical threats that faces the modern world: cyber attacks. Although cyber security has been an issue that competing industries worldwide have been facing directly for several years, the maritime industry has only just begun to analyze the threats of cyber attacks and breaches, and the resulting devastating impacts on the industry. The maritime industry has become extremely reliant on information and communication technology (ICT), yet it has failed to make cyber security a priority. As a result, the maritime industry has suffered many losses at the hands of cyber criminals and through unintentional cyber breaches, and continues to remain vulnerable to cyber attacks, manipulation of data, and threats to e-navigation.

Cyber security, or information security, has become a critical concern for every industry that relies on ICT, and the concern is the theft of digital information, the manipulation of information, and denial of service to authorized users. The prevalence of cyber attacks in recent years has escalated tremendously, affecting a wide range of industries. For example, throughout 2014 and 2015, cyber criminals attacked SONY Pictures Entertainment, Home Depot Inc., JPMorgan Chase & Co., Target Corp., Anthem, Inc., the Houston Astros, and also the White House and the Office of Personnel Management (OPM). While this is a small sampling of widely reported cyber attacks, it is clear that every industry that is reliant on ICT is susceptible to cyber attacks: entertainment, healthcare, finance, and the government, and the maritime industry is no exception. Although the maritime industry has largely remained out of the spotlight compared with competing industries, and has thus far failed to address this grave new threat, it is only a matter of time before a cyber attack in the maritime domain occurs with catastrophic effects.

Cyber Security Awareness in the Maritime Industry

In addressing cyber security and the risks associated with reliance on ICT, the maritime industry is decades behind the curve compared to office-based computer systems, and competing industries worldwide (Jones 2014). In 2011, the European Network and Information Security Agency (ENISA), issued the first ever European Union (EU) report on cyber security challenges in the maritime sector titled "Analysis of Cyber Security Aspects in the Maritime Sector." This report outlined the risks facing the maritime industry and analyzed how the industry should respond to such a threat (ENISA 2011). In 2014, the U.S. Government Accountability Office (GAO), issued a similar report, "Maritime Critical Infrastructure Protection: DHS Needs to Better Address Port Cybersecurity," which confirmed the threats of cyber attacks and cyber breaches facing industry, and ultimately found that the maritime industry had failed to make cyber security a priority (U.S. GAO 2014). Both the ENISA report and the GAO report analyzed and outlined the various systems that are at risk to cyber attacks and cyber breaches within the maritime sector. These include mission-critical systems onboard vessels, such as those used in communication, navigation, loading, and data storage. Systems at major ports and mainland computer systems at maritime companies were no less vulnerable. Any aspect of the industry that is reliant on ICT is vulnerable to a cyber attack. For example, propulsion, freight management, traffic control communications, terminal operating systems, and industrial control systems are all targets of cyber attacks. Additionally, the use of laptops, smart phones and USB keys are also vulnerable to cyber attacks and hacking. For example, when operating a vessel at sea, a USB key used to update onboard software can contain a virus that can subsequently jeopardize any and all interconnected systems on the vessel.

There are many perpetrators who use cyber attacks, to specifically target the maritime industry. For example, nation states regularly engage in cyber warfare, and the maritime industry is a prime target. Rival companies within the maritime domain also frequently engage in cyber attacks in an attempt to gain a competitive advantage. Confidential charter parties, rate information, and ship designs are just a few examples of valuable information that can be hacked and stolen by cyber criminals. Criminal organizations, pirates, terrorists, and independent and freelance hackers are also of great concern. Not all attacks are from outsiders however; corrupt industry insiders and sloppy employees are also sources of cyber breaches. Employees who fail to practice good cyber security hygiene leave their companies vulnerable and susceptible to cyber breaches.

Threats to E-Navigation

Many industries worldwide are concerned with cyber security and protecting customer information, privacy and data. However, the maritime industry has even greater responsibility and vulnerability due to its over-reliance on e-navigation. For example, due to the weakness of GPS signals, such signals can easily be manipulated, spoofed or jammed. In a spoofing attack, a GPS receiver is deceived by receiving counterfeit GPS signals, which then causes the receiver to estimate its position to be somewhere other than where it actually is. In relying on a fake or faulty GPS signal, mariners will respond by altering the course of the vessel, which will also affect vessels in the surrounding areas.

In addition to a spoofing attack, jamming attacks are also of concern for the maritime sector. A jamming attack is the intentional interference with GPS signals such that the signals are stopped, blocked or "jammed." Instead of providing false data as in a spoofing attack, in a jamming attack the GPS signals are blocked. In a jamming attack, multiple systems onboard the vessel can be affected. For example, the automatic identification system (AIS), electronic chart display & information system (ECDIS), voyage data recorder (VDR) and the vessel traffic service (VTS), are all affected when GPS is lost or blocked. Without GPS, manual navigation, visual data, and cryptic methods must be used to gauge location, communicate with surrounding vessels, and maneuver accurately. The electronic chart display and information system (ECDIS), a computer-based navigation system, is also at risk and vulnerable to cyber attacks, due to its interconnectivity to the Internet and standard communication platforms, which can allow cyber criminals unauthorized access. For example, a point of vulnerability is the introduction of a virus through a portable USB key, which can result in malicious software updates. The manipulation of ECDIS is of great concern as cyber criminals can cause widespread damage by altering charts and downloading, deleting or replacing files. To combat such attacks and manipulation of data, USB keys used in chart updating must be scanned for malware prior to use, disposed after each usage, and access to ECDIS entry points must be restricted.

Threats to e-navigation have become an operational problem for some maritime industry sectors and solutions are being discussed and considered. For example, it is imperative that our mariners understand alternate position sources, and rely on paper charts when failures occur. Ship owners and operators must also consider operational responses to the possibility of spoofing and jamming. There must also be a renewed initiative focusing on improved maritime training and education, which is of particular concern to the U.S. Coast Guard. As our mariners are over-reliant on e-navigation, there must be additional training and cyber security education to address the manipulation of GPS signals, anticipating such attacks and responding effectively and appropriately. Advanced technology and improved equipment can also help to detect and prevent cyber attacks. Nulling antennas, those resistant to jamming signals, must be utilized, as well as updated GPS receivers. Updated GPS receivers can identify non-GPS signals from their location and strength, thus alerting mariners to the possibility of spoofed information. While cyber threats to e-navigation are real and the consequences are disastrous, as discussed above there are steps that can be taken to help counter such threats.

Cyber Attacks in the Maritime Industry

While cyber attacks in the maritime industry have remained far from the front pages of the *New York Times*, such breaches and attacks have been happening for several years. For example, from 2011 through 2013, the Port of Antwerp was hacked by organized criminals who breached the port IT system to facilitate the smuggling of heroin and cocaine. Drug traffickers recruited hackers to breach the port IT systems that controlled the movement and location of containers (Bateman 2013). Also in 2011, pirates hijacked the Enrico Ievoli, in a premeditated attack, sponsored by the Italian mafia. Using information obtained online, the pirates knew the itinerary, the cargo, the crew, the location and that there were no armed guards aboard the vessel. Pirates have also been able to target specific crew for the purposes of kidnapping based on information they obtain online and through cyber attacks. Throughout 2015 and 2016, pirates also hacked into a global shipping conglomerate to facilitate their attacks on ships. It was determined that a malicious web shell had been uploaded onto the server which allowed the hackers access to valuable data and information. Pirates were able to board a vessel, locate by bar code specific sought-after crates containing valuables, steal the contents of that crate, and depart the vessel without incident (MarEx 2016).

The bunkering sector is also highly susceptible to cyber-attacks, and is often targeted by industry insiders. As the bunkering community relies on email communications for its transactions, cyber criminals are able to easily manipulate these communications to defraud shipping companies across the globe. For example, in 2014, World Fuel Services (WFS) was the subject of a cyber crime that resulted in a loss of $18 million (Stamford 2014). A cyber criminal impersonated the US Defense Logistics Agency (DLA) and engaged WFS to participate in a tender for a large amount of fuel. WFS supplied the cargo and submitted an invoice but the DLA had no record of the fuel tender. WFS had already purchased 17,000 metric tons and supplied the fuel but received no payment. Cyber crimes like this happen frequently in the bunkering community as the bunkering community has failed to make cyber security a priority.

In 2014, the U.S. Senate's Armed Services Committee issued a threat assessment report, focusing on cyber-attacks and the need for cyber security. The Committee reported that over a 12-month period there were 50 successful cyber intrusions on US Transportation Command Contractors (Transcom). Of those 50 cyber-intrusions, 20 qualified as advanced persistent threats (APT), and of those, Transcom was only aware of two such attacks. All of which were attributed to China and targeted at airlines or shipping companies. Foreign governments often seek to disrupt military resupply networks, making shipping companies, and operators prime targets for cyber attacks (Stamford 2014).

China's People's Liberation Army also specifically targets marine shipping providers through spear-phishing campaigns (Stamford 2014). Spear-phishing campaigns are automated programs that send spoof emails to companies and employees to secure access to confidential data. Most spear-phishing emails have the appearance of being sent from a bank or well-known organization, and include a corrupted link requesting the recipient click on the link to provide private or confidential information. If the recipient clicks on the link, the cyber criminal now has access to the recipient's system and data. There are over 80,000 victims of spear-phishing emails each day. Accordingly, employee training and cyber security awareness is of paramount importance.

In addition to protecting customer information, data and cargo, the maritime industry is also responsible for the environment and human life. Computer systems on oil rigs, for example, are hackable and vulnerable to cyber attacks and breaches. In 2013, an off-shore oil rig in Houston was shut down due to malicious software unintentionally downloaded by offshore oil workers. An example of a sloppy employee and the failure to maintain good cyber security hygiene, malware was brought aboard the rig by laptops and USB drives which incapacitated the computer networks on the rig. Additionally, infected files are often downloaded from online sources through satellite, including pornography and music piracy, which is of particular concern in the maritime industry. While this particular incident did not result in a well blowout, explosion, or oil spill, the potential for catastrophe is great if the maritime industry fails to make cyber security a priority.

Maritime Cyber Security in the Future: Regulations, Reviews and Audits?

As cyberattacks continue to plague the maritime industry, and the necessity of cyber security has become indisputable, the industry continues to navigate this challenge with little guidance and is unregulated in this regard. While it is undeniable that the maritime industry has been and will continue to be targeted by cyber criminals at every level, the maritime industry has just begun to address this threat and the disastrous consequences that will inevitably follow, as they can no longer be ignored.

The U.S. Coast Guard has taken the laboring oar in tackling this new challenge. In 2015, the U.S. Coast Guard launched a cyber security initiative which was a yearlong process to develop cyber security guidance for the maritime world. This initiative began on January 15, 2015, with a U.S. Coast Guard Public Meeting, "Guidance on Maritime Cybersecurity Standards." This meeting was held to discuss cyber security issues in the maritime domain, and was open to representatives throughout the maritime industry. As information sharing concerning cyber security issues in the maritime sector has been limited, the U.S. Coast Guard has relied on industry to weigh in on what cyber security challenges are being faced in the maritime industry, how these issues are being handled, and how deep Coast Guard oversight should go concerning potential cyber security regulations. After a period of analysis and information exchange, in June of 2015, the U.S. Coast Guard issued its "Cyber Strategy", which outlines its approach to defending cyber space, including a continued focus on risk assessment and risk management, and the strategic priority of protecting Maritime Critical Infrastructure, including ports, facilities, vessels, and related systems.

Shortly after the U.S. Coast Guard issued its "Cyber Strategy," the Maritime Safety Committee of the International Maritime Organization (IMO) held its annual meeting and cyber security regulations were discussed. The U.S. Coast Guard suggested the IMO develop voluntary guidelines for cyber security and proposed amendments to the International Ship and Port Facility Security (ISPS) Code were debated. The IMO stated that more time was needed to develop appropriate guidelines, and also noted that industry would need to take the lead in addressing cyber security challenges and guide the international maritime community. Ultimately, a correspondence committee was established to create a draft set of cyber security guidelines for the IMO to consider in 2016.

On October 8, 2015, the House Homeland Security Committee (2015), Border and Maritime Security Subcommittee held the first ever Congressional hearing to examine cyber security at our nation's ports. The committee's con-

CHAPTER 10

cern was that the U.S. government has fallen behind when it comes to cyber security at its ports. The witnesses who testified before the committee included Rear Admiral Paul Thomas, Assistant Commandant for Prevention Policy USCG, Gregory Wilshusen, Director, Information Security Issues, GAO, Randy Parsons, Director of Security Services, Port of Long Beach and Jonathan Sawicki, Security Improvement Program Manager, Ports of Harlingen and Brownsville, Texas. The testimony focused on the challenges of cyber security in the maritime industry, and the committee was concerned that U.S. ports have not addressed such issues and must protect against cyber breaches. Information sharing was of particular concern.

Shortly after the hearing, on November 2, 2015, the Strengthening Cybersecurity Information Sharing and Coordination in Our Ports Act of 2015 was introduced (H.R. Rep. 3878). The goal of this bill is to improve cyber security information sharing at U.S. ports through enhanced participation and reporting. The bill seeks enhanced participation by the Maritime Information Sharing and Analysis Center, including the Department of Homeland Security and the Coast Guard, and enhanced reporting by the National Maritime Security Advisory Committee to address cyber security situational awareness and information sharing. The bill also directs each Captain of the Port to establish a working group of members of Area Maritime Security Advisory Committees to facilitate the sharing of information about and development of plans to address port-specific cyber security vulnerabilities.

The hesitation to share information on cyber attacks and breaches has been extremely detrimental to the maritime industry. While it is indisputable that cyber attacks have been occurring throughout the maritime domain, the lack of awareness of these attacks and the failure to share information has hindered the ability of the industry to confront these challenges directly. Information sharing is necessary to develop the appropriate regulations, procedures and tools to combat these new threats. It is difficult to address cyber security issues throughout the maritime industry without knowing who is targeted, how they are targeted and the true extent of the damages caused. Industry working groups must be created to establish anonymous information sharing forums to facilitate and encourage the free exchange of information.

Conclusion

In addressing cyber security, the maritime industry is facing a new challenge. While cyber security affects every industry that relies on information and communication technology (ICT), the demands on the maritime industry are great, and the consequences of a cyber attack in the maritime domain are profound. In addition to protecting customer information, data and privacy, the maritime industry is responsible for protecting cargo, the environment and human life. Accordingly, although the maritime industry has been slow to respond to the growing demands of cyber security, it is responding and working toward greater awareness and protections. As it did with the first circumnavigation of the globe, to the development of LNG-powered containerships, virtual buoys and remotely operated vessels, the maritime industry will continue to evolve and meet the demands of the future.

REFERENCES

Bateman, Tom. 2013. "Police Warning After Drug Traffickers' Cyberattack." *BBC News*, October 16. http://www.bbc.com/news/world-europe-24539417.

European Network and Information Security Agency (ENISA). 2011. "Analysis of Cyber Security Aspects in the Maritime Sector." https://www.enisa.europa.eu/publications/cyber-security-aspects-in-the-maritime-sector-1/at_download/fullReport.

House Homeland Security Committee. 2015. "Protecting Maritime Facilities in the 21st Century: Are Our Nation's Ports at Risk for a Cyber-attack?" Border and Maritime Security Subcommittee. October 8. https://homeland.house.gov/hearing/protecting-maritime-facilities-in-the-21st-century-are-our-nations-ports-at-risk-for-a-cyber-attack/.

House of Representatives. Strengthening Cybersecurity Information Sharing and Coordination in Our Ports Act of 2015. H.R. Rep. No. 3878. (2015).

Jones, Stevens. 2014. "SAMI On Top Line Security Concerns." MarinkeLink.com, September 24. http://www.marinelink.com/news/security-concerns-line377865.aspx.

MarEx. 2016. "Case Study: Pirates Hack Cargo Management System." *The Maritime Executive*, March 1. http://maritime-executive.com/article/case-study-pirates-hack-cargo-management-system.

Stamford, Eric Martin. 2014. "Sophisticated Scams Highlight Growing Cyber Risk to Shipping." *TradeWinds*, October 10. http://www.tradewindsnews.com/weekly/346334/Sophisticated-scams-highlight-growing-cyber-risk-to-shipping.

U.S. Government Accountability Office (GAO). 2014. "Maritime Critical Infrastructure Protection: DHS Needs to Better Address Port Cybersecurity." GAO-14-459. http://www.gao.gov/products/GAO-14-459.

CHAPTER 11:
DEVELOPING RISK-BASED PERFORMANCE STANDARDS FOR CYBER SECURITY IN THE MARITIME TRANSPORTATION SYSTEM

Kimberly Young-McLear
U.S. Coast Guard Academy

Abstract

This chapter proposes strategies the U.S. Coast Guard can champion to develop risk-based performance standards for the Maritime Transportation System (MTS). The U.S. Coast Guard already has authority under the Maritime Transportation Security Act (MTSA) to regulate the MTS. Because MTSA primarily focuses on physical security and risks from cyber security can also disrupt the MTS, it is imperative new standards are developed to address these vulnerabilities. Private sector organizations including BIMCO, ABS, and Lloyd's Register in the maritime industry released guidelines for cyber security risk management. Lessons learned from the Department of Homeland Security and the Chemical Sector System can also provide insights on cyber security standards and strategies for collaborating with industry more effectively. Leveraging disciplines including systems engineering and risk management are essential to understanding risk management and developing holistic performance standards.

Keywords: Risk Management, MTS, Anti-Terrorism, Business Continuity, Systems Engineering

ISSUES IN MARITIME CYBER SECURITY

Introduction

In a global economic system, the world is more interconnected than ever with the advent of technology. A system is a "combination of interacting elements organized to achieve one or more stated purposes" (ISO/IEC 15288:2008). The Maritime Transportation System facilitates the safe transit of goods and has direct implications for the economy. In addition to the transport of goods, individual cruise lines move thousands of passengers at a time for recreational or vacation travel. Given the increase of vessel traffic and nature of weather, the MTS is vulnerable to maritime disruptions which may cause loss of life, loss of property, or an environmental catastrophe. Furthermore, these maritime disruptions may also be the result of a cyber-related attack or incident. The MTS currently does not have any regulatory standards for cyber security risks. As discussed in a later section, a few companies have developed guidelines, which are meant to assist owners and operators within the International Ship and Port Facilities Security Code (ISPS Code) and the International Safety Management Code (ISM Code). Both of these Codes have been promulgated by the International Maritime Organization (IMO). Its regulations are adopted and enforced worldwide. The U.S. Coast Guard is the lead agency responsible for regulating the MTS in the United States. Following 2015 GAO-16-116T study on Maritime Critical Infrastructure Protection, the U.S. Coast Guard has been working toward developing guidance for incorporating cyber-security into its maritime safety and security risk-based assessments. The Coast Guard has consulted with the general public, academia, industry, and other state/federal partners to determine potential cyber-security assessment guidelines to assist vessel and facility operators and owners manage risk. Systems engineering and risk management disciplines offer insights to how risk-based performance standards could be adopted for the MTS. Lessons can also be applied from the development of cyber-security standards in other critical infrastructure systems, such as in the chemical sector.

Systems Engineering Perspective of the Maritime Transportation System

Systems engineering is multidisciplinary (What is Systems Engineering 2016) which considers the business and technical requirements in developing a viable and resilient system. This discipline is useful for examining the Maritime Transportation System. The Transportation Systems Sector is one of 16 critical infrastructure sectors in the United States (Presidential Policy Directive-21). In particular, the Maritime Transportation System (MTS) is the marine portion of the transportation systems sector, which includes land based aspects of facility, rail, and highway based activities designed to support each other. The MTS is a complex "network of maritime operations that interface with shore-side operations as part of the overall global supply chains" (DHS 2004). Vulnerabilities can exist within the MTS through the network of interconnected maritime operations and must be secure to prevent malicious or human error cyber disruptions. These disruptions degrade the resilience of maritime operations which may have devastating economic impacts. Systems engineers have also contributed to research on the management of supply chains, disaster response management, information systems, and cyber security.

Role of Private Sector in MTS Business Continuity Planning

In the maritime industry, private businesses are at risk for cyber-attacks, which if occurs adversely affects the MTS. According to the Government Accountability Office (GAO), "cyber-attacks can potentially disrupt the flow of commerce, endanger public safety, and facilitate the theft of valuable cargo" (GAO 2014). Given the increase of computer-based technologies on vessels and in ports to facilitate the safe transit of more than $1.3 trillion in cargo annually (Maritime Cybersecurity 2004) and private sector owning the vast majority of critical infrastructure (Eckert 2005), the maritime industry must practice a cybersafe culture, which the U.S. Navy defines as a "change in organizational culture and crew proficiency required to [prevent] failure caused by a cyber attack (AFCEA 2015)." A cybersafe culture is central to effective business continuity planning (BCP), or increasing the likelihood of a business recovering from an attack or incident.

From a regulatory perspective of MTS continuity, the Coast Guard is the lead agency for enforcing maritime safety and security in domestic ports and on U.S. flag vessels. The Coast Guard is responsible for enforcing all other Maritime Transportation Security Act (MTSA) regulated elements. The Coast Guard, however, had not addressed cyber security regulations in its port and area facility security plans. Additionally, a GAO study concluded that the Coast Guard played a key role in informing maritime stakeholders of cyber-based threats (GAO 2014). Rear Admiral Paul Thomas, the Coast Guard's Assistant Commandant for Prevention Policy stated in a 2015 testimony that the Coast Guard must continually adapt its "regulatory regime" as computer-based technology is increasing in port and vessel operations (Thomas 2015).

Private Sector Maritime Cyber Security Guidelines

In 2016, three major maritime private sector organizations released guidance for cyber security. The American Bureau of Shipping (ABS), Lloyd's Register, and Baltic and International Maritime Council (BIMCO) captured best practices and recommendations for managing cyber security risks. There are commonalities among the guidelines. Given that all three guidelines were released within 2 months of each other is a strong implication that industry is being proactive and understands the benefits of adopting effective cyber security practices. Of likely concern of directive and restrictive regulation, these industry guidelines are an attempt to influence that regulatory process. Each of the three sets of industry guidelines provides a holistic, systematic perspective that incorporates risk management and capability maturity concepts.

"The Application of Cybersecurity Principles to Marine and Offshore Operations," developed by The American Bureau of Shipping (ABS), provides guidance for three levels of organizational maturity to be "implemented by ship and platform owners, operators and crews." ABS notes that it will continue to develop "to support self-assessment, self-test, and self-audit" for basic, developed, and mature organizational maturity levels. This allows owners and operators to allocate appropriate resources to cyber security practices to achieve a desired level of mitigation. Although this flexibility is advantageous, regulations will need to set a baseline level of mitigation so that compliance can be measured and evaluated. Resources should be first prioritized to address basic levels before developed or mature levels.

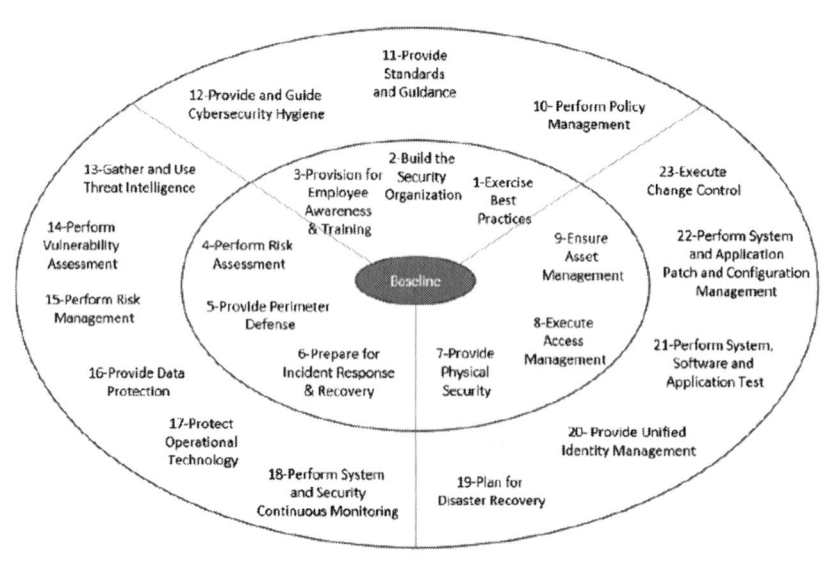

Figure 1: Basic and developed maturity levels (American Bureau of Shipping 2016)

Figure 1 above outlines six risk areas that owners and operators should consider when developing a risk management plan. Human-system, software, network and communications, data assurance, and cyber security should be holistically addressed through an appraisal process. ABS uses a systems engineering approach to define the concept of operations to achieve an integrated, risk-based approach to appraising or assessing cyber vulnerabilities, which are linked to international ISO and IEEE standards.

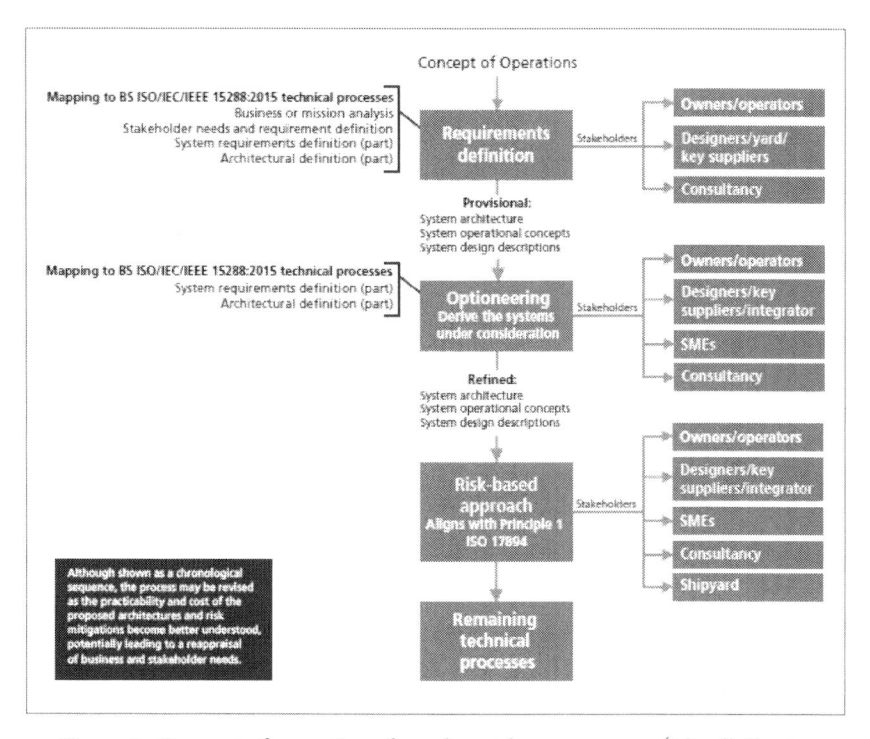

Figure 2: Concept of operations for cyber risk management (Lloyd's Register 2016)

From "Cyber-enabled Ships: Deploying Information and Communications Technology in Shipping—Lloyd's Register's Approach to Assurance," mapping to technical processes is depicted in Figure 2. This concept of operations ensures a risk-based approach to managing cybersecurity vulnerabilities by mapping standards from technical bodies, such as, IEEE, ISO, and IEC, through various programmatic and 3rd stakeholders.

BIMCO's guidelines as shown in Figure 3 was developed in partnership with more than 15 other organizations in the maritime and cyber security industries. The guidelines capture how to assess risk, reduce risk, and develop contingency plans from a holistic cyber security awareness and lifecycle model. Similar to ABS and Lloyd's Register, the BIMCO guidelines outline procedural controls, such as, training, software maintenance, anti-virus updates, administrator privileges, and removable media controls. The guidelines also call out specific cyber security awareness areas including cargo management systems, bridge systems, propulsion and machinery systems, power control systems, access control systems, passenger management systems, and guest entertainment networks.

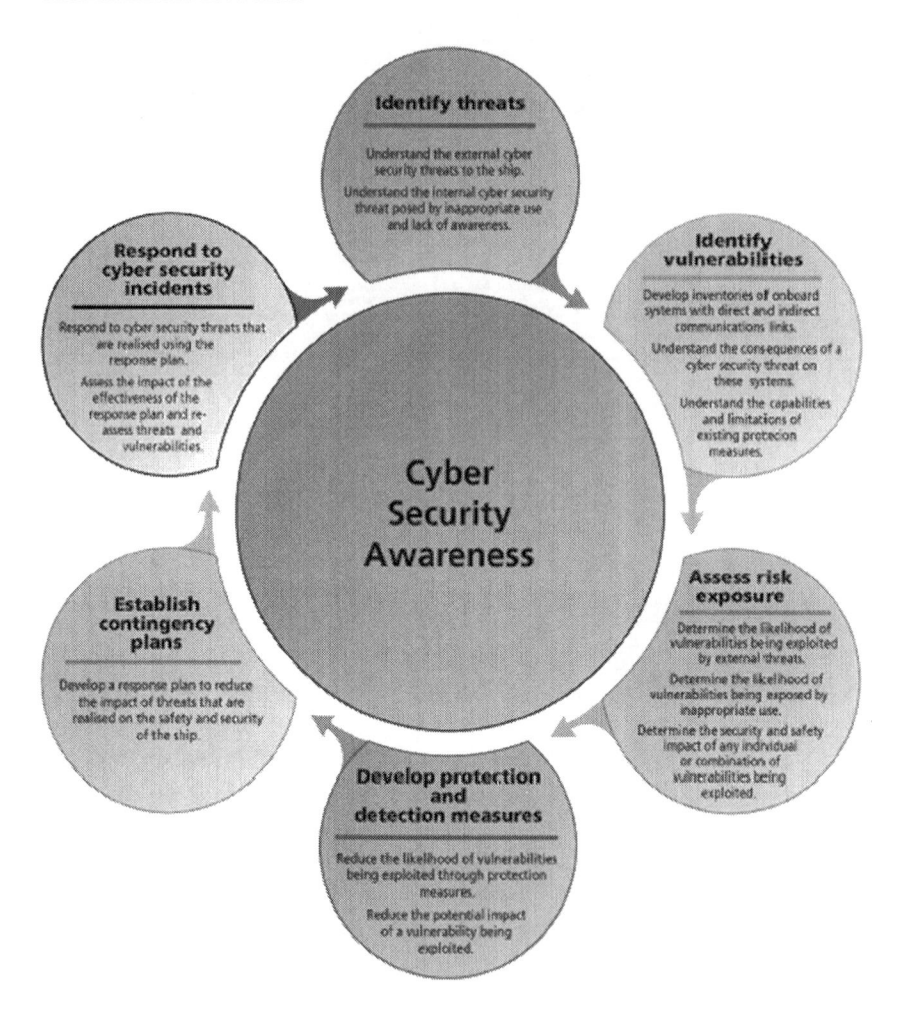

Figure 3: Cyber security awareness model (BIMCO 2016)

Moving Toward Risk-Based Standards

From a systems engineering perspective, it is imperative to include the users in the development of the requirements of the system. The ABS, Lloyd's Register, and BIMCO guidelines are a prudent starting point for developing risk-based performance standards because it increases the likelihood that it would be adopted across industry if a purely voluntary standard of practice is adopted. However, uniform regulation will mandate adoption and compliance to attain a minimum baseline.

The MTS and the chemical sector have commonalities with threats and vulnerabilities as well as facility ownership by private companies. The chemical sector, also owned primarily by the private sector, challenged the guidelines' applicability, maturity, and cost associated with implementing recommendations and conducting the audits. Given the pervasiveness of cyber technologies throughout these critical infrastructures, there are key lessons learned from how risk-based performance standards have been implemented. For example, in May of 2009, the Department of Homeland Security released risk-based performance standards guidance for the chemical critical infrastructure. The 18 standards included a focus on anti-terrorism and cyber security measures. The Department of Homeland Security Chemical Facility Anti-Terrorism Standards (CFATS) program required the government to review private facility security plans. According to a 2013 GAO study, there was a 7–9 year backlog to review facility security plan reviews. This presented challenges between the government and the private facilities. Without timely inspections and audits, the risk assessment of the facilities was inevitably delayed. This delay challenged the validity of the inspection program, reduced private sector confidence in government, and undermined the risk prioritization process of chemical facilities. In developing risk-based performance standards, it is imperative to identify the scope of inspection resources needed to review facility security plans as well as conduct site visits to avoid backlogs and delays. Working with private sector as the regulations are being created is also imperative. This ensures cooperation and may increase the likelihood of facilities self-reporting incidents, such as in Coast Guard marine safety port and vessel inspections. These key lessons learned from the chemical sector may assist the Coast Guard in developing guidance for maritime partners.

Ultimately, the government (U.S. Coast Guard) would be regulating cyber security standards in the MTS. Given the parallels between the cyber security risks and economic impact in the chemical sector and in the MTS, further research is needed to investigate how to best leverage both the cyber security guidelines released by the maritime industry and anti-terrorism standards for the chemical sector released by the Department of Homeland Security. A specific methodology the U.S. Coast Guard with industry, academia, and government stakeholders can take to develop risk-based performance standards include:

Phase 1 • Determine commonalities amongst maritime industry guidelines (ABS, Lloyd's, BIMCO)

Phase 2 • Determine which maritime industry guidelines can be enforceable standards

Phase 3 • Conduct a usability assessment to determine which chemical facility anti-terrorism standards would be favorable for the MTS private sector industry to adopt

Phase 4 • Conduct a usability assessment to determine which chemical facility anti-terrorism standards would be favorable for the MTS private sector industry to adopt

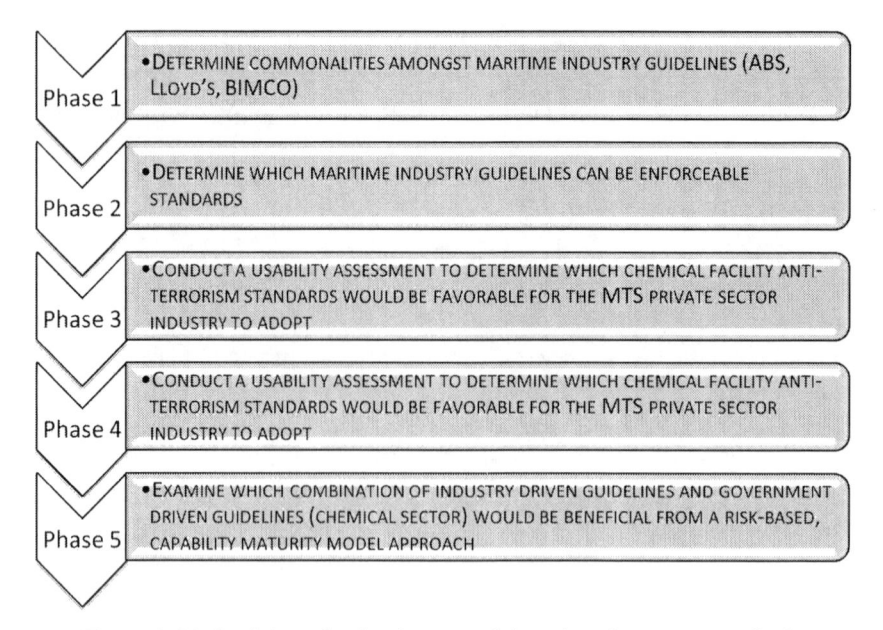

Phase 5 • Examine which combination of industry driven guidelines and government driven guidelines (chemical sector) would be beneficial from a risk-based, capability maturity model approach

Figure 4: Methodology for developing risk-based performance standards

Conclusion

The U.S. Coast Guard must continue to manage efforts toward developing risk-based performance standards while adhering to leadership that is inclusive of public–private partnerships. Following steps from Figure 4, these efforts will ensure maximum implementation success when the standards and regulations are ultimately created. The U.S. Coast Guard should continue to engage and collaborate with groups to include Information Sharing and Analysis Centers (ISACs) and other entities to reduce redundancy of efforts and also produce a more robust set of standards. A regulation in which the private sector was closely involved in its development has shown to increase the likelihood of success from both a "buy-in" implementation perspective and also from an enforceability perspective. Lack of sound cyber security practices throughout the MTS has environmental and economic implications. The U.S. Coast Guard is positioned well; both as the largest component within the Department of Homeland Security and also because of its legacy relationships with industry to be on the forefront of ensuring future performance standards are effective and appropriate.

CHAPTER 11

REFERENCES

AFCEA. 2015. www.afcea.org/events/navyday/15/documents/IDIndustry DayTFCAOverview_releasable.pdf (accessed January 15, 2016).

"Cyber-enabled Ships: Deploying Information and Communications Technology in Shipping—Lloyd's Register's Approach to Assurance." 2016. Lloyd's Register.

Eckert, Sue. 2005. "Protecting Critical Infrastructure: The Role of the Private Sector." In *Guns and Butter: The Political Economy of International Security*, ed. P. Dombrowski. Boulder, CO: Lynne Rienner Publishers.

ISO/IEC 15288:2008: Systems and software engineering—System lifecycle processes. https://www.iso.org/standard/43564.html (accessed January 15, 2016)

"Maritime Cybersecurity: A Growing, Unanswered Threat." 2014. The Maritime Executive. http://www.maritime-executive.com/article/Maritime-Cybersecurity-A-Growing-Unanswered-Threat-2014-10-24 (accessed October/November 24, 2014).

"The Application of Cybersecurity Principles to Marine and Offshore Operations." 2016. American Bureau of Shipping.

"The Guidelines on Cyber Security Onboard Ships" 2016. BIMCO.

U.S. Coast Guard. 2015. "Written Testimony of USCG Assistant Commandant for Prevention Policy RDML Paul Thomas for a House Committee on Homeland Security, Subcommittee on Border and Maritime Security hearing titled 'Protecting Maritime Facilities in the 21st Century: Are Our Nation's Ports at Risk for a Cyber-Attack?'" Reading, 311 Cannon House Office Building, Washington, D.C., October 8, 2015.

U.S. Department of Homeland Security. 2004. *Maritime Transportation System Security Recommendations for the National Strategy for Maritime Security*. Washington, DC. http://oai.dtic.mil/oai/oai?&verb=getRecord&metadataPrefix=html&identifier=ADA474574.

U.S. Government Accountability Office. 2014. *Maritime Critical Infrastructure Protection: DHS Needs to Better Address Port Cybersecurity*. GAO-14-459. http://www.gao.gov/products/GAO-14-459.

"What is Systems Engineering." International Council on Systems Engineering

Website. http://www.incose.org/AboutSE/WhatIsSE (accessed January 10, 2016).

CHAPTER 12:
CYBER SECURITY IN MARITIME DOMAIN: A RISK MANAGEMENT PERSPECTIVE

Unal Tatar and Adrian Gheorghe
Old Dominion University

Abstract

Maritime transportation security has been important for hundreds of years. Today, the maritime sector cannot survive without information technologies. The dependence on cyber infrastructure introduced many opportunities and innovative systems along with threats. In this study, we aim to draw a picture of the current state of cyber security in maritime domain from a risk assessment perspective to inform about the cyber security risks and vulnerabilities. Strategic and organizational level examination of cyber security in the maritime domain is the scope of this study. In order to identify the vulnerabilities and risks current literature was reviewed to extract the information.

Introduction

Increasing the reliance on cyber technologies brings threats as well as opportunities. Although the first cyber incident, the infamous Morris worm that seriously affected the security perception of the Internet occurred in 1988, it took almost a decade to handle cyber threats, vulnerabilities and risks as a national security issue. Studies on the critical information infrastructure protection and the publication of national cyber security strategies are two important indicators of cyber security efforts at national level. National cyber security is a new phenomenon with a history of almost 20 years, and this is very obvious when these two indicators are examined.

The term of critical infrastructure is first used within the Executive Order of President of United States in 1996 (Clinton 1996). Any physical or cyber infrastructure is called critical infrastructure if damage to that infrastructure has a harmful effect on economy of the country, social order and/or national security (Scrantom 2001). The purpose of the executive order was to introduce the term "Critical Infrastructure Protection", to define the problem and to establish interim commissions in order to recommend comprehensive strategies and amendments to the existing laws in order to protect critical infrastructures. The executive order mentioned two types of threats against critical infrastructures; physical threats and cyber threats. Although critical infrastructures existed long before the wide use of cyber technologies and internet prevalence, "Critical Infrastructure Protection" is defined as an important governmental notion because of the dominant use of cyber systems in infrastructures that serve society. There are two reasons for that. Firstly, cyber systems welcome a novel type of threats called cyber threats. Cyber threats are asymmetric in nature; an attacker can hide himself easily compared with conventional weapons. Cyber weapons are extremely cheap and prevalent. Consequently, cyber threats easily pave the way for harmful attacks against critical infrastructures. Secondly, cyber systems either established or increased already existing interdependencies among critical infrastructures. These interdependencies are considered the main cause of the cascading failures (Little 2002; Eusgeld, Nan, and Dietz 2011). In other words, a problem in one infrastructure may result in a subsequent failure in a separate infrastructure. Hence, countries started to think about protection of the critical infrastructure more seriously.

The first national cyber security strategy was published in 2000. The distribution of published national cyber security strategies shows that there exists significant acceleration after 2007, 2008, and 2010 (Tatar et al. 2014). Russian hackers attacked Estonia's critical information infrastructures in 2007 (Czosseck, Ottis, and Talihärm 2011). Russian hackers were on the stage again in 2008. They attacked Georgian government web sites just before the conventional war emerged from the conflict in South Ossetia (Korns and Kastenberg 2008). Stuxnet, the so-called first digital weapon, which targeted an Iranian nuclear plant in Natanz appeared in 2010 (Chen and Abu-Nimeh 2011). These three attacks are indicators of changes in perception specific to cyber threats and risk at a national level. The first event, the Estonian attacks, showed how a nation can be disrupted merely with cyber-attacks. The Georgian case is one of the first examples of aligning cyber-attacks with kinetic attacks as a force multiplier. In both of these cyber-attack campaigns, not the government officials but patriotic hackers were employed. Thus, non-state actors became visible in the cyber-warfare arena. The Stuxnet case was a game-changer in the cyber security world. Before Stuxnet, cyber-attacks to critical infrastructures were perceived as hypothetical claims. Stuxnet showed that the threat is real even when occurring in a cyber domain, as it had kinetic effects such as causing a malfunction to physical systems. Stuxnet is also accepted as the first example of a state sponsored cyber-attack against critical infrastructures. Shortly, after these three events, there were new rules for national cyber security.

- Cyber-attacks to critical infrastructures are not hypothetical threats anymore.
- Cyber-attacks to critical infrastructures may cause catastrophic consequences.
- Nation states are sponsors for cyber-attack campaigns
- Non-state actors also pose a serious threat for national critical infrastructure protection.

The maritime transportation sector is considered to be a critical infrastructure itself, or as a critical infrastructure under the transportation systems sector (Department of Homeland Security 2010). Maritime transportation is vital for the world's economy since it is the primary means of transportation for national and international trade. An interruption in the supply chain of food and oil may have catastrophic impacts on both the economy and society. All these are enough to recognize maritime transportation as a critical infrastructure.

Maritime transportation security has been important for hundreds of years. Today, the maritime sector cannot survive without information technologies. The dependence on cyber infrastructure has introduced many opportunities and innovative systems such as ECDIS and AIS along with threats. The cyber threat landscape makes the maritime domain important for examination and requires solutions to its problems.

Purpose, Scope, and Method of the Study

In this study, we aim to demonstrate a picture of the current situation of cyber security in the maritime domain from a risk assessment perspective in order to inform about the cyber security risks and vulnerabilities. Strategic and organizational level examinations of cyber security in the maritime domain are the scope of this study. In order to identify the vulnerabilities and risks, current literature was reviewed to extract this information. Since it is not in the scope of this study, another study is desirable to validate the findings.

Literature Review

There are many academic studies and policy reports on maritime security. However, these studies are focused on the physical aspect of the topic and there are only a limited number of works on cyber security of maritime domain. In this section of the study, a brief summary of the extracted information will be shared in chronologic order.

The European Network and Information Systems Agency (ENISA) published one of the first studies that handles cyber security aspects in the maritime sector. ENISA prepared the "Analysis of Cyber Security Aspects in the Maritime Sector" report to provide a baseline for better understanding key cyber security challenges in the maritime sector, identifying initiatives on cyber security in the maritime sector in national, European Union (EU), and the global level. Finally, further developed recommendations for enhancing safety, security and resilience of cyber based maritime capabilities. The target audience of the report is policymakers, all stakeholders in the maritime sector, and also initiatives for developing and implementing cyber security solutions. In order to assess vulnerabilities, ENISA conducted research and conducted interviews and administered questionnaires to experts from both the public and private sectors (Cimpean et al. 2010). The findings of this research were validated in a workshop by ENISA in 2011 and will be discussed in the next section of this study.

The Center for 21st Century Security and Intelligence in Brookings Institute published a policy paper which is titled "The Critical Infrastructure Gap: U.S. Port Facilities and Cyber Vulnerabilities" written by Commander Kramek of US Coast Guard 2 years after the ENISA report. Kramek (2013) gathered data by interviewing port security officials, government officials and other stakeholders. Kramek visited six ports in the United States which were selected according to their threat ranking, size, volume, and type of cargo and geographic location. The policy paper examines cyber security awareness and culture, prevention and preparedness capacity (i.e., conduct of vulnerability assessment), and level of cyber resiliency (i.e., having a written response plan) of the selected ports (Kramek 2013). A set of recommendations is listed at the end of the report.

In order to analyze the current situation, future predictions, and to develop recommendations to cope with the challenges, "The Future of Maritime Cyber Security" report developed a framework that utilizes three elements of cyber operations in the maritime environment: information, technology, and people. All three elements were employed to have a holistic view while examining constantly changing and increasingly demanding cyber threat environment (Fitton et al. 2015).

CHAPTER 12

US Government Accountability Office's report to Committee on Commerce, Science and Transportation of US Senate aimed to "identify the extent to [which] the Department of Homeland Security (DHS) and other stakeholders have taken steps to address cyber security in the maritime port environment" (Wilshusen 2015). The analysis was conducted by reviewing the laws, regulations, policies and reports. Besides these, the author of the report conducted site visits and did interviews to identify the roles and responsibilities of government organizations for addressing cybersecurity in maritime port environment. The report concludes that the US Department of Homeland Security should better address cyber security in the maritime domain (Wilshusen 2015).

The major documents on cyber security in the maritime domain were either policy papers or reports until 2015. The US Coast Guard published its cyber strategy which is different from previous documents and objectives. In order to embrace cyber space as an operational domain, three strategic priorities are identified: defending cyber space, enabling operations and protecting infrastructure. To ensure long term success in support of the strategic priorities, seven factors are recognized (US Coast Guard 2015). These are:

- recognition of cyber space as an operational domain,
- developing cyber guidance and defining mission space,
- leveraging partnerships to build knowledge, resource capacity, and an understanding of Maritime Transportation System cyber vulnerabilities,
- sharing of real-time information,
- organizing for success,
- building a well-trained cyber workforce, and
- making thoughtful future cyber investments (US Coast Guard 2015).

The Collaborative Cyber/Physical Security Management System (CYSM) is a European research project which developed a risk management methodology that relies on modelling and group decision-making techniques (Papastergiou, Polemi, and Karantjias 2015). Although output of the project is not a list of vulnerabilities, threats, risks, or a set of recommendations, we consider it important to mention this study since it is a specialized tool to identify these factors. Technical managers of the project published a paper for explaining the details and findings of the project (Papastergiou, Polemi, and Karantjias 2015). The developed tool which can be downloaded from the project's website free of charge, enables port operators to "(a) model physical and cyber assets and interdependencies; (b) analyze and manage internal/external/ interdependent physical and cyber threats/vulnerabilities; and (c) evaluate/ manage physical and cyber risks against the requirements specified in the ISPS Code and ISO27001" (Papastergiou, Polemi, and Karantjias 2015).

A more recent paper that tries to shed light on the vulnerabilities of maritime transportation system, particularly shipboard systems, oil rigs, cargo, and port operations was written by DiRenzo, Goward and Roberts (2015). The authors emphasize that maritime cyber security is not a well-studied area. Furthermore, the authors inform readers about how vulnerable maritime systems are in terms of cyber security by giving specific examples from previous incidents.

BIMCO, a leading international shipping association, published a guideline for cyber security of ships. This is the first document which focused on ships specifically. The document is based on National Institute of Standards (NIST) Cybersecurity Framework. The document is a practical guideline for assessing and reducing risks and developing contingency plans (BIMCO 2016).

Risk Management Based Evaluation of Maritime Domain

The risk management based approach is one of the primary tools for decision-making. For complex and interdependent systems such as Maritime Transportation Systems, assessing and quantifying the risk is important for decision-making on security investments and preparing a future strategy. With risk assessment, the heart of the risk management, it would be practicable to identify assets that need protection, vulnerabilities of the systems, the value of each asset, and implications when a system is attacked.

Almost all maritime cyber security documents up to now have adopted a risk management-based perspective (Cimpean et al. 2011; Fitton et al. 2015; Wilshusen 2015). A specific tool (Papastergiou, Polemi, and Karantjias 2015), and a guideline (BIMCO 2016), is prepared to make the application of risk management practices easy for stakeholders in the maritime domain.

There is not a universal definition of risk. In general terms, risk is a function of vulnerability, threat, and impact. Risk is defined in US Coast Guard's Cyber Strategy as "The potential for an unwanted or adverse outcome resulting from an incident, as determined by the likelihood that a particular threat will exploit a particular vulnerability, with the associated consequences" (US Coast Guard 2015). Vulnerability is defined as "a characteristic or specific weakness that renders an organization or asset (such as information or an information system) open to exploitation by a given threat or susceptible to a given hazard" (US Coast Guard 2015). BIMCO defines risk management as "the process of identifying, analyzing, assessing and communicating risk and accepting, avoiding, transferring or controlling it to an acceptable level considering associated costs and benefits of any actions taken" (BIMCO 2016). Cyber threats are actors or tools which can exploit the vulnerability of a system. Hackers, hacktivist groups, insider threats and states are examples of possible threat sources.

The aim of this study is to provide a list of vulnerabilities in the maritime domain at strategic and operational levels. The first group of vulnerabilities, also the root cause of some other vulnerabilities, is low level awareness and limited focus on maritime cyber security (Cimpean et al. 2011). Although cyber security studies have a history of 20+ years and an extensive literature, studies on the maritime domain is very limited because of the lack of demand. Policymakers and stakeholders are not fully aware of the threat. This vulnerability triggers another one, lack of the basic cyber hygiene factors (conducting user security trainings before using systems, conducting vulnerability assessments, and preparing response plans etc.) (Kramek 2013). The lack of awareness also causes human based vulnerabilities: insider threat and social engineering attacks.

The second type of vulnerabilities stems from governance issues. There is a fragmented governance structure in the maritime domain at both global and national/regional levels. At the global level, there are several organizations such as IMO, WCO, ICC IMB, and IMSC. However, lack of coordination between these organizations and stakeholders, different maritime zones which are subject to different laws and regulations and fragmentation in existing maritime policies has resulted in a fragmented governance structure which causes inadequate coordination, governance gaps and overlaps, difficulty in enforcing minimum cyber security requirements for the maritime domain, and difficulty in identifying roles and responsibilities (Cimpean et al. 2011; Kramek 2013). At the national level, privatization of ports causes a fragmentation of governance which results in a misperception where cyber security investments could be seen as a financial burden. The main reason behind this fact is the tendency in the private sector to view security as a cost (Fitton et al. 2015).

The third type of cyber vulnerabilities of the maritime domain is based on the complexity of the environment. The maritime environment has an intensive dependency on information technologies (IT) and is composed of interdependent systems. Besides technological complexity and interdependency, the maritime domain comprises of multiple stakeholders such as port authorities, ship industry, maritime companies, oil rigs, IT providers, customs, and security organizations (coast guard and navy). Complexity and interdependency of the maritime environment makes it difficult to secure the domain from cyber risks. In order to manage and cope with the risks that result from this complex interdependent structure, a holistic approach to maritime cyber security risk is a necessity. Nevertheless, cyber security issues are handled in an ad hoc manner (Cimpean et al. 2011).

The fourth and the last type of vulnerability is on the technical side. Highly advanced systems (i.e., radar systems, command centers, ECDIS, AIS, and etc.) are developed and networked without adequate cyber security measures (Fitton et al. 2015). There are two main steps which might end in hardware and software vulnerabilities including development and integration. During the development of technological systems, large groups of people are needed which causes the system to become vulnerable to communication errors. Another cause of having vulnerable systems is the fact that functionality is much more important than good practice and security. Increasing reliance on open standard technology is good for integration of interdependent systems. However, open standards are also open to adversaries who can possibly find a vulnerability of the system to exploit (Kramek 2013). Throughout the integration stage, automating and integrating separate systems decreases resiliency. The technology is designed to be secure in its primary function, but the designer may ignore securing its tertiary functions which are not adequately decided functions during the requirements analysis stage of development (Kramek 2013).

Conclusion

Cyber security in the maritime domain is not a well-studied area. There is only a limited number of introductory studies available. There are studies on ports and ships, however, a detailed study that covers vulnerabilities and risks and makes recommendations with a holistic, risk management perspective is missing (Kramek 2013; Papastergiou, Polemi, and Karantjias 2015; Wilshusen 2015; BIMCO 2016). Most of the cyber security literature is dedicated to technical side of the field. In order to trigger efforts of cyber security in the maritime domain, strategic level studies which inform policymakers and make them aware of the risk are required. A resilient maritime cyber infrastructure is possible only with efforts that adopt a multistakeholder approach. This is the only way to cope with the challenges posed by the fragmented structure of the maritime environment.

CHAPTER 12

REFERENCES

BIMCO. 2016. *Guidelines on Cyber Security Onboard Ships*. Denmark: BIMCO. https://www.bimco.org/News/2016/01/~/media/AEEEE215CBE3421F8F7493A6A1B0E521.ashx.

Chen, Thomas M., and Saeed Abu-Nimeh. 2011. "Lessons from Stuxnet." *Computer* 44 (4): 91–93.

Cimpean, Dan, Johan Meire, Vincent Bouckaert, Stijn Vande Casteele, Aurore Pelle, and Luc Hellebooge. 2011. "Analysis of Cyber Security Aspects in the Maritime Sector." European Network and Information Security Agency. file:///C:/Users/Nicole/AppData/Local/Temp/2011_ENISA_Analysis_of_cyber_security_aspects_in_the_maritime_sector_1%200.pdf.

Clinton, William Jefferson. 1996. "Executive Order 13010–Critical Infrastructure Protection." *Federal Register* 61 (138): 37347–37350.

Czosseck, Christian, Rain Ottis, and Anna-Maria Talihärm. 2011. "Estonia after the 2007 Cyber Attacks: Legal, Strategic and Organisational Changes in Cyber Security." *International Journal of Cyber Warfare and Terrorism* 1, no. 1: 24-34.

Department of Homeland Security. 2010. *Transportation Systems Sector-Specific Plan an Annex to the National Infrastructure Protection Plan 2010*. DHS. https://www.hsdl.org/?view&did=736911.

DiRenzo, Joseph, Dana A. Goward, and Fred S. Roberts. 2015. "The Little-Known Challenge of Maritime Cyber Security." Paper presented at the 6th International Conference on Information, Intelligence, Systems and Applications (IISA).

Eusgeld, Irene, Cen Nan, and Sven Dietz. 2011. "'System-of-Systems' Approach for Interdependent Critical Infrastructures." *Reliability Engineering & System Safety* 96 (6): 679–686.

Fitton, Oliver, Daniel Prince, Basil Germond, and Mark Lacy. 2015. "The Future of Maritime Cyber Security." Lancaster University. http://eprints.lancs.ac.uk/72696/1/Cyber_Operations_in_the_Maritime_Environment_v2.0.pdf.

Korns, Stephen W., and Joshua E. Kastenberg. 2008. "Georgia's Cyber Left Hook." *Parameters* 38 (4): 60.

Kramek, Joseph. 2013. "The Critical Infrastructure Gap: Us Port Facilities and Cyber Vulnerabilities." *The Brookings Institution*. (July 3). https://www.brookings.edu/research/the-critical-infrastructure-gap-u-s-port-facilities-and-cyber-vulnerabilities/.

Little, Richard G. 2002. "Controlling Cascading Failure: Understanding the Vulnerabilities of Interconnected Infrastructures." *Journal of Urban Technology* 9 (1): 109–123.

Papastergiou, Spyridon, Nineta Polemi, and Athanasios Karantjias. 2015. "Cysm: An Innovative Physical/Cyber Security Management System for Ports." In *Human Aspects of Information Security, Privacy, and Trust*. 219–230. Springer.

Scrantom, Timothy D. 2001. "USA Patriot Act: Implications for Foreign Financial Institutions and International Finance Centres." *Trusts & Trustees* 8 (1): 23–26.

Tatar, Unal, Orhan Calik, Minhac Celik, and Bilge Karabacak. 2014. "A Comparative Analysis of the National Cyber Security Strategies of Leading Nations." In *9th International Conference On Cyber Warfare and Security*, 211–218. UK: Academic Conference and Publishing International Limited. West Lafayette, Indiana.

US Coast Guard. 2015. *U.S. Coast Guard Cyber Strategy*. U.S. Coast Guard. https://www.uscg.mil/SENIORLEADERSHIP/DOCS/cyber.pdf.

Wilshusen, Gregory C. 2015. "Maritime Critical Infrastructure Protection: DHS Needs to Enhance Efforts to Address Port Cybersecurity." United States Government Accountability Office. GAO-16-116T. http://www.gao.gov/assets/680/672973.pdf.

CHAPTER 13:
AN ANALYSIS OF RESPONSE STRATEGIES USED TO MITIGATE DISRUPTIONS TO THE MARITIME TRANSPORTATION SYSTEM CAUSED BY A CYBER INCIDENT

Joseph Couch
U.S. Coast Guard

Background

Natural and man-made disasters, including disasters caused by a cyber incident could cause a significant disruption within the maritime transportation system (MTS), which would most likely result in an economic impact due to delays on the movement of goods in our nation's ports. "If or when there is a cyber incident in any given port area, our collective goal must be to continue safe and security operations with" minimum disruptions (Thomas 2015, 4). Following Hurricane Katrina, a task force was established to focus on issues impacting the recovery and restoration of the MTS. The recommendations of the Maritime Recovery Restoration Task Force (MR^{2}TF), in addition to the focus of Congress and DHS regarding the development of strategies to facilitate the resumption of trade and recovery of the MTS following a major disruption, led to an overhaul on how the U.S. Coast Guard, as Incident Commander, plans for and responds to a significant disruption to the MTS. To accomplish this task, the Coast Guard and its port partners have established Essential Elements of Information (EEI) that provide a consistent approach to MTS planning and response including a process for tracking and reporting EEI data regarding five general categories of the MTS. In addition, each COTP Zone has developed MTS Recovery Plans and established MTS Recovery Units (MTSRU) as part of the Planning Section of the Incident Command System organization.

A MTSRU (which include many port stakeholders) was established to assist Unified Command (UC) with planning MTS recovery and facilitation of the resumption of commerce and re-establishment of the basic functionality of the MTS. The MTSRU involves five primary objectives: (1) "Track and report on the status of the MTS;" (2) "Develop a clear understanding of critical recovery pathways;" (3) "Develop courses of action to support MTS recovery;" (4) "Provide an avenue of input to the response organization for" all MTS stakeholders; and (5) Identify and "develop long-term restoration issues and forward to UC" (US-Canada Maritime Commerce Resilience Initiative 2012). In recent years, these initiatives were tested and further refined as a result of port closures following major hurricanes, western river floods, and environmental disasters. Those types of emergency response incidents significantly impacted the MTS forcing incident commanders to manage and coordinate MTS recovery activities within their response organization. The same measures can be applied when responding to cyber incidents that could disrupt the MTS. To meet this need, federal agencies, like the U.S. Coast Guard has established MTS recovery procedures to address re-establishing the movement of goods and services in U.S. ports following any type of natural or man-made disaster, such as a cyber incident.

Why MTS Recovery Processes Can Be Applied to a Cyber Disruption

In the United States, critical incidents impacting the maritime domain are better managed when current MTS recovery functions are implemented within the response organization. The MTS recovery task was created by the U.S. Coast Guard following Hurricane Katrina and the devastating impact after this storm passed over the Gulf Coast area. Accordingly, the MTS recovery incident management concepts developed by the Coast Guard are directly linked to the response activities associated with this tropical storm. According to Henstra (2010), "recovery policies are meant to aid in restoring and rehabilitating a community after emergencies" (237). Because of Hurricane Katrina's emergency response needs to address MTS recovery activities, the U.S. Coast Guard formed a *Maritime Recovery and Restoration (MR^2) Task Force* to assist incident commanders with recovery efforts immediately following landfall of the hurricane along the Gulf Coast (O'Neil 2005). Given these pieces of information, the intent of this chapter is to highlight why the effectiveness of the response to Hurricane Katrina can be applied to man-made cyber incidents to help response organizations manage disruptions to the MTS.

The concept of MTS recovery began in 2005 immediately after Hurricane Katrina devastated the Gulf Coast states. The MR^2 Task Force was created to address the significant disruption to the MTS. The task force, headed by Rear Admiral Larry Hereth, identified "maritime issues affecting people, the environment, infrastructure, and the economy ... By working in concert with local, state, and federal partners as well as maritime industry and other stakeholders, the task force was able to identify those actions necessary for the short-term recovery and long-term restoration of the marine transportation system" (O'Neil 2005, 1). O'Neil (2005) reported the task force was a "naturally occurring part of the Coast Guard's response to Hurricane Katrina" and served as a "support element for the incident commanders engaged in response operations." Consequently, the Coast Guard MR^2 Task Force's role during the Hurricane Katrina incident led to the creation of new strategic guidance documents, and supported many concepts already addressed in national maritime-related strategies including the National Response Plan (NRP), which ultimately became the National Response Framework (NRF).

Prior to the Hurricane Katrina incident, the federal government enacted several laws, presidential directives, and many strategic policy guidance documents focused on maritime recovery planning. Two laws addressing MTS recovery include the (a) *Maritime Transportation Security Act (MTSA) of 2002*, and the (b) *Security and Accountability for Every Port (SAFE Port) Act of 2006*. Both laws required the Department of Homeland Security (DHS) develop plans "to ensure that the flow of cargo through United States ports is reestablished as efficiently and quickly as possible after a Transportation Security Incident" (TSI), including strategic guidance plans and protocols for the resumption of trade in the event of a MTS disruption (Maritime Transportation Security Act 2002). The MTSA 2002 and the SAFE Port Act of 2006 required development of the National Maritime Security Plan (NMSP), and the Maritime Infrastructure Recovery Plan (MIRP) addressing "the restoration of domestic cargo flow following a security incident that occurs under, in, on, or adjacent to waters subject to the jurisdiction of the United States" (Maritime Infrastructure Recovery Plan 2006). In addition, the NMSP and the MIRP were created because of the *National Security Presidential Directive (NSPD) 41* and the *Homeland Security Presidential Directive (HSPD) 13*. President Bush specifically addressed maritime infrastructure recovery in both directives. Bush (2004) conveyed "rapid recovery from an attack or similar disruption in the maritime domain is critical to the economic well-being of our nation." In spite of these critical documents, emergency response professionals agreed MTS recovery ideas learned during Hurricane Katrina needed to be incorporated in operational and tactical plans, and our nation's incident management practices.

The nation's economy depends on an effective functioning inter-modal transportation system, which involves cargo flow pathways through coastal ports as well as the extensive inland river system commodity flows that serves major sections of the country. International, national, regional, and local supply chain interdependencies with a maritime transportation nexus exist across other surface transportation modes, cyber-dependent systems, and critical infrastructure and key resources sectors. Therefore, the MIRP was created to address this concern and how recovery needs to be managed in the United States. The "MIRP contains procedures for recovery management and provides mechanisms for national, regional, and local decision-makers to set priorities for redirecting commerce, a primary means of restoring cargo flow" (Maritime Infrastructure Recovery Plan 2006). Furthermore, "the plan is employed when the DHS Secretary declares (a) an actual or threatened TSI occurred under, in, or adjacent to waters subject to the jurisdiction of the United States, or (b) an incident of national significance in accordance with the criteria established in the NRF and HSPD 5" (Maritime Infrastructure Recovery Plan 2006, 2). Additionally, the Maritime Infrastructure Recovery Plan 2006 identified four critical areas considered to be high priorities by the Secretary of Homeland Security: (a) "port cargo-handling capacity information is unknown or" not "readily available for use in" deliberation "regarding recovery of the MTS following a national TSI," (b) "there is no national communications network or information exchange between the federal government and the private sector focused on maritime recovery management," (c) "the Area Maritime Security (AMS) Plans do not adequately and uniformly address critical MTS recovery management planning elements," and (d) "salvage capability specific to national and regional recovery is unknown" (49–50). For these reasons, the U.S. Coast Guard updated tactical plans and developed national incident management procedures to address natural and man-made disasters (including cyber) that could significantly disrupt the MTS.

In 2006, the U.S. Coast Guard, in cooperation with port partners and stakeholders in the public and private sectors, established a common, scalable, all-hazards concept of operations planning framework nationwide. This plan helps to facilitate a standard process for accomplishing short-term recovery "(partial functionality) of the MTS following a substantial or catastrophic transportation disruption, including resumption of trade inside and outside" the incident area (US-Canada Maritime Commerce Resilience Initiative 2012, IX). Consequently, the MTS recovery concept was born to address the MIRP's high priorities discussed previously. The Coast Guard developed policy to address MTS recovery concepts, which helped frame the planning and management practices that enabled recovery of the MTS and restoration of trade. National Coast Guard policy addressed six major needs to enable recovery of the MTS, which involved (a) "an integrated government and industry recovery management organization," (b) "an integrated recovery communications system," (c) "a national plan for logistics support for cargo diversion," (d) "government awareness of cargo flows and inter-modal connectivity, and" (e) "federal funding mechanisms to support state and local preparedness" (U.S. Coast Guard 2014). Because of this policy guidance, U.S. Coast Guard command's implemented the concepts of MTS recovery in tactical plans and its incident management practices.

INSTITUTIONALIZATION OF MTS RECOVERY CONCEPTS

MTS recovery concepts are institutionalized by the Coast Guard incident management program managers. In addition, many of the ideas learned from the MR^2 Task Force were incorporated into Coast Guard contingency plans, and incident management protocols and practices. For example, the Coast Guard uses a *MTS Recovery Unit* in its incident management organization to help incident managers address significant disruptions of the MTS. Incidents affecting the MTS will be conducted following the guidance contained in the NRF and use the principles of the National Incident Management System (Marine Transportation System Recovery 2012). Therefore, Coast Guard policy requires its commanders establish MTS Recovery Units within the Planning Section of an Incident Command System (ICS) organization. Figure 1 depicts where the Coast Guard's MTS Recovery Unit is located in an ICS organizational diagram. Many qualified MTS Recovery Unit Leaders are responsible for leading the ICS recovery function during real and planned events. For example, the MTS recovery incident management actions were used during the 2010 major oil spill in the Gulf of Mexico, which produced MTS disruptions in the ports of New Orleans, Louisiana, and Mobile, Alabama. The U.S. Coast Guard set a high priority on keeping the ports and waterways open (Young 2011). Similar actions would be used in the United States during MTS disruptions caused by a cyber incident. During the response of a cyber incident, MTS recovery actions would be coordinated and executed at the local, regional, and national level of the U.S. government. "The Coast Guard, at the local, regional, and national levels, implemented its MTS recovery protocols and procedures consistent with the Coast Guard's doctrinal policy and implementation directives" (Young 2011, 1). In spite of these facts, the MTS Recovery Unit provides unique functional responsibilities to local incident commanders in an ICS organization.

THE MTS RECOVERY UNIT FUNCTION

The U.S. Coast Guard is the ideal agency to lead a response to a cyber incident affecting the MTS. Following major incidents in the maritime domain, "the Coast Guard has traditionally regulated navigation and other activities on the water, coordinated efforts to restore waterway usage, mitigated environmental impacts, and ensured the public and private sectors were adequately informed of maritime conditions" (National Capacity for Marine Transportation Recovery 2007, 1). In addition, as first responders, Coast Guard leadership positions used to manage maritime domestic incidents is provided by senior commissioned officers designated under law to perform unique tasks for the federal government. "As Captains of the Port (COTP), Federal Maritime Security Coordinators (FMSC), and Federal On-Scene Coordinators (FOSC), the Coast Guard is well positioned to coordinate the short-term recovery activities aimed at restoring the flow of commerce and other critical maritime activities within the ports" (National Capacity for Marine Transportation Recovery 2007, 1). Therefore, the MTS recovery incident management function became an important feature within an ICS organization to help Coast Guard incident commanders operating in a Unified Command.

If a domestic cyber incident caused a significant disruption to maritime transportation, incident commanders operating under Unified Command will set up a MTS Recovery Unit. Incident commanders shall ensure a MTS Recovery Unit is established within the Planning Section of the ICS response organization (Marine Transportation System Recovery 2012). Once established, the MTS Recovery Unit Leader is responsible for (a) tracking and reporting the status of the MTS, (b) understanding the critical recovery pathways, (c) recommending courses of actions to the incident commanders, and (d) providing all MTS public and private stakeholders with an avenue of input to the response organization (Marine Transportation System Recovery Unit Leader 2014). "The MTS Recovery Unit Leader prepares transportation data for the Situation Unit, and daily situation briefs applying core Essential Elements of Information" (Marine Transportation System Recovery Unit Leader 2014, 8–13).

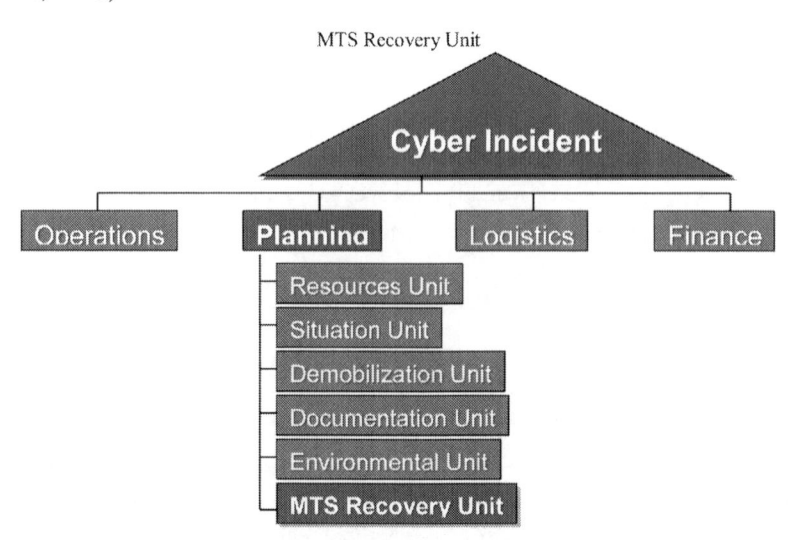

Figure 1: The MTS recovery unit depicted in an ICS organization

CHAPTER 13

Essential Elements of Information

The MTS recovery function is accomplished using five general categories of Essential Elements of Information (EEI), which address (a) waterways and navigation systems, (b) port area critical infrastructure, (c) port area vessels, (d) offshore energy, and (e) monitoring systems. For example, following a cyber incident that disrupted the MTS, the MTS Recovery Unit would track and report the status of critical waterways, critical aids-to-navigation, brides, bulk liquid facilities, and container cargo facilities under the port area critical infrastructure EEI category. The EEIs were designed to accomplish two objectives: (a) tracking overall MTS recovery progress, and (b) facilitating MTS recovery decision-making (Marine Transportation System Recovery 2012). "The specific data for each EEI provides the MTS Recovery Unit with a basis for analysis that can lead to development of recovery priorities and recommendations" (Marine Transportation System Recovery 2012, 8). For example, by comparing the nature of one waterway against another or one containerized cargo facility against another, the MTS Recovery Unit will be able to assess the relative importance or criticality of the components of the MTS. However, the MTS Recovery Unit will not be successful unless it interfaces with port stakeholders responsible for operating or controlling key transportation systems or critical infrastructure within the port area.

Importance of Stakeholders

The MTS Recovery Unit must directly interface with a significant number of public and private port stakeholders to obtain the information it needs to fully understand the status of the systems within the MTS. Recognizing the COTP's unique authority to direct vessel and port traffic, as well as the Coast Guard's objective to ensure resumption of commerce as soon as possible following a transportation disruption, the maritime industry is vital because these public and private entities provide the information needed to help prioritize and execute resumption of vessel traffic in the impacted port area (Marine Transportation System Recovery 2012). The MTS Recovery Unit must work closely with port partners to develop and execute plans that are aligned with industry concerns and capabilities. Therefore, the MTS Recovery Unit responsibilities associated with this ICS management function will require participation from a broad spectrum of agencies and entities.

According to the Marine Transportation System Recovery (2012), the MTS Recovery Unit's success will depend on the having the right mix of representatives from the government and the private sector. Consequently, Hurricane Katrina taught the federal government a lot about how to effectively respond and manage maritime disasters. "Although the size of the hurricane was extraordinary, much of the destruction and human suffering was attributed to failure of governments to adequately plan for a large-scale emergency of this kind" (Henstra 2010, 236). "Recovery policies that assist with prioritizing recovery operations helps ensure that sufficient resources are in place to manage disasters" (Henstra 2010, 240). Therefore, based on an evaluation of Hurricane Katrina, along with the facts, ideas, and examples provided in the paper, this article compared and contrasted the effectiveness of the response to this actual incident and discussed how the response and recovery activities learned from that natural disaster would be effective if applied to MTS disruptions caused by a cyber incident.

Conclusion

In summary, MTS recovery offers unique incident management processes, which help coordinate response activities associated with all types of disruptions of the MTS, including cyber-related disruptions. Since Hurricane Katrina, recovery and the restoration of trade have become the mechanisms for maintaining a healthy economy after a disaster has occurred in the United States. Recovery polices "are meant to aid in restoring and rehabilitating a community after emergencies" (Henstra 2010, 237). The MIRP is considered a critical national policy document focused on managing the restoration of trade after an incident has occurred. "The MIPR reflects the organizational constructs detailed in the NRF, as well as the use of the ICS and Unified Command procedures utilized to manage disruptive incidents requiring MTS recovery management" (Maritime Infrastructure Recovery Plan 2006, 2). For that reason, the U.S. Coast Guard, as the premier agency for managing domestic maritime incidents, established MTS Recovery Units within its ICS organization to assist incident commanders with MTS recovery-related incidents. Accordingly, Coast Guard emergency responders are now using the MTS recovery function within Unified Command when a cyber incident significantly disrupts the MTS. In addition, the MTS Recovery Unit leverages a port coordination team in each port area to address EEIs designed to help decision makers with addressing recovery priorities (Marine Transportation System Recovery 2012). Consequently, the recovery processes employed since Hurricane Katrina have increased our nation's capability to successfully address maritime disasters. "The Coast Guard effectively built resiliency within the MTS by (a) engaging the private sector, (b) establishing government roles, and (c) developing recovery policies, plans, and procedures at the local, regional, and national levels" (National Capacity for Marine Transportation Recovery 2007, 1). Because of Hurricane Katrina, one can conclude the MTS recovery incident management practices implemented by the Coast Guard have helped made the United States more effective at responding to domestic cyber incidents.

References

Bush, George W. 2004. *National Security Presidential Directive NSPD-41/Homeland Security Presidential Directive HSPD-13*. Washington, DC: The White House.

Henstra, Daniel. 2010. "Evaluating Local Government Emergency Management Programs: What Framework Should Public Managers Adopt?" *Public Administration Review* 70 (2) (March/April): 236–246.

Marine Transportation System Recovery. 2012. "Marine Transportation System Recovery." In *U.S. Coast Guard Atlantic Area Instruction 16001.1B*. Portsmouth, VA: U.S. Coast Guard Atlantic Area Command.

Marine Transportation System Recovery Unit Leader. 2014. *Incident Management Handbook COMDTPUB P3120.17B*. Washington, DC: U.S. Coast Guard. https://homeport.uscg.mil/ics.

Maritime Infrastructure Recovery Plan. 2006. *The National Infrastructure Recovery Plan*. Washington, DC: The Department of Homeland Security. http://www.dhs.gov/files/programs/editorial_0757.shtm.

Maritime Transportation Security Act of 2002. 107th Congress Public Law 295. U.S. Government Printing Office.

National Capacity for Marine Transportation Recovery. 2007. "Developing a National Strategy for Marine Transportation Recovery." In *The U.S. Coast Strategy for Maritime Safety, Security, and Stewardship*. Washington, DC: The U.S. Coast Guard. http://www.uscg.mil/strategy/docs/CGS-Final.pdf.

O'Neil, C. 2005. Press Release: *Coast Guard Forms Maritime Recovery and Restoration Task Force*. Washington, DC: U.S. Coast Guard. http://www.d8externalaffairs.com/go/doc/425/104549/Coast-Guard-forms-Maritime-Recovery-and-Restoration-Task-Force.

US-Canada Maritime Commerce Resilience Initiative. "U.S.–Canada

Beyond the Border—Perimeter Security and Economic Competitiveness Action Plan." December 2012.

U.S. Coast Guard. 2014. "Marine Transportation System Recovery Planning and Operations." In *Commandant U.S. Coast Guard Instruction 16000.28A*. Washington, DC: U.S. Coast Guard.

U.S. Department of Homeland Security. "Written Testimony of USCG Assistant Commandant for Prevention Policy RDML Paul Thomas for a House Committee on Homeland Security, Subcommittee on Border and Maritime Security Hearing Titled 'Protecting Maritime Facilities in the 21st Century: Are Our Nation's Ports at Risk for a Cyber-Attack.'" October 8, 2015. https://www.dhs.gov/news/2015/10/08/written-testimony-uscg-house-homeland-security-subcommittee-border-and-maritime.

Young, W. 2011. "Marine Transportation System Recovery." In *Deepwater Horizon* OSC Report. Washington, DC: U.S. Coast Guard.

CHAPTER 14:
CYBER SECURITY INDUSTRIAL CONTROLS

Jim Cooper
Port Authority of New York and New Jersey

ABSTRACT

September 11, 2001 provided the all hazards community a multitude of lessons. The response of our critical infrastructure operations was initially loosely defined and absent coordination. Over the years, we recognized the value in sharing our solutions and aligning our risks from one owner or industry to a neighbor's infrastructure, and reducing the shifting of risk. Our cyber community will benefit from learning the lessons of the physical space and applying these lessons to enable assessments of the cyber domain to include the level of awareness, integration and general approach of the critical infrastructure security operators to then address the evolving risks that are present in the cyber domain. The assessment should consider the relationship of our physical operations, information technology and the maturing partnership of these communities as required to secure our critical maritime infrastructure.

Awareness for cyber security has escaped from the shadows of the information technologists and is now a present topic of the general population, in our media, and discussed in the boardrooms of corporations. With such visibility, the cyber community has a laser like focus on our actions by many to include the armchair experts, organizational leadership and the threat actors. The visibility is similar to the experiences from the events of 9/11, which motivated the first responder community to adopt a collaborative workspace. Unfortunately, cyber and physical security domains most often operate in separate communities and the semantics and experiences of each are too often, not shared. The isolation has allowed our practitioners of our physical security, and of our information technologies, to exist as foreigners in each other's terrain. While the lessons from our response to the threats and consequences in the physical security arena are transferable to the cyber domain, we lack a repeatable model for knowledge transfer and in some cases, a common nomenclature for transferring the knowledge.

Keywords: awareness, all hazards, critical infrastructure, industrial controls, NIST CRF, NIST 800

Introduction

Since the 1800s, emergency responders from fire, police and the emergency medical services have worked together to evolve ideas, share lessons and improve our capabilities to respond, contain, and mitigate incidents. Over the recent years, the community of responders has adopted an all hazards approach to incident response. The all hazards term encompasses natural disasters, fire, crime, terrorism, and cyber events; this community of fire, police and EMS responders routinely train and fight together while absent the participation, and even the awareness of the cyber responders. The cyber community is further challenged by the stove piping that exists between the system operators and the information technologists. The first responders are the system owners and IT operators but they lack the commensurate response training and skill sets of other responder disciplines.

One apparent benefit to including the cyber community in our all hazards approach centers is the adoption of the National Incident Management System, which affords responders from disparate communities and domains the experience to operate within a common incident command framework.

Background

Over the past decade, the cyber threat to the maritime domain has evolved in sophistication while the response from the systems operators has lagged. In the maritime space, research and real-world examples have shown the inherent risks to the vessel positioning systems and the targeted threats against our container movements to include the documented tactics of drug trafficking in the 2014 CyberKeel report.

The cyber-attack described in the CyberKeel (2014) report has some exposure within the maritime community, yet how wide is the distribution across the sectors? How wide is the distribution of the report describing the attack against a German steel mill in 2010 (Lee, Assante, and Conway 2014)? While information sharing is a core lesson of the events from September 11, 2001, the lessons on sharing are not well known in the cyber domain and this contributes to the lack of threat awareness.

The cyber community is further challenged by the disparity that exists between the business systems and industrial control systems that are often vastly different. It is common to hear an IT department passionately debate the necessity of patching software vulnerabilities in business systems while in the industrial control space, experienced operators apply a defense in depth approach that considers physical access, technical access, and software vulnerabilities. In the control space, patching the operating system versus the programmable logic controllers presents unique operational risks that are less understood by some IT departments. At times, the IT departments lack experience in the analog domain yet insist all vulnerabilities, regardless of the reporting source or the affected domain, must be patched and patched quickly. The needs of these hybrid digital/analog systems are unique and for some operators, poorly understood. Given that the industrial controls operate on legacy platforms that likely have a number of legacy vulnerabilities, establishes the need for an informed system operator. Yet, the operators of the control systems lack a fundamental awareness of the cyber threat. Vulnerability management is more than software patching and requires a risk centric approach. The simple threat vector created from the use of a USB drive between a networked system and stand-alone industrial control system is poorly understood. The routine practice of moving log files from a standalone environment to a networked desktop creates a link between the internet and standalone system, defeating the model of security through obscurity. This disparity is not unique to the maritime sector and requires the attention of the broader community of critical infrastructure operators. Continuing to address cyber security as an IT issue, absent the involvement of the system operators, will provide opportunities for exploits and delay the timely detection of an advanced persistent threat.

When considering the impact of a cyber event in the maritime environment, we can embrace the studies that predict the impact of a port shutdown due to a cyber event by considering the impact of a labor action. In 2002, it was suggested that a 10-day labor lockout of West Coast ports would cost the U.S. economy an estimated $1 billion a day, the same lockout today is projected to cost $2.1 billion a day, according to a June study by Jeffrey Werling, executive director of the University of Maryland's Inforum (2016) forecasting project. A cyber incident will likely result in similar consequences of a labor action.

Lessons from Physical Security

The concepts and understandings of defense in depth are maturing in the physical space but less widely exposed in the IT domain and the community of industrial operators. In the critical infrastructure community, information systems are often managed by IT professionals while the control space is managed by mechanical system engineers who understand operations, and the control software is maintained by the software developer who may lack exposure to the operational risks. The lessons from the physical space spawned the well-known "See Something, Say Something" campaign. The challenge for a system engineer or operator to differentiate between a system "bug" from a concerted assault by a nation state is an under developed skill set. This is one lesson that is transferrable to the cyber domain and can accelerate the awareness of the operators in the industrial controls space.

The operators in the physical space, subsequent to the events of September 1, 2001, spent a number of years to learn and understand the threat space and to define an appropriate response, this work continues today. The initial security response to the events of 9/11 was emotionally driven and progressed absent a clear understanding and awareness of the threat space and how these threats manifest as risks. The communities of security professionals that are operating within the physical space are moving toward a risk/cost benefit model that equates the cost of proposed counter measures to a commensurate reduction in risk. Within the cyber community, the concepts of cost benefit versus risk buy down is not well understood, and at times, misplaced. A significant amount of capital is expended on security appliances, to detect and mitigate commodity malware; at the expense of the advanced persistent type threats. The IT community operating in the critical infrastructure domain, operate absent a thorough understanding of the threat space and may create an over reliance on technical solutions that convey a sense of security that does detect commodity threats, but the needed detection algorithms for detecting nation states targeting the operational community are less mature. Equally, the operators themselves have not developed an understanding of the threat space often subscribe to a vendor model of intrusion detection, absent a commensurate reduction in risk. While developing a security posture in the physical space, it is routine to establish physical boundaries of access through a defense in depth approach with the use of environmental controls, lighting, door locks, access control, gates, and patrols. Access control in the physical space is user centric, access is often established by time of day and day of week for what access is permitted, whereas the cyber access is typically established by role with

CHAPTER 14

some maturing in the concepts of least privileged—domain administrators, system administrators and then, everyone else. In the cyber space, we often find the legacy network topology lacks a defense in depth strategy and is instead a flat terrain with a broad attack surface. The legacy networks were designed and built to support user experience with less emphasis on the needed layers of security. A cyber security specialist assessing the cyber domain would first determine the threat landscape and attack surface. Once understood, they would then perform a rapid risk assessment to rack and stack the methods of exploits that afforded a threat actor access and then work to deny that actor access, determined by the risk priority. In contrast, we have varying levels of understanding and some professionals might look at the threat space more holistically and generalize a solution to deploy security appliances or applications to include antivirus, intrusion detection, and host detection. We must accept that the threat actors have or will gain access to the network; simply working on detection may report the subsequent loss of data, but not result in an adequate level of risk reduction. The separation of the cyber security professionals from the physical professionals denies an opportunity for collaboration and the sharing of mitigation lessons from the past decade. In the cyber community, we need to establish new partnerships of system operators and the IT departments to create the shared objective of cyber risk reduction. A number of critical infrastructure operators have established best practices that mitigate risk through a defense in depth approach and it is the sharing of these best practice models that might best inform the operational community and reduce risk.

Across the DHS sectors of critical infrastructure, we find similar disconnects within and across each of the sectors. The threat landscape of the maritime domain is not sector centric; the community has significant interaction with each of the sectors and shares many of the same threat vectors to include actors, objectives, and tactics. Yet, too often, we find the traditional stovepipes of excellence organically evolving and the notion of collaboration primarily limited to a specific sector. When considering the mission statement of the DHS National Infrastructure Protection Plan is to strengthen the security and resilience of the Nation's critical infrastructure by managing physical and cyber risks through the collaborative and integrated efforts of the critical infrastructure protection community, the inference of collaboration and integration is principally sector centric and inherent in the model is the limitation of cross sector experiences. The cyber threat space is modal agnostic and the lessons and best practices of each sector are not unique. Unlike the majority of the all hazards space, the cyber domain is common and spans these sectors.

Knowledge Gap

The approach to mitigating cyber risk is constrained by the experience and knowledge base of the system operators regarding security and risk including the threat, attack vectors, vulnerabilities in the threat space, and consequence. Our experience with known threats range from the Anonymous attacks on Government websites in 2011, to the Equation Group sophisticated attack on the attendee's at a Houston conference in 2009 (Kaspersky Lab 2016). The impact of the Northeast Blackout of 2003 provided significant lessons of the maritime domains influence of cross sector incidents. The ability to restore the "just in time" chain of critical commodities was inadequate during the blackout and conversations for how best to restore the fuel chain highlighted the expected challenges we would experience during a cyber event. Although not a cyber incident, the lessons are transferable to the cyber domain.

Evolving the DHS Lessons Learned Information portal with content that is relevant to the operators of industrial controls can augment the needed foundation of awareness. Today's comprehension of risk is further complicated with a misplaced trust in the air gaps for system security to the manufactures providing new systems—on antiquated computer platforms. The human factors of cyber security are not yet in the forefront of research. It is generally accepted that process designers can eliminate the human error of system failure if they remove the human from the loop. This is accomplished through system automation—the same venue targeted by cyber exploits. Although a 2006 study by Dr. Meshkati of the University of Southern California concluded that humans will remain in charge of an operational environment regardless of the advances in computer automation. As we adopt automation in process design, we require a similar focus on the human interface to ensure the operator can distinguish appropriate system behavior from the behavior of an exploit, and to make such a distinction on the first encounter. Too often, experience has shown that operators do not consider unusual system behavior as a security event and instead lean toward the unusual behavior as a consequence of a "buggy environment or software patch." The ability to infer a cyber incident from a system software or hardware failure is inadequate and such a discovery will likely require several system failures before the operator might consider a cyber incident. Most operators of our industrial control systems lack the requisite technical knowledge to identify cyber incidents from system problems.

The attack surface, to include support vendor office systems and laptops, extends globally; the threat scape which is global as well, is not adequately studied to establish the level of threat awareness by the operators of critical infrastructure. In known instances, the benefits of automation are not realized due to an operator's preferred reliance of human confirmation of the state of a valve versus a computer reporting a sensors detection of an open or closed valve. The operators of some transportation fuel farms continue to employ manual processes even while they have an elaborate network of sensors to detect valve states and tank levels—communicated via radio. These experiences are rarely shared within or across sectors. Over the past two decades, the pipeline industry has researched the impact of human operators on system failures and studied the integration of the operator with automation (Meshkati 2006), while the maritime sector has completed its own research and studies of system failure yet the sectors operate as silos, at the expense of the lessons from other sectors (Rothblum n.d.). When considering the threat vectors and human machine interface, cyber incidents are modal agnostic. An operator's skill sets, computer knowledge and reluctance to consider a cyber exploit instead of a software glitch is similar across sectors. The cyber ecosystem will benefit from cross sector collaboration of human factors research. It is generally accepted that critical infrastructure operators will always require a human in the decision process and will not defer to root automation for life safety or mission critical decisions. The attack surface and cyber threats are global in nature and modal agnostic yet our cyber strategy has followed our physical security strategy which is modal and sector centric.

Conclusion

President John F. Kennedy once uttered, "victory has a thousand fathers, but defeat is an orphan." How we engage the community of diverse cyber security specialists, IT professionals, and physical professionals can bring us closer to victory. In the cyber space, we are fortunate to adopt a defined framework to strengthen an organization's cyber security program including a defense in depth model. The framework was developed through the collaboration of a diverse group of impacted stakeholders from the public and private sectors. The "Cybersecurity Framework (CSF)" was developed and published by the United States National Institute of Standards and Technology's (NIST) Computer Security Division in 2014. The CSF provides the business information technology and industrial control systems the standards and best practices to manage security controls (NIST 2014). The implementation of the CSF provides a methodology for the two professions to target the familiar physical categories of people, systems, and hardware. When considering a program to enhance an organizations cyber security posture, the NIST Risk Management Framework (RMF) affords a cost-effective and repeatable approach to manage cyber security-related risks (NIST 2016).

A key challenge is establishing an organization's cyber security center ability to share the experiences of the IT professionals and the physical professionals. Each professional community has a poor understanding of the other community's interest and needs. The NIST (2013) Special Publication 800-53 "Security and Privacy controls for Federal Information Systems and Organizations" will help the two professions align efforts and begin the hard work of identifying and defining the security processes and controls.

When considering a program to enhance an organizations cyber security posture, the NIST RMF affords a cost-effective and repeatable approach to manage cyber security-related risk. The NIST 800-53 Security and Privacy Controls for Federal Information Systems and Organizations, will help the security and IT professional align efforts and begin the hard work of defining roles and responsibilities. The implementation of the NIST cyber framework provides a methodology for security and IT professionals to target the familiar physical categories of people, systems, and hardware. Within the cyber community, IT professionals focus on the applicability of intrusion detection sensors, the hardware; security software, and new appliances that are marketed to detect the threat—with less focus on the needs of the security professionals to establish the intent of the threat to disrupt, damage or exfiltrate data. Although intrusion sensors are part of a defense in depth strategy, the appliances required a skilled set of operators to understand the detections.

Security appliances and applications require experienced security analysts to monitor and interpret the events. The vendors of software or security appliances develop labels of threat severity for an exploit that is agnostic to the threat target. The widely used Malware Rating System (MRS) is consumer centric and modal agnostic. It is not necessarily relevant to operators of critical infrastructure and even less applicable to the industrial control space. These labels dismiss the potential for repurposed exploits, an area poorly understood by vendors and operators alike. When reviewing threat severities and response actions, how an exploit purposed for certain behavior can be repurposed is not given adequate consideration.

The lessons from the physical security space support demonstrate the benefit of a partnership of the operational users and information technologists. The new partnership should embrace an existing cyber security model that is risk centric and considers the cost benefit relationship of mitigating strategies to risk buy down. The NIST Cyber Risk Framework is one such model that offers a mature approach to improving cyber security.

CHAPTER 14

REFERENCES

Cyberkeel. 2014. "Maritime Cyber-Risks." Copenhagen, Denmark. October 15. http://www.cyberkeel.com/images/pdf-files/Whitepaper.pdf.

Inforum. 2016. University of Maryland. October 3. http://www.inforum.umd.edu/.

Kaspersky Lab. 2016. "Kaspersky Lab Equation Group: The Crown Creator of Cyber-Espionage." http://usa.kaspersky.com/about-us/press-center/press-releases/2015/kaspersky-lab-discovers-equation-group-crown-creator-cyber-espi.

Lee, Robert M., Michael J. Assante, and Tim Conway. 2014. "ICS CP/PE (Cyber-to-Physical or Process Effects) Case Study Paper." Industrial Control Systems. December 30. https://ics.sans.org/media/ICS-CPPE-case-Study-2-German-Steelworks_Facility.pdf.

Meshkati, Najmedin. 2006. "Safety and Human Factors Considerations in Control Rooms of Oil and Gas Pipeline Systems: Conceptual Issues and Practical Observations." *International Journal of Occupational Safety and Ergonomics* 12 (1): 79–93.

National Institute of Standards and Technology. 2013. "Security and Privacy Controls for Federal Information Systems and Organizations." NIST Special Publication 800-53. http://nvlpubs.nist.gov/nistpubs/SpecialPublications/NIST.SP.800-53r4.pdf.

National Institute of Standards and Technology. 2014. "Framework for Improving Critical Infrastructure Cybersecurity." Cybersecurity Framework Version 1.0. February 12. https://www.nist.gov/sites/default/files/documents/cyberframework/cybersecurity-framework-021214.pdf.

National Institute of Standards and Technology. 2016. "Risk Management Framework (RMF) Overview." August 25. http://csrc.nist.gov/groups/SMA/fisma/framework.html.

Rothblum, Anita M. n.d. "Human Error and Marine Safety." http://bowles-langley.com/wp-content/files_mf/humanerrorandmarinesafety26.pdf.

CHAPTER 15:
BASIC CYBER SECURITY IS EASY
(SO WHY IS IMPLEMENTING IT SO HARD?)

Jerry Doherty
Trackline Strategies

An effective cyber security posture is pretty basic; we have known the recipe for several years, and nothing about it is overly complicated or expensive: position current hardware defenses at the perimeter, keep the software that connects everything properly configured and up to date, have the wetware (read: people) that make use of everything remain aware of and vigilant against threats, and monitor network traffic for unusual behavior, so you will know when one of the "wares" has let you down and it's time to take action. Sure, things change and threats are constantly evolving, and there are varying strategies and techniques to layer the defenses, which means our cyber security posture needs to continually evolve, but that's background noise—a constant that applies to all aspects of business, and life. So, if cyber security is so easy, why are so few organizations good at it? Three reasons usually answer this question: failure to appreciate the risks, assessing the risks as technical ones instead of risks to business operations, and presuming a technology solution instead of a cultural one.

First Barrier: The Risk is not Real

It is my observation that the chief problem is organizations do not yet perceive the cyber security threat as real, or value a breach as costly as they should. This failure to appropriately assess the risks associated with cyber security is particularly puzzling given the string of recent high-profile cyber security breaches, including at Mossack Fonseca, The Hollywood Presbyterian Medical Center, Verizon, and MedStar Health in 2016, Anthem, Experian, Ashley Madison, Home Depot, VTech, and voter registration data in 2015, and eBay, JP Morgan

Chase, Sony Pictures, Target, and the Office of Personnel and Management in 2014. That cyber security breaches are on the rise, and that they are costly is not news. Yet there is still a sense among most organizations, and individuals, that it is a problem for the other guy. Why is this so?

A common way to assess a particular risk is to put a number on it by multiplying the perceived probability of an event occurring times the expected cost of the consequences should it occur. If either the probability of an event occurring or the consequences of that event are not valued very high, then there is not much perceived risk. When both have real values, perceived risk is moderate or even high. It's a useful way to model a very hard thing to measure. Such a risk number can be compared against other numbers that represent other risks, creating a prioritized risk assessment. Cyber security currently does not make it into this kind of list because few organizations are thinking of cyber systems as an integral part of their business processes. Read the latest 10K or 10Q from a few of your favorite companies, and see if they mention their cyber systems as a critical asset or cyber security as a business risk. Chances are good (too good) that they do not, which is odd given that few organizations (or shareholders) could imagine, let alone survive, going back to doing business without cyber systems. In many organizations, and even in most homes, devices outnumber people, and with each passing day we are figuring out new ways to connect old devices, in the name of increased productivity. Connected thermostats, anyone? "We're starting that meeting this morning an hour earlier than usual, so let's use our smartphone app to talk to the building and tell it to have the climate control adjusted before we get there." The Internet of Things is coming fast, which is to say that the number of attack surfaces is increasing fast and, therefore, the risks are similarly increasing. Still, while most folks can rationalize why Sony Pictures was targeted over the release of a film that depicts an assassination plot against the North Korean head of state, they cannot draw a parallel to their own organization. It may be reasonable to discount state-sponsored and terrorist cyber threats, but state-sponsored actors are perhaps the smallest segment of the threat spectrum. Much more prevalent are common thieves (your spam folder ought to substantiate that claim), and the recreational hackers—those who seek to breach systems just to see if they can. Still, the biggest source of potential threats, and the most pervasive, are the inside threats, which include software bugs, machine failures, human error, and intentional mischief perpetrated by disgruntled employees (think Edward Snowden), which can cause just as much harm as external threats, sometimes more.

The risks also are not perceived as real because the cyber infrastructure is evolving relatively slowly (one device at a time) and because each device, by itself, actually does have low probability and consequence values associated

CHAPTER 15

with it. This leads to the fallacious assumption that it cannot (or will not) happen here. The odds, however, are against that notion. Formerly stand-alone industrial control systems, with little or no built-in security features (such as thermostats and other environmental controls), are now connected to the Internet and, consequently, gaining access to that system does not require a physical intrusion to visibly sit at a control console. These breaches can now be done remotely and, instead of requiring dexterous and knowledgeable fingers, can be done with even more dexterous automated tools, with no obvious indication of it happening. Once that industrial control system is breached, it becomes a launch pad for breaching every other connected system, again with no obvious indications. What starts off in the thermostat moves to the video surveillance system, then the WiFi access point, then the network router, then the shared drive, which has the financial access credentials on it, and that leads to consequences. Awareness is the key to resolving this first barrier to effective cyber security. Organizations, and particularly the executive leadership team, need to embrace the facts that cybercrime is a chronic problem, there are many, many sources of threats, including from inside our organizations, and the number of attack surfaces is increasing by the day. Add to that the fact that any one breach can daisy-chain its way to high potential consequences. Probability? Even an unbiased casual observer has to conclude that it is not low. Consequence? Recent headline cases indicate that even modest breaches come with significant cost to reputation and brand, in addition to real dollars. Probability times consequence equals risk. Given this, let's presume your organization recognizes the threat as real. Now what?

Second Barrier: It is an IT Problem

Once an organization accepts that the risks associated with cyber security are real, it needs to next properly assign them as operational business risks, which require a cross-functional approach to address. The next obstacle along that path is that cyber security is viewed as just an IT problem. This common misconception is perpetuated by users and business leaders, and embraced by IT, because it just seems right; information technology is what introduced cyber security vulnerabilities, so it is natural to presume it is an IT's problem. Nothing could be more dangerously wrong. Overall cyber security is not an IT function any more than finance is an IT function. Though all financial information is stored in and processed on IT infrastructure, no one in their right mind would categorize finance as an IT function. Why do we let ourselves do so for cyber security? The threat of a cyber breach is a business risk. Until it is viewed as such, it will not get the organizational attention it needs to be addressed. It is not just a security problem, it is not just an IT problem, it is an imperative operational/business risk management challenge. We expect IT to provide technical security of the financial systems, and we can similarly expect IT to provide technical security of the cyber systems, but overall cyber security, like overall financial security, requires a cross-functional enterprise-wide solution—everyone needs to own it and be part of the solution. Otherwise, as the saying goes, they are part of the problem.

Most organizations that get this far do not have trouble with this mental shift initially but, because the discussion can quickly become complicated with tech-heavy jargon, actually pursuing this path seems hard and causes business leaders to want to push it back toward IT. That is because they are approaching it as if the solution will be a technical one instead of a cultural one.

Third Barrier: There is a Technical Solution

It is not the technology that is the biggest target. Recalling the threat sources, inside threats are by far the most prevalent, and users make up the largest vulnerability of inside threats. Everyone with access to your systems, including current and former employees, customers, and suppliers, is a potential attack surface for a cyber breach. People are the weakest link. You can have the most expensive and up-to-date hardware and software, but one careless user can render it all impotent with the click of a mouse. To implement effective cyber security, organizations must turn their users from weakness into strength by converting them from potential threats into vigilant defenses. This is a bridge too far for most organizations, because it requires a cultural change. Culture change is hard, so they look for a technical solution instead. There is not one. Technical tools can help, but as long as you have still got users, you have still got a problem. Even the most drastic technical solution of disconnecting your organization from the Internet will not work because employees will still get targeted at home or on their phones, and then walk the threats to work right through the front door. Cyber security only works when it is everyone's responsibility. There needs to be a cultural bias toward cyber security, just like there is with safety and financial reporting. Nearly every industrial workplace has a sign indicating the number of days since the last injury, as a means of reminding employees of the importance of safety, for themselves and for each other. Publicly traded companies impose quarterly quite periods on wide swaths of their employees, not just to comply with FCC rules about non-disclosure of financial data in the weeks prior to release of quarterly results (which could be met with a much smaller group of employees), but because doing so helps remind everyone involved of the importance of accurate and timely financial reporting. Cyber security demands nothing less—an enterprise-wide effort, regularly reinforced through a common cultural norm or practice.

CHAPTER 15

Change is hard under normal circumstances, but especially so when the members of an organization do not recognize the need for change, and what is asked of them is inconvenient. Security is inconvenient. It is also inefficient, at a time when we praise efficiency (and introduce IT tools to improve efficiency!) Efficiency is not good for security; security gets in the way of convenience. The battle lines are clearly drawn. All security comes at a cost to convenience; it is a trade-off we are willing to make when we see a value exchange. We once left the keys to our cars in the ignition, because that was the most convenient place to find them when you needed them. After a few people and businesses got their cars stolen, because it turned out that hanging in the ignition was also a convenient place for thieves to find the keys, we started taking the keys with us and locking our cars. We are not yet that far along in the value exchange with cyber security. We still find it too inconvenient to remember complex (read: secure) passwords for each application we use, so we use simple ones that are easy to remember and, therefore, easy for thieves to guess. But, worse than with our cars of old, we now have the ability to use the same key for multiple applications, so once thieves have one password, they may have them all. As a society, we are probably still a long way from getting everyone to recognize their role as cyber warriors. It is going to take several more high-profile data breaches to get the message to sink in. Will your organization be one of those who serve as a bad example, or are you ready to take action to protect yourself?

THE SOLUTION

Just like the recipe for basic cyber security, the recipe for implementing it is simple. All three barriers will be brought down, eventually, through awareness. Start a cultural change by building awareness throughout your organization about the threats of cyber security. This will not be hard to do, but it will take specific effort. It needs to be a part of your precious few subjects in employee outreach programs. Every time there is a cyber breach in your industry, or in any industry that your employees can relate to, your organization needs to push out a message to employees calling their attention to it, and asking if the same thing could happen here. Executives need to mention cyber threats in every one of their workforce engagements, and challenge employees as to what they can do to reduce their vulnerability. As awareness and validation of the cyber threats rises, introduce the concept of good cyber hygiene, which will help naturally dispel the myth that cyber security is an IT problem with an available technical solution. The more employees become aware of the threats and the steps they can take personally to reduce their and the organization's vulnerabilities, the more they will be empowered to do something. When they are at that point, introduces the link between cyber security and corporate social responsibility: taking care of employees, customers, and business partners.

This is not a one-time effort, though. Employees leave and new ones arrive, cyber threats evolve and the hygiene to combat them must also evolve. Like safety and financial reporting, cyber security needs to be regularly reinforced through organization-wide engagements that are story-based and engaging, not from slides on a screen, and those stories need to draw from recent headline-grabbing breaches to explore how that breach could happen here, and what can be done to prevent it. Organizations that rely on annual online training as a requirement for gaining network access are teaching their employees that cyber security is an annoyance to be avoided. Cyber thieves thank you. Do not be that organization. Instead, take a positive, proactive approach, now and for the foreseeable future. Your shareholders will thank you.

CHAPTER 16:
CYBER SECURITY AND THE MARITIME TRANSPORTATION SYSTEM: CLOSING THE FINANCIAL AND REGULATORY INCENTIVE GAP

Chris Conley
U.S. Coast Guard

Executive Summary

The Maritime Transportation System (MTS) is a key component of national power, and this largely privately operated system is at risk to cyber attacks presenting a clear threat to national security. The MTS accounts for $4.6 trillion in total economic value, provides import–export, and refining capacity for U.S. energy needs, and has the potential to be used as a conduit for weapons of mass destruction (WMD) (U.S. Public Port Facts 2015). Individually, the thousands of businesses that make up the MTS may not present sufficient risk that would justify changes to current cyber security practices that leave the system vulnerable. However, when considered in aggregate, the potential for a large-scale cyberattack or precision manipulation to facilitate movement of WMD demands policy changes and the development of risk markets to enhance the MTS to a defendable and resilient state.

There is significant evidence that cyber threats to the MTS are real and that cyber security practices of MTS industries are lacking. Part of this gap can be explained by a lack of awareness of cyber-related risks, the cost of securing cyber systems, and the view of security as a public good where individuals share in the benefits regardless of how much they spend. In essence, there is an incentive gap that prevents MTS stakeholders from instituting a more robust cyber security posture.

Closing the incentive gap should involve both financial and regulatory incentives. The development of financial risk modeling and cyber-risk insurance markets are fundamental to enhancing cyber security in the MTS. Additionally, as the regulatory agency for the MTS the US Coast Guard should transition from a voluntary to a mandatory compliance regime with third-party verification. Ultimately, combining market and regulatory forces will advance US national security by enhancing and protecting our critical infrastructure.

PART 1: CYBER AND THE MARITIME TRANSPORTATION SYSTEM

Overview of Critical Infrastructure and the MTS Policy and Responsibility

The Nation's critical infrastructure provides essential services that underpin American society creating the fundamental components for the safety and security of the American public enabling our economic vitality (Obama 2013). "Many of the nation's critical infrastructures have historically been physically and logically separate systems that had little interdependence" (Clinton 1998). As a result of advances in information technology and the necessity of improved efficiency, these infrastructures have become increasingly automated and interlinked. These same advances have created new vulnerabilities across an extensive landscape to a variety of threats including espionage, intellectual property theft, terrorism, and state-sponsored attacks (Clinton 1998). The challenge for the federal government in providing oversight to the security of critical infrastructure is that 90% of it is privately owned and operated with unique, often proprietary operating systems creating an extensive attack surface (Stouffer, Falco, and Scarfone 2011).

The Department of Homeland Security provides strategic guidance to public and private partners, promotes a national unity of effort, and coordinates the overall Federal effort to promote the security and resilience of the nation's critical infrastructure (U.S. Department of Homeland Security 2015a). Presidential Policy Directive 21 establishes 16 critical infrastructure sectors and assigns agency responsibility according to oversight responsibilities for the relevant sectors. The Transportation Systems sector includes the Marine Transportation System (MTS) and the US Coast Guard is the agency responsible for oversight of the MTS (U.S. Department of Homeland Security 2015a).

The MTS and its Link to National Security

The Marine Transportation System is a complex system that is both geographically and physically diverse in character and operation. The unique qualities of the MTS present extraordinarily complex challenges for those charged with ensuring its security. It is a network of maritime operations that interface with shore-based operations at intermodal connections and as part of global supply chains and domestic commercial operations (U.S. Department of Homeland Security 2010). The various operations within the MTS network have components that include vessels, port facilities, and waterways as well as related infrastructure such as railroads and highways that move goods and people essential to our economy and way of life (U.S. Department of Homeland Security 2010). The United States, like many other nations, works toward maintaining a balance between safe, secure ports and facilitating trade to promote economic growth.

The MTS is made up of assets, systems, and networks in both the physical and cyber domains with industrial control systems (ICS) and information technology systems embedded in nearly every aspect of the MTS. While not all encompassing, the major components include:

- 70 deep draft ports
- Container and bulk cargo vessels
- Oil and gas terminals and refineries
- 2,000 major terminals operated by the private sector
- Off-shore oil and off-shore renewable energy installations
- Navigation infrastructure that includes thousands of miles of main channels
- Ocean going vessels and passenger carriers (U.S. Department of Homeland Security 2010, 176–181).

To put things in perspective as to the importance of the MTS to our national economy and national security, consider that 90% of the world's trade is carried by sea (IMO Profile 2016). For the United States, 53% of imports and 38% of exports are transported via the MTS making it the single largest mode for transportation of goods (Chambers and Liu 2015). Furthermore, business activities related to waterborne commerce contributed approximately $4.6 trillion in total economic value to the U.S. economy in 2014 (U.S. Public Port Facts 2015). As a mode of transportation and additionally leisure, ferries carry over 100 million passengers per year while cruise ships host an average of 10 million passengers annually (Committee on the Marine Transportation System 2014).

Not only does the MTS present a clear feature of national power in terms of the economy, the U.S. military relies on the MTS to deliver military personnel, equipment, and supplies to military forces abroad. The MTS also plays an important role transporting raw materials and equipment to fuel war-fighting efforts both at home and overseas for our own forces and in support of allies. Military experts and economists cite the U.S.'s advantage in materials and manufacturing as the key enabler that was ultimately decisive in winning

World War II in both the European and Pacific theaters, and the MTS is a fundamental component of the U.S. economic machine (Goldsmith 1946). Considering the effect of globalization and the aforementioned reliance on the MTS for our economy and national production, it is highly likely that a secure and highly functional MTS will be necessary for the Unite States to succeed in a nation-state conflict with a major power or powers. Finally, a properly functioning MTS enables movement of our naval forces where they are highly constrained in their freedom of maneuver, and it encompasses key bases for the logistical support of carrier battle groups and submarine fleets, both key instruments of national power.

Cyber Systems in the Maritime Transportation System

Technology and network communications are embedded in nearly every aspect of the MTS. A prime example lies in container terminal operations where vessels with as many as 13,000 containers moor to off-load their cargo (Port of Los Angeles 2015). Computerized cargo and terminal management systems help operators on networked gantry cranes to offload containers, and then software systems direct equipment operators to place the containers in specific locations in the terminal for further transport (Kramek 2013). Stacking cranes then assemble the containers where they are later loaded onto final transport vehicles such as truck or rail. Wireless networks allow equipment operators to continually view and update information in the logistics database via handheld devices using radio frequency identification (RFID) tags (Kramek 2013). While it is clear that cyber-related systems are heavily embedded in this process, it is expected that in the future, even greater levels of technology and automation will be employed. In ports such as Rotterdam, the Netherlands and Hamburg, Germany, all three major aspects of container movement at the terminal: gantry cranes, ground movement, and stacking cranes, are handled by software and office workers that remotely control the cranes over their computers (Henriksson 2016; IHS 2016). Due to the cost of automation and labor issues, the United States retains a relatively antiquated terminal system, but with automation improving productivity rates by up to an estimated 80%, it is likely only a matter of time until technology advancements are more widely adopted by U.S. port operators thereby increasing the cyber attack surface at our ports (Why Are not America's Shipping Ports Automated? 2015).

Another significant aspect of the MTS where cyber systems are a fundamental component includes marine oil and gas terminals and port-based refineries that represent a key aspect of energy security with implications for the economy, public safety, the environment, and national defense. While the increase in domestic oil production has brought U.S. oil imports to their lowest levels in nearly 50 years, oil imports arriving via tanker continue to make up about 10% of our energy needs and historically have represented approximately 25% (U.S. Energy Information Administration 2014; Weitz 2015). Additionally, as an economic driver for the U.S. economy and a world supplier of oil, the United States exported 340 million barrels of oil overseas in 2012 and that number continued to rise as the United States ascended to the top oil producing country in the world in 2014 (Wile 2013; U.S. Energy Information Administration 2015). As with the container terminal, refining, storing, and moving petroleum products involves highly integrated industrial control systems. At the refinery, communication between cells is often linked via a fiber-optic network to the control center. Distributed control systems monitor temperature and pressure while alarm settings are configured to alert operators remotely to control the process from a distance (Moxa 2008). During movement and storage, differential pressure flow transmitters measure flow rate, and control-valves manage operating conditions such as flow pressure, temperature, and liquid level by opening and closing in response to signals received from controllers (Firoozshahi, Allahyari, and Haghdosti 2011). While there may be variations between facilities to the extent of ICS and centralized control, the reality is that cyber networks play an extensive role in managing oil terminal and related operations, and that role is expected to grow to increase productivity and safety (Energy API 2016).

The very security systems that have been established to secure port facilities are also vulnerable to cyber threats because they employ technology that is part of a networked system. For instance, the Port of Los Angeles uses 321 networked security cameras monitored by a single operator in a control room to identify and dispatch resources for potentially hazardous activity at the port (Marroquin 2009). Security cameras have become the eyes of the port security effort with ports such as Baltimore employing up to 500 cameras that feed video monitors via wireless communications (Kramek 2013). Additionally, access control to secure areas of the port are facilitated through the Transport Worker Identification Credential that employs biometrics and background checks to verify the identity of workers accessing facilities. These cards are often read by handheld electronic scanners that that provide a form of credentialing that is likely better than human verification but still create a vulnerability for fraud if the system can be externally manipulated as was demonstrated by hackers in 2014 (McGlone 2014). Vessel traffic services also play an important role in major ports to prevent collisions and groundings in confined and busy waterways while expediting ship movements for efficiency. These systems use land-based sensors such as radar, the automated identification system (AIS) or closed circuit video which output their signals to a central location where operators monitor and manage vessel traffic movement (Vessel Traffic Services 2014).

Finally, ocean-going vessels, the key feature of the Marine Transportation System, also employ an abundance of cyber-related systems that have become essential for vessel operations. For instance, the ship's autopilot coordinates data from many devices on the ship and interfaces with control and propulsion systems to keep the vessel on a predetermined course. Ships can also be monitored and controlled from shore-based facilities via satellite:

> The bridges of some modern vessels are now more likely to contain computer screens and joysticks than engine telegraphs and a giant ship's wheel. The latest supply ships serving the offshore oil and gas industry in the North Sea, for instance, use dynamic positioning systems which collect data from satellites, gyrocompasses, and wind and motion sensors to automatically hold their position when transferring cargo (also done by remote control) to and from platforms, even in the heaviest of swells (Ghost Ships 2014).

Modern ships have complex cargo operations connected through cyber space with even the vessel's cranes being moved by GPS. Industrial control systems use software to manage engine operations, and that software is updated automatically underway, sometimes without the master's knowledge (Paganini 2015). US Navy and Coast Guard vessels employ many of the same industrial control systems and automated features as commercial vessels and thus share in the broad attack surface created by the evolution of ship systems. In addition to that, they contain highly technical and destructive weapons systems. The future envisions an even broader role for technology aboard ships as efforts to move to crewless vessels abound with every aspect of vessel operation being enabled through cyber space (Ghost Ships 2014).

Threat Assessment and Vulnerabilities

Deciding on policy changes to protect the MTS must take into account both the cost to the maritime industry and the public as well as the benefits gained by a more robust cyber security position. Significant evidence has been provided in terms of the national economy, national protection, and the U.S.'s ability to wage war that the MTS, at least in its collective, is of significant national security interest. That being said, is the MTS at substantial risk that would warrant a dramatic shift to the status quo?

The cyber threats for critical infrastructure and the maritime transportation system are the same as for any information technology or industrial control system albeit with potentially devastating impacts. As previously discussed, the vast majority of critical infrastructure is privately owned and operated using proprietary systems that could be subject to intellectual property theft from market competitors or foreign governments. Terrorists or hacktivists could also exploit vulnerabilities in critical infrastructure to achieve effects similar to that of a kinetic attack that these groups may not have the resources to accomplish otherwise. State-on-state espionage also presents a threat to intellectual property as well as U.S. secrets and information essential to the secure operation of critical infrastructure. Finally, many experts believe that the first salvo in a large scale state-on-state military conflict will likely involve cyber attacks that degrade an enemy's ability and willingness to fight (Mazanec 2009).

Espionage and Intrusion: We Are Already Breached

In any cyber exploitation or attack involving critical infrastructure, the adversary often must gain an understanding of system operation through espionage. Probing cyber system operations relating to software, hardware, and communications facilitates future exploitation that may involve theft of intellectual property, terrorism, or an attack with outcomes similar to those of kinetic weapons. While theft of intellectual property is cited by U.S. leaders as possibly the greatest long-term threat to national security (IP Commission Report 2013), this paper focuses on the more immediate problems of cyber threats from terrorism and warfare and their impact on critical infrastructure as a national asset in military force projection, public safety, and a functioning economy. In testimony before Congress, the Commissioner of the US–China Economic and Security Review Commission stated that foreign intelligence or military services penetrate the computers that control our vital national infrastructure or our military, reconnoiter them electronically, and map or target nodes in the systems for future penetration or attack (Wortzel 2010). Malicious code is often left behind to facilitate future entry (Wortzel 2010). In the case of China, they are very transparent about their desire to gain the advantage against critical infrastructure going as far as publishing academic research articles on how to take down the U.S. power grid through a cascade-based attack (Wang and Rong 2009). Broadly speaking, cyber intrusions into critical infrastructure continue to grow with report data illustrating a 400% increase in the reporting of vulnerabilities to the Industrial Control Systems Cyber Emergency Response Team (ICS-CERT) from 2010 to 2011. In the first half of FY 2015, ICS-CERT responded to 108 cyber incidents impacting critical infrastructure in the U.S., and this figure may underrepresent the actual number of incidents as only 25% of reports came from owners and operators while the remaining reports were made by federal partners, researchers, and media (U.S. Department of Homeland Security 2015b). The intrusion figure also does not take into account yet undiscovered intrusions and vulnerabilities that are likely to exist (Sistrunk 2015).

To complicate matters further, it is difficult to assess the level of intrusion into critical infrastructure, and cyber systems are also vulnerable to attacks for which they may not be the intended target. Gaining knowledge of how to attack critical infrastructure does not necessarily require probing of systems in the United States that would give officials and security personnel an understanding of what may be known by adversaries. As a manufacturer of information technology components, China has demonstrated the ability to preload malware into hardware as reported by DHS. "In one example, Zombie Zero was implanted in the software of scanner hardware manufactured in China as part of an attack targeted against shipping and logistics industries" (Sood and Enbody 2014). Both Russia and China also have the ability to purchase commercial ICS components to learn how to exploit these systems without alerting cyber monitors. There is also the issue of containment and unintended consequences that are a peculiar aspect of cyber attacks where malware, once released into cyber space, can have detrimental impacts far beyond the intended target. The well-known Stuxnet virus that caused damage to nuclear centrifuges in Iran also infected tens of thousands of unrelated computers and reportedly 14 industrial control systems in Germany (Schneier 2010). The virus was designed to become dormant if it did not find its intended target (Schneier 2010), but the level of penetration into so many systems should yield some concern at the prospect of collateral damage or unintended consequences. One can suspect that not all such malware will be designed in this manner or be successful in becoming inert if they are designed to do so.

Are U.S. Vulnerabilities Overstated?

It is very difficult to determine how much is known about the operation of our critical infrastructure cyber systems by adversaries or potential adversaries. Many experts in risk psychology argue that the language used by cyber security experts to describe risk to critical infrastructure is over-dramatic leading to exaggerated assessments of U.S. vulnerability (Quigley, Burns, and Stallard 2015). Additionally, cyber security expert Dr. Herbert Lin (2012) notes that, "threat assessment in cyber space is an inherently more uncertain endeavor than for more traditional domains of potential conflict. Under circumstances of information scarcity and faced with potential threats, there are many influences on analysts to make worst-case assessments" (337). These worst-case assessments often overestimate adversary operational skills and intent leading to overreaction by policy makers to address the worst-case scenario instead of the most probable scenario (Lin 2012). Even with the possibility that we may be overestimating adversary capability, policy-makers and industry are struggling to find the right security balance considering the potential national security and economic impacts. While dedicating resources to guarantee 100% security from a worst-case attack scenario may not be possible or supported by a cost benefit analysis, the lack of even adequate cyber security measures in critical infrastructure demonstrates that the United States must make significant improvements to defend against the most probable scenarios the United States is already facing.

Assessing MTS Cyber risks: Adversary Capability

The greatest potential for damage and acute national economic impact to critical infrastructure come from state actors and terrorists while other threats such as insiders can also have detrimental effects (Andress and Winterfeld 2013). In defining risk for the MTS critical infrastructure, a simple formulation of risk is a function of probability and severity. Probability is based on both the capability of adversaries and their intent which varies significantly based on the objectives to be achieved, the relationship between the interests of the attackers and their target, and deterrent factors for how the United States might respond if an attack can be accurately attributed (Lin 2012). In terms of capability, the following examples demonstrate how adversaries might attack or compromise technology in the MTS:

- 2011–2012: A criminal organization hired hackers to facilitate a shipment of a large cargo of drugs in the port of Antwerp by hacking into the port's management systems to facilitate movement of the illicit cargo manifested as bananas from South America. The simple software and hardware hacks that used USB keyloggers and more sophisticated purpose built devices allowed traffickers to send in drivers and gunmen to steal particular containers before the legitimate owner arrived. Even after the port become aware of the initial breach and installed a firewall, the criminals penetrated physically into the port and installed wireless bridges on the operating computers creating open access to the operating system that enabled pickup and transfer of the drugs (New Cyber Frontiers).

- July, 2013: A radio navigation research team from The University of Texas at Austin subtly coerced a 213-foot yacht off its course, using a custom-made GPS device. The team transmitted spoofed GPS signals that overrode civilian GPS obtaining control over primary and backup GPS with no alarms on radar, gyro, or compasses alerting the vessel operator. The team then gained navigational control of vessel and redirected its course (UT News 2013).

- August, 2011: A cyber attack on the Iranian shipping line ISIRL "damaged all the data related to rates, loading, cargo number, date and place, meaning that no-one knew where containers were, whether they had been loaded or not, which boxes were onboard the ships or onshore" (Jensen 2015a, 35). The attack disrupted IRISL communications and resulted in cargo being sent to the wrong destination and the loss of cargo creating significant financial losses. "A similar attack on a major international container line would have a crippling effect on the supply chains of thousands of international companies" (Jensen 2015a, 35).

- August 15, 2012: Hackers unleashed a computer virus into the systems of Saudi Aramco and Qatar's RasGas to initiate what is regarded as among the most destructive acts of computer sabotage on a company to date. The virus erased data on three-quarters of Aramco's corporate PCs destroying between 30,000 and 55,000 computers (MacKenzie 2012). As with the previous example, a similar attack on major port facilities and shipping lines could have devastating consequences.

- 2014: The FBI released information on Ugly Gorilla, a Chinese attacker who invaded the control systems of utilities in the United States. While the FBI suspects this was a scouting mission, Ugly Gorilla gained the cyber keys necessary for access to systems that regulate the flow of natural gas (Riley and Robertson 2014). This example is particularly troubling considering the number of port-based petroleum refining facilities and oil and gas terminals where such an attack could have significant safety, economic, and environmental impacts.

- 2014: A malicious actor infiltrated a German steel facility using a spear phishing email to gain access to the corporate network and then moved into the plant network. According to German authorities, the adversary showed knowledge in ICS and was able to cause multiple components of the system to fail. This specifically impacted critical process components to become unregulated which resulted in massive physical damage (Lee, Assante, and Conway 2014). While not directly related to the marine environment, the large industrial components of the MTS such as shipboard engine rooms and gantry cranes create a large attack surface for this type of threat.

As demonstrated in the instances above from both the Marine Transportation System and other critical infrastructure breaches that are transferrable to the MTS, our critical infrastructure cyber systems are vulnerable to the capabilities of both state and non-state actors. For the case that occurred in the port of Antwerp with the manipulation of manifests to facilitate movement of illicit cargo, the issue of border security at our ports that might allow for the movement of weapons of mass destruction (WMD) including chemical or biological agents into the United States presents a serious threat. While managing container integrity within U.S. port facilities is just one part of broader efforts such as the Proliferation Security Initiative and Container Security Initiative to prevent the spread of WMD, the Antwerp example offers an indication of the vulnerability of the MTS to attackers with the skills, imagination and resources. The other attacks or demonstrations in isolation may not rise to the level of national security threats in themselves. At issue here is the function of severity in determining risk and whether these types of attacks can be scaled up and across the critical infrastructure sectors to wreak nationwide havoc with effects similar to that of conventional weapons, to initiate extensive economic harm, or as a key enabler for adversaries to conduct warfare against the United States or allies.

Assessing MTS Cyber risks: Adversary Intent

Adversary intent is much more difficult to divine in most instances (Lin 2012). State actors contemplating a cyber attack on U.S. critical infrastructure could expect significant retaliation by cyber and conventional military means as well as diplomatic and economic sanctions. Since the United States is arguably the most connected and reliant on technology systems for warfighting, safety, and its economy, it is very likely that any cyber attack would be asymmetric in nature lending the advantage to the attackers. Despite this asymmetry, it is hard to imagine that a state with the capability to attack would conduct a cyber attack on critical infrastructure for the purpose of generating significant damage or economic turmoil unless their intent was a large-scale confrontation. For example, were China to attack U.S. ports, they would compromise their ability to ship their goods to the United States which accounts for 5% of their GDP which would have a negative impact on their own economy in addition to the negative consequences that the United States may impose in retaliation for such an offense (World Bank Group 2014). However, this leaves an unanswered question as to the reason for the dedicated effort of state actors to conduct the extensive levels of espionage that have gone into identifying and exploiting vulnerabilities in United States critical infrastructure including in the maritime domain as was the case when a commercial ship on contract to the United States military was the target of an intrusion by suspected Chinese military hackers (Paganini 2015). The exact levels of intrusion from state actors specifically into the MTS is unknowable as it is a near certainty that public security organizations and private enterprise could not possibly be aware of intrusions that have yet to be, or may never be discovered. More importantly in assessing adversary intent is the extensive economic and military advantage that could be gained by significantly disrupting the system which is extensive in the case of the MTS.

The significantly larger asymmetric nature of terrorist groups who may want to attack critical infrastructure or manipulate the MTS to facilitate attacks with WMD increases chances that these groups would be willing to engage in a cyber attack. The threats from these groups do not in aggregate approach the levels of damage presented by nation states but with much less to lose from the asymmetric nature of these attacks, it is likely that these groups may be more willing to attack via cyber means (Andress and Winterfeld 2013). Overall, when balancing the risk posed by these two potential adversaries, state actors have greater capability but are less likely to pursue an attack while non-state groups that may not be able to effect the same level of destruction, are more likely to attempt such an attack.

The risk to the MTS from cyber attack is real though difficult to quantify. As previously demonstrated, cyber espionage and probing of our critical infrastructure is rampant from both state and non-state actors. Adversaries and researches have additionally demonstrated the capability to attack critical infrastructure in the MTS and in other critical infrastructure sectors with techniques that can be transferred to the MTS. The severity of an attack on the MTS could be substantial particularly if the MTS were used as a conduit to support infiltration of WMD. Additionally, cyber attacks on port facilities could have severe economic impact with $12.6 billion per day in economic loss (Martin Associates 2015), and shortages of critical supplies and food, 19% of which is imported (Jerardo 2016). Cyber-attacks on oil and liquefied natural gas terminals could have severe economic and environmental impacts. An adversary's ability to take control or disrupt the operations of cargo, oil, and high capacity passenger vessels has the potential to cause death, injury, and destruction and would eliminate trust in the integrity of vessel operating systems grinding the MTS to a halt (Kramek 2013). Cyber threats to the MTS are real, and the United States must have an effective defense, a program of deterrence, and resiliency in our ports, ships, and waterways to ensure the integrity of our critical infrastructure.

Existing Cyber Security Practices and Regulations for the MTS

Current Cyber Security Posture in the MTS

It is inherently challenging to assess the status of cyber security in the MTS. The MTS is highly dynamic, expands beyond national boundaries, and is widely varied from the operation of ships to oil and gas terminals. Additionally, it encompasses both public and private elements that connect and extend to other critical infrastructure sectors through the supply chain. Further, the variables in facility size, purpose, location, and operation including proprietary and customized systems create a seemingly limitless landscape to assess. A survey of the MTS can give a broad evaluation of the security environment but obviously cannot account for the practices of every terminal operator, port facility, or ship. With that in mind, there is evidence that the overarching cyber security environment is lacking. In a report released in October 2015, the Government Accountability Office (GAO) concluded:

> Until DHS and other stakeholders take additional steps to address cyber security in the maritime environment—particularly by conducting a comprehensive risk assessment that includes cyber threats, vulnerabilities, and potential impacts—their efforts to help secure the maritime environment may be hindered. This in turn could increase the risk of a cyber-based disruption with potentially serious consequences. (Wilshusen 2015)

This report largely focused on the lack of overall consideration and oversight of cyber security citing the absence of cyber security as a component in maritime security plans, varying levels of information-sharing mechanisms, and an assessment that the Port Security Grant Program did not provide guidance for cyber security-related proposals eligible for funding.

Beyond the GAO report, other research and maritime industry cyber security groups such as Cyberkeel point to a gap in the current security posture. Based on reports from 2014, there is generally low awareness of cyber security incidents among maritime industry leadership and awareness of the incidents often does not translate into follow-on action (Kramek 2013). Industry leaders often see cyber security as a technical matter for the information technology (IT) department or chief information officer, and cyber attacks are often viewed as theoretical in nature with a low threat to their own operations (Cyberkeel 2014b). The disconnect between cyber security threats and perceived risk is further evidenced by a survey conducted of six commercial ports where information technology departments were separate and distinct from security departments and not a single port had a dedicated cyber incident response plan nor were cyber response plans contained within the broader risk management plans (Kramek 2013).

In addition to an overall understanding of the cyber security threat, actual cyber practices in the maritime industry reveal poor cyber hygiene. In a white paper published by Cyberkeel, the group revealed that 37 out of 50 of the largest container shipping carriers display vulnerabilities via simple tests that:

- Identify the potential to insert code, possibly malicious code, through company websites.

- Garnered open source information on company hardware that can reveal already known vulnerabilities (Cyberkeel 2014b).

While these weaknesses may not be exploitable, they have the potential to reveal overall cyber security capabilities presenting these companies as potential targets, and again, this was from the most basic of tests. The firm also conducted a spot check that revealed that, "37% of maritime companies had not applied the appropriate patch from Microsoft, leaving their webservers open for denial of service attacks as well as potential remote code execution" (Cyberkeel 2014a). A Brookings Institute report additionally notes that of the six commercial ports surveyed, most are not taking basic cyber hygiene steps such as requiring users to receive cyber training prior to being granted network access (Kramek 2013). While Cyberkeel's survey is international, it is also revealing noting that eight carriers, controlling 38% of global trade, allow "password" as a password to access sensitive eCommerce applications with two carriers allow "x" as a password (Jensen 2015b, 7). Furthermore, the world-wide aspect of the report does not eliminate the potential risk to the United States as the ability to manipulate cargo manifests or disrupt the U.S. economy persists. Through research and evidence from more than 100 attacks and attempted penetrations for 2014 and 2015 in the Transportation Sector, the MTS is not currently demonstrating, nor are regulators demanding adequate levels of cyber security to safeguard MTS infrastructure (U.S. Department of Homeland Security 2014, 2015c).

Legal Authorities and Regulation of Cyber Security

The Maritime Transportation Security Act (MTSA) 2002 and Security and Accountability for Every Port (SAFE) Port Act 2006 provides the legal basis by which the Coast Guard regulates port facilities and vessels. The laws do not specifically use the term "cyber" but the 33 CFR 105.300 and 104.305 address security vulnerability assessments using the term: "radio and telecommunications systems, including computer systems and networks" which should capture all cyber-related functions. Furthermore, in the absence of a law specifically related to cyber security and critical infrastructure, there is some debate as to whether the Executive Branch under existing law has the statutory authority to oversee cyber-related elements in critical infrastructure (Liu, Rollins, and Theohary 2014). MTSA requires the Coast Guard to conduct vulnerability assessments to identify and evaluate critical assets and infrastructure and their associated threats (Maritime Transportation Security Act 2002). Among the list of areas in which the Coast Guard is to identify weaknesses, there is no specific mention of information technology. However, the list does include requirements for procedural policies and communications which would capture the cyber aspects of port and vessel operations if it is construed to include the exchange of information and data between both people and equipment. MTSA also requires facilities and vessels to create and update security plans for approval by the Coast Guard. These security plans are to identify and ensure the availability of security measures necessary to deter a transportation security incident which is defined as significant loss of life, environmental damage, transportation, or economic disruption (Maritime Transportation Security Act 2002).

The Government Accountability Office noted that in the Coast Guard's 2012 biennial maritime risk assessment, it did not include cyber-related threats and vulnerabilities (Wilshusen 2015). In reviewing the updated risk assessment, the GAO continued to assert that the Coast Guard did not identify vulnerabilities to cyber-related assets though it did identify some cyber threats and potential impacts (Wilshusen 2015). In October 2015, the GAO additionally stated that the Coast Guard did not address cyber-related risks in its guidance for developing port area and port facility security plans (Wilshusen 2015). As of June 2016, the U.S. Coast Guard has drafted but not yet released Guidelines for Cyber Risk Management that provide overarching practices to enhance cyber security but compliance will remain voluntary.

Part 2: Closing the MTS Cyber Security Incentive Gap

Based on the cyber threats articulated throughout this paper, and the assessment of the vulnerabilities and current cyber security practices, there appears to be insufficient incentives that would drive industries and regulators of the MTS to achieve an adequate cyber security posture. Closing the incentive gap will require a significant change to current practices and the three actions outlined below should be used to help close this gap:

1) Financial incentives to industries in the MTS

2) Mandatory compliance enforced by regulators

3) Third-party verification

This list of proposals is not comprehensive for all that needs to be done to fully secure cyber systems in the MTS, but it does offer a framework for initial action to close the current incentive gap.

Financial Incentives

Financial Modeling

The Coast Guard and MTS industry groups should communicate in financial terms the cost-benefit-analysis of cyber security investments. While demonstrating the direct linkage of financial incentives for individual maritime businesses presents a challenge due to the sheer volume and diversity of companies within the MTS, there is an ongoing effort to provide modeling to show that investments in protection and resiliency of cyber systems do protect a company's bottom line. University of Southern California's National Center for Risk and Economic Analysis of Terrorism Events (CREATE) has developed a comprehensive framework for estimating the total economic consequences of maritime cyber threats that includes the effect of cyber resilience tactics that can reduce business interruption losses. They have developed a user-friendly software system called the Economic Consequences and Analysis Tool (E-CAT) that provides rapid estimates for threats to the U.S. economy that include wide ranging impacts that extend beyond the MTS both up and down the supply chain as well as the impact of resiliency measures (Rose 2016). Using the E-CAT model in a simulated 90-day disruption at Port Arthur/Port Beaumont, a standard estimate of economic impact was $1.3 billion, but taking into account all of the economic linkages, the loss grew to $14.8 billion in the absence of resilience. However, when factoring multiple resilience tactics, the total economic loss was reduced by 67% to $4.8 billion (Rose 2016). Rose (2016) provides similar cyber resiliency tactics with associated impacts on economic loss to both the business directly affected as well as those up and down the supply chain to demonstrate that there is a measurable financial incentive to enhancing a company's cyber security posture. Economic modeling and simulations that help businesses of the MTS to conduct cost–benefit analysis of enhancing their cyber security should be expanded to provide companies with individual assessment tools. Additionally, models such as the E-CAT could be developed to factor the economic impact of defensive and mitigation measures to cover the full spectrum of cyber security.

Cyber-risk Insurance Markets

Cyber-risk insurance markets can also provide an avenue to provide MTS industries with financial incentives to enhance their cyber security, and this line of insurance has grown dramatically in recent years with estimates of global premiums rising from $1 billion in 2013 to $2.5 billion in 2015 according to a Cybersecurity Ventures (2015) market report. However, there are significant challenges that must be overcome to develop a mature and sustainable cyber risk insurance marketplace and industry stakeholders, insurers, governments, and international organizations world-wide have significant obstacles to overcome in achieving a viable market.

Currently, most existing marine and commercial general liability policies exclude cyber-related risks. Insurance policies that cover ships and cargo-handling facilities have instituted cyber attack exclusion clauses where losses caused directly or indirectly by computer or software attack as a means of inflicting harm such as loss, damage or liability would be excluded from coverage (Marsh 2014). Furthermore, commercial general liability insurance that covers physical damage or loss of profit was never designed to cover loss of profits or other intangibles that might be associated with a cyber attack (Bonner 2012). This limitation was emphasized by the well-publicized 2011 Sony data breach when the insurance company Zurich won a decision from the New York Supreme Court that it was not liable to reimburse Sony for cyber-related third-party claims including lawsuits against Sony for loss of data and privacy breach (Bishop 2014). In general, damage caused by intrusions, attacks, or other losses must be covered by a specific cyber policy that generally covers liability, business interruption, third-party coverage, and the cost of information technology notification and forensics (Kirkpatrick 2015).

Significant among other common insurance policy exclusions are war-risk and terrorism exclusions. War-risk and other such exemptions may allow insurers to distinguish between which types of claims can be paid if attacks are deemed to be politically motivated or militaristic (Douse 2010). There are provisions in the U.S. war risk insurance program that encompass war and terrorist risks involving waterborne commerce and the law establishes a structure for federal reinsurance if markets cannot provide coverage on a reasonable basis (Benzie 2003). However, that coverage is limited to marine transit while goods are being shipped (Benzie 2003). Similar federal reinsurance programs are also found in the Terrorism Risk Insurance Act (TRIA) passed in 2002 that was renewed most recently in 2015 as the Terrorism Risk Insurance Program Reauthorization Act (TRIPRA). Following the terrorist attacks on the United States on 9/11 and concerned about the limited availability of terrorism coverage in high-risk areas, Congress passed TRIA so that in the event of a major terrorist attack, the insurance industry, and federal government would share losses according to a specific formula (Terrorism Risk and Insurance 2015). It is unclear whether TRIPRA would cover a cyber terror attack and whether such an attack would meet the definition under TRIA of a violent act that is dangerous to human life, property or infrastructure (Terrorism Risk Insurance Act of 2002; Lang and Mullen 2013). Additionally, a business would most probably need an underlying cyber risk policy for TRIPRA to apply (Terrorism Risk Insurance Act of 2002; Lang and Mullen 2013).

Further complicating the issues of both war and terrorism exclusions and federal reinsurance is the problem of attribution that plagues cyber attacks. It will be inherently difficult to determine whether an attack came from a state actor, terrorist group or some other actor qualifying as neither. Determining whether an attack came from state actors officially sanctioned by the state or a third party at the behest of or complicity of a state could be an insurmountable obstacle under the current structure. There will be the additional challenge of determining who is qualified and capable to conduct the forensics and apply attribution to both persons and in the case of terrorism, motives that include political ends. There is an opportunity for Congress and the federal government to expand TRIA and the war risk insurance plan to act as an insurer of last resort that would facilitate the issuance of coverage from risks of cyber-war and cyber-terrorism where insurers are challenged to accurately assess risks due to the lack of data that would allow them to determine the likelihood or severity of loss. Based on the previously discussed assessment of capability and intent to inflict damage and harm, cyber war and cyber terror attacks on critical infrastructure present the most significant national and economic security threats in the cyber domain. The United States would enhance its national security as a byproduct of a cooperative relationship between industry and insurance providers to raise overall cyber security and should be willing to act as a financial backstop in the case of a massive cyber attack to encourage development of this market.

As previously noted with the rapid rise in cyber risk insurance policies, the demand for cyber coverage and willingness of insurers to begin insuring this risk has grown substantially and is projected to grow threefold to $7.5 billion by 2020 (Morgan 2015). Despite the growth and expected expansion in cyber risk insurance policies, it is unclear whether the current construction of policies will meet the needs of MTS operators or other critical infrastructure sectors. In a somewhat bizarre scenario, the vast majority of marine insurance policies exclude losses related to cyber security breach or failure under clauses such as CL380, and most cyber risk policies exclude coverage for physical losses, damage, and injury or death that are caused by accidental or malicious cyber activity which could very well occur in a shipping or industrial environment (Cyber Marine Attacks). Most of the cyber risk insurance policies focus coverage related to data breach, third-party suits, financial loss, theft, and fraud as these areas have cost U.S. businesses significantly as demonstrated by the high-profile Sony hacks of 2011 and 2014 and the Target and Home Depot breaches of 2013–2014 that cost $248 million and $43 million respectively (Weiss and Miller 2015). Further illustrating the limitations of cyber risk insurance for marine operators, Christine Marciano of Cyber-Data Risk Managers notes that "few cyber policies cover physical injuries or damage caused by an attack that started online, but then caused actual physical damage in the real world" (Kirkpatrick 2015). A review of insurance policy information available through insurance company websites provides further evidence of the exclusion of computer-based attacks for marine policies and the primacy of data breach expenses, theft, third-party costs, and business interruption in available cyber-risk policies. This is not to say that there are no such policies in existence as some insurers have begun to offer cyber-gap insurance specialty lines (Cyber Marine Attacks). However, comprehensive coverage for the full spectrum of marine operators' risk to the cyber threat does not appear to be mature or prolific.[1]

There is recognition among stakeholders that the cyber-risk insurance market can provide financial incentives that will improve overall cyber security. The Department of Homeland Security (DHS) National Protection and Programs Directorate (NPPD) has engaged academia, infrastructure owners and operators, insurers, chief information security officers, and risk managers to find ways to expand the cyber security insurance market's ability to address this emerging risk area (U.S. Department of Homeland Security 2016). The NPPD sought input from these same stakeholders on the market's potential to encourage businesses to improve their cyber security in return for more coverage at more affordable rates (U.S. Department of Homeland Security 2016). The NPPD has also explored how a cyber-incident data repository could foster both the identification of emerging cyber security best practices across sectors and the development of new cyber security insurance policies that reward businesses for adopting and enforcing those best practices (U.S. Department of Homeland Security 2016). Insurers issuing cyber-risk policies also attempt to gauge a potential client's cyber security posture as a method of assessing risks including intensive and comprehensive applications that can cover details of remote network access, firewalls, security audits, and a host of other cyber security-related issues (Yates and Varholak 2013). Insurer Lloyd's of London has also taken steps to better develop the cyber-risk market announcing in January 2016 that it is partnering with catastrophe modeling and consulting firm AIR

---

1   Authors review of the following company websites accessed on March 11, 2016; Zurich https://www.zurich.com.au/content/dam/marine/cargo_insurance/marine_cargo_insurance/annual_marine_cargo_policy.pdf; Zurich Security and Privacy Protection Policy, 2014; Allianz https://www.allianz.com.au/openCurrentPolicyDocument/POL309BA/$File/POL309BA.pdf; Beazely https://www.beazley.com/specialty_lines/professional_liability/tmb/information_security_privacy_.html; https://www.beazley.com/specialty_lines/professional_liability/tmb/beazley_breach_response.html; https://www.beazley.com/specialty_lines/professional_liability/tmb/information_security_privacy_.html.

Worldwide and RMS Cambridge Centre of Risk Studies to develop common core data and risk-modeling to accurately calculate risk (Lloyd's 2016).

There are significant benefits from businesses and insurers working collaboratively to reduce and insure cyber risks. The measures that businesses may be incentivized to institute to improve their cyber security and consequently their financials are indifferent to who is attacking so that the same efforts that prevent what are generally considered insurable losses such as theft, accidents, and business interruptions will also enhance protection across the broader sector to improve national and economic security. While many challenges remain including developing standard definitions, sovereignty of state laws, and the global nature of threats, businesses, insurers, governments, and inter-governmental organizations are working to help forge a viable cyber-risk market. Considering that an estimated $2.5 billion in cyber-risk insurance premiums were collected in 2015 worldwide and business losses were estimated to be $400 billion, improved cyber security to reduce business vulnerabilities must be a component of a viable market (Cybersecurity Ventures 2015; Gandel 2015).

Mandatory Regulatory Compliance

There is no current law that provides for comprehensive cyber security requirements or activities for operators in the MTS, nor is there a comprehensive cyber security law that compels the private industry that operates critical infrastructure to take action to be able to defend and recover from cyber-threats. The most recently passed legislation to address cyber security was the Cybersecurity Act of 2015, but it focusses mainly on requirements for information sharing and reporting vice mandating actions to secure networks associated with critical infrastructure (White and Halpert 2016). Specific legislation that addresses port cyber security has passed the U.S. House of Representatives with language that requires area maritime security plans and facility security plans to include a mitigation plan to prevent, manage, and respond to cyber security risks, but it has not yet been passed by the Senate (H.R. 3878). Across different critical infrastructure sectors, regulations directly related to cyber security vary with most agencies providing only voluntary guidance (Hogan and Newton 2015). Other agencies such as the Nuclear Regulatory Commission (NRC) require operators to implement a cyber security program that provides high assurance that safety, security, and emergency preparedness functions are protected from cyber attacks, and the NRC conducts inspections to verify this condition (Wiggins, Erlanger, and Harris 2015). There is a legitimate argument that efforts to regulate cyber security will be ineffective because it may provide a false sense of security. The fear is that industries might view achieving the minimum security standards as the goal resulting in inadequate protection instead of designing a system that encourages a rise to the highest levels of security (Bucci, Rosenzweig, and Inserra 2013). This issue can be partly overcome through other incentives such as the financial incentives previously mentioned that should help to drive high levels of security through the competitive marketplace. Another difficulty of cyber security regulations is that they could be a hindrance to both business and security by anchoring security requirements in the technology at the time the law was written. However, regulations need not be prescriptive and must allow for flexibility with changes in technology and cyber threats. This has been demonstrated by the NRC that issues performance-based regulations that describe the outcome to be achieved but not necessarily how to accomplish that outcome (Wiggins, Erlanger, and Harris 2015), and the NRC has been deeply involved with this issue since 2006 when it included cyber security requirements as a subsection within the nuclear power plant physical security requirements.

Ultimately, existing incentives have been insufficient to drive cyber security postures to a level that is both adequately defendable and recoverable and thereby risking national security. As was aptly noted by James Lewis of the Center for International and Strategic Studies in 2005, "National defense and homeland security are public goods where individuals share in the benefits irrespective of how much they spend or if they spend at all. Markets are inefficient at supplying goods and services in situations where groups of people must work together to achieve a good outcome but the incentive for investment and cooperation is low. In these situations, the private sector will not produce an optimal outcome" (821). While there are financial and market incentives that can be illuminated and fostered to improve cyber security, ultimately, some form of federal regulation and oversight will likely be required as well.

The Maritime Transportation Security Act provides the authority to for the Coast Guard to regulate the MTS and with broad language that does not exclude cyber threats. The fundamental and overriding feature of the act provides the Coast Guard with extensive authority to require operators in the MTS to take action to deter a transportation security incident and the authority is indifferent to the nature of the threat vector. The Coast Guard should amend existing regulations to require compliance with its Guidelines for Cyber Risk Management making them compulsory instead of voluntary as will be the case when they are released. While issuing regulations understandably draws concern over the impact to business operations, the draft guidelines available to the author for review contain a mere six pages of actionable items which is not insignificant but should be manageable considering the threat to business operations due to inadequate cyber security. The guidelines focus on a cyber security regime of: identify, protect, detect, respond, and recover and offer an overall strategy that operators in the MTS can employ. As previously mentioned, the complexity and diversity of cyber systems in the MTS precludes a prescriptive approach to regulation. However, the framework of the risk management guidelines provides for what must be done in broad terms leaving the specifics of how to accomplish this task appropriately to individual operators. Currently, the Coast Guard enforces security of the MTS by outlining requirements and reviewing and approving plans submitted by operators. The Coast Guard further compels compliance by conducting on-site assessments of port facilities and ships to ensure that operators are adhering to the approved security plans. This same approach should incorporate cyber vulnerabilities as well as protective and resiliency measures and assure compliance using third-party verification.

Verifying Cyber Security Compliance

Both financial and regulatory measures to close the cyber security incentive gap will require verification and inspection of cyber systems in the MTS. In the private sector, verification of security will likely be driven by the financial incentives of reducing financial losses as recognized by individual businesses, and it will also be a component of insurance companies' desire to reduce risk exposure through audits and the application process. In the regulatory arena, there are several issues that confound verification but they are not insurmountable, and regulatory schemes should leverage verification by third parties for the foreseeable future.

The Coast Guard, as the regulatory agency for the MTS, does not currently possess nor has it undertaken the development of a cyber security regulatory workforce with the capacity to provide verification of cyber security standards across the thousands of vessels, port facilities, and other infrastructure that constitute the MTS. The sheer size of the MTS landscape and scope of verifying cyber security as a regulatory agency is evident by considering just the two ports of Los Angeles and Long Beach that combined host 53 terminal operators, support 163,000 port-related jobs, and serve as ports of call for approximately 4,000 ships annually (Port of Los Angeles; Port of Long Beach 2016). Service-wide, the Coast Guard has grown its workforce to 70 cyber-specialists, but those specialists' expertise is spread across the Coast Guard's three lines of effort on cyber including defending cyber space and enabling Coast Guard operations in addition to protecting critical infrastructure (Knox 2015). In its Cyber and Human Capital strategies, the Coast Guard emphasizes and articulates the need to grow cyber competency within the organization to help protect the nation's critical infrastructure. While the Coast Guard continues to grow its cyber workforce and make programmatic changes for the management and sustainment of cyber expertise in alignment with the cyber and human capital strategies, neither strategy leads to the significant changes to the workforce that would be required were the Coast Guard to take on inspection and verification of cyber security standards across the MTS. This is understandable considering the current budget and commitment of the existing workforce to the Coast Guard's 11 statutory missions that receive a constant demand signal. In addition to that, the Coast Guard has not yet moved to a mandatory compliance regime that would push the organization to make a dramatic workforce shift. Even if the Coast Guard moves from voluntary to mandatory compliance and outsources verification requirements to third parties, it will still need cyber competencies within the workforce capable of validating third party findings.

As it stands, there is already a "nationwide shortage of highly qualified cyber security experts and the federal government in particular has fallen behind in the race for this talent" (Booz Allen Hamilton 2015). This cyber talent gap does not take into account the additional personnel resources that would be needed to enforce cyber regulations so for the near term, third-party verification should be used to validate cyber security measures in the MTS. Beyond the lack of the federal government's ability to provide cyber regulators, another benefit of third-party verification is the increased likelihood of cyber security expertise, adaptability free of government bureaucracy, and familiarity with emerging technology trends. There are several options for third-party verification that could be explored with stakeholders for determining feasibility and assuring compliance. Existing cyber security firms that already perform security services for businesses within the MTS provide one avenue to accomplish verification as many companies offer cyber security audit services that could be tailored to meet regulatory requirements. The readily available resources of this method of verification make it an attractive option, yet there could be an issue regarding conflict of interest if businesses are relying on their existing cyber security firms for verification. There are also possible proprietary conflicts among security audit providers if competing security companies are used.

Another option for third-party verification may be to create an industry cyber-classification society similar to that of the classification societies with which the maritime industry is already familiar. Classification societies originated in the 18th century when marine insurers based at Lloyd's coffee house in London developed a system for the independent technical assessment of the ships presented to them for insurance cover as a means of assessing the risks of insuring individual voyages (International Associations of Classification Societies 2011). Since that time, classification societies have matured as intermediary and independent institutions that serve a public role in contributing to safety and prevention of marine pollution. The objective of classification societies is to verify the structural integrity of a ship's hull and other essential components through the application of rules and technical knowledge and by verifying compliance with international and statutory regulations on behalf of flag Administrations (International Associations of Classification Societies 2011). In addition to verification of these standards during construction, classification societies provide for vessels to maintain certification over their economic lifetime only if they satisfy requirements during annual surveys to assure continued compliance (Furger 1997).

Developing cyber-classification societies for the MTS presents both opportunities and challenges. The classification societies that developed out of the 18th century were organically generated by the need of marine insurers to assess the risks they were underwriting so imploring the creation of cyber security societies may be difficult. One way in which the federal government might aid in developing a cyber-classification society would be to seed the initial investment required to create the societies whose funding would ultimately be transferred to the business owners through fees as is done for ship classification societies. Another barrier to the creation of cyber security classification societies is the vast scope of cyber security challenges in the MTS and competition for talent from other enterprises when compared with the relatively narrow shipbuilding industry. It is projected that by 2020, the global information technology security workforce shortfall will be 1.5 million worldwide (Suby and Dickson 2015), and by way of comparison, the U.S. shipbuilding industry employed just 111,905 workers in 2015 (Spring 2015 Industry Study 2015). The challenges of the talent gap in cyber security specialists are pervasive and will be a challenge to any solution to provide for cyber security verification. However, this should not be used as an excuse for inaction considering the significant risks posed to our economic and national security by cyber threats.

Finally, classification societies have a history of working across international organizations, nations, regulators, insurers, and industry to act as an intermediary to promote safety, and this attribute makes the model a particularly attractive option for cyber security verification in the MTS. The security of our marine transportation system is highly contingent on the global nature of the shipping system where foreign flagged vessels are loaded at overseas port facilities with foreign-produced goods, call on U.S. ports, and transit though the maritime global commons. The United Nation's International Maritime Organization (IMO) of which the U.S. is a member, acts as the global standard-setting authority for the safety, security, and environmental performance of international shipping (Introduction to IMO 2016). The International Association of Classification Societies has observer status in the IMO and may develop unified interpretations that clarify the intent and application of international standards or establish rules that will allow industry to meet goal-based standards (International Associations of Classification Societies 2011). Marine insurers also use the determinations of classification societies as an essential standard for underwriting risk (Ott 2014), and this is currently a deficient component of the current cyber insurance market as previously noted. Creating cyber classification societies could be a key factor in promoting enhanced cyber security in the MTS by facilitating self-governance within industry, informing risks and adherence to standards for insurers and regulators, and providing a connection to the international maritime community to apply sound cyber security practices worldwide.

Conclusion

Cyber threats to the MTS represent a risk to national security, and the current defense and resiliency in this largely private domain with extensive implications for the national economy and safety is insufficient to the current challenge. While there are numerous initiatives that are being undertaken that include information-sharing on threats and vulnerabilities, expanding federal grant guidelines to make cyber-related projects eligible for grant funding, and Coast Guard-promulgated Cyber Risk Management Guidelines that provide a framework for a cyber security program, there appears to be an incentive gap that would drive industry to a status where it could defend and recover from the most probable of cyber-attacks. There are several reasons for the incentive gap including the voluntary nature of compliance, the cost of enhancing security in cyber systems, the lack of a perceived threat, and proprietary systems that industry understandably seeks to protect. The 9/11 Commission concluded in its report that a significant reason for our inability to thwart the attack was a failure of imagination (National Commission on Terrorist Attacks Upon the United States 2004). In the cyber domain, imaginations run the full spectrum, but preventing, mitigating, and recovering from cyber attacks on the maritime transportation system will require a significant change to the current structure. It is not likely a lack of imagination that is stifling action to secure the MTS but the enormity and complexity of the challenge across tens of thousands of businesses operating in the MTS, government regulators and numerous other stakeholders. When considering that we are already behind where we need to be, and the rapid advance of technology and interconnectedness that is quickly approaching including the "internet of things," a massive effort must be undertaken to drive the MTS to a defendable and resilient state.

References

Andress, Jason, and Steve Winterfeld. 2013. *Cyber Warfare: Techniques, Tactics and Tools for Security Practitioners*. Waltham, MA: Elsevier, 46.

Benzie, Helen M. 2003. "War and Terrorism Risk Insurance." *John's J. Legal Comment*. 18: 432. http://scholarship.law.stjohns.edu/cgi/viewcontent.cgi?article=1160&context=jcred.

Bishop, Stewart. 2014. "Sony Units Denied Coverage For Suits Tied To Cyberattack." *Law360*. (February 21). http://www.law360.com/articles/512263/sony-units-denied-coverage-for-suitstied-to-cyberattack.

Bonner, Lance. 2012. "Cyber Risk: How the 2011 Sony Data Breach and the Need for Cyber Risk Insurance Policies Should Direct the Federal Response to Rising Data Breaches." *Washington University Journal of Law & Policy* 40: 270.

Booz Allen Hamilton. 2015. "Cyber In-Security II: Closing the Federal Talent Gap." Partnership for Public Service. (April): 1. http://ourpublicservice.org/publications/download.php?id=504.

Bucci, Steven P., Paul Rosenzweig, and David Inserra. 2013. "A Congressional Guide: Seven Steps to US Security, Prosperity, and Freedom in Cyberspace." *Heritage Foundation Backgrounder* 2785: abstract.

Chambers, Matthew, and Mindy Liu. 2015. "Maritime Trade and Transportation by the Numbers." Bureau of Transportation Statistics. http://www.rita.dot.gov/bts/sites/rita.dot.gov.bts/files/publications/by_the_numbers/maritime_trade_and_transportation/index.html (accessed October 21).

Clinton, William. 1998. "Presidential Decision Directive 63." Washington, DC: The White House. http://fas.org/irp/offdocs/pdd/pdd-63.htm.

Committee on the Marine Transportation System. 2014. "Components and Functions." Marine Transportation System Fact Sheet. (October 9). http://www.cmts.gov/downloads/CMTS_MTS_Fact_Sheet_9.15.14_FINAL.pdf.

Cyberkeel. 2014a. "Marine Cyberwatch." April. www.cyberkeel.com.

Cyberkeel. 2014b. "Virtual Pirates at Large on the Cyber Seas." Maritime Cyber-Risks. (October 15): 3. http://cyberkeel.com/images/pdf-files/Whitepaper.pdf.

"Cyber Marine Attacks: The Next Shot Across the Bow?" 2016. *Kennedys: Legal Advice in Black and White*. (February 2). http://www.kennedyslaw.com/article/cyber-marine-attacks-the-next-shot-across-the-bow/.

Cybersecurity Ventures. 2015. "The Cybersecurity Market Report." (October). http://cybersecurityventures.com/cybersecurity-market-report-q3-2015/.

Douse, Christopher M. 2010. "Combating Risk on the High Sea: An Analysis of the Effects of Modern Piratical Acts on the Marine Insurance Industry." *Tulane Maritime Law Journal* 267. https://litigation-essentials.lexisnexis.com/webcd/p?action=DocumentDisplay&crawlid=1&doctype=cite&docid=35+Tul.+Mar.+L.+J.+267&srctype=smi&srcid=3B15&key=19c595ff3bdc8ce848bb7050aa44a4a7.

Energy API. 2016. "Advancements in Technology Reduce Environmental Impact." http://www.api.org/policy-and-issues/policy-items/exploration/environmental_impact (accessed March 25).

Firoozshahi, Amir, Hossein Allahyari, and Farzad Haghdosti. 2011. "Intelligent and Innovative Valve Control DCS-Based in Large Tank Farm Oil Terminal." Paper presented at the 3rd International Conference on Advanced Computer Control (ICACC). Harbin. (January): 209–213.

Furger, Franco. 1997. "Accountability and Systems of Self-Governance: The Case of the Maritime Industry." *Law & Policy* 19 (4): 445–476.

Gandel, Stephen. 2015. "Lloyd's CEO: Cyber Attacks Cost Companies $400 Billion Every Year." *Fortune*, Finance. (January 23). http://fortune.com/2015/01/23/cyber-attack-insurance-lloyds/.

"Ghost Ships." 2014. *The Economist*. (March 8). http://www.economist.com/news/technology-quarterly/21598318-autonomous-cargo-vessels-could-set-sail-without-crew-under-watchful-eye (accessed November 5, 2015).

Goldsmith, Raymond W. 1946. "The Power of Victory: Munitions Output in World War II." *Military Affairs* 10 (1): 69–80.

Henriksson, Björn. 2016. "Automated Container Terminals Are Taking off." Generations. http://new.abb.com/marine/generations/technology/automated-container-terminals-are-taking-off (accessed April 1, 2016).

Hogan, Michael, and Elaine Newton. 2015. "Supplemental Information for the Interagency Report on Strategic U.S. Government Engagement in International Standardization to Achieve U.S. Objectives for Cybersecurity." *NISTIR 8074* Volume 2 (draft). (August): 1–81. http://citeseerx.ist.psu.edu/viewdoc/download?doi=10.1.1.699.2193&rep=rep1&type=pdf.

H.R. Rep. 114-379. "H.R.3878 - Strengthening Cybersecurity Information Sharing and Coordination in Our Ports Act of 2015." 114th Congress. https://www.congress.gov/bill/114th-congress/house-bill/3878 (accessed June 1, 2016).

IHS. 2016. "New Automated Rotterdam Container Terminal Shows How Far US Lags." http://www.joc.com/port-news/terminal-operators/apm-terminals/new-automated-rotterdam-container-terminal-shows-just-how-far-us-lags_20150502.html (accessed April 1).

"IMO Profile." 2016. Business.un.org. https://business.un.org/en/entities/13 (accessed October 21, 2015).

International Associations of Classification Societies. 2011. "Classification Societies: What, Why and How?" (March): 5. http://www.iacs.org.uk/document/public/explained/Class_WhatWhy&How.PDF.

"Introduction to IMO." 2016. International Maritime Organization. http://www.imo.org/en/About/Pages/Default.aspx (accessed April 6, 2016).

"IP Commission Report." 2013. The Commission on the Threat of American Intellectual Property by The National Bureau of Asian Research. (May). http://www.ipcommission.org/report/IP_Commission_Report_052213.pdf, 4.

Jensen, Lars. 2015a. "Challenges in Maritime Cyber-Resilience." *Technology Innovation Management Review* 5 (4): 35.

Jensen, Lars. 2015b. "Maritime Cyber Risks and Resilience—What Is Real, What Is Fiction?" Proceedings of TOC Europe, Ahoy, Rotterdam, Netherlands. (June 11): 7. http://www.tocevents-europe.com/images/speaker-presentations/Lars-Jensen.pdf.

Jerardo, Alberto. 2016. "Import Share of Consumption." United States Department of Agriculture. http://www.ers.usda.gov/topics/international-markets-trade/us-agricultural-trade/import-share-of-consumption.aspx (accessed March 7, 2016).

Kirkpatrick, Keith. 2015. "Cyber Policies on the Rise." *Communications of the ACM* 58 (10): 21–23. http://cacm.acm.org/magazines/2015/10/192376-cyber-policies-on-the-rise/fulltext.

Knox, Jodie. 2015. "Coast Guard Commandant on Cyber in the Maritime Domain." *Coast Guard Maritime Commons* (blog). June 15. http://mariners.coastguard.dodlive.mil/2015/06/15/6152015-coast-guard-commandant-on-cyber-in-the-maritime-domain/.

Kramek, Joseph. 2013. "The Critical Infrastructure Gap: US Port Facilities and Cyber Vulnerabilities." *Policy Paper*. The Brookings Institution (July 13): 1–50. https://www.brookings.edu/research/the-critical-infrastructure-gap-u-s-port-facilities-and-cyber-vulnerabilities/.

Lang, Molly E., and John F. Mullen. 2013. "Is TRIA for Cyber Terrorism?" *Insurance Journal News*. (October 21). http://www.insurancejournal.com/magazines/features/2013/10/21/308184.htm.

Lee, Robert M., Michael J. Assante, and Tim Conway. 2014. "German Steel Mill Cyber Attack." Industrial Control Systems. (December 30). https://ics.sans.org/media/ICS-CPPE-case-Study-2-German-Steelworks_Facility.pdf.

Lewis, James Andrew. 2005. "Aux Armes, Citoyens: Cyber Security and Regulation in the United States." *Telecommunications Policy* 29 (11): 821–830.

Lin, Herbert. 2012. "Thoughts on Threat Assessment in Cyberspace." *I/S: A Journal of Law and Policy for the Information Society* 8 (2): 337–355.

Liu, Edward C., John Rollins, and Catherine A. Theohary. 2014. "The 2013 Cybersecurity Executive Order: Overview and Considerations for Congress." Congressional Research Service. R42984. 1–24.

Lloyd's. 2016. "Lloyd's Leads Development of Core Data Requirements for Cyber Insurance." *News release*. (January 19). Lloyd's News and Insights. https://www.lloyds.com/news-and-insight/press-centre/press-releases/2016/01/lloyds-leads-development-of-core-data-requirements-for-cyber-insurance.

MacKenzie, Heather. 2012. "Shamoon Malware and SCADA Security—What Are the Impacts?" Tofino Security. (October 25). https://www.tofinosecurity.com/blog/shamoon-malware-and-scada-security---what-are-impacts.

Maritime Transportation Security Act of 2002. 116 STAT 2069 § 701-70102. Public Law 107–295 (107th Congress).

Marroquin, Art. 2009. "Eying Port of L.A. Security." *Daily Breeze News.* (December 5). http://www.dailybreeze.com/general-news/20091205/eying-port-of-la-security.

Marsh. 2014. "The Risk of Cyber Attack to the Maritime Sector." (July 24). https://uk.marsh.com/NewsInsights/Articles/ID/41229/The-Risk-of-Cyber-Attack-to-the-Maritime-Sector.aspx.

Martin Associates. 2015. "The 2014 National Economic Impact of the U.S. Coastal Port System." American Association of Port Authorities. (March). http://aapa.files.cms-plus.com/SeminarPresentations/2015Seminars/2015Spring/US Coastal Ports Impact Report 2014 methodology - Martin Associates 4-21-2015.pdf.

Mazanec, Brian M. 2009. "The Art of (Cyber) War." *Journal of International Security Affairs* 16 (Spring): 84.

McGlone, Tim. 2014. "Cybersecurity Options Lag Behind Hackers." Government Technology. (May 13). http://www.govtech.com/security/Cybersecurity-Options-Lag-Behind-Hackers-Abilities.html.

Morgan, Steve. 2015. "Cyber Insurance Market Growing from $2.5 Billion in 2015 to $7.5 Billion by 2020." *Forbes.* (December 24). http://www.forbes.com/sites/stevemorgan/2015/12/24/cyber-insurance-market-storm-forecast-2-5-billion-in-2015-projected-to-reach-7-5-billion-by-2020/#348899253ffe.

MOXA. 2008. "Oil Refinery Remote Monitoring and Control System under Hazardous Environments." Moxa Application. (April 11). http://www.moxa.com/applications/success_stories_Oil_Refinery_Remote_Monitoring_and_Control.htm.

National Commission on Terrorist Attacks upon the United States, Thomas H. Kean, and Lee Hamilton. 2004. The 9/11 Commission report: final report of the National Commission on Terrorist Attacks upon the United States. (Washington, D.C.): National Commission on Terrorist Attacks upon the United States.

"New Cyber Frontiers (Antwerp Port Case Study)." 2015. Magal-S3: Security, Safety, Site Management. http://www.magal-s3.com/content

Managment/uploadedFiles/White_Papers/Cyber_For_ICS_Antwerp_Case_web.pdf (accessed November 16).

Obama, Barack. 2013. "Critical Infrastructure Security and Resilience." Presidential Policy Directive PDD-21. (February 12). Washington, DC.

Ott, Christian. 2014. "The Importance of Class in Marine Insurance, Claims, and Legal Liabilities." Skuld. (August). http://www.gia.org.sg/pdfs/Industry/Marine/MKSS/SS28_Presentation_ChristianOtt.pdf (accessed April 20, 2016).

Paganini, Pierluigi. 2015. "Hacking Ships: Maritime Shipping Industry at Risk." *Security Affairs*. (March 31). http://securityaffairs.co/wordpress/35504/hacking/hacking-maritime-shipping-industry.html (accessed November 5, 2015).

Port of Long Beach. 2016. "Facts at a Glance." http://www.polb.com/about/facts.asp. Compilation of data (accessed March 22).

Port of Los Angeles. 2015. "Port of Los Angeles to Start Construction on Yusen Terminal Improvements." *News release*. (June 29). https://www.portoflosangeles.org/newsroom/2015_releases/news_062915_Yusen_Terminal.asp.

Quigley, Kevin, Calvin Burns, and Kristen Stallard. 2015. "'Cyber Gurus': A Rhetorical Analysis of the Language of Cybersecurity Specialists and the Implications for Security Policy and Critical Infrastructure Protection." *Government Information Quarterly* 32 (2): 108–117.

Riley, Michael, and Jordan Robertson. 2014. "UglyGorilla Hack of U.S. Utility Exposes Cyberwar Threat." *Bloomberg.com*. (June 13). http://www.bloomberg.com/news/articles/2014-06-13/uglygorilla-hack-of-u-s-utility-exposes-cyberwar-threat.

Rose, Adam. 2016. "Economic Consequence Analysis of Maritime Cyber Threats." National Center for Risk and Economic Analysis of Terrorism Events. (February 24). University of Southern California.

Schneier, Bruce. 2010. "The Story Behind the Stuxnet Virus." *Forbes*. (July 10). http://www.forbes.com/2010/10/06/iran-nuclear-computer-technology-security-stuxnet-worm.html.

Sistrunk, Chris. 2015. "Has Your ICS Been Breached? Are You Sure? How Do

You Know?" *Power*. (June 1). http://www.powermag.com/has-your-ics-been-breached-are-you-sure-how-do-you-know/?pagenum=1.

Sood, Aditya K., and Richard Enbody. 2014. "U.S. Military Defense Systems: The Anatomy of Chinese Espionage." *Georgetown Journal of International Affairs*. (December 19). http://journal.georgetown.edu/u-s-military-defense-systems-the-anatomy-of-cyber-espionage-by-chinese-hackers/.

"Spring 2015 Industry Study Final Report - Shipbuilding." 2015. The Dwight D. Eisenhower School for National Security and Resource Strategy. Washington DC: Fort McNair. http://es.ndu.edu/Portals/75/Documents/industry-study/reports/2015/es-is-report-shipbuilding-2015.pdf.

Stouffer, Keith, Joe Falco, and Karen Scarfone. 2011. "Guide to Industrial Control Systems (ICS) Security." *NIST Special Publication* 800 (82): 1.

Suby, Michael, and Frank Dickson. 2015. "The 2015 (ISC) 2 Global Information Security Workforce Study." *Frost & Sullivan in Partnership with Booz Allen Hamilton for ISC2*: Executive Summary. https://www.isc2cares.org/uploadedFiles/wwwisc2caresorg/Content/GISWS/FrostSullivan-(ISC)%C2%B2-Global-Information-Security-Workforce-Study-2015.pdf.

"Terrorism Risk and Insurance." 2015. Insurance Information Institute. June. http://www.iii.org/issue-update/terrorism-risk-and-insurance.

Terrorism Risk Insurance Act of 2002. § Section 1, Title 1, Sec. 102 (107th Congress).

U.S. Department of Homeland Security. 2010. *Transportation Systems Sector-Specific Plan an Annex to the National Infrastructure Protection Plan*. Washington, DC.

U.S. Department of Homeland Security. 2014. "ICS-CERT Year in Review: Industrial Control Systems Cyber Emergency Response Team." https://ics-cert.us-cert.gov/sites/default/files/Annual_Reports/Year_in_Review_FY2014_Final.pdf.

U.S. Department of Homeland Security. 2015a. "Critical Infrastructure Security." https://www.dhs.gov/topic/critical-infrastructure-security (accessed October 26, 2015).

U.S. Department of Homeland Security. 2015b. "Industrial Control Systems

Cybersecurity Emergency Response Team." *ICS-CERT Monitor.* (May/June). https://ics-cert.us-cert.gov/sites/default/files/Monitors/ICS-CERT_Monitor_May-Jun2015.pdf.

U.S. Department of Homeland Security. 2015c. "Monitor (ICS-MM201512)." (November/December). https://ics-cert.us-cert.gov/monitors/ICS-MM201512.

U.S. Department of Homeland Security. 2016. "Cybersecurity Insurance." https://www.dhs.gov/cybersecurity-insurance (accessed March 14, 2016).

U.S. Energy Information Administration. 2014. *Short-term Energy Outlook (STEO).* (September). http://www.eia.gov/forecasts/steo/archives/sep14.pdf.

U.S. Energy Information Administration. 2015. "Independent Statistics and Analysis." International. http://www.eia.gov/beta/international/ (accessed November 2).

"U.S. Public Port Facts." 2015. Port Industry Information. American Association of Port Authorities. http://www.aapa-ports.org/Industry/content.cfm?ItemNumber=1032 (accessed October 21).

UT News. "UT Austin Researchers Successfully Spoof an $80 Million Yacht at Sea." (July 29). https://news.utexas.edu/2013/07/29/ut-austin-researchers-successfully-spoof-an-80-million-yacht-at-sea (accessed December 4, 2016).

"Vessel Traffic Services." 2014. Navigation Center. July 300. http://www.navcen.uscg.gov/?pageName=vtsMain (accessed December 15, 2015).

Wang, Jian-Wei, and Li-Li Rong. 2009. "Cascade-Based Attack Vulnerability on the US Power Grid." *Safety Science* 47 (10): 1332–1336.

Weiss, N. Eric, and Rena S. Miller. 2015. "The Target and Other Financial Data Breaches: Frequently Asked Questions." *Congressional Research Service.* R43496. https://fas.org/sgp/crs/misc/R43496.pdf.

Weitz, Peter. 2015. "US Oil Imports by Country." In Black Weitz. January 12. http://inblackandweitz.com/us-oil-imports-by-country/.

White, Sydney M., and Jim Halpert. 2016. "Cybersecurity: 2015." DLA Piper. (February 11). https://www.dlapiper.com/en/us/insights/publications/2016/02/cybersecurity-2015s-top-legal-developments/.

"Why Aren't America's Shipping Ports Automated?" 2015. Priceonomics. (September 30). http://priceonomics.com/why-arent-americas-shipping-ports-automated/.

Wiggins, James, C. Erlanger, and T. Harris. 2015. "Regulatory Efforts to Improve Cyber Security." *US Nuclear Regulatory Commission.* http://pbadupws.nrc.gov/docs/ML1315/ML13156A251.pdf (accessed March 16).

Wile, Rob. 2013. "Here Are the Countries Drinking Up 'Saudi America's' Sweet Oil [MAP]." *Business Insider.* (October 9). http://www.businessinsider.com/us-oil-export-destinations-2013-10.

Wilshusen, Gregory C. 2015. *Maritime Critical Infrastructure Protection: DHS Needs to Enhance Efforts to Address Port Cybersecurity.* No. GAO-16-116T. http://www.gao.gov/assets/680/672973.pdf.

World Bank Group, ed. 2014. *World Development Indicators 2014.* World Bank Publications. http://data.worldbank.org/data-catalog/world-development-indicators.

Wortzel, Larry M. 2010. "China's Approach to Cyber Operations: Implications for the United States." *Testimony before House Committee on Foreign Affairs, Hearing on "The Google Predicament: Transforming US Cyberspace Policy to Advance Democracy, Security, and Trade."* (March 10).

Yates, Brooke, and Katie Varholak. 2013. "Cyber Risk Insurance - Navigating The Application Process." Sherman and Howard LLC. (June 6). http://shermanhoward.com/publications/cyberriskinsurance-navigatingtheapplicationprocess/.

CHAPTER 17:
INFORMATION SHARING FOR MARITIME CYBER RISK MANAGEMENT

*Dennis Egan, Darby Hering, Paul Kantor,
Christie Nelson, and Fred S. Roberts*
Rutgers University

1. Executive Summary

Effective and timely sharing of cyber risk management information among all stakeholders in the Maritime Transportation System (MTS) is vital to maintaining a safe, secure, and resilient MTS. To develop information sharing protocols across this complex system, we must consider the layers of cyber risk management, including communication and technology, economic, and legal and regulatory aspects. Our research addresses the following questions: *What is the most appropriate role for the U.S. Coast Guard (USCG), and how does guidance for physical security relate to cyber risk management needs? What organizational systems could best support the needed sharing? What kinds of incentives could be used to encourage participation, particularly from private industry? What information needs to be shared, and when? What technologies could be used to enable and safeguard the information sharing?* In this white paper, we discuss the approach taken by the CCICADA-Rutgers team to address these topical questions. Our research process included interviews with experts, literature reviews, and taking a leadership role on the Port of New York and New Jersey Area Maritime Security Committee cyber subcommittee. We present our initial findings based on the interviews conducted and documents read, and we conclude with a set of recommendations related to each topical question.

2. The Background

At the March 2015 Maritime Cyber Security Symposium held at CCICADA/Rutgers University, one of the important themes was that the ability to share information in an effective and timely manner with all stakeholders in the MTS is essential in keeping the MTS safe, secure, and resilient. At the Symposium, VADM Charles Michel of the USCG laid out six research challenges. This paper deals with one of those challenges:

Information Sharing—How would a framework for network analysis be developed to support optimal information sharing with partners to address maritime cyber issues?

In June 2015, a Maritime Cyber Security Research Summit was organized at California Maritime Academy to investigate these six research challenges. Working groups were formed to address each of the challenges and this led to a report (Clark and Roberts 2015).

After the report, three more focused research questions were posed by USCG-FAC (Office of Port and Facility Compliance). This report deals specifically with the following one of those questions:

Information Sharing—Develop Information Sharing Protocols to meet the needs of industry and government.

To address this challenge, the CCICADA team set out to investigate methods to achieve rapid and useful information sharing in a way that both large and small players in the MTS can participate. In particular, how can we entice larger content providers to take the lead on information sharing within the MTS on cyber issues? We sought to explore ways to incentivize environments that are both transparent and candid in the sharing of information.

As part of the research, we also sought to investigate ways to categorize what information about the latest cyber threats and countermeasures should be shared and with whom. To answer this question we looked to understand the types of information that need to be shared rapidly as well as the types of information that do not impose an immediate threat. One example we set out to investigate is how and when to share reports on "near misses."

CCICADA also set out to understand what organizational structures for information sharing between government and industry in the MTS and between private sector MTS entities make the most sense to better understand:

- What information sharing leverage can be gained from existing organizations such as the Maritime Information Sharing and Analysis Center (M-ISAC) and Area Maritime Security Committees (AMSCs) or the National Cybersecurity and Communications Integration Center (NCCIC) or the International Maritime Organization (IMO) or NATO's Center for Combined Operations from the Sea (CJOS)?

- How is information sharing performed in other sectors such as those facilitating financial services, utilities, and oil and natural gas?

- Can we find good systems for use of real-time machine to machine interfaces such as the Security Information and Event Management (SIEM) software that can automatically collect, filter, correlate, vet, and distribute threat analysis and trends?

Finally, CCICADA sought to analyze the roles of the USCG in cyber risk management information sharing, roles such as developing standards for sharing systems, exchanging best practices, or enforcing sharing regulations. Can we learn a great deal from USCG reporting procedures for physical security risks, and translate those into good reporting procedures for cyber security risks?

This was an ambitious agenda for a project of a few months, and this paper reports on our preliminary findings and recommendations. There is a great deal that still needs to be done.

This report is organized into five topical areas:

The role of the USCG and extending physical security to cyber security—cyber risk management:

- Organizational systems for information sharing
- Motivation and barriers for sharing information
- What information to share, and what to share rapidly versus slowly
- Technologies to support information sharing

A Comment on Terminology: In this paper, we use the terms "cyber security" and "cyber risk management" somewhat interchangeably. We tend to favor the latter terminology since we feel that management of cyber risk is a key to maximizing cyber security.

3. Context

In the maritime cyber security arena one may identify five kinds of adversarial threats or risks. One is TCOs (Transnational Criminal Organizations) which might disrupt cyber systems with goals such as hijacking, concealing contraband transport, or, potentially, hostage-taking. A second class of threats would originate with Violent Non-State Actors (VNSA) such as Al Qaida or ISIS/ISIL. While these might exploit some of the same technologies, they may have goals quite different from the essentially economic goals of TCOs (Sanctions Wiki 2014). Maritime cyber-systems are subject to attack by nation-states, either as part of a declared war, or part of an undeclared military contact, such as the encounters in the South China Sea. So-called "hacktivists," cyber specialists acting in extreme ways in support of a cause, may create havoc and cause damage to call attention to a social or political issue. Finally, there may be cyber attacks for purposes of corporate espionage. For each of these, the response requirements, both in terms of velocity, and of appropriate responding agents, may be quite different. And this, in turn will affect the architecture and technology, as well as the legal structure for information sharing. It should be emphasized that careless cyber behavior or misuse of cyber systems is a major cause of cyber system failures with potential consequences as serious as those of deliberate attacks, and information sharing about the consequences of such behavior or misuse is also covered by our findings and recommendations.

3.1. Maritime Cyber Risk Management a Novel Challenge

The problem of information sharing for maritime cyber risk management has little in common with many marine security issues. Because of this, there are not strong analogies. One key issue of maritime safety and security is hull breach. The defense is waterproof bulkheads. But the analogous approach—shutting off cyber communications, removes their value completely. Whether the cyber system is GPS, or other computer controlled systems, their key contribution to maritime activity is their ability to bridge long distances and maintain nearly instant situational awareness. Therefore, nothing analogous to a "complete lockdown" seems feasible. As to the cause of hull breach, other than rare failures due to extreme weather, and those due to poor maintenance, the key cause is obstructions, which are more or less fixed in space. In contrast, cyber space does not offer "chartable hazards," as bad actors can rapidly change their IP addresses or obscure them completely. The closest analogy to physical world hazards seems to be the notion of a "campaign," in which similarities in the specific technologies and messages serve to "locate" an attack in some abstract "ocean of possible attacks." Assembling information in terms of those characteristics seems to come closest to the historical approaches to maritime safety and security. Whether that abstract ocean of threats can be usefully presented remains to be explored.

3.2. Layers of Interaction

The problem of organizing for maritime cyber risk management seems to have three distinct aspects that must be considered. These arise because the players are of very different types: governments and their agencies, commercial shipping and cruise firms, the onboard captains, and crew and the ports and associated personnel. Their coordinated efforts to advance Maritime Cyber Risk Management appear to involve at least three different "layers" of concern: communications and technology arrangements, economic considerations, and legal and regulatory matters.

CHAPTER 17

Communication and Technology Arrangements

In the **communication and technology layer,** we find the problems of collecting information (about attacks and signatures) and of distributing warnings and remedies. The key considerations for this layer are of two kinds: technical capability, and architectural design. Technical capability limits the roles that individual parties may play. Architectural design asks questions such as: what channels should be used to communicate? Is the organization peer-to-peer or centralized? How does the architecture deal with varying levels of security and classification of information? What are the trust mechanisms?

The entire MTS comprises many players with outright conflicting interests, ranging from simple commercial competition to declared hostilities. How will access to shared information be limited (if at all) in consideration of these conflicts? Centralized control requires a trusted center. This can be accomplished for a single nation, but is much harder for a plurality of nations. Centralized control also puts "all the eggs in one basket" so that an attack on that control center can have widespread impact, worse than would be realized in a distributed or peer-to-peer system. There is some recent research on building decentralized systems that can enforce trust without putting all the eggs in one basket (Minsky 1991; Minsky and Leichter 1995; Minsky and Ungureanu 2000).

Economic Considerations

The **economic layer** represents not only the fact that multiple players are in competition with each other, but also the sheer costs of being a participant. Many maritime activities work on a narrow economic margin, and the costs of being an effective participant in a sharing system may be out of reach. As soon as some players are excluded, however, the entire system loses much of its value, and the outcasts are ripe for an attack that could affect many players across the maritime system. From an economic perspective every organization must watch its "bottom line." As the SONY attacks[1] showed, the entertainment industry, which had felt that cyber-security concerns were limited to IP issues, can be harmed in other ways. It has been reported that since that attack, that industry as a whole has become more interested in information sharing.

In addition to the false confidence that one will not be a target, if a firm reports that it has been hacked, it may lose the confidence of the public and suffer overall harm much greater than was caused by the specific attack. Each corporation or business is asked to weigh the potential downstream benefit to all of its competitors against its immediate loss by revealing the attack. This

---

1   Attributed to North Korea. See http://www.nytimes.com/2015/01/19/world/asia/nsa-tapped-into-north-korean-networks-before-sony-attack-officials-say.html?_r=0.

layer brings us face to face with all the complexities of maintaining competitive advantage when the threat is ubiquitous and invisible.

Legal and Regulatory Matters

A third layer is the **legal or regulatory layer**. In the United States (and many other countries), cooperation among firms, which might have the effect of reducing competition, and therefore raising consumer costs, is tightly regulated. Since cyber risk management is a cost, and cooperation or sharing will lower those costs, such sharing is in danger of falling under the regulations. While there are proposed (limited) legislative remedies (Senate Committee on Intelligence 2015), the problem is a significant one. Conceivably there may one day be an extension of the seafarer's obligation to assist persons, to an obligation to assist systems (Davies 2003).

4. The Research Process

We drew information from several kinds of sources as we compiled findings and developed recommendations for this white paper. The process is described briefly below, and was aimed to organize and synthesize the information into specific recommendations for consideration.

4.1. Interviews

Our best sources of information were numerous interviews with experts. We reached out to all of the participants in Working Group Team #6 of the Maritime Cyber Research Summit held at the CSU Cal Maritime Safety and Security Center, June 16–17, 2015. A summary of Working Group 6's findings and recommendations can be found in Clark and Roberts (2015). That working group focused on Information Sharing, and many of its findings and recommendations led to the topic of the present white paper. We were able to interview a majority of the Working Group 6 participants, who in turn gave us additional contacts to interview. Besides that key set of sources, we interviewed other senior USCG officers specifically charged with developing cyber risk management policies and guidelines, as well as some people in the private sector with specific expertise in areas such as maritime law and Information Sharing and Analysis Centers (ISACs), and also representatives of other government agencies such as the FBI, port security, NYPD, and other law enforcement agencies.

In all we conducted approximately thirty interviews. Most interviews were conducted by a pair of project team members who used an interview guide, took notes and later combined their notes into a single interview summary. Since we did not ask permission of the interviewees to attribute specific quotes or ideas to them, in the following, we refer to (Interviews 2015-6) when we present a finding based on one or more interviews.

4.2. Literature Review

We also reviewed selected documents related to cyber risk management information sharing. Some of these are listed in the Reference section of this white paper. These include relevant legislation and regulations, government reports, security guidelines, best practices and standards, and academic research on technologies, incentives and risk related to information sharing. The documents cited in the Reference section are a tiny fraction of the literature available on this topic.

4.3. Port of New York and New Jersey AMSC Cyber Subcommittee

Another source of information for the project was the knowledge and experience gained from our leadership role in the Cyber Security Subcommittee of the Area Maritime Security Committee (AMSC) for the Port of New York and New Jersey. This subcommittee, formed and officially chartered in 2015, is chaired by the USCG with Rutgers University/CCICADA as a co-chair along with Stevens Institute/Maritime Security Center and the NYPD. Meeting and working with this subcommittee brought us into contact with numerous USCG personnel, commercial partners, and representatives of law enforcement concerned with maritime cyber risk management in the region surrounding the Port of New York and New Jersey. Through meetings and conversations we were able to begin to understand issues related to cyber risk management information sharing by commercial companies (some of whom are competitors with one another), planning for cyber risk management exercises, and cyber risk management training needs. Since a primary activity of each AMSC is to create a security plan (the AMSP), a natural part of the maturation process for AMSCs is to create a subcommittee to address cyber risks. At the time of this writing, about one-third of the AMSCs have chartered cyber security subcommittees (Interviews 2015-6).

4.4. Process for Organizing and Synthesizing Information

After gathering information by conducting interviews and reading relevant documents, the project team systematically worked to organize and synthesize the information. Project team members were asked to summarize major takeaways from the interviews and literature in bullet points, and to categorize these bullet points by placing them under one (or more) of the five substantive topics under information sharing that we used to organize our project. As mentioned in Section 2, the five topics are:

- The role of the USCG and extending physical security to cyber security—cyber risk management
- Organizational systems for information sharing
- Motivation and barriers for sharing information
- What information to share, and what to share rapidly versus slowly
- Technologies to support information sharing

These clusters of bulleted items formed the basis of the findings in Section 5 of this report. Documents and/or interviews are cited in support of the findings. The findings in turn lead to the recommendations in Section 6.

5. Findings

In this section, we present selected findings that provide context for the recommendations given in Section 6. Throughout this section, we link the discussion points to the recommendation(s) they produce using the notation [R 6.x.y] to indicate relevance to recommendation 6.x.y, for example.

5.1. The Role of the USCG and Extending Physical Security to Cyber Security—Cyber Risk Management

The USCG has an extensive set of guidelines and regulations for physical security. Developing cyber security—cyber risk management guidelines for the MTS seems to be a natural extension of that role for the USCG. It was suggested in interviews that the USCG could develop cyber risk management guidelines for facilities similar to 33CFR105 and continue, similarly, to develop guidelines for vessels (Maritime Security 2010; Interviews 2015–6). Since there are many diverse players in the MTS, and they have competing interests, these guidelines should be written at a "high" level—specifying the characteristics of a cyber risk management plan, not detailed prescriptive requirements (Interviews 2015–6).

5.1.1. Resources for Planning Cyber Risk Management

There are numerous resources for planning cyber risk management, but most were not developed specifically for the maritime sector. Examples include the NIST framework (NIST 2014), the NIST 800 series (NIST 1990–2015), the

ISO/IEC series, the Center for Internet Security Controls for Effective Cyber Defense Version 6.0, and the BIMCO recommendations (BIMCO 2016). The ISO/IEC 27,000 series provides international best practice recommendations on security management (ISO/IEC 2013). The Center for Internet Security (CIS) Controls for Effective Cyber Defense Version 6.0 provides ways to defend against the most common and dangerous cyber attacks (CIS 2015). The NIST 800 Series provides security guidelines, policies, and procedures for federal government IT systems and organizations (NIST 1990–2015). The BIMCO recommendations are specific to a segment of the maritime sector, and carefully address cyber security for onboard systems (BIMCO 2016). [R 6.1.1, 6.1.2, 6.1.3]

The NIST guidelines are perhaps the most widely known, and provide an example framework of a process for developing cyber risk management plans (NIST 2014), called the Cybersecurity Framework (CSF). The NIST Cybersecurity Framework was developed to support protection of critical infrastructure resources. It includes a list of steps to take (and repeat) to develop and refine a cyber security plan. Additionally, the NIST Framework Core contains a list of "Functions, Categories, Subcategories and Informative References" that describe common cybersecurity activities. As described by NIST (NIST 2016), "The goal of the framework is to minimize risks to the nation's critical infrastructure, such as the transportation, banking, water and energy sectors. The executive order directed NIST to work with stakeholders across the country to develop the voluntary framework based on existing cybersecurity standards, guidelines and best practices." In creating this framework, NIST was "extremely collaborative with the public sector" (NIST 2016, 2). However, even this framework is not a perfect document. CSF is referenced in several documents as a living document, and when requesting feedback on the framework through a response analysis, respondents felt that it needed frequent updating (suggested yearly), and that it should be done by either NIST or a neutral third party. It is important to note that CFR is "consistent with voluntary international standards" (NIST 2015), which is important in the maritime international setting. If the USCG decides to issue guidelines for cyber security plan development, this could inform part of the guide.

The DHS Cyber Resilience Review (CRR) process uses the NIST guidelines. The CRR predates the NIST CFR, and although not a perfect matchup, closely aligns with the NIST framework. Included in the CRR self-assessment package is a document that maps the CRR to the CFR (DHS CRR).

ICS-CERT (Industrial Control Systems Cyber Emergency Response Team), operated through DHS, offers self-assessment tools as well. Though there are many of these tools available, the ICS-CERT Cybersecurity Evaluation Tool (CSET), is a free well supported option. DHS offers approximately 60 YouTube videos showing how to utilize this tool (CSET 2015). CSET has several approaches for self-assessment: a questions-based approach (recommended for most assessments), a standards-based approach (for regulated industries, presenting requirements as they are written in the standards), and a cyber security framework-based approach (a risk-based cyber security evaluation using a customized question set). This self-evaluation allows users to customize their assessment based on need: regulated industries have requirements available to assess built into the tool and there is also an option to select a desired security level (low, moderate, high, very high). Questions in the self-assessment are based on 27 different categories such as access control, physical security, training, maintenance, etc., each with many subtopics.

5.1. Cyber Risk Management Audits

Guidelines that are for protocols for ports, companies, etc. should not become a basis for auditing individual approaches. But companies may welcome government guidelines. NIST could be one such starting point (Interviews 2015–6; BIMCO 2016). [R 6.1.1]

Because cyber risk management lacks a specific physical presence, there is little functional connection, beyond physically securing (e.g., requiring two-person authentication) access points to cyber systems. None of our interviewees discussed issues such as physical protection against GPS spoofing, and other threat-specific physical measures. Therefore, it seems that physical security and cyber risk management might be better linked through audit systems currently in place or third-party audits, and companies should not rely solely on external audits (Interviews 2015–6; BIMCO 2016). [R 6.1.4, 6.1.5]

One example of audits is those performed by the Bureau of Safety and Environmental Enforcement (BSEE). BSEE regulates and inspects all oil and gas operations on the outer continental shelf (if oil rigs are in transit, they are regulated by the USCG). BSEE does not write a company's hazards plan; it is developed by the operators themselves and is approved by a third party. BSEE assesses how well the companies meet these plans. These plans focus primarily on physical security, but in the future they may include some cyber risk management as well. It is important to note that BSEE might be a reasonable entity to conduct cyber auditing for oil and gas operations. However, as of the time of our interview, BSEE had never conducted a cyber audit.

There are additional regulations, 33CFR Subchapter H (Maritime Security 2010), relating to maritime vessels. These regulations focus on owner/operators of Mobile Offshore Drilling Units (MODU), foreign cargo vessels greater than 100 tons, U.S. self-propelled vessels greater than 100 tons (except

CHAPTER 17

commercial fishing vessels), passenger vessels with more than 150 passengers, or other types of passenger vessels carrying more than 12 passengers when including at least one passenger for hire, and certain types of barges, tankships, and temporary assist vessels, but do not apply to warships. There are compliance audits for various types of security and safety topics including drills and training. Audits are performed through Vessel Security Assessments, and owners or operators must have a Vessel Security Plan. This is another area to which cyber components could be added. Amendments to the Vessel Security Plans are approved by the Marine Safety Center and may be added by the USCG or the vessel owner or operator. These regulations also apply to facilities. The regulations for facilities include access control, systems and equipment maintenance, handling cargo, training, drills and exercises required, monitoring, procedures for incidents. This is yet another area in which cyber regulations, training, drills, etc. could supplement the existing plans. Amendments to a Facilities Security Plan are approved by the Captain of the Port (COTP) and may be initiated by the COTP or the owner/operator. [R.6.1.4]

Although not designed for auditing, the BIMCO guidelines may also provide suggestions for components to integrate into these cyber risk management audits, assessments, trainings, and drills.

5.1.3. Metrics

Our interviews made it clear that all are concerned with the cyber-threat to the MTS. However, there are not currently any agreed-upon measures in place to assess "how secure" any part of the system is. Similarly, there are no measures in place to assess "how insecure" or "at risk" parts or subsystems may be. Clearly there is need for metrics to determine the cyber secure status of ports, vessels, container handling systems, etc. The Maritime Resource Center in Middletown, RI provides one example of an organization that is beginning to develop such metrics, through their proprietary methodology for assessing vessel and marine terminal cyber risk management. The primary use of such metrics for that organization is for use in their educational programs for mariners. However, many other uses can be envisioned, for example, in cyber risk management audits. The Department of Energy's Cybersecurity Capability Maturity Model (C2M2 2014) provides a complementary approach focused on assessing an organization's implementation and management of cyber risk management practices. Information Sharing and Communication is one of ten cyber security domains for which an organization can use C2M2 to assess the maturity of its processes. An effort to develop performance-based standards and the metrics to measure achieving those standards focused on maritime cyber risk management could be very important. [R 6.1.5]

5.1.4. Training and Exercises

The 33CFR103.515 specifies the USCG role to coordinate with the Area Maritime Security (AMS) Committee to conduct and participate in exercises to test the effectiveness of the AMS Plan. The AMS Plan should include a cyber component, and exercises should increasingly include tests of the effectiveness of the cyber risk management plan. Strategies for incentivizing sharing (together with new technologies) could be tested at upcoming or future USCG cyber risk management exercises (Interviews 2015–6). These exercises could be held in conjunction with physical security exercises since we know a cyber attack may be brought about by physical damage or vice versa. The AMS Committee for the Port of Pittsburgh held the first such exercise in 2013. Exercises can range in scope from tabletops and workshops to full-scale, simulated, coordinated cyber attacks. In the latter case, access to a cyber range may be useful (Interviews 2015–6). [R 6.1.6, 6.1.7, 6.1.8]

Conventional education (in Technical Schools, Community Colleges, Colleges and Universities) moves slowly. Today, some do not even realize that cyber is a threat (Interviews 2015–6). It may be that the national or international coverage of dramatic problems contributes more to the essential awareness of cyber threats than does any formal program of education. Therefore, educational efforts should be of two kinds: "Slow:" the development and dissemination of courses and training materials suitable for players at all levels from port managers to mariners, and "Fast:" effective media campaigns to build upon any major attacks (or near-misses) as they occur, to increase awareness, and motivate players to engage with the training materials, and/or the sharing organizations. Building awareness and capability requires training tailored to components of the maritime system (Interviews 2015–6; BIMCO 2016). The private sector and non-profit organizations have an important role to play in such training (Interviews 2015–6). This might coincide with rolling out new cyber guidance from the USCG. [R 6.1.9, 6.1.10]

5.1.5. Collaboration with Other Government Agencies

The USCG has a unique position in the MTS as part of the U.S. Government. In developing guidelines and technical standards for cyber risk management information sharing, the USCG has the opportunity to collaborate with other government agencies (such as NIST, ODNI, Cyber Command, NavSea, and DHS CERT). In support of these opportunities for enhanced information sharing, strengthened by the USCG presence at the NCCIC described below in Section 5.2, further research is needed into the most appropriate role for the USCG in (1) pushing best practices for cyber risk management to the private sector (versus just posting the information), and (2) developing regulations for sharing information about cyber attacks, vulnerabilities, and defenses with the private sector (Interviews 2015–6). [R 6.1.11, 6.1.12, 6.1.13]

5.2. Organizational Systems for Information Sharing

The question of how to organize systems for information sharing had by far the richest source of information, as there are several model organizations, and there is a strong consensus that some combination of those models will form the basis for any effective program of cyber risk management for the MTS. Key findings seem to be that: (1) industry players, based on their resources, will play roles of varying intensity in the organizations that are developed; (2) to permit all needed kinds of cooperation some organizations should be non-governmental, while others are governmental and perhaps even multinational; (3) issues of trade secrets, proprietary information, public embarrassment, lack of technical (IT) skills of even a basic nature, and national security will limit the willingness of players to share information, and must be countered with an array of incentives, as discussed in Section 5.3 below; (4) there are significant technical challenges in developing protocols for rapid sharing, and in coping with the expected flow of information, as participation expands to include all the parts of the MTS (see Section 5.5 below).

With such diverse organizations in the MTS, a range of organizational, technical, and incentive systems will be needed. To ensure timely dissemination to appropriate players, some of our interviewees emphasized the importance of a tiered approach to information sharing (Interviews 2015–6). [R 6.2.2]

5.2.1. Enhancing USCG Presence at the NCCIC

Organizationally, partnering with effective national organizations can help the USCG to a running start. By increasing its presence at the NCCIC, the USCG would expand its opportunities to coordinate with NCCIC partners and report cyber risk management alerts, trends and mitigation strategies across the USCG, commercial partners, and other appropriate government agencies. We understand from interviews that the USCG currently has one member of the CG Cyber Command onsite at the NCCIC, and we are recommending this presence be extended to a 24 × 7 capability (Interviews 2015–6). [R 6.2.1]

Through interviews and related research we learned that the NCCIC is able to receive and analyze Protected Critical Infrastructure Information (PCII),[2] a category of Sensitive but Unclassified (SBU) information that is protected

---

2   For more information on the Protected Critical Infrastructure Information (PCII) Program see: https://www.dhs.gov/protected-critical-infrastructure-information-pcii-program.

from FOIA disclosure and regulatory use to encourage reporting of information important to the security of the nation's critical infrastructure. Furthermore, through the new DHS Automated Indicator Sharing (AIS)[3] program, the NCCIC is able to receive cyber threat indicators from private industry, perform automated analyses and tasks such as removing personally identifiable information (PII) or anonymizing the sender, and distribute the indicators to federal departments or private industry, as appropriate. This kind of two-way, machine to machine sharing accelerates the pace at which DHS, and therefore the NCCIC, is able to receive and provide cyber measures and signatures. Finally, the NCCIC works with a variety of DHS training and assessment tools available to Critical Infrastructure and Key Resources sectors. These tools include the previously mentioned Critical Resilience Review (CRR)[4] available as self-assessment or DHS-facilitated evaluation, and the National Cybersecurity Assessment & Technical Services (NCATS)[5] through which a variety of cyber assessment services (such as architecture reviews and red-team, blue-team penetration testing) are available at no cost to stakeholders. [R 6.2.1]

5.2.2. Re-developing the Maritime ISAC

Partnering with effective private sector organizations will be needed to bring competing firms and competing nations into an effective overall system. Relying solely on a governmental organization might limit information sharing among private sector partners (and international partners), and this leads to the idea of a re-development of the Maritime ISAC to provide an industry-focused community for information sharing (Interviews 2015–6). [R 6.2.3]

Reflecting the economic layer of interaction, organizations differ in the resources they can direct to cyber risk management. Some interviewees suggested a Maritime ISAC with membership levels that provide and require different levels of information and capability (FS-ISAC 2015; Interviews 2015–6). A fast-acting ISAC is needed to complement periodic, face-to-face information sharing (supported by the AMSCs) since some cyber threats and attacks must be met in real time (Interviews 2015–6). There seem to be variously: very tight agreements among small numbers of large players with major budgets (Interviews 2015–6); more broad sharing, such as ISACS; and smaller players with low or no cyber budget or expertise. To include the full range of MTS stake-

---

3 For more information on Automated Indicator Sharing (AIS) see: https://www.us-cert.gov/ais.

4 For more information on Critical Resilience Review CRR) see: https://www.us-cert.gov/sites/default/files/c3vp/crr-fact-sheet.pdf.

5 For more information on the National Cybersecurity Assessment & Technical Services (NCATS) see: https://www.us-cert.gov/ccubedvp/federal.

CHAPTER 17

holders, some models are: ISACs; fusion centers; neighborhood watch as developed by the FBI Office in Los Angeles. Incremental development can start with key players and expand, perhaps using AMSC cyber risk management subcommittees as an initial step that the USCG is able to support immediately while industry partners evaluate the viability of developing and running an ISAC (Interviews 2015–6). [**R** 6.2.2, 6.2.3, 6.2.4, 6.2.7]

Reflecting both the economic and legal layers of interaction, sharing agreements may require: anonymity; authenticated messaging; and no FOIA access (FS-ISAC 2015). The FS-ISAC model (particularly its technical systems guaranteeing submission anonymity) is a possible model for a new Maritime ISAC (FS-ISAC 2015). [**R** 6.2.3]

Again at the legal layer, multinational membership adds challenges. The FS-ISAC may provide a model. (FS-ISAC 2015; Interviews 2015–6). National laws on cyber vary greatly (Interviews 2015–6). Since some important information is classified, it seems reasonable that a proposed Maritime ISAC ultimately be a cleared organization (FS-ISAC 2015; Interviews 2015–6). The ISAC could interface with other government agencies to ensure appropriate notification of organizations as part of the membership/access levels. [**R** 6.2.3, 6.2.4]

At the technical layer of interaction, shipboard systems and concerns are specialized, and, for example, legacy supervisory control and data acquisition (SCADA) systems and their network connections, in particular, need to be assessed for cyber risk (Konon 2014). To have effective communication among those with true common interests, the ISAC or similar organization might maintain a subgroup focused on shipboard systems, perhaps guided by the BIMCO publication (BIMCO 2016). [**R** 6.2.4]

Any industry-led information sharing platform (such as M-ISAC) and the USCG information sharing platform (such as a part of the NCCIC) must themselves share critical cyber risk management information regarding cyber threats. This leads to the idea that the M-ISAC maintain a presence at NCCIC, as is done by the FS-ISAC, the Aviation ISAC, the MultiState ISAC, and others (Senate Committee on Intelligence 2015; FS-ISAC 2015). [**R** 6.2.3]

5.2.3. Enhancing Cyber Incident Reporting Capability

As defined in 33CFR101.305, maritime security plans require that "activities that may result in a transportation security incident" be reported to the U.S. Coast Guard National Response Center (NRC). Some of our interviewees have suggested that the USCG either develop the capability or partner with an organization (such as the NCCIC[6]) to receive centrally information about cy-

---

6 For more information on the National Cybersecurity and Communications Integration Center (NCCIC) see: http://www.dhs.gov/topic/cybersecurity-information-sharing.

ber risk management incidents and suspicious activities (Interviews 2015–6). The analysts at this organization could send relevant alerts to the affected maritime community members (Interviews 2015–6) establishing a regulated two-way path and ensuring the USCG has all relayed information. In the meantime, we heard in interviews with port officials that there is confusion regarding whom they should contact in the event of a cyber incident (Interviews 2015–6). The NRC maintains a hotline[7] for "anyone witnessing an oil spill, chemical releases or maritime security incident," but there have not yet been thresholds guiding which types of incidents should be reported to the NRC (versus more locally, perhaps at an AMSC Cyber Subcommittee meeting). In fact, there are currently no regulations on reporting cyber incidents unless it reaches a Transportation Security Incident (TSI) level incident for the USCG, where TSI is defined as "any incident that results in a significant loss of life, environmental damage, transportation system disruption, or economic disruptions to a particular area."[8] No industry cyber incidents have ever reached the TSI level. We understand that the USCG Office of Port & Facility Compliance (CG-FAC) is updating its breach of security requirements soon to include thresholds for reporting (Interviews 2015–6). [**R 6.2.2, 6.2.8**]

5.2.4. ENHANCING AMSC CYBER INFORMATION SHARING

Currently, the AMSCs enable public and private partners in a geographic port area to meet periodically (often quarterly), discuss current concerns in the area, and build relationships of trust necessary for information sharing.[9] To maintain these relationships and extend them into cyber space, each AMSC could follow the example of the Port of Pittsburgh AMSC, the Port of Northern California AMSC, the Port of New York and New Jersey AMSC and others and create a cyber security subcommittee (Interviews 2015–6; Senate Committee on Intelligence 2015). As noted previously, about one-third of the AMSCs already have chartered a cyber security subcommittee. As the NY/NJ AMSC and others have done, all AMSCs could consider sharing cyber risk management information through the USCG HOMEPORT Portal (Interviews 2015–6). [**R 6.2.5, 6.2.6**]

In many of our interviews, we were told that a large number of entities of the MTS do not have the resources to hire employees with sufficient back-

---

7   The hotline phone number can be found on the NRC homepage: http://www.nrc.uscg.mil/ (accessed March 23, 2016).

8   http://www.uscg.mil/d8/msuBatonRouge/mtsa.asp.

9   For discussion of the importance of trust for information sharing, see: European Network and Information Security Agency (ENISA), 2010. Incentives and Challenges for Information Sharing in the Context of Network and Information Security.

ground to understand anything beyond the most rudimentary aspects of good cyber hygiene, and certainly not information about evolving cyber attacks, cyber vulnerabilities, and cyber defense. Some of these entities are represented on various AMSCs. More work is needed to understand organizational structures for information sharing that will develop ways to communicate cyber issues to the large number of MTS entities without technical expertise. [R 6.2.7]

5.2.5. MULTINATIONAL MARITIME ORGANIZATIONS: CJOS, AAPA, AND IMO

Complex nation-specific laws on cyber related issues along with concerns of sharing information about national security with other nations makes multinational information sharing very challenging. "Maritime operations to counter illegal activity at sea are difficult to coordinate between nations, governing bodies, security organizations, and armed forces. Responsibilities, jurisdiction, co-ordination, information and intelligence exchange, as well as the command and control of units conducting or supporting law enforcement operations are a maze of classifications, information systems, hierarchies and varied forces. ... None of the groups alone can provide all the necessary capabilities and coordination needed to succeed against threats" (NATO OTAN 2013). To help facilitate this information sharing and other security issues, the North Atlantic Treaty Organization (NATO) developed a NATO Memorandum of Understanding to create the Combined Joint Operations from the Sea (CJOS) Center of Excellence. CJOS was established on June 28, 2006, and includes 13 nations: Canada, France, Germany, Greece, Italy, the Netherlands, Norway, Portugal, Romania, Spain, Turkey, the United Kingdom, and the United States. CJOS is located in Norfolk, Virginia and is the only NATO accredited COE in the U.S. The purpose of CJOS is to "support the transformation of joint maritime expeditionary operations in assistance to NATO" (United States Department of State 2006).

The CJOS Memorandum of Understanding states that external security is the responsibility of the host nation (US) and internal security is the responsibility of the CJOS Director, following NATO and US security regulations. Relating to information sharing, the nations involved in CJOS are responsible to safeguard the security of any classified information provided in the course of the CJOS mission. Confidentiality is expected to remain intact even if the MOU is terminated or withdrawn.

CJOS held Maritime Security Conferences (MSC) from 2008 to 2012 which built on the idea of information sharing. In 2012 they found that "a bottom-up approach is more likely to be supported than an international governance model. ... The outcome of MSC 2012 identified widespread agreement that there is a need for information sharing and, for this sharing to occur, there needs to be a shift from the current 'need to know' mentality to a culture of 'need to share'" (Combined Joint Operations from the Sea Centre of Excellence 2016). [**R 6.2.9**]

Another international maritime organization, the American Association of Port Authorities (AAPA), is a trade association representing more than 130 deep draft public ports in the United States, Canada, Latin America and the Caribbean. The AAPA provides education, services, and advocacy for its members, which also include more than 300 associate and sustaining members such as inland river ports and firms doing business with corporate member ports. Some of the education opportunities available in 2015 included an intensive Marine Terminal Management Program, a Port Security Seminar and Exposition, a Cybersecurity Seminar, and a workshop on Shifting International Trade Routes. Along with the newsletters and surveys the AAPA publishes, they maintain a list of Port Industry Best Practices,[10] which includes categories of resources such as Emergency Preparation Response and Recovery, that could potentially be a forum for sharing port cyber risk management guidelines. Just as BIMCO recently issued detailed Guidelines for Cyber Security Onboard Ships (BIMCO 2016), some of our interviewees said that it might be appropriate for the AAPA to develop similarly focused guidelines for port facility cyber risk management (Interviews 2015–6). [**R 6.2.10**]

The International Maritime Organization (IMO), an agency within the United Nations, has 171 member states and three associate members responsible for regulating shipping. The main role of the IMO "is to create a regulatory framework for the shipping industry that is fair and effective, universally adopted and universally implemented. In other words, its role is to create a level playing-field so that ship operators cannot address their financial issues by simply cutting corners and compromising on safety, security and environmental performance" (International Maritime Organization 2016a).

In the IMO's 2014 year in review, The Maritime Safety Committee and the Facilitation Committee agreed to include on their agendas the topic of cyber security for the following year (2015).[11] This came about after Canada presented a paper on the topic to the 39th session of the IMO facilitation committee in September of 2014. "The Canadian presentation called for voluntary guidelines on cyber security practices to protect and enhance the resilience of electronic systems of ports, ships, marine facilities, and other parts of the mar-

---

10 For more information see: http://www.aapa-ports.org/Issues/content.cfm?ItemNumber=1262&navItemNumber=543.

11 For more information see http://www.imo.org/en/MediaCentre/HotTopics/yearreview/Pages/2014-Security-and-facilitation.aspx.

itime transport system. It is understood to have suggested that cyber issues are brought into the coverage of the International Ship and Port Facility Security Code (ISPS)" (Gooding 2014). The committee agreed, "recognising it as a relevant and urgent issue for the Organization, in order to guarantee the protection of the maritime transport network from cyber threats" (Gooding 2014).

In a January 2016 IMO letter, describing trends affecting the organization in order to help develop their strategic framework for 2018–2023, cyber risk was at the top of the list. The following excerpt was taken from this letter: "The increasing trend in the use of cyber systems benefit the maritime industry, but their use also introduces great risk. From a security perspective cyber systems may be exposed to deliberate, malicious acts from individuals who may attempt to control, disable, or exploit cyber systems. From a safety perspective, non-targeted malware, innocent misuse of systems, and simple technical errors may impact vital systems related to ship and propulsion control, navigation-related technologies, industrial ship control technologies including propulsion, steering, ballast water management, electrical systems, heating, ventilation, air conditioning systems, cargo pumps, cargo tracking and control, ship stability control systems, fire detection and protection, gate access control and communication and monitoring systems, alarm systems and various hazardous gas alarm systems, pollution and other safety and environmental monitoring" (International Maritime Organization 2016b). However, the IMO does not yet publicly state what measures they will be taking. [R 6.2.11]

5.3. Motivation and Barriers for Information Sharing

The question of how to incentivize sharing among players in the MTS is a central one. As with any sharing scheme, information sharing for cyber risk management faces the "problem of the commons." Several industries appear to be further along in developing solutions, and their models provide guides. In complexity, MTS is closest to international finance, and the economic and security concerns of many kinds of organizations, and of competing nations, are involved. Positive incentives (motivations for information sharing) could include technical support and timely sharing of information or insurance industry pressure (through rate reductions) to encourage participation. They also could include believable guarantees of protection from (1) action by competitors (2) legal action and (3) FOIA pressures by competitors, NGOs, and activist groups. Negative incentives (overcoming barriers to sharing) might include regulations and penalties for non-reporting. It may be possible to test some models in USCG exercises.

Providing incentives for sharing could be particularly important as the industry begins to take the small, initial steps that will lead to enhanced maritime cyber risk management. For example, we are recommending that the U.S. Government require "landlord" port operators[12] to incorporate maritime cyber risk management standards into the leases they issue to terminal operators for the right to use the ports. Landlord ports are highly autonomous and can easily implement requirements of this nature into their leases without waiting for a legislative or regulatory process, but terminal operators may then decide to "port shop" for easier restrictions, thereby hurting the ports working to improve cyber risk management. For this reason, regulation requiring these standards at all ports is needed to ensure a "level playing field" in cyber risk management, preventing terminal operators from being able to avoid cyber standards by relocating to a more "lax" port operator (Interviews 2015–6). [**R** 6.3.1]

Realization of this kind of legislation or regulation will likely take some time, however, and in the interval, there are opportunities to motivate early adoption of enhanced cyber risk management practices. As port operators negotiate leases with tenants, they could, for example, offer discounted rates to tenants that agree to comply with cyber risk management standards (Interviews 2015–6). Furthermore, the terms of port leases could be opportunities for requiring tenants' participation in a reinstantiated M-ISAC. [**R** 6.3.1, 6.3.2, 6.3.3]

Players in this industry bring greatly different resources to the problem (Interviews 2015–6). The arrangements made at the E-ISAC (Electrical Industry) and the FS-ISAC (Financial) may yield useful models for development of incentives (FS-ISAC 2015; Interviews 2015–6). We heard in interviews, for example, that the FS-ISAC maintains a Gmail Listserv for communicating threat information to its members, and the fact that U.S. Law Enforcement is not allowed to join the list encourages foreign-based partners to participate (Interviews 2015–6). Other models are the FBI, which shares at industry conferences, and the Oil and Gas industries (ONG-ISAC) (FS-ISAC 2015; Interviews 2015–6). It seems clear that incentives will have to be tailored to be effective for the various classes of players. [**R** 6.3.4, 6.3.5, 6.3.6]

Since information sharing benefits all, but costs the contributors, distillers, and disseminators, there is a risk of "free-riding" (Gal-Or and Ghose 2005). However, the MTS is an interdependent system of systems, and a major cyber event somewhere in the system will likely disrupt the business of all parties and potentially affect the reputation of the whole industry (Interviews 2015–6). [**R** 6.3.5]

Some useful incentive models may be found in other domains, such as the WHO's provision of subsidized vaccine targeted to countries reporting

---

12 See the American Association of Port Authorities (AAPA) Glossary of Maritime Terms for definitions of "landlord" versus "operating" ports. http://www.aapa-ports.org/Industry/content.cfm?ItemNumber=1077 (accessed March 23, 2016).

outbreaks of bacterial meningitis (Laxminarayan, Reif, and Malani 2014). It is possible that compliance in sharing will be motivated by the insurance industry, although it has not yet taken any positions on this issue (Interviews 2015–6). [R 6.3.6]

The European Network and Information Security Agency (ENISA) found, in a research effort regarding information sharing for network and information security, that stakeholders felt "Economic incentives stemming from cost savings" were of the highest importance for information sharing, whereas "Economic incentives from the provision of subsidies" or "Economic incentives stemming from the use of cyber insurance" were of low importance (ENISA 2010). That is, the participants identified the most important incentive for participating in an Information Exchange (IE) such as an ISAC to be the cost savings they would realize from more efficiently allocating the information security resources of the group. The challenge remains, however, to prove that participation in an IE does bring these savings and efficiencies (ENISA 2010). [R 6.3.6]

A different kind of incentive for information sharing, ranked third in importance out of ten incentives for information sharing by the participants in the ENISA research Delphi exercise, is the "presence of trust amongst IE participants." Although trust is perhaps more difficult to quantify than cost savings, many interviewees highlighted operating and sharing information within a community of trusted partners (such as the current AMSCs) to be a critical component of the current security arrangements (Interviews 2015–6). Furthermore, a 2001 study of organizations accustomed to sharing information, conducted by the U.S. General Accounting Office (GAO), found, "All of the organizations identified trust as the essential underlying element to successful relationships and said that trust could be built only over time and, primarily, through personal relationships" (U.S. GAO 2001). [R 6.3.6]

5.4. What Information to Share, and What to Share Rapidly versus Slowly

What information should be shared? The Cybersecurity Information Sharing Act of 2015 proposes requirements for communication of "cyber threat indicators," defining these as the "information necessary to describe or identify." The Act identifies eight categories of cyber threats, and could be the framework of a strategy describing what to share regarding threats. More broadly, the FS-ISAC structures its sharing according to incidents, threats, vulnerabilities, and resolutions/solutions (FS-ISAC 2015). We learned that information shared with the MS-ISAC may include: advisory notices, tactical information, and known malicious IPs (Interviews 2015–6). Information to share will include: vulnerabilities, TAXii[13] information, botnet information, malicious IP addresses, near misses, incidents, threats, resolutions/solutions, and the seven key Netflow fields (Interviews 2015–6). Once again the economic layer is in play here as only a select set of private sector companies and law enforcement agencies have the resources to dedicate groups of highly skilled people to analyze this information. [R 6.4.1]

A Red | Yellow | Green Traffic light protocol to code sensitivity of information could be useful (FS-ISAC 2015; Interviews 2015–6; BIMCO 2016). [R 6.4.2] The Port of NY and NJ has developed what many regard as a "best practice" for sharing sensitive but unclassified information with private sector partners (Interviews 2015–6). The process involves individual invitations to a closed door meeting where participants' identification is checked at the door. Participants are typically long-standing AMSC members, and the meeting is chaired by the COTP. If further dissemination of information beyond the meeting is deemed necessary, the USCG vets the information to remove the sensitive material. This process may include a USCG legal advisor if necessary. Once fully vetted, the information is posted to the HOMEPORT portal.

Targeted small briefings, including classified briefings, are vital, but these typically lag events by weeks. Faster sharing is needed. "Slow" sharing can also be done at industry conferences or with AMSCs (Interviews 2015–6). Research may be needed to automate the filtering and classification of the large volume of information (Interviews 2015–6). Additional research is needed into the cyber risk management industry issue of filtering the large volume of information available on cyber incidents. More information is not always better, and it can be difficult to filter through the noise to understand what is actually a malicious attack. Key players need to avoid information overload, which can cause actual events to be overlooked as noise. This is not just a maritime cyber issue, but a cyber risk management industry issue in general. [R 6.4.2, 6.4.4]

Because information about potentially catastrophic near misses may unduly influence cyber risk management investment decisions (Dillon and Tinsley 2015), one can ask whether these events should be examined to determine whether resilience was key to the "miss." This could help others to learn from the disseminated information. [R 6.4.3]

5.5. Technologies to Support Information Sharing

Technology presents several challenges: Each player must have adequate resources to share and receive information, to protect sensitive information, and

---

13 For more information see https://securityintelligence.com/how-stix-taxii-and-cybox-can-help-with-standardizing-threat-information/.

CHAPTER 17

to respond promptly enough. The players have to agree on protocols for reporting problems, attacks, and countermeasures. The responsible coordinating bodies must also have technologies for receiving and filtering streams of information, prioritizing them, and classifying them for controlled dissemination to the players.

Rapid sharing requires standardized reporting, etc. The FS-ISAC employs the STIX and TAXii systems that are being developed by a community led by DHS. STIX and TAXii may be relevant for standardized reporting, but they are not software tools. Full implementation should not require vendor specific software (FS-ISAC 2015–6; Interviews 2015–6). Several existing protocols/systems could be evaluated to see whether they are appropriate for MTS use (Interviews 2015–6). For example, the FS-ISAC uses technical systems developed by Soltra, a DTCC and FS-ISAC company. The systems include a threat intel server, SoltraEdge, to aggregate and distribute information about threats (peer-to-peer and firm-to-firm) and the SoltraNetwork that connects these servers in a hub and spoke manner. [R 6.5.1. 6.5.3]

Utilizing STIX and TAXii is the new DHS Automated Indicator Sharing (AIS)[14] program. As previously mentioned, this is now used by the NCCIC. AIS is a two-way sharing program, which does not need a human in the loop to share information; the information is shared from machine-to-machine, either from the NCCIC to partners or from partners/industry to the NCCIC.

The National Cybersecurity Protection System (NCPS) is a system of systems providing capabilities to defend the federal government's information technology infrastructure. NCPS broad cyber security capabilities include detection, analytics, information sharing, and prevention. For example, its analytics capabilities include Secure Information and Event Management (SIEM), Packet Capture (PCAP), Enhanced Analytical Database (EADB) and flow visualization, and Advanced Malware Analysis. These or related technologies may prove useful in developing information to share within certain segments of the MTS.

The ability to anonymously submit reports of cyber risk incidents or near misses could allow firms to share information without fears of harming their reputations or incurring regulatory penalties. An example of this kind of anonymized incident reporting is found in the collaboration of the American Bureau of Shipping (ABS) and Lamar University to develop and maintain an online Mariner Personal Safety (MPS) database[15] for tracking maritime injury and close call reports (Interviews 2015–6). Another example of data anonymization is the Vocabulary for Event Reporting and Incident Sharing

---

14 https://www.us-cert.gov/ais.

15 http://ww2.eagle.org/en/rules-and-resources/safety-human-factors-in-design/mariner-personal-safety.html.

(VERIS) framework[16] used by Verizon to gather information for its annual Data Breach Investigation Report. Finally, the FS-ISAC has as one of its Cornerstones that information is able to be submitted anonymously through its technical systems (FS-ISAC 2015). The M-ISAC or other consortium of MTS partners could increase participation in information sharing by identifying or developing an independent data anonymization platform for sharing cyber risk management incidents and false alarms (Interviews 2015–6). [R 6.5.4]

Recent research in decentralized "trust systems" also may prove helpful (Minsky 1991; Minsky and Leichter 1995; Minsky and Ungureanu 2000). [R 6.5.2]

6. Recommendations

6.1. The Role of the USCG and Extending Physical Security to Cyber Security—Cyber Risk Management

6.1.1. We strongly endorse the ongoing USCG effort to develop cyber risk management guidelines analogous to the physical security requirements found in 33CFR Subchapter H.

6.1.2. Since there many diverse players in the MTS, and they have competing interests, we recommend guidelines for the MTS be written at a "high" level—specifying the characteristics of a cyber risk management plan, not detailed technical prescriptions.

6.1.3. We recommend the NIST Framework (NIST 2014) as a guide for the process of developing cyber risk management plans covering facilities, NIST (1990–2015) as a resource for federal government IT system security, and BIMCO (2016) as a resource for developing cyber risk management guidelines specific to vessels.

6.1.4. We recommend that physical security and cyber risk management be more strongly linked, reflecting the likelihood that a cyber attack may be manifest by physical damage or vice versa. This may be facilitated through the audit systems currently in place, such as found in 33CFR Subchapter H (Maritime Security 2010), in addition to self-audits. These may also be integrated into current vessel and facility drills, exercises, and trainings. The BIMCO Guidelines may offer some insight of topics to include.

6.1.5. We recommend a research effort to develop cyber risk management performance-based standards and metrics to be used by the USCG in security audits, educational programs, and other applications. Again, these can be added

---

16 http://veriscommunity.net/index.html.

as additional content into pre-existing vessel and facility drills, exercises, and training.

6.1.6. We recommend the USCG develop and roll out the capability to assess and communicate the cyber readiness of the MTS and its components.

6.1.7. We recommend that the USCG increase its effort to coordinate and lead regular cyber risk management exercises in collaboration with the AMSCs and in conjunction with physical security exercises. Exercises should range in scope and complexity as appropriate from tabletops to full-scale simulated cyber attacks perhaps facilitated by access to a cyber range.

6.1.8. We recommend cyber risk management exercises as opportunities for evaluating proposed organizational structures, performance-based standards and technologies for information sharing within the USCG, and between the USCG, its commercial partners, and other government agencies.

6.1.9. We strongly endorse ongoing USCG efforts to provide guidelines for training to raise awareness of cyber risk management threats for members of the AMSCs. Considerations such as who pays for the training and who develops, delivers and receives the training need to be worked out.

6.1.10. We recommend cyber risk management training tailored to specific components of the maritime system be developed to coincide with, and enhance understanding of, new cyber guidance from the USCG.

6.1.11. We recommend that the USCG expand collaboration with other government agencies (such as NIST, ODNI, Cyber Command, NavSea, and DHS CERT) to develop technical standards for cyber risk management information sharing.

6.1.12. We recommend further research into the appropriate role of the USCG in pushing best practices for cyber risk management to the private sector.

6.1.13. We recommend further research into the appropriate role of the USCG in developing regulations for sharing information about cyber attacks, vulnerabilities, and defenses with the private sector.

6.2. Organizational Systems for Information Sharing

6.2.1. We recommend the USCG enhance its presence at the NCCIC into a 24 × 7 capability for coordinating with NCCIC partners and reporting cyber risk management alerts, trends and mitigation strategies to the USCG, commercial partners, and other appropriate government agencies.

6.2.2. We recommend that the USCG lead an organization (such as a branch of the NCCIC) for sharing cyber risk information with its MTS partners, which may include several tiers of information corresponding to the type of information to be shared (automated reports of probes versus discussion of possible trends over time, etc.) with appropriate groups of partners such as ISACs, fusion centers, AMSCs, local FBI offices, and state and municipal law enforcement units. We note that not all MTS partners will be able to participate at all levels of sharing (limitations may be technical, economic, or based on national policy).

6.2.3 We recommend that private industry within the MTS develop and lead an industry-focused organization (such as a re-instantiated M-ISAC) for sharing cyber risk information, providing an arms-length relation to the USCG-led organization in Recommendation 6.2.1. Investigation of the business and technical models employed by existing organizations such as the FS-ISAC, E-ISAC, ONG-ISAC, and A-ISAC, particularly as relates to supporting anonymous sharing of information, may provide a good starting point for this organization.

6.2.4. We recommend that the industry leaders of the M-ISAC establish membership levels that vary according to the member's size (ability to contribute financially) and industry sector (terminal operator, oil and gas import/export, international shipping, etc.).

6.2.5. We strongly endorse the requirement, proposed in the Strengthening Cybersecurity Information Sharing and Coordination in Our Ports Act of 2015 (Senate Committee on Intelligence 2015), that each AMSC create a cyber risk management working group or subcommittee, and we recommend the subcommittee meet at least quarterly.

6.2.6. We recommend a system-wide coordination effort to develop a compilation of mission, focus, and operation found at existing AMSC cyber security subcommittees, with results to be shared across AMSCs.

6.2.7. We recommend further research on the best organizational structures for sharing information with components of the MTS that do not have any information-technology-trained personnel.

6.2.8. We recommend that all MTS partners report cyber security incidents, including near misses, to the USCG National Response Center (NRC) until an alternative organization (perhaps the NCCIC) is identified and reporting requirements are specified in the cyber risk management guidelines referred to in Recommendation 6.1.1.

6.2.9. As found by CJOS, "a bottom-up approach is more likely to be supported than an international governance model". We recommend this approach be utilized, emphasizing buy-in from international industry partners as much as possible rather than regulations.

6.2.10. We recommend the American Association of Port Authorities (AAPA) develop cyber risk management guidelines for port facilities, similar to the BIMCO (2016) guidelines for ships.

6.2.11. We recommend that the USCG continue to work with the IMO and monitor their international efforts to establish cyber risk management guidelines.

6.3. Motivation and Barriers for Information Sharing

6.3.1. We recommend that the USCG advise Congress that legislation and/or regulation is needed that requires "landlord" port operators to incorporate maritime risk management standards in their leases to terminal operators and "operating" ports to adopt the standards themselves.

6.3.2. We recommend that "landlord" port operators offer discounts to terminal operators that agree to adopt cyber risk management standards before legislation requires it.

6.3.3. We recommend that "landlord" port operators require their tenants to be members of the M-ISAC once it is reinstantiated.

6.3.4. We recommend that the new M-ISAC communicate threat information among its membership in a way that does not involve the U.S. Government or Law Enforcement in order to encourage participation by non-U.S. firms. We note that U.S. firms may be required to also (separately) report threat or incident information to an appropriate U.S. authority.

6.3.4. We recommend that industry partners working to establish new information sharing agreements evaluate the incentives used to avoid pitfalls such as free-riding and withholding critical information from competitors in the FS-ISAC, ONG-ISAC, and E-ISAC.

6.3.5. We recommend research efforts focused on the following:

> 6.3.5.1: In-depth interviews with all participants in an AMSC to identify the specific barriers to investment in information sharing faced by these MTS partners. Incentive plans, such as identification of a third-party anonymization service for reporting incidents, can then be proposed to target these specific, MTS-centric barriers.

Issues in Maritime Cyber Security

6.3.5.2: The legal challenges of global cyber risk management information sharing and incentives.

6.3.5.3: Methods to achieve rapid and useful information sharing in a way that both large and small players in the MTS can participate, and in particular on how one can entice larger content providers to take the lead on information sharing.

6.4. What Information to Share, and What to Share Rapidly versus Slowly

6.4.1. We recommend that categories of information to be shared could be taken from existing sources that include: the TAXII, STIX, and CybOX specifications,[17] the FS-ISAC categories of information for submission (Incidents, Threats, Vulnerabilities, and Resolutions/Solutions), the threat types listed in CISA 2015, and data elements known to be shared by MTS entities with organizations such as the MS-ISAC.

6.4.2. We recommend development of standardized protocols for managing the sensitivity (as relates to confidentiality and to timing) of information to be shared. Examples of protocols in use that could serve as models are the USCG RGA (Red, Green, and Amber) scheme, and the approach employed at the Port of NY/NJ.

6.4.3. We recommend that reports of "near misses" be shared together with an analysis of the apparent reason(s) the attack was unsuccessful.

6.4.4. We recommend additional research into the cyber risk management industry issue of filtering the large volume of information available on cyber incidents (noise versus malicious). More information is not always better; instead, the research should focus on what is the most important critical information that key players need to avoid information overload, which can cause actual events to be overlooked as noise. Additionally, different MTS partner organizations, and different roles, positions, and levels within these organizations, will require different kinds of filters to ensure the right information reaches each party.

6.5. Technologies to Support Information Sharing

6.5.1. We recommend research to evaluate how existing technical protocols for

---

17 https://www.us-cert.gov/Information-Sharing-Specifications-Cybersecurity.

information sharing (such as TAXii/STIX) are currently at use by MTS partners, such as the MS-ISAC, and how their use could be more widely adopted where needed. Rather than identify products from a single vendor, technical recommendations should identify industry protocols supported by multiple software products.

6.5.2. All of the models discussed so far have one or two central nodes, which present a single point of failure (SPOF). We recommend further research seeking distributed models that can deal with the complexities of the MTS without presenting a SPOF.

6.5.3. We recommend a research effort aimed at analyzing the many vehicles for sharing to see what role they may play in a comprehensive information sharing strategy for the MTS. Examples include: HOMEPORT, SharePoint, briefings (internal, other agencies, etc.), DHS Communities of Practice, forums, and automated network monitoring systems.

6.5.4. We recommend that the MTS industry research available anonymization platforms and technologies that could allow commercial partners to share cyber risk information such as incidents and false alarms without fear of negative publicity. Examples include: the online Mariner Personal Safety (MPS) database led by American Bureau of Shipping and Lamar University, the Vocabulary for Event Reporting and Incident Sharing (VERIS) framework[18] used by Verizon to gather information for its annual Data Breach Investigation Report, and the technical systems used by the FS-ISAC to allow anonymous submission of threat information by its members.

REFERENCES

BIMCO. 2016. "The Guidelines on Cyber Security Onboard Ships." https://www.bimco.org/~/media/Products/Manuals-Pamphlets/Cyber_security_guidelines_for_ships/Guidelines_on_cyber_security_onboard_ships_version_1-1_Feb2016.ashx (accessed January 12, 2016).

C2M2. 2014. "Cybersecurity Capability Maturity Model." Version 1.1. Department of Energy, February. http://energy.gov/sites/prod/files/2014/03/f13/C2M2-v1-1_cor.pdf (accessed February 2, 2016).

CIS. 2015. "CIS Critical Controls for Effective Cyber Defense." Version

---

18 http://veriscommunity.net/index.html (accessed March 21, 2016).

6.0. Center for Internet Security. http://www.cisecurity.org/critical-controls.cfm.

Clark, Bruce G., and Fred Roberts. 2015. "Summary Report of Findings: Maritime Cyber Security Research Summit." July 2015.

Combined Joint Operations from the Sea Centre of Excellence. 2016. http://www.cjoscoe.org/.

CSET Cybersecurity Evaluation Tool. 2015. "13 CSET 6.2 Questions Screen." (video). January 15. https://www.youtube.com/watch?v=CfRDUqA5WnI.

Davies, Martin. 2003. "Obligations and Implications for Ships Encountering Persons in Need of Assistance at Sea." *Pacific Rim Law & Policy Journal* 12: 109–141.

Department of Homeland Security (DHS). Cyber Resilience Review. https://www.us-cert.gov/sites/default/files/c3vp/crr-fact-sheet.pdf.

Dillon, Robin, and Cathy Tinsley. 2015. "Near-Misses and Decision Making Under Uncertainty in the Context of Cybersecurity." *Improving Homeland Security Decisions*, forthcoming.

European Network and Information Security Agency (ENISA). 2010. *Incentives and Challenges for Information Sharing in the Context of Network and Information Security.*

Financial Services Information Sharing & Analysis Center (FS-ISAC). 2015. "Operating Rules." https://www.fsisac.com/sites/default/files/FS-ISAC_OperatingRules_2015.pdf (accessed January 12, 2016).

Gal-Or, Esther, and Anindya Ghose. 2005. "The Economic Incentives for Sharing Security Information." *Information Systems Research* 16 (2): 186–208.

Gooding, Nicholas. 2014. "IMO is Being Warned of 'Scary' Potential of Maritime Cyber Attacks." AllAboutShipping.co.uk. (October 25). http://www.allaboutshipping.co.uk/2014/10/25/imo-is-being-warned-of-scary-potential-of-maritime-cyber-attacks/.

International Maritime Organization. 2016a. "Introduction to the IMO." http://www.imo.org/en/About/Pages/Default.aspx.

International Maritime Organization. 2016b. "Letter from J.G. Lantz to Kitack Lim." 16707/IMO/C 116 Rev. 1-2014. (January 15). http://www.imo.org/en/About/strategy/Documents/Member%20States%20-%20tdc/United%20States%20-%20Input%20to%20TDCs.pdf.

Interviews. 2015–6. Interviews with Industry and Other Experts, Conducted by CCICADA.

ISO/IEC. 2013. International Organization for Standardization, ISO/IEC 27001 – Information Security Management.

Konon, Jennifer. 2014. "Control System Cybersecurity: Legacy Systems are Vulnerable to Modern-Day Attacks." Proceedings of the Marine Safety & Security Council. *The Coast Guard Journal of Safety at Sea* 71 (4): 45–47.

Laxminarayan, Ramanan, Julian Reif, and Anup Malani. 2014. "Incentives for Reporting Disease Outbreaks." http://journals.plos.org/plosone/article?id=10.1371/journal.pone.0090290 (accessed January 12, 2016).

Maritime Security. 2010. "33CFR Subchapter H pt. 101-107 (2010)." https://www.gpo.gov/fdsys/granule/CFR-2010-title33-vol1/CFR-2010-title33-vol1-part101 (accessed January 12, 2016).

Minsky, Naftaly H. 1991. "The Imposition of Protocols Over Open Distributed Systems." *IEEE Transactions on Software Engineering* 17 (2): 183–195.

Minsky, Naftaly H., and Jerrold Leichter. 1995. "Law-Governed Linda as a Coordination Model." *Object-Based Models and Languages for Concurrent Systems*. 924: 125–146. Springer.

Minsky, Naftaly H., and Victoria Ungureanu. 2000. "Law-Governed Interaction: A Coordination and Control Mechanism for Heterogeneous Distributed Systems." *ACM Transactions on Software Engineering and Methodology (TOSEM)* 9 (3): 273–305.

National Institute for Standards and Technology (NIST). 1990–2015. Special Publications SP-800 Computer Security. http://csrc.nist.gov/publications/PubsSPs.html#SP%20800.

National Institute for Standards and Technology (NIST). 2014. "Framework for Improving Critical Infrastructure Cybersecurity, Version 1.0." http://www.nist.gov/cyberframework/upload/cybersecurity-framework-021214.pdf (accessed January 12, 2016).

National Institute for Standards and Technology (NIST). 2015. "Overview of the Cybersecurity Framework." http://www.nist.gov/cyberframework/upload/cybersecurity_framework_coast_guard_maritime_public_meeting_2015-01-15.pdf.

National Institute for Standards and Technology (NIST). 2016. "Cybersecurity Framework Comments Reveal Views on a Framework Update, Increased Need to Share Best Practices and Expand Awareness." http://www.nist.gov/itl/acd/cybersecurity-framework-comments-reveal-views-on-a-framework-update.cfm.

National Institute for Standards and Technology (NIST). 2016-2. "Analysis of Cybersecurity Framework RFI Responses." http://www.nist.gov/cyberframework/upload/RFI3_Response_Analysis_final.pdf.

North Atlantic Treaty Organization Allied Command Transformation (NATO OTAN). 2013. "Delivering Maritime Security in Global Partnership." October 29. http://www.act.nato.int/article-2013-2-14.

Sanctions Wiki. 2014. "Transnational Criminal Organizations–TCO." July 2. http://www.sanctionswiki.org/TCO.

Senate Committee on Intelligence. 2015. "Cybersecurity Information Sharing Act of 2015." S. Rept. 114-32, 114th Cong. https://www.congress.gov/bill/114th-congress/senate-bill/754 (accessed December 19, 2015).

Strengthening Cybersecurity Information Sharing and Coordination in Our Ports Act of 2015. H. Rep 114-379. 114th Cong. https://www.congress.gov/bill/114th-congress/house-bill/3878/text (accessed January 12, 2016).

United States Department of State. 2006. "Memorandum of Understanding Concerning the Establishment, Administration, and Operation of the CJOS COE." http://www.state.gov/documents/organization/75818.pdf.

United States General Accounting Office (U.S. GAO). 2001. "Information Sharing: Practices That Can Benefit Critical Infrastructure Protection." GAO-02-24. Washington, DC.

CHAPTER 18:
HOW DO WE PROMOTE THE USE OF SOUND CYBER RISK MANAGEMENT PRINCIPLES?

Eduardo V. Martinez, Jessica L. Adkisson, Jacob A. Babb, Eric S. Casida, Frank S. Hooton, Gabriel Nunez, Jeremy Quittschreiber, and Joshua S. Weishbecker
American Military University

EXECUTIVE SUMMARY

With the many types of Maritime Cyber Security Threats in existence, there comes the need to prepare and contain such risks in a responsible and intelligently aggressive manner. This white paper is designed to assist in this endeavor, through a culmination of extensive research conducted by an American Military University/American Public University (AMU/APU) Research Team which met from November 2015 to January 2016. The team consisted of members of several governmental and civilian entities associated with the Department of Homeland Security, Department of Defense and the National Security Agency, including seasoned military personnel from the Navy, Coast Guard, Air Force, Army, Marine Corps, and the National Guard. Researcher biographies follow the conclusion. Tasked to answer the objective "How do we promote the use of sound cyber risk management principles?" their combined experience and insight greatly added to the direction and focus in providing the very best research and recommendations to the leadership of the United States Coast Guard (USCG). In this research, the AMU/APU team placed a great deal of emphasis on locating ways in which both an organization and the United States as a whole can be affected by Cyber Security Threats, which is ultimately the initial step that must be completed in any risk management strategy. The team then determined the most important elements of those risks and how to mitigate them. These techniques were then integrated into the mitigation process within a promotion and resiliency strategy to ensure that, in the event of a cyber attack, the USCG would be able to successfully and efficiently respond. Based on an extensive literature review, the team evaluated the risks, possible ways to mitigate them, and how to respond through promotion methods by organizing them into five areas designated as pillars: Awareness, Flexible and Adaptable Security, Domain Understanding, Enabling the Mission, and Risk Management. Each of these pillars stands for a very important aspect that must be understood and implemented unanimously, for if one is not constructed as an equal with the others, the entire structure falls apart. Through this design and understanding, promotion of sound cyber risk principles is obtained.

Introduction

For many Americans, the advent of technology and its advantages are taken for granted. Their own personal safety and security tends to be something they rarely ever consider until something goes wrong and it is far too late to respond. It is the responsibility of those who have the ability to protect the American public and, ultimately, the world, from the growing threat that a cyber system poses not only to our National Security, but our own individual security.

For many consumers, the worldwide web, ancillary networks and anything coupled to it should be simple, efficient and as convenient as possible. This means the consumers desire what many industry leaders have begun to provide such as a single username and password across all possible combination of devices and systems that can accomplish nearly everything they might achieve on a daily basis. This concept of simplicity seems like a great idea; however, the more things become connected to a single or non-complex pathway activity, the higher the risk involved. Why does centralizing and simplifying things within the cyber realm create greater security risk? Simply, attackers ultimately need to figure out less ways to access all necessary information to plan and carry out not only a cyber intrusion, but also a physical attack on a target of their choice. Today, that also means attacks on the even easier target of individual personnel. In the intelligence realm, we all know that the easiest way to solve a problem is to break it into several pieces like a puzzle. Then, you begin to put it together, one piece at a time rather than trying to create an entire picture in a single event. When cyber security becomes involved, it no longer takes an entire team to acquire this information efficiently. Theoretically, a single individual who has even the smallest amount of background in computers and a $500-dollar laptop can do the same task in a matter of hours. That individual could hack into targeted agencies' or companies' records, which are now all stored digitally, such as the Office of Personnel Management which was hacked in the summer of 2015 (Weintraub Schifferle 2015). Those hackers potentially can gain access to nearly every starting point they may need, for an attack on every individual's personal information who has ever applied to work with the government.

For individual agencies, such as the United States Coast Guard (USCG), which this study was specifically prepared for, this means preparing for added responsibilities outside maintaining protection for the homeland as their mission states. They must also protect their personnel, personnel's information, and their own agency from both the threat of external attack, as well as that of an insider. In addition, it compels the USCG to inform companies of potential Maritime Cyber Security Threats. If any one area fails to maintain security, the entire system can collapse. Therefore, all agencies, individuals, and governments need to heighten their focus on risk management and security from not only physical threats, but also cyber attacks. These issues also affect the Maritime Transportation System in their daily management of international shipping. Yet, despite technological advances, the critical infrastructure concerns are threatened, to some degree, by these same mediums. Cyber communication is now the most pertinent medium of communication of any type, in any situation. Electronic communication involves the spread of information, and information is power. It is for this reason that all security professionals also recognize that we not only have to be pro-active to manage all these risks, but also figure out ways to be resilient in the event an attack is successful. "Resilience" is the ability to rebound after something difficult. In the event a cyber attack, for example, shuts down all critical infrastructures, how does an agency proceed? Will the agency fall apart or will it return even stronger and bounce back quickly? To best accomplish the goal of an enhanced and strengthened recovery, this team proposes weaving a layered security network, recognizing the cyber environment as its own vulnerability and using everyone to safeguard against threats. To do this, we must focus on promoting three primary things: teamwork, resilience and risk mitigation. By utilizing and incorporating, different departments and agencies, the common goal of cyber security and safety in mind is achieved.

RESOURCES AND METHODOLOGY

The AMU Team utilized scholarly resources that can be verified such as government publications, peer reviewed articles and other such verifiably accurate documents.

In a two-month timeframe, the team's initial research commenced with a comprehensive list that was presented to the team by AMU instructors and then new information developed by different members of the team selected to a specific area. The qualitative case study approach was selected to best accommodate the available resources, time constraints of the project, and the intended end user environment, while at the same time allowing for the most complete analysis available from the team.

Pattern Matching

- Cyber Security is a problem in other realms; therefore, cyber is a problem in the maritime environment, as the same concerns exist as in other environments. This realization requires educated assumptions as to how they correlate in order to best deal with the threats.

Content Analysis

- Key concept grouping
- Indexing, categorized by grouping key words and themes

Cases Studies/Secondary Analysis

- Use already published records and data
- Use current and valid data: the cyber world is fast moving and evolving in nature.

By addressing our methodology in this manner, we have also organized our analysis and recommendations in a logical way, which answer many of the questions an intelligently aggressive leader is looking for on a topic such as risk management.

Status of the Research, Gaps, and Areas of Research That Need Further Exploration:

New and Emerging Field and Focus

- Additional research is needed, especially in maritime area
- Landscape of the threat is changing constantly
- Many of the answers have not yet been found and the questions are in constant evolution

CHAPTER 18

Evolving and Growing Field

- Many gaps in research due to fast evolving variables-The technology is advancing faster than the response to it
- Qualitative/creative thought is needed to attempt to predict the future of the cyber world and possible threat from it

Trends of Attack Types

- Criminal, terrorism and espionage remain as constantly expanding vital concerns. Additional research and information sharing remain, to determine whether or not there is any relationship between them or if they are distinctly separate.

Promoting Cyber Security

- Focusing on training all personnel on the importance of Cyber Security principles (first step in any risk management is awareness).
- Limiting or not allowing personal use of networks at work.
- Encouraging teamwork not only among members of the same entity, but also between other entities.

The Cyber Environment

- Extremely porous with multiple vulnerabilities.
- A wide variety of players and customers with diverse levels of structure and security needs.
- Will require a multiphase approach to engage and minimize the threat environment.

The task facing the cyber domain will require dealing with the five primary pillars built upon a base of policy and regulations and enacted by the Maritime Transportation System.

1) **Awareness:** All players must realize there is a threat. Research indicated that the threats posed and the level of sophistication of adversaries is much greater than personnel realize. These threats are rapidly evolving and there must be continual education, awareness, and training.

2) **Flexible and Adaptive Security:** with a wide range of players at mul-

tiple levels, the security will need the ability to spread across a varied cyber-scape and be scalable to be usable to all levels of players.

3) **Domain Understanding:** there must be an understanding of the terrain of the cyber-scape and the range of threats in that terrain. The Maritime domain is extremely varied and vast in its scope, as such it presents a broad range of potential threat domains. To effectively deal with each domain, the awareness of the challenges to each must be understood by those in it.

4) **Enabling the Mission:** it is critical that all parties are enabled, possibly at different levels, to deal with the threats in the cyber environment. From sensitive and secure to completely un-vetted, players must have information available to them at varying levels of security that enables the cross sharing of information.

5) **Risk Management:** not all threats can be stopped, or immediately dealt with. There must be a planned risk management structure to deal with the range of threats and manage them at an acceptable level.

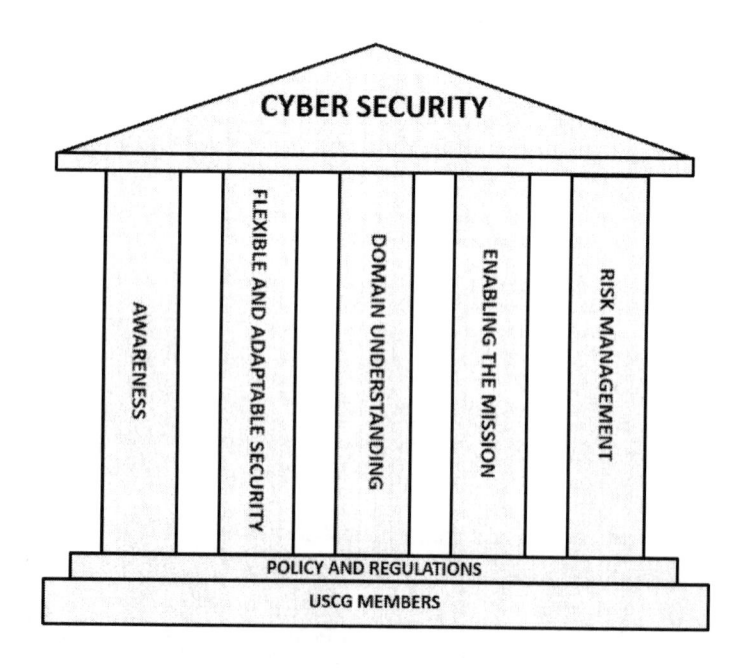

CHAPTER 18

PROMOTION METHOD SOLUTIONS:
Building upon the pillars of understanding cyber security, we move toward promotion methods. We will examine some of the areas, which the research indicates needs to be addressed and further explored.

The concept of understanding cyber security naturally leads to the principles of risk management in the above referenced category, and if followed helps make others aware of how important it is. There are only a few ways of promoting anything—essentially filling some need (monetary, country, self-image, fear of job loss, devotion to duty, or a cause). How do we promote? The act of motivation deals with a goal, purpose, or encouragement toward an outcome.

TYPES OF MOTIVATION: POSITIVE AND NEGATIVE

Positive Motivation

Recognition or Awards: National, local (i.e., best new cyber safety practice, etc.) in the industry through newsletters, magazines, or industry awards.

Creation of a New Field: Prestige needs to be added to the security offices within the shipping companies and ancillary companies that support them. This recognition will alert the shipping companies and their customers of which entities are pursuing best practices in countering the threats through education directed action and training in real-time scenarios. Prestige needs to be added to the IT/Cyber/Intel sector, as this is a fast evolving field with major ramifications.

Education:

- Establish a Cyber Department and stronger IT support
- Attract qualified graduates via signing bonuses and special pay programs
- Use combination of online learning modules, specific courses, and conferences to develop and maintain cyber proficiency
- Send personnel to cyber specific schools, such as Defense Cyber Investigations Training Academy (DCITA)

Creation of an Innovation Hotline or email address: This would be from inside a company, identifying real vulnerabilities and hopefully giving actionable solutions toward fixing them (give a quarterly award for best idea, best vulnerability identified best solution, etc.), obviously praise and reward these people.

Rewards: for turning in malicious emails (monetary, passes, free computer/electronics, etc.).

Raising Awareness: Dissemination of a weekly cyber newsletters or email:
- This is what we are seeing
- This is what happened
- This was the repercussion or punishment of this incident

Technology: Networks
- Quarantine Zone (or firewall) of sorts that establishes a set boundary between inside and outside a network.
- The ability to scan outside data/incoming messages before they are allowed access—should be primarily automated and less human initiated.
  - The use of WIFI and Bluetooth technology makes these steps difficult, but they need to happen.

Technology: Bio-scan versus Password Access
- Using fingerprint scanning or facial recognition for system access instead of passwords.
  - We need to remember too many passwords that are getting longer and more complex.
  - You must change them too frequently and not able to reuse past ones.
  - This forces them to be written somewhere or forgotten (causing vulnerability and access delays).
  - This gives approved personnel easier secure access, prevents aggravation.
  - Less chance of creating workarounds (such as writing down of passwords) = less insider threats.

CHAPTER 18

Negative Motivation

Accountability:

- Widely based on punishment or fear (jail, civil lawsuit, job loss, discharge)
- Repercussions for failure to follow procedure:
- Forced to take a class or present a class to co-workers on an incident/best practices (raising awareness which is the most important)

Overall Environment Including Port Security:

- Mandate regular patch updates from all vendors and those who have access
- Mandate sharing of data on cyber attacks with partners who track and combat such efforts
- Offer preferred status to companies/partners with high standards of Cyber Security, for example ones with dedicated Cyber Security and Data Breach reporting officers. This enhances port security operations
- Enact data monitoring and the possible disabling of access for employees deemed high risk or disgruntled, and those who show signs of meeting that criteria

Facilitating this study we have a combination of active duty and prior service members representing Navy, Army, Air Force, Marine Corps, Coast Guard, National Security Agency, Custom and Border Patrol and other civilian agencies giving us a wide range of standard operating procedures and tactics used throughout the US Government and Cyber Security environment. AMU/APU involvement is a valuable partner in any future research or study, especially those relating to Department of Homeland Security and Department of Defense procedures and policy implementation.

What Remains in the Near Future

Cyber security professionals are in great need, not just in the maritime industry, but also in all aspects of government and businesses, in general. There are not enough security professionals to keep pace with the ever-advancing technology. Per Lee and Rotoloni (2015) "... there is a significant lack of trained security experts, which will result in a shortfall of as many as 1.5 million workers by 2020" (4). Trying to overcome these shortfalls is no easy task because it takes years for students to obtain a degree in cyber security and another year of on-the-job training to become proficient in defending networks, "When Intel hires a new computer science or engineering graduate for a security position, it takes about one year to train ... them for their work ..." (Lee and Rotolini 2015, 5). Training these professionals is a long-term solution that will eventually provide much needed cyber security. Furthermore, the USCG should work toward increasing their cyber security staff to include hiring more contractors to fill in personnel gaps.

One way to enhance cyber security, is simply to educate employees about the safety practices that should be taken while working online, "companies need to focus on educating their employees about security issues—teaching them about the dangers and consequences of phishing, unencrypted data and lax reactions" (Lee and Rotolini 2015, 6). Most employees may not be aware that what they do at work could be a potential risk to the network; therefore, unintentionally compromising the work place. Simply connecting to the internet while in the Maritime environment could open the door to dangerous attacks as many network connections are already infected with a variety of attacks waiting to strike:

> ... cyber experts who manage to get afloat report worrying signs of general defencelessness. At a recent conference, an expert noted that after every visit, he would routinely destroy his laptop, such was the level of "contamination" by viruses and the like in the systems he saw afloat. Pirated software is routinely reported in ships' navigational systems and devices brought aboard by crew or contractors are often badly contaminated (Grey 2016).

By incorporating cyber awareness training, employees will be less prone to accidentally compromising the network. Furthermore, this is a simple way to enable employees to be security assets rather than a hindrance to network security. If need be, "companies should, if appropriate, clamp down on personal use of employer-issued devices to minimize threats or monitor the use of consumer devices inside the workplace" (Lee and Rotolini 2015, 6). Removing these devices would maintain security in the workplace. Employers could utilize a safe space for devices to be used and lockboxes to safely and securely store their devices before re-entering their place of work.

Additionally, the national government needs to realize vulnerabilities within their port infrastructures and provide the necessary training and manpower to successfully improve their cyber security (Jensen 2015, 38). Cyber attacks on ports can have damaging effects on the economy because such attacks can cause crippling delays in the importing and exporting of goods to different countries. Various delays due to natural disasters highlight the economic importance of the shipping industry, "northeast ports lost an estimated $50 billion—$1 billion in cargo delays alone—because of Hurricane Sandy in 2012" (Walters 2015, 1). Furthermore, "in 2002, the key ports on the western coast of the United States were shut down for ten days due to a labour dispute ... it was estimated that this had a cost to the United States economy of $1–2 billion USD per day due to disrupted supply chains" (Jensen 2015, 35). Complex cyber attacks by technically advanced adversaries can produce these same results or worse if the industry cannot adapt to new threats.

This realization cannot occur too soon as the threat level is increasing at a dramatic rate. "The number of incidents reported by federal agencies to the federal information security incident center has increased by nearly 680 percent" (United States Government Accountability Office 2012). Further, not only is the number of attacks increasing, but the sophistication and caliber of the attackers is also ramping up to unseen levels of participation. Harris (2015) uses Iran, North Korea and Russia to demonstrate this. In particular, Iran carried out a "massive attack on U.S. bank websites in 2012," North Korea hacked Sony Pictures, and Russia's "use of cyber offensive operations has been documented both in Georgia in 2008 and more recently with Russia's invasion of Crimea in 2014" (Harris 2015, 4).

The effect of the increased sophistication and capability can be felt in industries once considered ultra-secure such as national and state power grids in the U.S. and abroad, as well as closely protected information used in multinational corporate negotiations:

1) In Florida, a hacker caused a large power outage as he strayed while attempting to map the infrastructure (Harris 2008).

2) Overseas breaches of utility computer systems have occurred, with some resulting in power outages and attempted extortion, as ransoms were demanded. The attacks were validated by the CIA. Access is believed to have occurred through the Internet (Harris 2008).

3) There is "... a huge increase in focused attacks on our national infrastructure networks." The source(s) of the attacks is/are believed to be foreign (Harris 2008).

4) As executives travelled through China, spyware was uploaded into their networks. "cyber counterintelligence" threat (Harris 2008).

The data are clear, the cyber world is no longer secure, and the elements making it less secure are increasing exponentially with expanding sophistication.

Governments need to take charge of their cyber security programs and to seriously address cyber security issues. This should include working closely with departments and to develop interagency cooperation in terms of finding solutions to cyber security problems. "Domestically, agencies should address continuity and simplicity in identifying cyber threats, such as the definition and severity of threats, attacks, and solutions, while avoiding the creation of catch-all regulation that hinders business" (Walters 2015, 3). While a catch-all regulation could potentially hinder business, the body of research indicates a broader approach such as a common regulation standard. This would be a good start to dealing with the threat, but such a step would be very complicated, involve multiple jurisdictions and governments, and need to cover a wide range of systems. Thus, a single common environment is viewed as unlikely in the near term. "Cyberspace is currently seen as 'ungovernable, unknowable, makes us vulnerable, is inevitably threatening and is inhabited by a range of threatening and hostile actors'" (Barnard-Wills and Ashenden 2012).

The situation is extremely challenging and complicated even when looking at a single country such as the United States. With a single governing body, and examining only the requirements for the Maritime areas, protecting the nation demands a collective effort that involves cooperation between all agencies and departments. Together, the collaborative efforts of the United States can create comprehensive security measures to protect America's ports. When such an effort is extended to groups of countries, with multiple governing bodies, and their multiple requisite agencies, the challenge grows exponentially. Further, the creation of comprehensive measures will require a change in thinking and assessment in order to reach such measures, which may add to the challenge. The problem can be described as "not being the lack of cyber security promotion but the inability of policy makers to fully understand the risk of cybersecurity as being the problem" (Kelic et al. 2013). To be sure, the cyber security world is evolving into a new and intelligently aggressive concept that bears observation in order to anticipate changes that support new ideas in the future.

To truly secure America's ports, there also must be coordination internationally, "increasing cyber information includes working with international partners because cyber attackers may enter U.S. port networks by any available means" (Walters 2015, 3). Working with other countries and international organizations can truly make the maritime industry's cyber security reforms successful. Attempts are underway in Europe to address the challenge in broad reach. "The European Union has been tinkering with, but has not yet implemented, a single Europe wide Data Protection Directive that would equalise regulation across all EU States" (Lackie 2016). If those involved in the Maritime industries can agree and execute a streamlined cyber security strategy, then it would be much easier to monitor threats and share vital information pertaining to cyber attacks. Doing this would involve the International Maritime Organization and although it would take a few years to come to an agreement on a streamlined cyber security strategy, it would be a step in the right direction (Jensen 2015).

Sound maritime cyber security is a complex task that will require the help of all those involved in this industry. Not to do so could prove catastrophic. "Imagine shutting down a port. Imagine running a ship aground. These are the kinds of implications we're worried about" stated Todd Humphreys, a GPS expert at the University of Texas (Roberts 2013). Such scenarios may seem extreme, but when looking for a baseline view of where overall industry standards currently stand outside the maritime environment, even among such sensitive areas as government personnel records, a recent audit found the mandatory regulations were not being adhered to. A recent comment indicated "... only 75 percent of OPM's critical systems had valid authorizations in accordance with FISMA [Federal Information Security Management Act] regulations, and in January, an inspector general audit of OPM that deemed the agency's cyber security sufficient relied on unverified data simulation ..." (Kanowitz 2015). When it comes to the maritime environment, and ships, a recent study indicates the scenario is even worse. A Copenhagen-based firm believes even larger vulnerabilities exist from unpatched Microsoft servers that could allow attackers to exploit and take control of the servers. Microsoft had released patches in April but spot checks revealed 37% of the servers have not been patched (Network World 2015). This further leads to a confirmation that not enough is being done to deal with the potential threat, especially in non-mandated areas of concern. In short, there may be entire systems that are already potentially severely infected in the Maritime Industry environment.

Conclusion

The UCSG should work toward getting those involved in the maritime industry to formulate a comprehensive cyber security strategy that will streamline maritime defenses. Doing so will allow for adequate information sharing which will make it easier to identify weaknesses and potential cyber attacks that could have crippling effects. The effect of the attacks are already being felt, "2015 saw some of the biggest data hacks to date costing the global economy some US$400bn, highlighting the inability of companies to properly guard valuable entrusted data" (Lackie 2016). Furthermore, information technology professionals and security experts need to stay informed when it comes to new ways terrorists and criminals implement new viruses and ways to hack network systems. Learning these new advancements can help provide them with the necessary knowledge to develop tools to combat cyber terrorism and cyber criminal activity. If the Maritime environment follows suit with traditional non-maritime corporate environments, there will likely be a new broad acceptance of the position of Cyber Security Officer finding its way into business leadership. This move will likely come as a result of the high cost, and potential liability, according to Lackie's (2016) research, which theorizes the new reporting officers will become a part of organizational fabric across the globe as enforcement steps up in response to the expense. The enhanced enforcement already exists in the Unites States and is expected soon in Europe.

References

Barnard-Wills, David, and Debi Ashenden. 2012. "Securing Virtual Space: Cyber War, Cyber Terror, and Risk." *Space and Culture* 15 (2): 110–123.

Grey, Michael. 2016. "Cyber Attacks—Coping with New Threats to the Maritime World." *Seatrade Maritime News.* http://www.seatrade-maritime.com/news/americas/cyber-attacks-coping-with-new-threats-to-the-maritime-world.html (accessed November 12, 2016).

Harris, Shane. 2008. "China's Cyber Militia." *NextGov.com.* http://www.nextgov.com/defense/2008/05/chinas-cyber-militia/42113/ (accessed November 13, 2016).

Harris, Shane. 2015. "China Reveals Its Cyberwar Secrets." *The Daily Beast*, March 3. http://www.thedailybeast.com/articles/2015/03/18/china-reveals-its-cyber-war-secrets.html.

Jensen, Lars. 2015. "Challenges in Maritime Cyber-Resilience." *Technology Innovation Management Review* (April): 35–39.

Kanowitz, Stephanie. 2015. "Old Technology, Poor Governance to Blame in OPM Breach, Report Finds." *FierceGovernmentIT.com.* http://www.fiercegovernmentit.com/story/old-technology-poor-governance-blame-opm-breach-report-finds/2015-07-20?utm_medium=nl&utm_source=internal&mkt_tok=3RkMMJWWfF9wsRons6jJdO%252Fhm jTEU5z14uQrW6CylMI%252F0ER3fOvrPUfGjI4DSsVrM6%252BT FAwTG5toziV8R7LMKM1ty9MQWxTk.

Kelic, Andjelka, Zachary A. Collier, Christopher Brown, Walter E. Beyeler,

Alexander V. Outkin, Vanessa N. Vargas, Mark A. Ehlen, Christopher Judson, Ali Zaidi, Billy Leung, and Igor Linkov. 2013. "Decision Framework for Evaluating the Macroeconomic Risks and Policy Impacts of Cyber Attacks." *Environment Systems and Decisions* 33 (4): 544–560.

Lackie, Lara. 2016. "2016—The Rise of the Cyber Security and Data Breach Reporting Officer." *Itsecurityguru.org*. January 7. http://www.itsecurityguru.org/2016/01/07/2016-the-rise-of-the-cyber-security-and-data-breach-reporting-officer (accessed November 12, 2016).

Lee, Wenke, and Bo Rotolini. 2015. "Emerging Cyber Threats Report 2016." Georgia Institute of Technology, (October): 1–17. http://www.iisp.gatech.edu/2016-emerging-cyber-threats-report.

Network World. 2015. "Maritime Cybersecurity Firm: 37% of Microsoft Servers on Ships Vulnerable to Hacking." *Network World*. (May 4). http://www.networkworld.com/article/2917856/microsoft-subnet/maritime-cybersecurity-firm-37-of-microsoft-servers-not-patched-vulnerable-to-hacking.html.

Roberts, John. 2013. "Exclusive: GPS Flaw Could Let Terrorists Hijack Ships, Planes." *Fox News Technology*. (July 26). http://www.foxnews.com/tech/2013/07/26/exclusive-gps-flaw-could-let-terrorists-hijack-ships-planes.html.

United States Government Accountability Office. 2012. "Cybersecurity: Threats Impacting the Nation." GAO-12-666T. http://www.gao.gov/assets/600/590367.pdf.

Walters, Riley. 2015. "The U.S. Needs to Secure Maritime Ports by Securing Network Ports." *The Heritage Foundation*, No. 4353 (February): 1–3.

Weintraub Schifferle, Lisa. 2015. "OPM Data Breach—What Should You Do?" Consumer Information (blog). Federal Trade Commission, June 4. http://www.consumer.ftc.gov/blog/opm-data-breach-what-should-you-do.

PART III:
Technical

CHAPTER 19:
ECONOMIC CONSEQUENCE ANALYSIS
OF MARITIME CYBER THREATS

Adam Rose[1]
University of Southern California

INTRODUCTION

The development and implementation of cyber technology is accelerating at a rapid pace in all facets of society. This is especially the case in areas of national defense in general and in the maritime domain in particular, where the U.S. Coast Guard is charged with the safety and security of vessels and ports. While cyber systems enhance defense and business operations, they also make them more vulnerable to disruptions because of the increased network dependency.

One major example pertains to cyber threats affecting seaports. The impacts are wide-ranging and not just confined to individual ships, cargo damage, or port operations. They cause a disruption in downstream supply chains for U.S. businesses depending on imports in their production process. They also cause a shortfall in the supply to satisfy consumer demands. Analogously, they disrupt upstream supplychains relating to U.S. exports whose production is halted because goods cannot be transported overseas. Actual events and simulation studies have indicated losses of tens of billions of dollars from various broader impacts of port disruptions (see, e.g., Cohen 2002; Park 2008; Rose

---

1 The author is Research Professor, Price School of Public Policy, University of Southern California (USC), and Faculty Affiliate, Center for Risk and Economic Analysis of Terrorism Events (CREATE), USC. This research is funded by U.S. Department of Homeland Security under Grant Award Number 2010-ST-061-RE0001-05. The author acknowledges the collaboration of Dan Wei on previous port studies, the collaboration of Fynn Prager, Zhenhua Chen, and Sam Chatterjee on the development of the Economic Consequence Analysis Tool (E-CAT) and the research assistance of Joshua Banks and Noah Miller. This summary also received valuable input from members of the Maritime Cyber Economic Consequence Analysis Working Group (see Appendix A). The author is, of course, responsible for any errors or omissions.

and Wei 2013; Werling 2014). Cyber disruptions could have similar outcomes.

However, those affected do not stand by passively. They undertake various types of post-disaster resilience by using remaining resources more efficiently and recovering more quickly. These responses take place with regard to cyber capability and also other aspects of port operations such as the use of excess capacity and ship re-routing. Other resilience tactics are implemented up and down the supply chain, such as use of inventories, conservation, input substitution, and lining up new suppliers).

The purpose of this chapter is to set forth a comprehensive framework for the estimation of total economic consequences of maritime cyber threats. This includes a categorization of threats and how they directly affect port operations. It includes a characterization of the major types of indirect, or ripple, effects this may cause. It also includes the specification of cyber resilience tactics that can reduce business interruption losses.

We will utilize this framework to develop a rapid estimation capability for the economic consequences of maritime cyber threats. The author has recently led a research team that developed a user-friendly software system capable of providing such rapid estimates for a dozen diverse threats to the U.S. economy (Rose et al. 2017). This decision support system is known as the Economic Consequence Analysis Tool, or E-CAT. The next step is to incorporate several types of cyber threats into this platform.

The enhanced software system is intended to help high-level decision-makers in the maritime cyber domain assess the severity of various threats in real time. This is a key aspect of a benefit–cost analysis or risk-benefit analysis, as the benefits of reducing threats are essentially the averted negative consequences. This holds both for pre-event mitigation efforts and post-event resilience.

Note this research does not address all maritime cyber-related issues. It does not assess the value of compromised military operations due to cyber threats. It does not assess the full extent of property damage. And, it does not address pre-event mitigation.[2] Consequences are measured in gross terms in the absence of resilience and in net terms in its presence. The metrics used are standard macroeconomic indicators of business interruption, such as gross domestic product (GDP), personal income, and employment.

THE ROLE OF ECONOMIC CONSEQUENCE ANALYSIS IN THE USCG CYBER STRATEGY

The design of this project takes direction from the *United States Coast Guard Cyber Strategy* (USCG 2015), which presents the Coast Guard's vision for op-

---

2 The reader interested in pre-event mitigation of cyber threats is referred to Farrow (2015) and Sinha et al. (2015).

erating in the cyber domain.

To begin, Economic Consequence Analysis (ECA) is at the heart of the document's definition of a Cyber Incident—"An occurrence that actually or potentially results in adverse consequences to an information system or the information that the system processes, stores, or transmits and that may require response action to mitigate the consequences" (USCG 2015, 41, based on US DHS 2015).

Our focus will be on Cyber-Dependent Critical Infrastructure—"Critical Infrastructure where a cyber security incident can result in catastrophic regional or national effects on public health or safety, economic security, or national security" (USCG 2015 41, based on Executive Order 13636, 2013). The *Cyber Strategy* document notes that the Coast Guard is responsible for protecting the Maritime Transportation System (MTS), which consists of 360 sea and river ports that service $1.3 trillion in annual cargo. Other US government agencies have also stressed the importance of this system. For example, a Government Accountability Office report (GAO 2014) emphasized the critical role of port cyber security to the continued full operation of these ports.

A decision-support system that estimates the economic consequences of maritime cyber threats fits into several of the Coast Guard's strategic priorities. For example, in relation to the priority of Protecting Infrastructure, one of the objectives is to "identify existing cyber security risk assessment tools, and, where appropriate, tap them for Coast Guard use and share them with the maritime industry" (USCG 2015, 32). A related objective is to "Modify Maritime Security Risk Assessment Model (MSRM) to incorporate cyber risks, or identify a similar tool that performs the same function (USCG 2015, 32). In terms of the cross-cutting goal of Ensuring Long-Turn Success, there is a stated need for communicating in real time (USCG 2015, 36).

Although most of the *Cyber Strategy* document focuses on pre-event mitigation, significant portions do address post-event activities, which we discuss further below under the heading of *resilience*. Although this term is not defined in isolation, network resilience is defined as—the ability of a network to: (1) provide continuous operation, (i.e., highly resistant to disruption and able to operate in a degraded mode if damaged; (2) recover effectively if failure does occur; and (3) scale to meet rapid or unpredictable demands (USCG 2015, 43; based on US DHS 2015). The terms *response* and *recovery* are closely connected to resilience in both *Cyber Strategy* and in the E-CAT Tool. For example, Cyber Strategy defines recovery as—"The activities after the incident to restore essential services and operations in the short and medium term and fully restore all capabilities in the longer term (USCG 2015, 43, based on US DHS 2015).

ISSUES IN MARITIME CYBER SECURITY

A GENERAL FRAMEWORK FOR MARITIME CYBER CONSEQUENCE ANALYSIS

An overarching framework of analysis combining maritime physical, cyber, and economic systems would contain the following key elements:

1) Categorization of Major Systems dependent on cyber networks
   - Coast Guard
     - command centers
     - ases
     - intelligence units
     - individual vessels
     - surface support equipment
   - Ports
     - operations centers
     - loading facilities
     - emergency response centers and equipment
   - Ships (by origin, destination, and type)
   - Cargo (by origin, destination, and type)
   - Supply-Chain (upstream and downstream for each type of cargo)
   - Regional Economy (by size and economic structure)

2) Cyber Landscape, covering the role of cyber in each of the major systems. For starters, this could be a logistical analysis.

3) Cargo Movement, covering docking needs, handling equipment, and cargo characteristics perishability, fertility, strategic importance. Again a logistical analysis would be in order.

4) Type of Threat (including various subcategories)
   - Natural
   - Human Intentional
   - Human Accidental
   - Technological Failure

These various components could then be organized into a *Threat-Consequence Matrix*, which displays the degree to which various types of cyber threats/

incidents have potential impacts on the various systems. It would be useful to distinguish direct and indirect threat affects. Indicators of impacts would include degree of function and capacity. For starters, one could use qualitative designations ranging from low to high impacts (see also the discussion of an Economic Consequence Enumeration Table below.

The CREATE Economic Consequence Analysis Framework

CREATE's expanded framework for economic consequence analysis (ECA) of terrorist attacks and natural disasters is shown in Figure 1-1. It has been formulated to account for several standard and new considerations that affect bottom line impacts (Rose 2009a; 2015a).

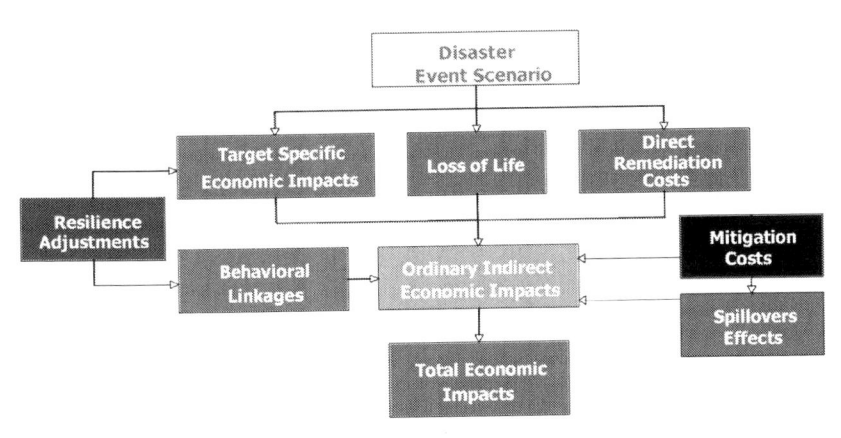

Figure 1. Economic assessment framework overview

For many years, the estimation of losses from disasters focused almost entirely on standard target-specific (Direct) Economic Impacts and Loss of Life, and, to some extent, Ordinary Indirect Effects in terms of multiplier (quantity supply-chain), general equilibrium (multimarket quantity and price interactions) or macroeconomic (aggregate behavioral) effects.

The first major refinement to these standard economic consequences is the inclusion of *Resilience*, which refer to actions that mute business interruption and that hasten recovery. Rose (2004; 2009b) has proposed as an operational metric of resilience: the avoided losses resulting from implementing a given resilient tactic as a proportion of the maximum potential losses for a given event in the absence of that tactic. Rose et al. (2009) measured the resilience of the New York Metropolitan Area economy to the 9/11 World Trade Center attacks at 72% as a result of business relocation as a resilient response. This stemmed from the fact that 95% of the businesses, comprising 98% of the employment, in the World Trade Center area did not shut down but rather relocated their operations, mainly within the New York Metro Area (the losses are simply due to the time lags in the relocation).[3]

In the past decade, the major extension of economic consequence analysis has been to include *Behavioral linkages*. A prime example is the "fear factor," which refers to changes in risk perception that translate into changes in economic behavior and may amplify damages instead of reducing them. Rose et al. (2009) measured the effect of the nearly 2-year downturn in air travel and related tourism in the United States following 9/11 at $85 billion, which accounted for over 80% of the estimated business interruption losses stemming from the event. A recent study by Giesecke et al. (2012) of a potential RDD ("dirty bomb") attack on the financial district of Los Angeles would lead to social amplification of risk and stigma effects that could exceed the conventional "resource loss" effects by 14-fold.

The framework includes three other aspects necessary for a comprehensive analysis, the implications of which are often misinterpreted. The first is *Remediation*, which is typically not part of traditional economic impact analysis and has a conventional role in hazard loss estimation as simply repair and reconstruction. In the case of a terrorist attack, this can take on a much larger role, especially if the attack is caused by an insidious chemical, biological, radiologic, or nuclear (CBRN) agent. For example, Baker (2008) found that the cost of remediation for a radionuclide attack on a reservoir of a small city of 100,000 was equal to the sum of the property and business interruption losses because of the extensive spread of the contamination and the high standards of remediation set by U.S. EPA.

Second, *Mitigation*, public and private actions prior to the event that reduce impacts, also enters the picture of a comprehensive economic consequence framework in its move toward a full-blown counterpart to benefit–cost analysis (BCA). The interesting consideration here is the interpretation by many that remediation and mitigation have benefits stemming from their direct expenditures alone (aside from the standard benefits of avoided losses). This perspective is often criticized because it appears to ignore the basic principle that resources are expended in the course of implementing remediation or mitigation, and that these resources typically must be diverted from productive use elsewhere. Of course, if the economy is not at full employment (the typical situation), or, at the regional level, where in-migration of new workers is likely, then indirect effects can be included, as admitted by most authorities on BCA (see, e.g., Boardman et al. 2001). ECA, on the other hand, does not make an a priori judgment on this question and simply explores whether the em-

---

[3] Others have used this metric as well to measure resilience (see, e.g., Kajitani and Tatano 2009). For a broader view of cyber resilience, see Linkov et al. (2013).

ployment adds, detracts, or is neutral with respect to the bottom-line, e.g., its impact on GDP. The answer has a great deal to do with whether the economy is initially at full employment, but is also influenced by whether higher-order effects of resource diversion are larger or smaller than those associated with mitigation or remediation.

Third, the mitigation effort can generate various types of "non-market" *Spillover* effects in the form of congestion, delays, inconveniences, changes in property values, changes in the business environment, and changes in the natural environment. These are difficult to measure, but have been found to be significant in both negative and positive directions, e.g., closed-circuit television surveillance is minimally intrusive, and its improvement in the business environment due to the public feeling safer from both terrorism and ordinary street crime can outweigh the intrusion on privacy (Rose et al. 2014).

The presence of *Resilience* and *Behavioral Responses* imparts significant variability to the economic consequences of terrorism in relation to attack mechanisms and targets. Simple rules of thumb cannot be used as in the relatively mundane areas of general economic impact analysis. Computable general equilibrium modeling is relatively superior to other model forms because of its ability to incorporate resilient actions (see, e.g., Rose and Liao 2005) and the behavioral consequences of changes in risk perceptions (see, e.g., Geisecke et al. 2012).

ILLUSTRATION OF APPLICATION OF THE ECA FRAMEWORK TO PORT AND SUPPLY CHAIN DISRUPTIONS

We now provide an illustration of the application of the CREATE ECA Framework to the estimation of the economic consequences of two simulated port disruptions. The first pertains to a 90-day closure of Port Arthur/Port Beaumont (PA/PB) complex due to a shipping accident (see Rose and Wei 2013; 2013) and a 2-day closure of the Ports of Los Angeles and Long Beach (LA/LB) due to a tsunami (Rose et al. 2016). Overall, the examples relate to a more comprehensive and enlightened view of the value of America's ports.

The standard approach to estimating the economic impacts of a port is to determine the direct impacts of the port's operation (lost operating revenue) and then apply some form of multiplier. More recently, it has been popular to use both supply-side and demand-side impact multipliers, but just applied just to port operations alone. Accordingly, the calculations would proceed as follows:

- PA/PB: $220 million × 5.9 = $1.3 billion
- LA/LB: $1465 million × 5.9 = $8.6 billion

However, the standard approach misses the value of the cargo and contribution to the rest of the economy. Thus, the prior analysis would grossly understate the economic consequences of these port shutdowns. At the same time, many analyses that include supply-chain aspects fail to take into account resilience, or the ability to mute the negative consequences by using remaining resources more efficiently or recovering more quickly, which have the effects of offsetting some of these negative impacts.

A more comprehensive view of the economic consequence domain is presented in Figure 2, which displays the major linkages in tracing port disruptions from closure and damages beginning with direct economic impacts through short-run and longer-run impacts across five analytical stages of a disaster scenario (see also Rose et al. 2016).

The analysis begins with the tsunami event, which first translates into a risk of a port shutdown, cargo damage, and isolated terminal downtime for extended periods of time. At the port level, this leads to disruption of imports, exports, and port onsite activities and operations. Various resilience measures can be implemented to mute impacts at the outset, including rerouting the traffic to other ports, diversion of exports for use as import substitutes, use of inventories by port customers, relocating activities within the ports, and rescheduling of activities once the port reopens by working overtime or extra shifts.

At the level of the macroeconomy, impacts stem from increased scarcity of intermediate and final commodities, and reduction in final demand associated with a decline in exports. Both supply-side and demand-side impacts must be taken into account when evaluating total economic impact. Supply-side impacts affect customers of imported goods down the supply chain, while demand-side impacts affect these customers' suppliers up the supply chain. Intermediate goods imports are subject to both supply and demand impacts. Firms using imports as intermediate inputs to production, as well as successive rounds of downstream customers are subject to supply shortfalls. In addition, curtailment of production by import-using businesses also reduces the demand for intermediate commodities produced by successive rounds of upstream suppliers within the region, or nation. Curtailments of "final" (finished) imported goods supplies only impact end-users (consumers, government, and purchasers of capital equipment) without generating forward or backward linkage effects, and are simply added to the total macroeconomic impacts.

The shutdown of port operations limiting export shipments is characterized as an impact on suppliers, since downstream customers are outside the region and thus do not affect California's GDP. Conversely, disruption of export commodities reduces the demand for inputs to the production of these goods. First-round suppliers in turn reduce their demand, triggering a cascading

CHAPTER 19

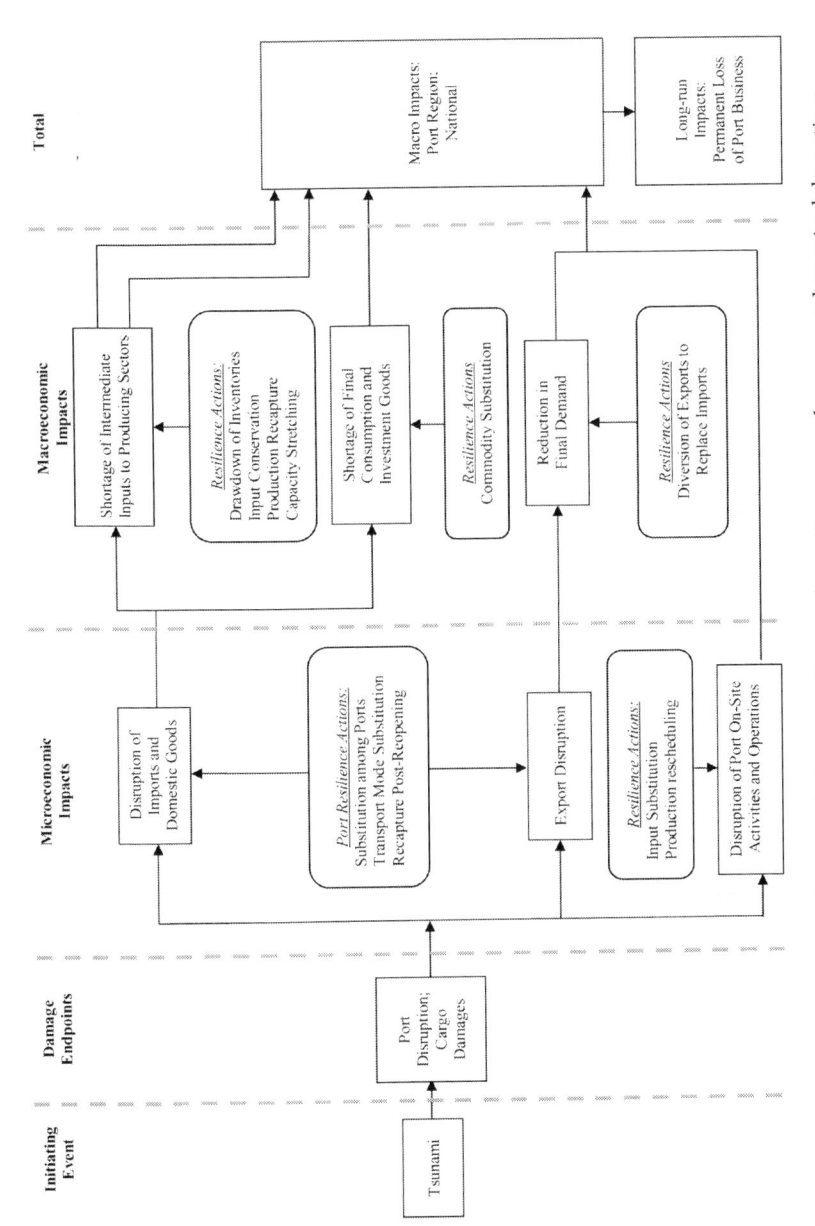

Figure 2. Estimating total economic impacts of a port disruption, cargo damages and terminal downtime

decline in upstream production activities, analogous to that experienced by imports. The sum of all of these impacts is a multiple of the original initiating shock; hence, the term "multiplier" effect (both price and quantity) characterizes the manner in which these reactions yield macroeconomic impacts.

The estimation of the numerous macroeconomic linkages and resilience offsets can also be illustrated by the results of the simulated 90-day disruption at Port Arthur/Port Beaumont (see Rose and Wei 2013). Again the standard estimate of economic impact on the U.S. economy presented at the outset of this section was $1.3 billion. Taking all of the linkages into account raises this estimate to $14.8 billion in the absence of resilience. However, there are many resilience tactics that are applicable as shown below, listed in terms of their percentage ability to reduce gross macroeconomic consequences, such as GDP. These range from very small gain from accessing the Strategic Petroleum Reserve or conservation efforts by the petrochemical industry dominating the surrounding economic region to rather high levels of resilience from imports ship rerouting in production rescheduling (making up lost production at a later date when input supplies are restored). Note that the total amount of resilience in terms of the reduction business interruption losses is 67% of the base estimates (the individual tactics are not fully additive, but contain some overlaps and offsets of their own). Thus, the bottom-line estimate of the economic consequences of the Port Arthur/Port Beaumont shutdown is $4.8 billion, or 3.7 times the initial estimate.

Strategic Petroleum Reserve	2.4%
Ordinary Inventories of All Goods	17.0
Conservation by Customers	3.0
Import Ship Rerouting	23.1
Export Diversion (to Replace Imports)	7.0
Production Rescheduling (Recapture)	25.4
Total Resilience (not additive)	67.0%

Economic Consequence Analysis Tool (E-CAT)

Reduced-Form Analysis

The state-of-the-art modeling approach for economic consequence analysis is computable general equilibrium (CGE) analysis. This approach models the economy in terms of the multimarket responses of individual producers and consumers in response to price changes, government policies, and external shocks, subject to constraints on labor, capital, natural resources. It essentially models the economy as a set of interconnected supply chains (see, e.g., Dixon and Rimmer 2002). CGE models have been used extensively for the analysis of economic consequences of terrorism and natural disasters (see, e.g., Rose, Oladosu, and Liao 2007; Rose et al. 2009; Dixon et al. 2010; Geisecke et al. 2012; Sue Wing et al. 2016). CGE models are especially complex, involving thousands of equations representing production, consumption, and trade activities. These models are typically on the utilization of those without extensive backgrounds in economics in general and CGE modeling itself.

A "reduced-form" capability refers to a simplified version of a more complex model that can readily be operated by users with a limited amount of knowledge of economics and with a rapid turnaround. Examples of these models have been developed by Dixon and Rimmer (2013) and by Rose et al., 2017) for computable general equilibrium (CGE) models and by Rose et al. (2011) for macroeconometric models.

In the development of the Economic Consequence Analysis Tool (E-CAT) by Rose et al (2017), for a given scenario, the CGE model is run hundreds of times for variations in key variables. This provides the "synthetic" data for statistical regression equations that are the reduced form. The dependent variable is a major consequence type (e.g., GDP losses or employment losses), while the independent variables explain these losses to the extent possible.

Three factors should be considered in performing this analysis. First is the soundness of the theoretical underpinnings. This is guaranteed to a great extent by the fact that the synthetic data are generated by CGE models, which have been vetted on both a theoretical and empirical plane. CGE models reflect the behavioral responses of businesses and households within an economy to changes in prices, as well as taxes, regulation and other external shocks, within the constraints of labor, capital, and natural resource assets. CGE models are based on economic theory relating to producer and consumer choice and the workings of markets. They are able to estimate not only the direct responses but also indirect ones leading to total economic impacts, or consequences, referred to as "general equilibrium." In this modeling approach these impacts relate to price and quantity interactions in upstream and downstream markets. CGE models are constructed on the basis of a comprehensive set of economic accounts for production, household and institutional sectors, as well as some parameters, such as price and substitution elasticities, from the literature.

The soundness of the CGE model helps to ensure that results are likely to be reasonably accurate. However, we should note that accuracy depends on more than just sound theoretical underpinnings and internal consistency of the model, but also is affected by the key variables that are included or omitted. For each threat Rose et al. (2017) consider 16 categories of direct impacts that might be relevant and then quantify those that are likely to have significant effects on the results. This "Enumeration" approach is discussed in the following chapter. The third consideration is ease-of-use. While the complexity of the underlying CGE model is a plus, the opposite requirement is needed here. The reduced form regression equations include a limited number of variables that are transparent and for which numerical values can readily be obtained. The user thus need only plug these variables into the estimating equation, and a simple multiplication by parameter values yields the value of the dependent variable (see Chen et al., 2017). In E-CAT the reduced-form equations have been constituted in a user-friendly spreadsheet format to facilitate this application in an even more facile fashion.

E-CAT Model Construction

E-CAT is constructed in seven steps, as outlined in Figure 3. In Step 1, "Enumeration Tables" for as many as 16 categories of impacts for each threat are filled out according to upper and lower bounds identified from searches of relevant historical data of prior threat incidents, related literature, and/or expert judgment. In Step 2, lower and upper bound Direct Impact numerical values are estimated for each of the Enumeration Table categories that are determined to be above the "Low Influence" threshold.

Figure 3. Seven-step E-CAT research framework

In Step 3, unique sets of User Interface Variables are identified for each threat and grouped under the following categories: Magnitude, Time of Day, Duration, Economic Structure, Location, Other, Behavioral Avoidance, Behavioral Aversion, Resilience Recapture, and Resilience Relocation. Randomized draws of 100 User Interface Variable combinations generate uniformly distributed values between range boundaries for the Magnitude variable and different options for the other variables relevant to each threat. These 100 random draws are then converted to CGE inputs via a series of linkages.

In Step 4, CGE model simulations are run for each of the 100 random draw scenarios. The identified relevant Direct Impact values are then input into the CREATE CGE model of the US economy (USCGE), which captures the combined and interactive effects of these impacts through price changes and substitution effects across multiple economic institutions—58 sectors, 9 household groups, government institutions, and international traders. GDP and employment impacts for up to the first year of consequences are generated for each of these 100 scenarios, and where relevant the Economic Structure of the impacted region is also factored in by scaling the national average results across three different example regional economy structures to render 400 unique GDP and employment results.

In Step 5, multivariate regression analysis is conducted to estimate the influence of each of the User Interface Variables on the dependent variables of GDP and employment impacts, respectively. This analysis produces a reduced-form equation on the basis of Ordinary Least-Squares and Quantile regression analysis, allowing for estimates of mean, 5th percentile, 25th percentile, 50th percentile, 75th percentile, and 95th percentile results.

In Step 6, these reduced-form equations are combined to model the mean response and uncertainty surrounding the GDP and employment results for any given combination of User Input Variables. Uncertainty distributions are determined by user inputs of the parameters of a triangle distribution (i.e., a lower-bound, a mid-point, and an upper-bound) for the Magnitude variable, alongside user inputs of the other variables for that particular threat. Validation criteria and methods applicable to CGE modeling are also implemented.

In Step 7, the coefficients from the reduced-form equations are input into the E-CAT Tool. The Tool is designed to be a user-friendly interface with which to explore the deterministic and probabilistic results of the reduced-form analysis of the CGE modeling for each threat. Users first select a threat and the level of detail for the results they would like. The resulting E-CAT Tool User Interface provides an Input Area, whereby the user selects values for each of the relevant User Input Variables, and an Output Area, and where economic impact results for GDP and employment are presented in both tabular and graphical formats and with respect to both point estimates and distributions.

E-CAT Software

Following Rose et al. (2017), this section introduces the design of the E-CAT user interface tool. The tool is based on Excel and Visual Basic for Application (VBA). Three different economic consequence options are developed for each type of threat, including a point estimate (option 1), interval estimate (option 2), and uncertainty distribution (option 3). Step-by-step instructions are presented in the User's Guide in Rose et al. (2017).

The conceptual framework of the E-CAT user interface tool is illustrated in Figure 4. The analytical function of E-CAT is structured in four layers. The master user interface is designed in layer 1, which functions as the gate for various options. The different user options are designed in layer 2, which functions as the major platform for both data input and output visualization. User input information is translated from contextual format into numerical format and is then calculated based on the corresponding reduced-form coefficients stored in layer 4. User option 3 differs from option 1 and 2 in that an additional step for Latin-hypercube sampling (LHS) procedure is added in layer 3 to present the output uncertainty in various forms of probability distribution.

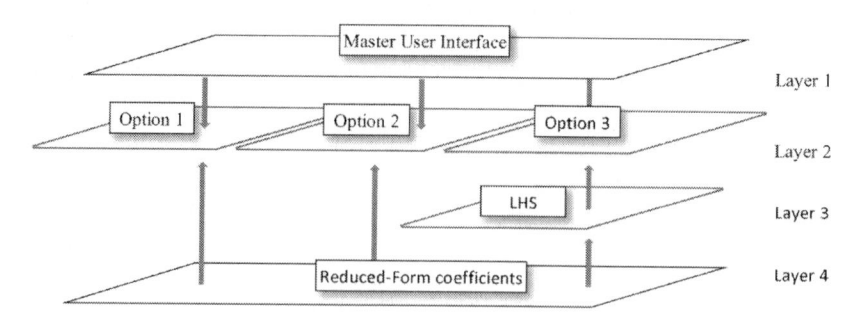

Figure 4. E-CAT user interface tool structure design

The designs of the various functional pages of E-CAT are introduced as follows. The master user interface page, as illustrated in Figure 5, is designed for the user to specify the types of threat and option of output estimation. The current version of E-CAT is able to conduct economic consequence analysis for the following categories of threats: human pandemic, nuclear attack, animal disease, earthquake, flood, tornado, and aviation system disruption. Three output estimation options are provided for each threat. When a user specifies the type of threat as "human pandemic" and the output option type as "point estimate", a point estimate page, as illustrated in Figure 6, is presented automatically. After the consequence analysis, the user can return to the main menu to select another threat or output option type by clicking the "Main Menu" button on the top right of each option page. The result can also be printed automatically when clicking the "Print Results" button. In addition, a "Reset Default" button is added in case the user wants to reset all the settings.

Figure 5. E-CAT user interface for threat type and option selection

CHAPTER 19

We illustrate the use of E-CAT with an application to estimating the consequences of a human pandemic following Prager, Wei, and Rose (2017) and Rose et al. (2017). The point estimate presented in Figure 6 allows the user to calculate the economic consequences of a selected threat type in terms of GDP and employment losses based on a single magnitude input variable as well as other user input variables, such as "time of day", "duration", "resilience", "location", etc. The area for user input is highlighted in yellow color boxes, whereas grey boxes are not applicable for the specified threat type. For instance, in the case of Option 1 for the human pandemic scenario, the user is provided with five selection options in terms of magnitude, duration, behavioral-avoidance, behavioral-aversion, and resilience-recapture. The magnitude variable requires an input of numerical values within the given range as suggested, whereas other variables provide various options of categorical selection from a dropdown list. For instance, the "time of day" variable allows the user to choose either a daytime or a nighttime. The "duration" variable allows user to choose either a 6-month period or a 9-month period. The "resilience" variable provides the user with three options: no resilience, lower-bound resilience and upper-bound resilience, whereas the two variables denoting behavioral effects only provide a "Yes or No" option for the user. Any change of an input variable would lead to an immediate update of results presented in the white color area. Outputs are presented in both numerical terms and cumulative distribution graphs. The numerical outputs of the mean estimates and estimates at various quantile levels are presented in terms of both level change and percent change, respectively.

As shown in Figure 6, without considering behavioral effects and resilience, in a human pandemic case where 60 million people are infected during a 6-month period, the mean GDP loss is $66.08 billion dollars, which is around 0.41% decline of U.S. national GDP, and the mean employment loss is 1,071 thousand jobs, which is equivalent to a 0.83% reduction in jobs nationally. Behavioral effects in terms of avoidance and aversion, and resilience in terms of production recapture could substantially alter the bottom-line. For instance, the mean estimate of GDP loss is amplified significantly to $79.88 billion dollars if the behavioral-avoidance option is switched on in this case. However, if lower-bound resilience-recapture is selected, the mean estimate of GDP loss then reduces to $55.33 billion dollars. If an upper-bound resilience-recapture is selected, the mean estimate of GDP loss then reduces to $35.76 billion dollars.

Option 2 of the E-CAT user interface (not shown) provides interval estimate, which allows the user to calculate economic consequences of a selected threats in terms of GDP and employment losses based on the given range of magnitude, together with other user input variables.

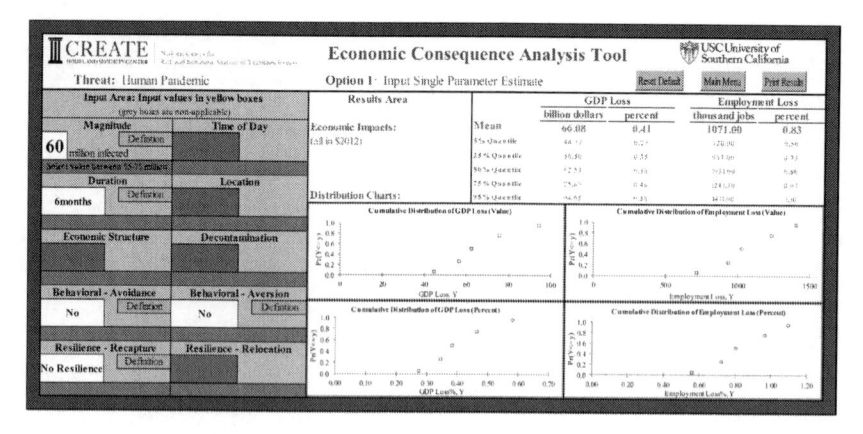

Figure 6. E-CAT user interface option 1 (human pandemic)

The uncertainty distribution estimate as illustrated in Figure 7 provides the user with an option to calculate GDP and employment losses based on a triangular distribution of the magnitude inputs, with interactions to other user input variables. In this option, the user is able to specify the magnitude values in terms of lower, middle and upper bounds. In addition, the user could also specify attributes, such as duration, behavioral-avoidance, behavioral-aversion, and resilience-recapture. Numerical estimates of GDP and employment losses are displayed automatically in the output area. In addition, the cumulative frequency distribution charts and the relative frequency distribution charts of the mean estimates of GDP and employment losses are updated automatically.

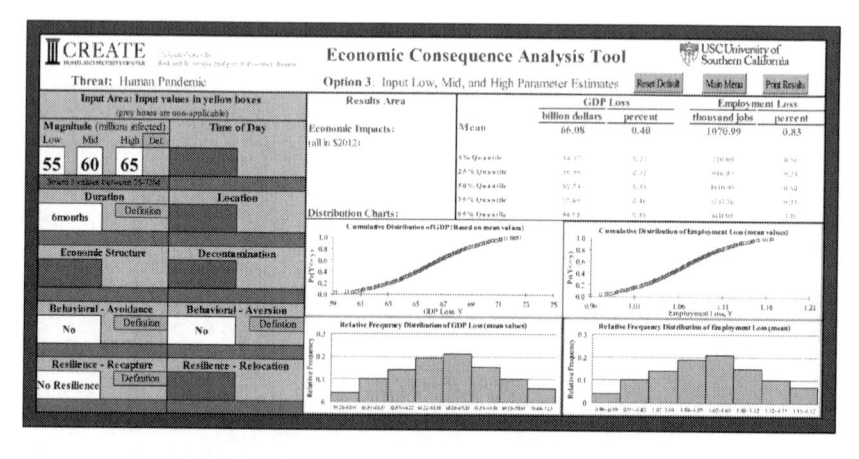

Figure 7. E-CAT user interface option 3 (human pandemic)

CHAPTER 19

Measurement of Cyber Resilience

This section summarizes the measurement of cyber resilience. This is done for several aspects of the cyber domain: (1) various types of cyber communications systems, (2) the electricity network, (3) manufacturing of communications equipment, and (4) provision of various cyber support services. The analysis is performed both on the supplier-side and customer-side. It includes not only the direct impacts but those rippling through the supply chain both upstream and downstream. The analysis is based on the Resilience framework developed by Rose (2009b), and applied to several disaster scenarios in general and to electricity and supply-chain analyses in particular (see, Rose, Oladosu, and Liao 2007; Rose et al. 2016; Rose and Wei 2013).

To date, there is no comprehensive study of cyber resilience. There are several in-depth studies on protecting cyber systems from various types of attacks, including criminal, malicious, and terrorist threats. However, nearly all of these represent pre-event mitigation to reduce the chance of an attack or minimize the direct effect of an attack. Resilience, in our study, refers to actions taken in response to disruption from a disaster, i.e., post-disaster recovery (Rose 2004; 2009b). Of course, resilience is a process, and its capacity can be enhanced prior to disaster, but nearly all of these enhancements are not implemented until after the disaster strikes. Examples are the stockpiling of critical inputs or backup generators, even though they will not be utilized until a disaster strikes. We also note that resilience differs with respect to suppliers of cyber equipment/services and their customers. In general, resilience on the supplier-side often involves expensive redundancies of equipment and systems, while many of the resilience options on the customer-side are relatively inexpensive, and in fact can pay for themselves (e.g., prioritization of access to limited bandwidth, substitution of satellite-based phones).

Table 1 summarizes studies on the microeconomic resilience options, or tactics, for businesses on the supplier and customer-sides. By microeconomic level, we are referring to the operation of the individual business, in contrast to meso-level resilience, which pertains to the operation of the entire industry or market, and macro-level resilience, which pertains to the entire regional or national economies. Following Rose (2004; 2009), the resilience options can be either "inherent" or "adaptive." The former represents resilience capacities that already exist or are planned in the economic system and would simply be accessed to increase the functioning of business activities (to the extent possible) in response to disruptions. The latter refers to any expansion of resilience from improvisation or regulatory and administrative changes. The categories of resilience emanate from economic production theory, which is the conceptual basis for analyzing how business transforms various inputs into the goods or services it produces (Rose 2015b). The column headed "Possible Action" refers to specific resilience tactics that represent the build-up of resilience capacity prior to disaster or the response after the disaster strikes. The column "Cost of Resilience" provides a rough approximation of the cost of actually implementing resilience. The "Effect of Resilience" provides a general indication of the extent to which it can reduce business interruption losses. The cost and benefits of resilience tactics could only be quantified where some evidence could be found.

A major source of customer-side cyber resilience is the existence of multiple communications systems that support input substitution. Fiber-optic, or hard-wired, systems are the most vulnerable to earthquakes in terms of physical damage and repair, followed by cellular data systems dependent on cell towers. Satellite providers are the most reliable in the face of an earthquake threat, but are not as widely used and have vulnerabilities of their own with respect to technological accidents and solar weather. Of course, substitution potential must be tempered by possible delays in re-routing and limits to system capacity, which have frequently caused systems to crash in the aftermath of disasters (cf., Altman 2012; Vantage Point 2013).

At the outset, we note that the literature and interviews by other researchers (e.g., Wein 2015) emphasize that critical facilities are far better prepared in terms of resilience than ordinary business enterprises. For example, NASA has extensive redundant and back-up systems.[4] Of course, recent experience indicates that no system is fail-safe. Another overriding issue is that technological advances have simultaneously made us more vulnerable and more able to respond to cyber disruptions.

The first cyber resilience tactic presented in Table 1 is Conservation. Examples include reducing nonessential usage, restricting nonessential access, and recycling cyber equipment. For example, removing non-essential access increases the ability and speed of responding to a cyber breach or general disruption by reducing the number of access points (CyberSheath 2014). Note that Conservation is an especially attractive resilience tactic, since it often pays for, or even more than pays for, itself. However, it is limited in scope in terms of being able to reduce business interruption from disasters in other or related contexts (see, e.g., Rose and Lim 2002; Rose, Oladosu, and Liao 2007) and there is every indication that this applies to the cyber realm as well.

Input Substitution has extensive possibilities in the case of cyber. It ranges from increased flexibility of systems to various substitutes and back-up capabilities (see, e.g., Sheffi 2005; Chongvilaivan 2012). Flexibility refers to both

---

[4] We distinguish these two terms as follows: "Redundant" refers to an entire system, and is usually applied to the supplier-side. "Back-up" refers to select components, and is usually applied to the customer-side.

supply procurement and to the conversion of inputs into final goods and services. The former entails investment into business-relationships between corporate management and suppliers, which leads to combined efforts toward quickly overcoming supply-chain disruptions.[5] "Multi-sourcing" is a classic example. Conversion flexibility relates to machinery and processes, which facilitates adjustment in resources and employees as necessary (Zolli and Healy 2012). There are many examples of back-ups, including portable electricity generators. More dramatic examples include the use of "Cells on Wheels" following Hurricane Sandy, in which Verizon deployed several mobile cell-towers throughout New York City in response to a number of conventional cell towers going down (Richtel 2009).

Import Substitution refers to bringing in goods and services in short supply from outside the region. It pertains primarily to the manufacturing of cyber equipment and various supply-chain effects. Setting up alternatives in advance, or at the minimum, researching options, can ensure smoother substitution of inputs following a disaster. Of course, it can be constrained by damage to transportation infrastructure, as often results from natural disasters.

Inventories refer to stockpiling critical inputs for the production of cyber equipment, other supply-chain inputs, and cyber systems. Sheffi (2005) notes the classic example of Nokia being much better prepared for a disruption of semi-conductor supply inputs than its major competitor, Ericsson, and thus was able to significantly increase its market share in the aftermath of the disruption. Note that the cost of inventories is not the actual value of the goods themselves, but simply the carrying costs. The goods themselves are simply replacement for the cost that would have been incurred had the ordinary supplies been forthcoming. That said, it should be further noted that carrying costs of electronic goods are typically much higher than other goods, as they depreciate quite quickly (and carrying cost is more than interest and storage costs, but also the cost of the obsolete inventory itself). Some companies, such as Dell, circumvent these carrying costs with a "Made-to-Order" business model, in which they typically hold only 4 days' worth of inventory, ordering more as they receive orders. At first glance this would seem a less-resilient business model, more prone to supply chain disruption, but during the semi-conductor shortage in 1999 their "Made-to-Order" direct consumer marketing model allowed Dell to steer its customers toward products that it had on hand and products that were less affected by the shortage. Alternatively, this could be termed a marketing strategy to promote resilience.

Excess Capacity overlaps to a great extent with system redundancy, which is primarily a supply-side resilience tactic. Typically, it is viewed as a rather expensive option, as, for example, in the case of back-up transformers for elec-

TABLE 1. MICROECONOMIC RESILIENCE OPTIONS FOR BUSINESSES

Category	Action/Investment	Cost of Resilience	Effect of Resilience	Source
Conservation				
• Reduce non-essential use	Data consolidation	More than pays for itself	Significant	
• Remove non-essential access	Remove non-essential administrator access	More than pays for itself (453 hours * (IT-wage+Mngr-wage) * 3 days + 260 hours * (IT-wage+Mngr-wage) * 7 days)	Significant (increases ability and speed of responding to a breach by reducing access points)	CyberSheath (2014)
Input Substitution				
• Paper records, traditional couriers	Re-contract	Low to moderate cost	Significant at small scale	
• Enhance flexibility of input combinations	Supply procurement flexibility	Low (investment in aligning corporate-supplier relationship)	Significant (quickly overcome disruptions through greater cooperation between businesses)	Chongvilaivan (2012)
	Process conversion flexibility	Low (investment in standardized processes, identical machinery)	Significant (ease of relocation)	Chongvilaivan (2012)
• Wireless-to-wired, and wired-to wireless internet and phone access	Use text messaging or social media	Low to moderate cost	Significant	
	Cells on Wheels (COWs)	Moderate (price of device + long term storage; transportation)	Moderate to large	Richtel (2009)
	Satellite phones	Low to moderate	Moderate to large	Verizon (2015)

CHAPTER 19

	Femtocells (small mobile cellular base station)	($189–$300 monthly rental charges; $6–$9 per minute)	(most reliable method of communication)	Ricknas (2010)
	Cellular signal boosters	Low (small scale: <$100)	Moderate to large (restores cell coverage; improved battery life for devices)	SureCall (2015)
	Voice over IP telephone lines	Low (small scale: ~$850; large business: ~$3,500; industrial scale: ~$4,000)	Moderate to large (restores cell coverage; improved battery life for devices)	Chacos (2012)
		Low to moderate ($2,500–$15,000 for initial installation + $40–65 per line/month)	Moderate (requires an internet connection)	Kremlacek (2012)
Import Substitution				
• Mutual aid agreements	Cooperative agreement	Low	Low to moderate	
• Re-routing of goods/services	Data-center failover	Low (slowdown in internet services)	Moderate	

Category	Action/Investment	Cost of Resilience	Effect of Resilience	Source
• Supply-chain management	Multi-sourcing strategy	Loss of quantity discount; higher admin costs; reduces strength of established partnerships; competition leads to lower costs	Moderate	Linthorst and Telgen (2006)
Inventories (Stockpiles)				
• Pool resources	Cooperative agreements	Very low	Significant	

341

• Stockpile products and other essentials	"Safety" stock	Low (carrying cost only; but higher than normal as cyber equipment depreciates quickly)	Significant (safety net for disruption of supply; lower costs when purchased in bulk)	Sheffi (2005)
• Stockpile product inputs	Build-to-order (stockpile inputs and parts instead of finished products)	Low (revise how business operates; some loss of economies of scale)	Significant (allows business to more efficiently use stockpile to meet customer demand while input supply chains are reestablished)	Papadakis (2006) Chongvilaivan (2012) Sheffi (2005)
	Direct consumer marketing (in conjunction with built-to-order model)	Low (cost of training/updating marketers and customer service staff)	Significant (allows promotion of products that were unaffected by supply chain disruption)	Sheffi (2005)
• Batteries	Install battery storage	Low ($250/kwh capacity; base 100 kwh, expandable up to 10 mwh)	Moderate (allows for 100 kwh–10 mwh worth of electricity to be stored)	Kassner (2015)
Excess Capacity				
• Maintain in good order	Maintain in good order	Low	Significant	
• System redundancy	Redundant Array of Independent Disks (RAID); on- or off-site	Low ($200–$15,000+ for RAID setup; for off-site add cost of storage)	Significant	Khasymski et al. (2015)
	E-mail and work mirroring software; off-site	Low ($150 + $0.50 per user/month to $1,000 per server + $0.50 per user/month)	moderate (requires a working internet connection)	Gros (2003)

Category	Action/Investment	Cost of Resilience	Effect of Resilience	Source
	Cloud-based backup servers; off-site	Low (cloud server: $0.024–$0.061 per GB/month and $0.0036 per 100,000 transactions)	Moderate (allows for easy connection to data if relocation is necessary)	Microsoft (2015) Dell Servers (2015a)
	Tape backups; off-site	Low ($1,500–$25,000 per month)	Moderate (much cheaper than RAID storage)	Gros (2003)
	Distributed data centers with data center "failover" capabilities	Low	Moderate	Wein (2015)
• Maintain capacity	Multiple internet service provider (ISP) contracts	Low (cost of additional ISP contracts + $275–$4,000 for routing equipment)	Moderate (maintains internet connections in the case of an ISP losing connectivity)	Barracuda (2015) Amazon (2015)

Category	Action/Investment	Cost of Resilience	Effect of Resilience	Source
• Maintain service	Uninterruptible internet service premiums	Low (usually 5% or less)	Low to moderate (ISP will prioritize returning service to the business over other customers)	

Input Isolation

• Decrease dependence — Permanent and temporary

• Segment production — Identify less essential cyber needs

Relocation

343

• Physical move	Arrange for facilities in advance	Low to moderate	Large	Rose et al. (2009)
• Telecommuting		Low	Moderate	
Production Recapture				
• Overtime/extra shifts		Low (overtime pay)	High (production and sales are not lost)	Park et al. (2011)
• Restarting procedures	Uninterruptible power supply (UPS) with generators	Low ($4,000–$15,000; plus cost of fuel and generators for as long as power is down)	Moderate	Datacenter UPS (2015b) Bruschi et al. (2011) Liebert Corporation (2004)
Technological Change				
• Change processes	Increase flexibility			
• Alter product characteristics				
Management Effectiveness				
• Succession/continuity	Train; increase versatility	Low	Low to moderate	Casey et al. (2015)
• Increased awareness/information sharing	Cyber security framework	Low (<175 full time employee hours to implement, otherwise resources are free)	Low to moderate	NIST (2014)
	Homeland Security Information Network	Low (average cost of $43.80 per month per user)	Low (provides a platform to share sensitive information,	IT Dashboard (2015)

tric power systems. However, cloud-based backups are a relatively inexpensive option. Another possibility related to this resilience tactic is the development of uninterruptible internet service contracts, which could give firms the option to pay a small fee for being priority customers in the event of shortages in internet access. Furthermore, multiple overlapping contracts with different internet service providers (ISPs) could provide higher day-to-day speeds and larger bandwidth, while providing redundancy toward maintaining service should one or more ISPs experience service loss following a disaster (Amazon 2015; Barracuda 2015). Finally, we recognize the inherent redundancy in the internet—data network system here. For example, Facebook data centers can failover, data can be rerouted.

Input Isolation is referred to in the technical earthquake literature by its complement—"Importance" (see ATC 1991). It pertains to the ability to separate aspects of the production process from dependence on lifeline utilities, including cyber systems. The Cybersecurity Framework, a federally developed set of guidelines for cyber standards and practices, provides resources to identify which aspects are essential and nonessential (NIST 2014). Input Isolation obviously applies to many aspects of agriculture with respect to electricity and communications, but it is increasingly less of an option as our economy advances in terms of technological sophistication. While it is typically inherent in the system or production process, it can also be applied in the aftermath of the disaster through improvisation.

Relocation is a tactic that increases resilience by physically moving the business' operations to a location away from the affected area. This requires not only the arrangement for alternate facilities with sufficient capacity, but is also facilitated by the standardization of processes and operations to allow for movement. Relocation would also include tele-commuting if the nature of the business allows for it.

Production Recapture refers to the ability to make up lost production by working extra shifts or over time after communication services and other capabilities are restored. It might involve replacement of expensive equipment that has been damaged, but otherwise the cost is only that of overtime pay for workers (Park et al. 2011). It is further facilitated by hastening the restarting of services such as electricity and internet access. This in turn can be promoted by other resilience tactics, such as uninterruptible power supplies (UPS), a form of input substitution, which provide an emergency power source until back-up generators can be started or central power service is restored.

Technological change is a tactic that can increase resilience capacity by imparting additional flexibility into production systems both before and after the earthquake hits (Zolli and Healy 2012). It can also refer to important improvisations in the way goods and services are produced in the aftermath of a disaster.

Management-effectiveness refers to any improvements in decision-making and expertise that improve functionality, primarily by using existing scarce resources more efficiently. Much of it refers to improvisation, but some relates to established emergency-management plans and information services. The Cybersecurity Framework is one such service that provides a platform for information to be shared between businesses on current threats and the tools available to counter and rebound from these threats. Typically, it is a relatively inexpensive option with costs limited solely to the implementation of the framework.

SYSTEMS ANALYSIS OF ECONOMIC CONSEQUENCES OF MARITIME CYBER THREATS

Figure 8 presents the many components of a system and their interconnections to estimate the economic consequences of maritime cyber threats. It begins with the specification of major characteristics of these threats and then includes the data set inputs and modules that use these inputs to perform a sequence of calculations leading to the estimation of Total Economic Consequences (TECs). Rectangles represent input data, diamonds represent the calculation modules, and ovals represent model outputs.

The initial input into the System is the specification of the Threat by type and key characteristics, the major categories of which are listed in the right-hand margin of the figure. Broader features of the Maritime Context, data from CART on ports and shipping and a history of cyber incidents, and information on the Cyber Role in this context are fed into an Incident Determination Module that calculates the major Incident Features, the major categories of which are listed in the right-hand margin again. The arrows connecting these three sets of input data are solid ones indicating that they are always included in the estimation. Another set of data that feeds into this Module, but on an optional basis, is information on Interdiction/Mitigation that can dampen the frequency and severity of each incident.

Incident Features along with two other sets of data are fed into the Direct Economic Consequences (DECs) Module. The first is an Enumeration Table, which lists the various categories of direct impacts that are applicable to a given Threat/Context combination. Because of the importance of Cyber-related considerations, information on the Cyber Role is again included. The Direct Economic Consequences Module then yields a set of estimates of DECs. Here, there are two optional enhancements of the analysis through the inclusion of Behavioral Response (e.g., fear arising from CBRN threats) and Microeconomic Resilience (e.g., ship-rerouting, use of inventories to cushion the economic shock of a disruption of critical input materials, production rescheduling). The first of these input data sets typically exacerbates the DECs, while the latter typically mutes them. These two aspects along with the Mitigation and Interdiction represent the major policy levers the Coast Guard has to reduce economic consequences.

The Direct Economic Consequences are then fed into the E-CAT Module along with data on Supply Chains and a Logistical Model to estimate Total Economic Consequences. Here again, there are options, including first Meso/Macro Resilience (e.g., price changes that spur resource reallocation or reliance on imported supplies from other regions through other transport modes). The system also can incorporate the effects of Repair and Reconstruction, relating to longer-term recovery of the economy. Both of these optional features reduce the level of TECs.

Note that the optional features of the System are intended to enable the user to analyze TECs in the absence of any external influences, such as private and public policy responses, behavioral reactions, resilience, and recovery activities. This enables the user to examine the influence of each of these factors that significantly affect TEC one at a time to gauge their relative effectiveness. Including the costs of these various influences, so as to be able to gauge their relative cost-effectiveness, is a key to developing an overall Risk Management Strategy.

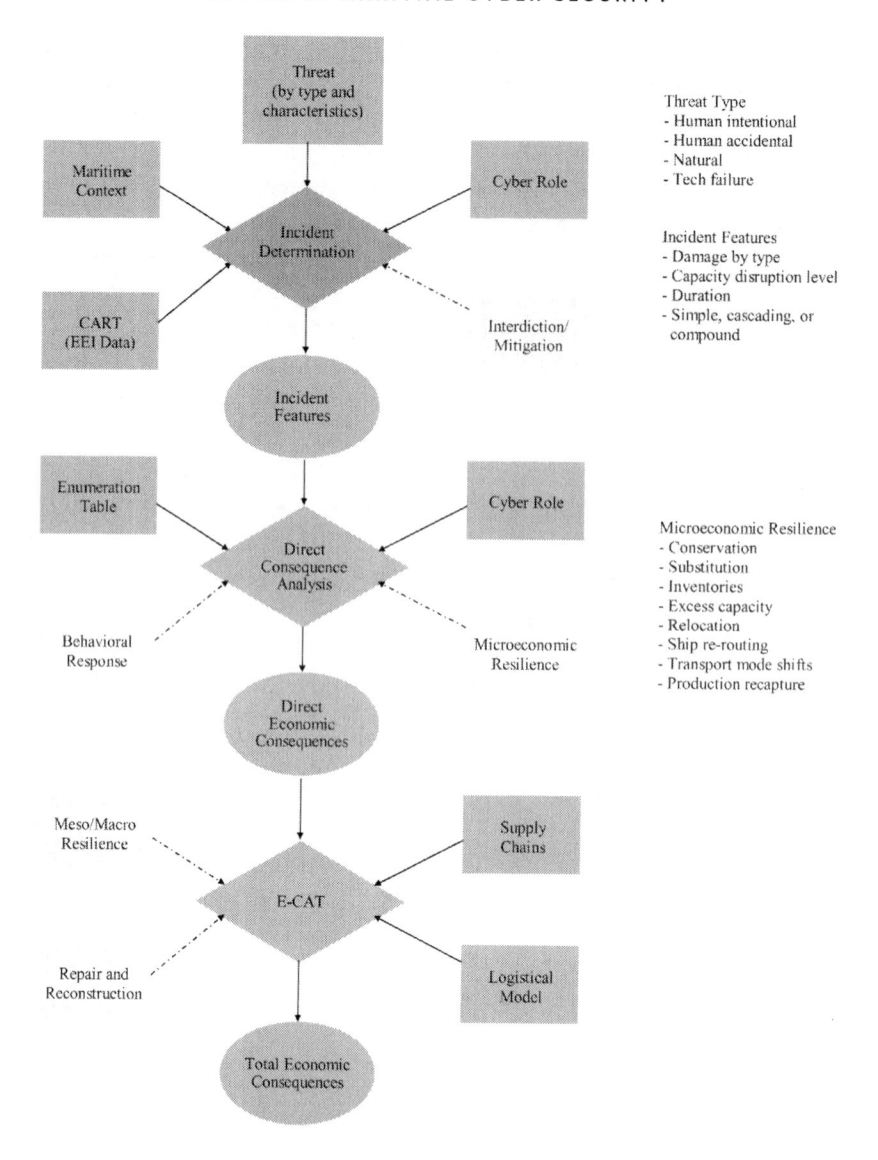

Figure 8. System for estimating economic consequences of maritime cyber threats

Conclusion

This article has summarized the current state of the author's research on economic consequence analysis (ECA) of maritime cyber threats. It has identified the role of ECA in the overall U.S. Coast Guard's cyber strategy, and has outlined a framework for integrating the two. It has summarized the well-established CREATE ECA Framework and illustrated its application to prior studies of port disruptions. These studies have demonstrated the need for a comprehensive framework that includes proper attention, not only to standard features of traditional economic impact analysis, but also to aspects of resilience, behavioral linkages, and remediation of damages.

The white paper also presented a summary of the recently developed Economic Consequence Analysis Tool (E-CAT), which is intended to provide rapid estimates of economic losses from more than 30 types of threats, including those related to the cyber domain and transportation system disruptions. Finally, we presented a summary of research on numerous resilience tactics applicable to the recovery from cyber threats.

The goal of this on-going research is to incorporate into E-CAT the capability to rapidly estimate economic consequences of various maritime cyber threats. These will be chosen and their characteristics determined in collaboration with the USCG. The E-CAT methodology will be adapted to the special needs of this objective. The product to be transitioned to the USCG will essentially be a decision-support capability that will enable high-level decision-makers to better allocate resources across numerous threats.

APPENDIX: MEMBERS OF THE MARITIME CYBER ECONOMIC CONSEQUENCE ANALYSIS WORKING GROUP

Name	Affiliation
Captain Bruce Clark	Cal Maritime Academy
Paul Kantor	CCICADA
Joseph Couch	USCG Atlantic Area
Evi Dube	LLNL
David Moskoff	USMMA
LTC Ernest Wong	Army Cyber Institute United States Military Academy
Randy Sandone	Critical Infrastructure Resilience Institute (CIRI)

Craig Moss
Adam Rose
Dan Wei
Zhenhua Chen

Oak Ridge National Lab
CREATE, USC
CREATE, USC
CREATE, USC

Text References

Applied Technology Council (ATC). 1991. *Seismic Vulnerability and Impacts of Disruption of Lifelines in the Coterminous United States.* Report ATC-25. Redwood, CA: Applied Technology Council.

Baker, M., Jr., Inc. 2008. *Municipal Water Distribution System Security Study: Recommendations for Science and Technology Investments,* Final Report to the U.S. Department of Homeland Security, Washington, DC.

Boardman, Anthony, David Greenberg, Aidan Vining and David Weimer. 2011. *Cost–Benefit Analysis,* Upper Saddle River, NJ: Prentice Hall.

Cohen, Stephen S. 2002. *Economic Impacts of a West Coast Dock Shutdown.* Berkeley, CA: University of California at Berkeley. http://www.brie.berkeley.edu/publications/ships%202002%20final.pdf.

Dixon, Peter B., Bumsoo Lee, Todd Muehlenbeck, Maureen T. Rimmer, Adam Z. Rose, and George Verikios. 2010. "Effects on the U.S. of an H1N1 Epidemic: Analysis with a Quarterly CGE Model." *Journal of Homeland Security and Emergency Management* 7 (1): Article 7.<not cited>

Dixon, Peter B. and Maureen T. Rimmer, 2002. *Dynamic General Equilibrium Modeling for Forecasting and Policy.* Emerald Group, Bingley, UK.

Dixon, Peter B. and Maureen T. Rimmer. 2013. "Validation in Computable General Equilibrium Modeling. In: Peter B. Dixon and Dale W. Jorgenson (Eds.), *Handbook of Computable General Equilibrium Modeling,* Vol. 1A. North-Holland, pp. 1271–1330.

Executive Order 13636. 2013. "Improving Critical Infrastructure Cyber Security." Office of the President, Washington, DC.

Farrow, Scott. 2015. *Integrating Cyber Losses into the Standard Microeconomics of the Consumer and Firm: Defining Losses in the Gordon and Loeb Model.* Baltimore: University of Maryland. http://economics.umbc.edu/files/2016/01/WP_15_03.pdf.

Geisecke, James, William J. Burns, Anthony Michael Barrett, E. Bayrak, Adam Z. Rose, Paul Slovic, and M. Suher. 2012. "Assessment of the Regional Economic Impacts of Catastrophic Events: A CGE Analysis of Resource Loss and Behavioral Effects of a Radiological Dispersion Device Attack Scenario." *Risk Analysis* 32: 583–600.

Government Accountability Office (GAO). 2014. *Maritime Critical Infrastructure Protection: DHS Needs to Better Address Port Cybersecurity.* Report GAO-14-459., Washington, DC.

Kajitani, Yoshio, and Hirokazu Tatano. 2009. "Estimation of Lifeline Resilience Factors based on Empirical Surveys of Japanese Industries." *Earthquake Spectra* 25 (4): 755–776.

Linkov, Igor, Daniel A. Eisenberg, Kenton Plourde, Thomas P. Seager, Julia Allen, and Alex Kott. 2013. "Resilience Metrics for Cyber Systems." *Environment Systems and Decisions* 33 (4): 471–476.

Park, JiYoung., Joongkoo. Cho, and Adam Rose. 2011. "Modeling a Major Source of Economic Resilience to Disasters: Recapturing Lost Production," *Natural Hazards* 58(2): 163-82

Park, JiYoung 2008. "The Economic Impacts of Dirty Bomb Attacks on the Los Angeles and Long Beach Ports: Applying the Supply-Driven NIEMO (National Interstate Economic Model)." *Journal of Homeland Security and Emergency Management* 5 (1): Article 21.

Prager, Fynnwin, Dan Wei, and Adam Rose. 2017. "Total Economic Consequences of an Influenza Outbreak in the United States." *Risk Analysis* 37: 4–19.

Rose, Adam. 2004. "Defining and Measuring Economic Resilience to Disasters." *Disaster Prevention and Management* 13 (4): 307–314.

Rose, Adam. 2009a. "A Framework for Analyzing and Estimating the Total Economic Impacts of a Terrorist Attack and Natural Disaster," *Journal of Homeland Security and Emergency Management* 6: Article 4.

Rose, Adam. 2009b. *Economic Resilience to Disasters.* Community and Regional

Resilience Institute Report No. 8. Oak Ridge, TN: Oak Ridge National Laboratory.

Rose, Adam. 2015a. "Macroeconomic Consequences of Terrorist Attacks: Estimation for the Analysis of Policies and Rules." In *Benefit Transfer for the Analysis of DHS Policies and Rules*, eds. C. Mansfield and V.K. Smith. Cheltenham, UK: Edward Elgar, 172–200.

Rose, Adam. 2015b. "A Methodology for Incorporating Cyber Resilience into Computable General Equilibrium Models." CREATE, USC.

Rose, Adam, Misak Avetisyan, and Samrat Chatterjee. 2014. "A Framework for Analyzing the Economic Tradeoffs between Urban Commerce and Security," *Risk Analysis* 34(5): 1554-79.

Rose, Adam, Bumsoo Lee, Gbadebo Oladosu, and Garrett R. Beeler Asay. 2009. "The Economic Impacts of the 2001 Terrorist Attacks on the World Trade Center: A Computable General Equilibrium Analysis." *Peace Economics, Peace Science, and Public Policy* 15: Article 6.

Rose, Adam and Shu-Yi Liao. 2005. "Modeling Resilience to Disasters: Computable General Equilibrium Analysis of a Water Service Disruption," *Journal of Regional Science* 45(1): 75-112.

Rose, Adam and Dongsoon Lim. 2002. "Business Interruption Losses from Natural Hazards: Conceptual and Methodology Issues in the Case of the Northridge Earthquake," *Environmental Hazards: Human and Social Dimensions* 4: 1-14.

Rose, Adam, Gbadebo Oladosu, and Shu-Yi Liao. 2007. "Business Interruption Impacts of a Terrorist Attack on the Electric Power System of Los Angeles: Customer Resilience to a Total Blackout." *Risk Analysis* 27: 13–31.

Rose, Adam., Fynnwin Prager, Zhenhua Chen, and Sam Chatterjee. 2017. *Economic Consequence Analysis Tool (E-CAT)*. Singapore: Springer.

Rose, Adam, and Dan Wei. 2013. "Estimating the Economic Consequences of a Port Shutdown: The Special Role of Resilience." *Economic Systems Research* 25 (2): 212–232.

Rose, Adam, Dan Wei, and Noah Dormady. 2011. "Regional Macroeconomic Assessment of the Pennsylvania Climate Action Plan," *Regional Science Policy and Practice* 3(4): 357-79.

Rose, Adam, Ian Sue Wing, Dan Wei, and Anne Wein. 2016. "Economic Impacts of a California Tsunami." *Natural Hazards Review* 17 (2).

Sheffi, Yossi. 2005. *The Resilient Enterprise*. Cambridge, MA: MIT Press.

Sinha, Arunesh, Thanh Nguyen, Debarun Kar, Matthew Brown, Milind Tambe, and Albert Xin Jiang. 2015. "From Physical Security to Cybersecurity." *Journal of Cybersecurity* 1 (1): 19–35.

Sue Wing, Ian, Adam Rose, Dan Wei, and Anne Wein. 2016. "Impacts of the USGS ARkStorm Scenario on the California Economy." *Natural Hazards Review*.17(4): A4015002-1.

U.S. Coast Guard. 2015. *United States Coast Guard Cyber Strategy*. Washington, DC.

U.S. Department of Homeland Security. 2015. *National Cybersecurity Assessments and Technical Services: Capability Brief*.

Werling, Jeffrey. 2014. *The National Impact of a West Coast Port Stoppage*. Washington, DC: National Association of Manufacturers.

Wein, Anne. 2015. "Personal communication," U.S. Geological Survey, August 24.

Zolli, Andrew, and Ann Marie Healy. 2012. *Resilience: Why Things Bounce Back*. New York: Free Press.

Cyber Resilience References

Altman, Lou. 2012. "Satellite Communications: The Myths, Costs & Capabilities." (March 5). http://www.continuityinsights.com/articles/2012/03/satellite-communications-myths-costs-capabilities.

Amazon. 2015. "Peplink Balance 20 Dual-Wan Router (pricing)." http://www.amazon.com/Peplink-Balance-20-Dual-WAN-Router/dp/B0042210U6/ref=sr_1_12?s=pc&ie=UTF8&qid=1373925222&sr=1-12&tag=viglink20237-20.

Barracuda. 2015. "Link Balancer, Advanced Internet Link Load Balancing." https://www.barracuda.com/products/linkbalancer/models#SUB.

Bruschi, John, Peter Rumsey, Robin Anliker, Larry Chu, and Stuart Gregson. 2011. *Best Practice Guide for Energy-Efficient Data Center Design*. Washington, DC: Department of Energy. http://energy.gov/sites/prod/files/2013/10/f3/eedatacenterbestpractices.pdf.

Casey, Tim, Kevin Fiftal, Kent Landfield, John Miller, Dennis Morgan, and Brian Willis. 2015. *The Cybersecurity Framework in Action: An Intel Use Case*. Santa Clara, CA: Intel Corporation. http://www.intel.com/content/www/us/en/government/cybersecurity-framework-in-action-use-case-brief.html.

Chacos, Brad. 2012. "VoIP Buying Guide for Small Business." *PC World Magazine*, April 14. http://www.pcworld.com/article/260859/voip_buying_guide_for_small_business.html?page=2.

Chen, Brian X. 2013. "F.C.C. Seeks Ways to Keep Phones Alive in a Storm." *New York Times*, February 5. http://bits.blogs.nytimes.com/2013/02/05/f-c-c-revisits-communications-failures-after-hurricane-sandy/.

Chongvilaivan, Aekapol. 2012. "Thailand's 2011 Flooding: Its Impact on Direct Exports and Global Supply Chains." *ARTNeT Working Paper Series*, No. 113. https://www.econstor.eu/dspace/bitstream/10419/64271/1/715937650.pdf.

CyberSheath Services International. 2014. "The Role of Privileged Accounts in High Profile Breaches." May. http://lp.cyberark.com/rs/cyberarksoftware/images/wp-cybersheath-role-of-privileged-accounts-6-2-14-en.pdf.

Dell. 2015a. "Dell PowerEdge Servers." http://www.dell.com/us/business/p/servers?~ck=bt.

Dell. 2015b. "Datacenter UPS." http://accessories.us.dell.com/sna/category.aspx?c=us&l=en&s=bsd&cs=04&category_id=7071.

Department of Homeland Security (DHS). 2013. "National Initiative for Cybersecurity Careers and Studies." https://www.dhs.gov/news/2013/02/21/dhs-launches-national-initiative-cybersecurity-careers-and-studies#.

Department of Homeland Security (DHS). 2015. "Homeland Security Information Network - Critical Infrastructure." May. https://www.dhs.gov/critical-infrastructure-0.

FEMA. 2015. http://m.fema.gov/get-tech-ready-additional-tips.

Goldman, D. 2012. http://money.cnn.com/2012/10/29/technology/mobile/cell-phone-sandy/.

Green Nicola, Tim Bentley, and David Tappin. 2014. "A Multi-Level Analysis of Telework Adoption and Outcomes within Organisations Following a Natural Disaster." Ergonomics, Work & Health Ltd, New Zealand. http://www.ergonomics.org.nz/LinkClick.aspx?fileticket=S_FdRN6qWpc%3D&tabid=39.

Gros, M. 2003. "Taking Care of Business—Small, Midsize and Large Companies Have Different Disaster-Recovery Needs and Budgets. The CRN Test Center Details a Wide Range of Solutions to Help Your Customers Weather the Storm." *Computer Reseller News*. http://go.galegroup.com/ps/i.do?id=GALE%7CA108267908&v=2.1&u=usocal_ain&it=r&p=AONE&sw=w&asid=e4066cd2123764c27c43f5afc6f2ba82.

IT Dashboard. 2015. "DHS—Homeland Security Information Network (HSIN)." https://itdashboard.gov/investment?buscid=134.

Kassner, Michael P. 2015. *Tesla's Powerpack Proposes Battery Powered Data Centers*. London: Datacenter Dynamics. http://www.datacenterdynamics.com/critical-environment/teslas-powerpack-proposes-battery-power-for-data-centers/93974.fullarticle.

Khasymski, Aleksandr, M. Mustafa Rafique, Ali R. Butt, Sudharshan S. Vazhkudai, and Dimitrios S. Nikolopoulos. 2015. "Realizing Accelerated Cost-Effective Distributed RAID." In *Handbook on Data Centers*, eds. A. Khasymski and M. Rafique. New Paltz, NY: Springer, 729–752.

Kremlacek, Randy. 2012. "How Much Does a Business VoIP Installation Actually Cost?" *TeleDynamic*, May 22. http://www.teledynamic.com/blog/bid/139621/How-Much-Does-a-Business-VOIP-Installation-Actually-Cost.

Liebert Corporation. 2004. *Choosing the Right UPS for Small and Midsize Data Centers: A Cost and Reliability Comparison*. Columbus, OH: Liebert Corporation. http://www.upsystems-inc.com/sites/default/files/resources/cost-and-reliability.pdf.

Linthorst, Merijn M., and Jan Telgen. 2006. "Public Purchasing Future: Buying from Multiple Suppliers." In *Advancing Public Procurement: Practices, Innovation and Knowledge-Sharing*, eds. K. Thai and G. Piga. Boca Raton:

PrAcademics Press, 471–482. http://www.utwente.nl/bms/iebis/staff/linthorst/67_linthorst_telgen_edited_acc.pdf.

Microsoft. 2015. "Backup Pricing." http://azure.microsoft.com/en-us/pricing/details/backup/.

National Institute of Standards and Technology (NIST). 2014. "Framework for Improving Critical Infrastructure Cybersecurity." http://www.nist.gov/cyberframework/upload/cybersecurity-framework-021214.pdf.

Papadakis, Ioannis S. 2006. "Financial Performance of Supply Chains after Disruptions: An Event Study." *Supply Chain Management* 11 (1): 25–33.

Richtel, Matt. 2009. "Inauguration Crowd Will Test Cellphone Networks." *New York Times*, January 18. http://www.nytimes.com/2009/01/19/technology/19cell.html.

Ricknas, Mikael. 2010. "Femtocell Prices Have Dropped Below $100, Says Vendor." *PCWorld*, March 30. http://www.pcworld.com/article/192855/article.html.

Samuelson, Tracy. 2013. "After Sandy, Questions Linger Over Cellphone Reliability." *NPR*, April 29. http://www.npr.org/sections/alltechconsidered/2013/04/29/179243218/after-sandy-questions-linger-over-cellphone-reliability.

SureCall. 2015. "Cellular Signal Boosters for Commercial." http://www.surecall.com/product/cellphonebooster/15/0/0/CommercialBoosters.

Vantage Point. 2013. http://apps.fcc.gov/ecfs/document/view?id=7520956711.

Verizon. 2015. "Satellite Phone FAQs." http://www.vzwsatellite.com/faqs.

CHAPTER 20:
SECURING THE INTEGRITY OF YOUR CONTROL SYSTEMS

Mate J. Csorba and Nicolai Husteli
DNV GL—Maritime, Norway

This chapter focuses on bridging the existing gap between information technology (IT) systems security and maritime cyber security by establishing an approach covering risk assessments, reliability simulations, and last but not least simulator-based software testing. In contrast to IT security challenges, when assessing or testing the security of Industrial Control Systems (ICSs), including maritime control systems, commercial-off-the-shelf tools become less applicable, one has to work with proprietary implementations and protocols, and generally in a more fragile networked environment. A statement echoed often—by system vendors or asset owners—is that a simple network scan might bring down ICS equipment. Hence, a more holistic and well-founded approach is required to mitigate cyber risks in maritime activities.

INTRODUCTION

Cyber/IT professionals have been combating threats to information technology and have been trying to keep communication networks and data secure for decades. Cyber security incidents have, however, until recently not penetrated the maritime world significantly; or at least they have not raised public attention. The need for guidance materials from the industry on cyber security is increasing as the maritime transportation industry, which is carrying approximately 90% of international commerce, is increasingly reliant on cyber systems. Breaches in cyber security can have significant impact not only on security, but on the safety of operations and business perspectives as well. Yet, regulations and recommendations have left the cyber security aspects of safety critical systems as something needing optional rather than mandatory effort to "harden" or increase robustness and resiliency. Hence, developers and vendors of these systems were reluctant to expend resources to address this critical dimension. Focus on security has been reduced due to the fast development of technology and increasing degree of automation in the maritime sector. For example, there is an increasing amount of communication and IT devices operational in ports connected to the Internet without considering their impact on security, and sometimes even without the real need to being connected. The vulnerabilities that arise in these communication systems might also compromise the underlying infrastructure, e.g., database systems and other services, and can lead to security breaches, such as the cyber breach at the port of Antwerp in Belgium (Bateman 2013).

Control systems providing safety functions may be designed to target certain Safety Integrity Levels (SIL) ranging between 1 and 4, as defined by the IEC 61508 standard (IEC 2010). These levels, however, will only hold as long as the software operating the control system is intact and functions correctly. As control systems today are increasingly networked, maybe even Internet-connected, the attack surface of these systems has increased dramatically. One current trend in control system communications is pointing toward all Ethernet and Internet Protocol (IP)-based communication, which will lead to an advent of reuse of existing malicious techniques to which IP networks worldwide have been exposed. Similarly, wireless communications are also being installed on-board of vessels, for various purposes. An example of industrial wireless communication in drilling control systems is a top-drive controller that employs wireless Profibus communication, a solution delivered by the German company Schildknecht AG. Another aspect of wireless networks on-board is their use for providing Internet access to crew or passengers, which might open security holes, unless proper network segregation is present.

Intentional (e.g., malicious attacks) or unintentional (e.g., untested patches) changes to control system software can violate safety and lead to potentially disastrous events. Accordingly, a more general categorization of threat vectors include: (1) hackers and hacktivists (intentional outsiders); (2) disgruntled employees and other insider threats (intentional insiders); (3) hardware and software errors (unintentional inherent); (4) software and firmware upgrades and patches (unintentional introduced). When critical infrastructure protection is addressed, the sectors typically considered are the nuclear industry, or the utilities and various types of telecom providers. Nonetheless, similar types and amounts of industrial control systems are present—increasingly so—in the maritime industry, e.g., in ports, in a variety of transport vessels, offshore supply and drilling vessels and offshore oil and gas installations.

CHAPTER 20

On the organizational plane, stakeholders are more concerned with downtime, i.e., availability, than with security. Accordingly, the best way to reach out to senior executive levels in an asset owner company is to devise a possible attack scenario on the company asset that has direct and significant economic consequences, or affects the company's liabilities in some other way. Regarding vendors and developers, they have to comply with a variety of regulations and certifications, however, cyber security has had little focus or is currently on optional operational detail (ENISA 2011).

At the same time, patching and bug fixing, i.e., correcting discovered vulnerabilities in ICS products—is not quite straightforward. The patching treadmill is difficult to handle in a remote environment, off-shore, or might even be impossible in case of legacy hardware installed on-board, which might have reached its end-of-life. Sometimes, patching difficulties can endanger the crew and the environment, as was the case when an infection, caused by the Slammer malware, resulted in flooding an otherwise isolated network on-board a rig in the Mexico Gulf (Shauk 2013). Generally, an attainment of a 3-month patching cycle is considered doing extremely well where an average patching frequency in ICSs might typically be around 12–18 months.

To better identify risk and vulnerability, the rest of this paper is organized as follows: first, in the remainder of this section, we look at some vulnerabilities that became public recently, as well as at some Norwegian incidents in particular. In the second section of this chapter, we introduce a consolidated approach and provide a methodology to mitigate risks in the maritime and offshore oil and gas industries. Next, fundamental building blocks of the approach are presented which discuss risk assessments; a description of possible simulation studies is discussed; and finally, an evaluation of robustness testing is conducted.

VULNERABILITIES EXPOSED

Holes in cyber security in an industrial environment—including the maritime and offshore industries—can have several aspects; organizational, educational, policy related, or technical. We refer to technical deficiencies as vulnerabilities, which might be present in any kind of computing nodes (e.g., operator stations, programmable logic controllers (PLCs), various servers and clients, etc.) or in the networking equipment (firewalls, routers and switches, all sorts of gateways and communication cards, etc.) interconnecting them. Specifically in the maritime environment there are few cyber security incidents that have been made public where detailed information is available. The energy industry, and in particular the oil and gas sector are however, increasingly becoming targets for various forms of cyber-attacks. These documented attacks have taken the form of espionage attempts, piracy, or acts of terrorism. In recent years, some vulnerabilities have reached the public media and attracted the attention of various maritime authorities and agencies, such as the International Maritime Organization, and the United States Coast Guard.

Maritime control and auxiliary systems are equally exposed to cyber security threats as drilling or process control systems. Research reports significant flaws in technologies, for example in navigation, as remote monitoring and automation has become increasingly widespread in these systems. These incidents have reportedly resulted in incidents such as tilting an oilrig off the coast of Africa, or standstill of control systems during relocation of a rig due to malware infections (Wagstaff 2014). Changing a vessel's direction via GPS spoofing was successfully demonstrated in 2013 by a team from the University of Texas at Austin. Serious flaws in the Electronic Chart Display and Information System (ECDIS) were documented (Dyryavyy 2014) that might lead to incidents such as the grounding of US Navy warship USS Guardian in the Philippines, which was supposedly caused by incorrect charting (Commander U.S. Pacific Fleet 2013). Researchers from the anti-virus vendor Trend Micro demonstrated weaknesses of the Automatic Identification System (AIS), effectively being able to tamper with or even shut down communication between a ship and the port authority using a $100 off-the-shelf radio kit (Balduzzi, Wilhoit, and Pasta 2013). While it is possible to detect spoofing (Katsilieris, Braca, and Coraluppi 2013), it will probably take some time for regulations to catch up.

Specific Incidents from Norway

In August 2005, the IT staff at a Norwegian oil and gas company noticed that network traffic through a firewall was increasing suspiciously and was not initiated by an unusual spike of web browsing activity of the employees. At the same time, several personal computers behaving strangely were reported to the IT department's ticketing system. Engineers quickly identified the cause of the suspicious behavior as a computer worm, which was a variant of Zotob (Microsoft 2005)—a piece of malware that exploits vulnerabilities in Microsoft operating systems and was, among other capabilities, able to force an infected computer to restart. The number of infected hosts increased from one to 157 during a period of 17 days and affected 185 clients and servers altogether. The investigations that followed concluded that the malware infection was caused by a third-party laptop connected to the internal network. However, the problem of needing to re-start certain hosts was not the highest risk from the worm, as its set of built-in tools included a "call-home" function and the possibility to include the infected host into a botnet (a combination of robot and network) that could have been utilized for criminal purposes. Consequences of the incident could have been as far-reaching as a complete halt in offshore oil and gas production, disturbance in refineries, and possibly disruption in the ability to purchase gasoline across Norway. After the cause of the problems was identified it took a taskforce 50 h to stabilize the network again. Difficulties included that some hosts were providing critical services, in other words it was very difficult to stop them even temporarily. Yet, it was an incredibly lucky end to a cyber security incident as there was no major impact due to downtimes (Johnsen, Ask, and Røisli 2007).

The January 2013 attack on the gas facility partially owned by the Norwegian state oil company in In Amenas, Algeria, also showed that Norwegian interests in the oil and gas sector could be an attractive target for terrorist organizations (Statoil 2013). A year later, the Norwegian National Security Authority (NSM) indicates in its annual report that it registered 15,815 security incidents in national networks that year, of which 50 serious infiltration attempts against critical industrial networks were traced (an increase from 46 in 2012 and 23 in 2011). The report concludes that enterprises are lacking the necessary awareness of their own vulnerabilities, and while commercial security products are widely used, they are only capable of handling cyber security threats where the vulnerability is known beforehand. In addition, several nation states have reportedly built up significant capabilities to execute operations in the cyber domain. In their 2014 report on national cyber security status, NSM suggests that third-party organizations could be used to strengthen cyber security of critical infrastructures and industrial communication systems (NSM 2014).

As control systems have incorporated more and more computerized remote operations there has been a corresponding dramatic increase in the number of cyber security incidents as the focus of these attacks has shifted from regular IT infrastructure to control systems. In order to address this issue, in 2005 the Norwegian Oil and Gas Association formed a workgroup, including members from the Petroleum Safety Authority of Norway (PSA) and the Norwegian Petroleum Directorate (NPD), to establish recommendations and guidelines for information security in industrial control and support systems and networks. As a result, the first version of recommendation OLF104 was published in June 2006 (Norwegian Oil and Gas Association 2006). OLF104 was founded upon the ISO 27001 standard. The recommendation documents best practices for information security and provides guidelines for implementation. Inspections resulting from OLF104 were first conducted in spring 2007 and discovered discrepancies in network segregation, staff competence, documentation, and confusion in the procedures for handling communication errors.

In connection with the recommendation, in 2012, a voluntary self-assessment schema was released by PSA to vessel and rig owners operating on the Norwegian continental shelf. Participants assessed their level of preparedness on a scale from zero to four, zero being the worst score. The assessment schema covered well-defined topics in cyber security, grouped into 16 Information Security Baseline Requirements (ISBRs). The voluntary responses to the self-assessment schema were collected and PSA released anonymized statistics. In the results specifically for drilling rigs, the overall average result was 2.4, with 1.5 being the worst score average on one of the ISBRs. In particular, the collected results highlighted discrepancies in the understanding of security risks, and in usage of security-critical systems, as well as in disaster recovery plans.

A Consolidated Approach

Mitigating cyber security risks in the maritime environment requires a composite approach encompassing a few critical building blocks. Simply installing arbitrary IT defenses without proper planning and configuration, conducting security audits without an appropriate baseline, or testing security defenses without thorough gap-analysis might all be inadequate. To cover the variety of aspects affecting ICS security a step-wise approach is required, as shown in Figure 1. Successful problem analysis and framing starts with deciding, what are the specific objectives for the review or study to be conducted. The asset(s) that will be part of the investigation need(s) to be defined and placed in a business context. An appropriate baseline for information security has to be agreed upon with the asset owner, followed by a listing of relevant requirements and national and international standards. Next, the cyber security assessment can be engaged with a review against the baseline within the defined frame and context. All the necessary information has to be captured by reviewing documentation, and conducting interviews with the stakeholders. This phase ends with the identification of gaps.

When the gaps are identified, prioritization can occur using the risk picture. The gaps are ranked based on their importance and possible impact. The risk assessment phase is further detailed in Section 3. As the findings of the assessment are prioritized, possible mitigation actions can be identified, developed and deployed. After mitigating actions have been put in place to close the identified gaps, evaluation and different forms of verification can start. Depending on the gap analysis, a reliability analyses for component parts, or the entire system can be conducted by various means including test simulation as presented, in section four below. To technically verify mitigating actions will be effective during real operations, the overall robustness of the integrated hardware and software system can be tested, something further discussed in

CHAPTER 20

section five. Finally, based on the residual risk picture, a feedback loop is possible, where the process revisits the assessment phase to further evaluate the asset at hand until the desired state of cyber security is reached.

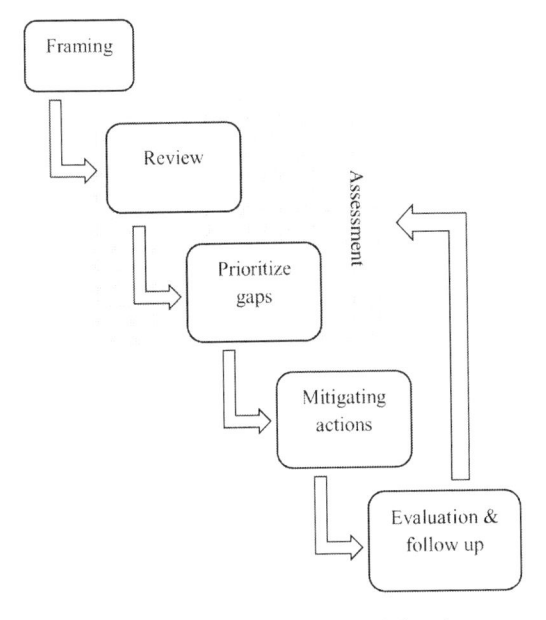

Figure 1. Mitigating cyber security risks, a consolidated step-wise approach

RISK ASSESSMENT

Industrial control systems' cyber security risk assessment standardization and regulatory work was launched mainly in the United States, related to smart-grids, and gradually spread out to the European Union and other countries. For example, the NIST 800 series is a set of publications rather widespread in the United States. The purpose of NIST 800 is to provide guidance for securing ICSs, including supervisory control and data acquisition (SCADA) systems, distributed control systems (DCSs), and other systems performing control functions. The documents provide an overview of ICSs and typical system topologies, identify typical threats and vulnerabilities to these systems, and provide recommended security countermeasures to mitigate the associated risks.

When it comes to risk assessment of networked systems, the European Telecommunications Standards Institute (ETSI) has also developed standards and recommendations, such as the ETSI TS 102 165-1, that is a "Method and Proforma for Threat, Risk, Vulnerability Analysis," applicable in particular for communication protocol vulnerabilities. Some countries have started the process of standardization and regulation in the oil and gas and maritime sectors as well. A number of organizations have put their own standardization work on hold, and have focused on a single series standards, ISA/IEC 62443—formerly known as ISA-99—addressing Industrial Automation and Control Systems (IACS) Security. There seems to be a trend toward the application of this standard across industries, hence it is highly relevant for the maritime industry to consider this standard as well. In addition, the ISO/IEC 27002 is also well established, and is the basis for many national regulations. This standard "established guidelines and general principles for initiating, implementing, maintaining, and improving information security management within an organization." Although relevant, it is targeted at the general IT policy level.

An asset owner has the ability to decide on the desired extent of cyber security risk reduction and choose the corresponding approach. Figure 2 presents the different approaches asset owners may decide to take to reduce cyber risks, where the extent of possible risk-reduction grows top to bottom depending on the approach chosen.

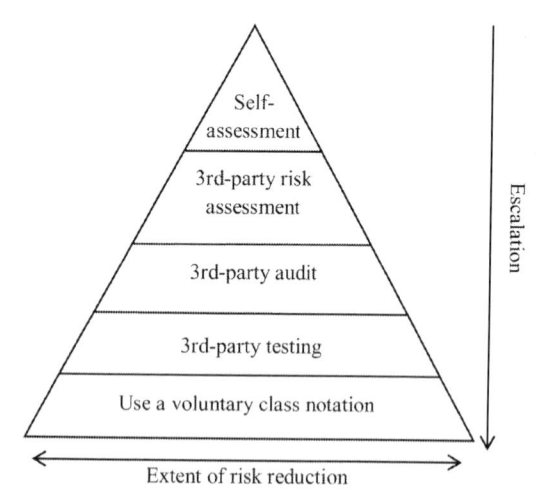

Figure 2. Approaches to cyber-risk reduction

The most basic approach to reduce cyber-related risks of an asset is to conduct a self-assessment. There are some available tools and methods to support cyber self-assessments, such as the OLF104 recommendation applied on the Norwegian continental shelf (Norwegian Oil and Gas Association 2006), or the CSET tool developed by the US Department of Homeland Security (ICS-CERT 2011). These tools can give detailed guidance for conducting a risk assessment by the asset owner. Alternatively, company standards and requirements, if available, can be used as a basis for the self-assessment. If an asset owner does not have the required competence or capacity to conduct an assessment, a third party can be involved to conduct an extended assessment of risks.

A total third-party cyber security audit is the next level of addressing cyber risks, and constitutes a more extensive approach. An audit will go deeper into cyber-dependencies and can be a sound basis for further actions, such as simulations for supporting decisions and technical verification and testing—techniques further described in subsequent sections. Det Norske Veritas/Germascher Lloyd (DNV GL) uses its proprietary tools, Synergy Life Risk Management, and BowTieXP during the audit process. The former tool is an enterprise-level software for managing operational risk and consists of software solutions for managing Asset Integrity and, Quality, Health, Safety and Environmental (QHSE) issues, as well as asset optimization for the gas, electric, pipeline, and water treatment industries. Risk management process support is based on the principles of, for example, ISO 31000 or ISO 27000, and enables uniformly applied risk-based decision-making. Whereas, the latter tool is used in risk assessments applying the BowTie method and visualizing complex risks. A BowTie diagram is generally used to visualize risks within a single, easy to understand picture. Proactive and reactive forms of risk management are clearly displayed as the left and right sides of a diagram shaped as a bow-tie, thereby presenting an overview and allowing analyses of multiple scenarios.

When weak spots in cyber security of an asset are identified, or in case there are uncertainties regarding the current status of systems (e.g., deployed equipment, or configuration) on an asset, technical verification can go beyond desktop exercises and map in what condition the system under consideration is. This is when third-party testing can help. Prior to launching testing of onboard communication networks and control systems, an audit can provide the information necessary about the system(s) under consideration to prepare for testing. Lastly, to gain the broadest risk-reduction a cyber security class notation could be used, possibly involving third-party verification as well. A cyber security class certificate could then be granted to a vessel if it complies with the specific rules how to obtain such a certificate, defined by the class society as a contractual document including both the requirements and the acceptance criteria. Although, to the author's knowledge such class notation does not exist today, it is expected that classification societies will catch up with the emerging cyber threat and develop new rules for new compliance/class notations. However, due to its expected cost burden, such a notation will probably be voluntary when introduced, unless public regulatory bodies will provide additional drivers for formal certification.

Simulation Studies

Reliability, Availability and Maintainability (RAM) modeling and simulation is originally an approach to assess the capabilities of a production system. This analysis can be done in either a design phase or after the system is in operation. When applied to investigate a system's production capabilities, RAM simulation can identify possible causes of losses and highlight alternative system implementations to mitigate the degrading factors. In this process, a RAM study supports decision-makers in a cost–benefit analysis. The factors normally taken into consideration in a RAM study are the configuration, operation, failure, repair, and maintenance of a system. Lifetime analysis of the system's production is then facilitated by simulating the system in a test bed setting that duplicates its components, configuration, operational, and maintenance philosophy. Borges, Hickey, and Gilmour (2014) provides an example of an offshore oil and gas application.

Predicting the performance of an entire system throughout its lifetime is a challenging task at least, if not impossible considering the pace at which technology is changing. Dependencies and technologies change reasonably often in complex systems, hence simulation techniques are promising tools to predict the performance of a system, but should be performed periodically to keep up with changing conditions. Simulation methods can also target cyber security analysis by providing insight into the effects of different configurations of mitigating actions and protection mechanisms proposed for installation. Using simulation the performance of cyber protection mechanisms can be studied, as well as their sensitivity to changes in configuration and setup, as a way of predicting optimization of the protective measures and taking into account given constraints as necessary. A sensitivity manager module can be employed during RAM simulation to identify the optimum design configuration, and in maintenance and operational strategy, e.g., remedial and upgrade patch management. This module allows for the investigation of multiple scenarios to ensure optimum performance (Borges and Hickey 2015).

As a RAM simulation framework we use ExtendSim—a tool by the California-based Imagine That Inc., which is a discrete-event, continuous, agent-based and discrete rate simulation engine (David 2011). While ExtendSim has frequently been applied for traditional RAM analysis, it is also suitable in a cyber security and information security context for supporting decision-making. A further set of toolboxes for cyber security analyses includes Reliability Workbench with the appropriate modules, such as AttackTree—both available from Isograph Inc. "Attack trees" enable graphical representation of how attacks might succeed, and allow a probabilistic analysis of which attacks are most likely to succeed. By specifying parameters of attacks, such as frequency, cost and consequence, analysis of feasible attacks can support planning of the next phase of decreasing cyber risks by performing real-world operational testing of system robustness.

Testing Robustness

Strengthening the robustness of systems—i.e., their ability to operate correctly and reliably across a wide range of operational conditions—can significantly improve their security. Testing the robustness of control system software has been proven to improve safety of offshore operations and reduce downtime. Hardware-In-the-Loop (HIL) testing has enabled comprehensive evaluation of control systems' functionality and failure handling capabilities without risk to personnel, equipment or the environment (Maslin 2014). This includes safety-barriers that otherwise cannot effectively be tested before they are needed (Korn 2013). Independent third-party HIL testing is a well proven methodology applied in the automotive, avionics and space industries, originally introduced by NASA in its Independent Verification and Validation (IV&V) facility for testing mission-critical software components (Bole et al. 2013). HIL testing has also been used in programs such as the United States Future Combat System and the Joint Strike Fighter programs.

Marine Cybernetics Advisory, a unit of DNV GL located in Trondheim, Norway—has been applying the HIL testing methodology since 2002 to test advanced marine control systems, such as dynamic positioning, power management, steering propulsion and thrusters, blow out prevention, and drilling systems. This test approach can be complemented with the verification of software cyber security to ensure integrity of control systems. Testing systems that provide open services and nodes in the telecommunication backbone (e.g., web servers, backbone routers, etc.) for vulnerabilities has a long history in cyber security research. A plethora of tools exist for penetration testing of networked equipment, e.g., including a free Linux distribution containing several hundred dedicated tools called Kali Linux (Offensive Security 2015). However, SCADA and control system networks are a relatively new arena for testing, and focus has only recently been shifted to critical infrastructure protection. Testing the communication networks serving human/machine interface (HMI) systems and other control systems often requires novel and custom tools due to proprietary restrictions and closed-source solutions, in contrast to the more open architectures such as those available via the Internet. Although various solutions exist targeting known vulnerabilities of communication systems, such as malware and virus scanners, vulnerability scanners, and intrusion detection systems (IDSs), they are severely dependent on updates or training (in case machine intelligence is involved). Therefore, it is equally important to scan communication systems for unknown vulnerabilities and to verify their robustness.

A dynamic analysis method known as "fuzzing" can be used to test communication protocol stacks in ICSs for unknown vulnerabilities (Sutton, Greene, and Amini 2007). The method of fuzzing relies on fuzzy logic and hopes to trigger completely unexpected behavior in the software under test. However, testing software, in this case communication protocols, in an entirely random way would be quite ineffective. That is the reason why state-of-the-art "fuzzers"—toolboxes for fuzzing—combine techniques such as grammar rules that specify which parts of a protocol to fuzz, and various strategies for generating packets in an efficient way, systematically simulating invalid communication. To illustrate what kind of hidden vulnerabilities can be uncovered using the method of fuzzing, the message exchange relevant for the infamous Heartbleed bug is shown in Figure 3.

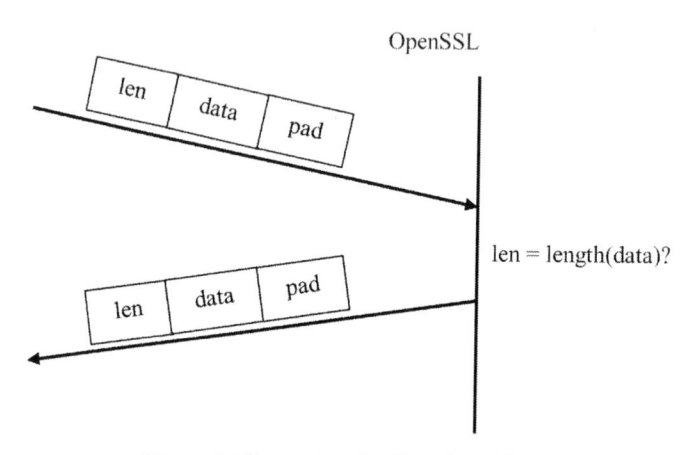

Figure 3. Illustrating the Heartbleed bug

The Heartbleed bug was a serious security bug found in 2014, by two researchers independent of each other, and at least one of them could have been using the method of fuzzing (Vassilev and Celi 2014). The bug was found in a cryptography library and open source implementation of the Secure Socket Layer protocol, OpenSSL, widely used in the open Internet and impacted a huge amount of servers in a variety of systems. The essence of the software error, as shown in Figure 3, is improper input validation. The vulnerable server might receive a maliciously crafted request that contains a length and a data field, with some padding in the end. Then, as a form of heart-beat protocol, the OpenSSL implementation has to reply with an equivalent length amount of bytes, and the same data as in the request. However, whether the requested length amount of bytes matches the actual length of data is not validated. In the specific case of the Heartbleed bug, an intelligently fuzzed, but valid OpenSSL request with a length field larger than the length of data could have been enough to expose the vulnerability. Accordingly, using the technique of fuzzing, i.e., crafting requests where the fields and values are modified following some strategy, and then observing the responses, it is able to uncover these types of software errors, or in general the lack of robustness.

Fuzz testing of communication units is one useful technique, but not sufficient in itself. Execution of a security tests must be based on the information collected from documentation and the topology of the plant. Information can be gathered during risk assessments, detailed in Section 3. A security audit conducted prior to testing can facilitate a more detailed overview over the systems at hand and will likely lead to a more thorough testing result.

Contracted vendors often use remote login to provide support. During a penetration test, the security of remote login solutions is evaluated to determine whether there is proper network segregation. Appropriate network segregation is key in order to properly seal off the control network from less critical networks, such as an office network. Segregation of networks is deemed insufficient if we can reach a probe installed in the critical part of the network from the other side of a segregation point. There are various ways to conduct a so-called penetration test, and both commercial and open-source tool support is available, one example is arguably the most widely accepted tool, Metasploit, offered by Rapid7—a publically traded U.S. firm located in Massachusetts (Rapid7 2015).

Maintaining the proper user rights is paramount in order to limit unauthorized access to critical networks and control systems. Generally, all types of access should be forbidden unless explicitly granted. The ability to achieve authorization bypassing, privilege escalation, and login locking have to be checked and frequency of occurrence verified. To check for weak passwords, brute force or dictionary-based password guessing can be conducted, although, it might save a lot of effort to do a table-top assessment of passwords, where the existing passwords and related policies are openly examined during an audit. Passwords analyzed in this process might of course be changed and updated following an audit where this information is shared.

The general robustness of field communication can be evaluated by load testing and network storm simulation. During a network storm simulation, switches, and controllers are flooded with telegrams and their capability of handling the overload is tested. Test cases can be tailored to the specific environment or predefined ones can be used, such as the ones in the Embedded Device Security Assurance (EDSA) certification from the ISA Security Compliance Institute—a non-profit consortium based in North Carolina, USA—developed to verify supplier compliance (ISA Security Compliance Institute 2011). Furthermore, firmware and applied patches have to be screened as part of vulnerability scanning. Thorough scanning of the networks as a means to verify that the documentation of topology is corresponding to the implementation is a base requirement. If passive network monitoring is possible, any suspicious traffic anomalies have to be examined.

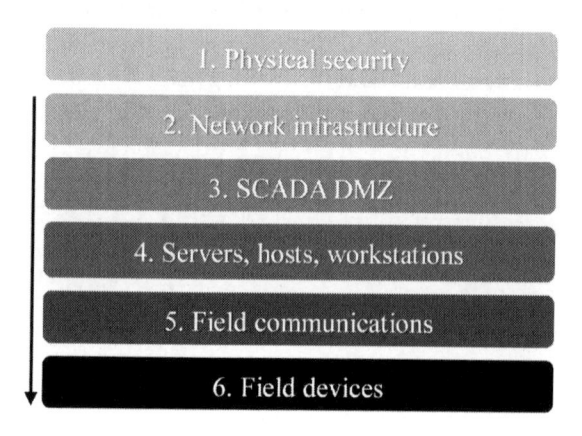

1. Physical security
2. Network infrastructure
3. SCADA DMZ
4. Servers, hosts, workstations
5. Field communications
6. Field devices

Figure 4. Levels of threats to industrial communication

Also, to verify redundancy concepts, disconnection tests can be conducted. During a disconnection test links between parts of a control system are intentionally shut down to understand the extent subsystems are capable of autonomous operation. When connections are reestablished the system's ability to recover can be verified. Concluding a successful security test, a report is presented summarizing the findings. Applying all the above techniques during testing can cover five of six layers from high to the lowest levels of communication security (Figure 4.). Ideally, robustness testing can be conducted in two phases. First, as part of vulnerability assessment, supporting an assessment or audit. Second, as technical verification of the mitigating actions put in place for an asset. Both test elements are crucial for hardening the robustness and security of industrial communication networks.

Conclusions

As cyber security threats are increasing in numbers and are entering the maritime domain, an up-to-date methodology is required to secure safe operations at sea. The challenge of maintaining or improving the integrity and resilience of cyber-physical systems, including safety-critical control systems, requires taking into account cyber security and safety at the same time. Regulations and recommendations targeting cyber security are being worked out by authorities, e.g., USCG, industrial associations, e.g., IADC, and also class societies such as DNV GL.

Possible alternatives asset owners have to mitigate cyber security risks, ranging from self-assessments, guided assessments, and audits, to third-party testing, and voluntary class notations were presented as well as a consolidated analysis approach, consisting of framing, risk assessment, mitigation, and follow up was outlined. Furthermore, an integrated approach for handling software and software updates is essential for successful operations of vessels in the future. Combining HIL testing and cyber security testing can contribute to increased safety and security in the maritime and offshore industries. Not all tests can, however, be integrated into tools that can be completely automated. The experience of a human "tester" is still necessary to discover and investigate specific holes in cyber defense. Future research in this area should seek to develop a holistic control system integrity management approach that addresses the software development cycle, software testing and cyber security as an aspect of overall robustness, ultimately to support maritime and off-shore asset owners in mitigating cyber security risks.

References

Balduzzi, Marco, Kyle Wilhoit, and Alessandro Pasta. 2013. "Hey Captain, Where's Your Ship? Attacking Vessel Tracking Systems for Fun and Profit." 11th Annual HITB Security Conference in Asia, Kuala Lumpur, Malaysia.

Bateman, T. 2013. "Police Warning After Drug Traffickers' Cyber-Attack." *BBC News Europe*, October 16. http://www.bbc.com/news/world-europe-24539417.

Bole, Brian, Christopher Teubert, Cuong ChiQuach, Edward Hogge, Sixto Vazquez, Kai Goebel, and George Vachtsevanos. 2013. "SIL/HIL Replication of Electric Aircraft Powertrain Dynamics and Inner-Loop Control for V&V of System Health Management Routines." Annual Conference of the Prognostics and Health Management Society 4: 1–12.

Borges, Victor and Colin Hickey. 2015. "Uncertainty Analysis Incorporated to Lifecycle Cost Analysis in the Oil and Gas Industry." ABRISCO 2015, Rio de Janeiro, Brazil.

Borges, Victor, Colin Hickey, and John Gilmour. 2014. "Optimisation of Deepwater Maintenance Strategy Taking into Account Reliability and Financial Aspects." Deep Offshore Technology International Conference. Aberdeen, UK.

Commander U.S. Pacific Fleet. 2013. "Command Investigation into the Grounding of USS Guardian (MCM 5) on Tubbataha Reef, Republic of the Philippines That Occurred on 17 January 2013." http://webcache.googleusercontent.com/search?q=cache:3WciUqDqJ4oJ:www.cpf.navy.mil/foia/reading-room/2013/06/uss-guardian-grounding.pdf+&cd=1&hl=en&ct=clnk&gl=us.

Dyryavyy, Yevgen. 2014. "Preparing for Cyber Battleships—Electronic Chart Display and Information System Security." NCC Group Whitepaper. 3–10.

European Network and Information Security Agency (ENISA). 2011. "Analysis of Cyber Security Aspects in the Maritime Sector." 1–25.

ICS-CERT. 2011. "Cyber Security Evaluation Tool (CSET)." https://ics-cert.us-cert.gov/Assessments.

International Electrotechnical Commission (IEC). 2010. "Functional Safety of Electrical/Electronic/Programmable Electronic Safety-related Systems." (IEC 61508). http://www.iec.ch/functionalsafety/standards/page2.htm.

ISA Security Compliance Institute. 2011. "Embedded Device Security Assurance." Version 2.0, http://isasecure.org.

Johnsen, Stig, Rune Ask, and Randi Røisli. 2007. "Reducing Risk in Oil and Gas Production Operations." 1st Annual International Conference of IFIP WG 11.10 on Critical Infrastructure Protection. Hanover, NH.

Katsilieris, Fotios, Paolo Braca, and Stefano Coraluppi. 2013." Detection of Malicious AIS Position Spoofing by Exploiting Radar Information." 16th International Conference on Information Fusion, Istanbul, Turkey.

Korn, Milton. 2013. "In the Loop." Proceedings of the Marine Safety & Security Council, the Coast Guard Journal of Safety at Sea, Winter 2013–2014.

Maslin, Elaine. 2014. "Testing Subsea BOPs to the Limit." Off-shore Engineer. http://www.oedigital.com/energy/item/5461-testing-subsea-bops-to-the-limit.

Microsoft. 2005. Microsoft Security Advisory 899588.

The Norwegian National Security Authority (NSM). 2014. "Rapport om Sikkerhetstilstanden." https://nsm.stat.no/publikasjoner/rapporter/rapport-om-sikkerhetstilstanden/.

Norwegian Oil and Gas Association. 2006. "Recommended Guidelines for Information Security." https://www.norskoljeoggass.no/en/Publica/Guidelines/.

Offensive Security. 2015. "Kali Linux." https://kali.org.

Rapid7. 2015. "Metasploit." http://www.rapid7.com/products/metasploit.

Shauk, Zain. 2013. "Malware Threatening Offshore Rig Security." (February 25). http://fuelfix.com/blog/2013/02/25/malware-on-oil-rig-computers-raises-security-fears/.

Statoil ASA. 2013. "The In Amenas Attack." Statoil investigation report. http://www.statoil.com/en/NewsAndMedia/News/2013/Pages/12Sep_InAmenas_report.aspx.

Sutton, Michael, Adam Greene, and Pedram Amini. 2007. *Fuzzing: Brute Force Vulnerability Discovery*. Upper Saddle River, NJ: Addison-Wesley.

Vassilev, Apostol, and Christpoher Celi. 2014. "Avoiding Cyberspace Catastrophes Through Smarter Testing." *Computer* 47 (10): 102–106.

Wagstaff, Jeremy. 2014. "All at Sea: Global Shipping Fleet Exposed to Hacking Threat." (April 23). http://www.reuters.com/article/us-cybersecurity-shipping-idUSBREA3M20820140424.

CHAPTER 21:
MAN AND MACHINE: HOW VISUAL ANALYTICS CAN ENABLE INSIGHTS IN MARITIME CYBER SECURITY

Sungahn Ko, Abish Malik, Guizhen Wang, and David S. Ebert
Purdue University

Abstract

The maritime sector is crucial for the wellbeing of the modern society. With the maritime industry adapting sophisticated systems that are increasingly connected to cyber space (e.g., cargo handling, container tracking, vessel navigation, propulsion systems), the industry has become increasingly susceptible to cyber attacks. These attacks can have potentially disastrous economic consequences, and in the worst case, can lead to a loss of lives. There is a need for solutions that enable maritime operators and analysts to identify deviant behaviors, and make quick and accurate decisions from the massive cyber data generated by the maritime systems that can be indicative of cyber attacks. This chapter provides an overview of the role of visual analytics in the maritime cyber security domain, along with a discussion of the role of human analysts and how their domain expertise can be harnessed in the increasingly automated maritime cyber security space. The authors provide several examples of visual analytics solutions that facilitate extracting insights from maritime data.

Introduction

The maritime sector contributes greatly to the global economy. The U.S. deep-draft seaports and supporting firms employed more than 8 million people and contributed approximately $2 trillion to the U.S. economy in 2008 (USCG 2008). The maritime industry has adopted new technologies that are increasingly connected to cyber space in order to improve their day-to-day operations (e.g., vessel navigation and propulsion systems, freight management, cargo handling, traffic control communication systems). While these technologies have greatly enhanced the maritime operations, they have exposed the industry to new cyber threat vectors that can disrupt their daily operations and lead to potentially significant local, regional, and national economic losses.

Maritime stakeholders have taken note of these threats in recent years. The newly established U.S. Coast Guard maritime cyber strategy (Michel 2015) outlines the Coast Guard's cyber priorities to include the following objectives: (a) Defending cyber space and identifying vulnerable cyber systems, (b) Enabling operations by delivering cyber resilient capabilities, and (c) Protecting critical infrastructure by performing cyber risk assessments. These priorities establish the need for maritime stakeholders to effectively monitor their cyber systems for deviations from normal behaviors. There is also a need for analysts to create simulation models that model real-world processes and behaviors for evaluating the severity of different threat vectors and testing for system resilience.

One of the major bottlenecks in these analyses, however, is the size and complexity of the data sets generated by cyber systems. As modern data sets increase in size and complexity, it becomes increasingly difficult for analysts and decision makers to identify deviant behaviors that can be indicative of cyber attacks. This is especially challenging since these data can exhibit multiple aspects of the dimensions of *big data*—large in terms of variety (e.g., multi-source, multiscale, multivariable), velocity (e.g., streaming, frequent updates), and volume. However, as Hamming (1962) points out, the purpose of computing must be to gather insights into the dynamic processes of the world instead of the mere generation of numbers. The wealth of information generated by cyber systems creates a dearth of attention for analysts that require them to allocate their attention efficiently in order to prevent them from getting overburdened from the information overload (Simon 1971). To address these challenges, the field of visual analytics has emerged to assist analysts and decision makers in their use of data for effective monitoring, analysis, and decision-making.

Visual analytics has been defined as the science of analytical reasoning facilitated by interactive visual interfaces (Thomas and Cook 2006; Keim et al. 2010). The use of visual analytics techniques facilitate and enhance the cognitive abilities of analysts and decision makers by maximizing the utilization of their perceptual and cognitive capabilities in an integrated visual analysis and exploration environment (Eick and Wills 1993; Stasko, Görg, and Liu 2008; Zhao et al. 2011; Kim et al. 2013). Consequently, visual analytics can provide insights to domain experts into the massive data generated by cyber systems. Analysts can utilize visual analytics to design resilient maritime systems, and also to provide situational awareness tools for detecting anomalies in the data that can be indicative of cyber attacks. Additionally, such visual analytic systems, when used in conjunction with other threat vectors (e.g., significant weather events), can greatly improve the situational awareness of emerging or emergent situations.

This chapter provides an overview of the role that visual analytics can play in the maritime cyber security domain. The main areas covered in this chapter include the following:

- Monitoring maritime cyber systems for aberrant behaviors: We present a visual analytics system that enables the monitoring of large-scale high-dimensional cyber network systems.

- Cyber attack modeling and simulation tools for maritime resiliency analysis: We demonstrate the utility of visual analytics in designing resilient critical maritime infrastructure systems by presenting a visual analytics system that allows analysts to create multiple "what-if" scenarios to evaluate the impacts of multiple cyber-threat vectors.

- Economic risk assessments due to disruptions caused by cyber attacks: We discuss how maritime stakeholders can utilize visual analytics to perform economic risk assessments in the event of port closures due to cyber attacks.

BACKGROUND

In recent years, there has been much work done in applying visual analytics principles in different domains. Data analysis processes can often be complex and deal with multiple variables that exhibit complex relationships. Visual analytics integrates data visualization techniques along with human–computer interaction methods to assist users with their decision-making tasks. It allows domain experts to utilize their domain knowledge to augment the underlying machine learning and statistical processes. Compared with monotonous numerical data analysis methods, visual analytic methods allow users to quickly understand their data through intuitive and interactive data visualization techniques.

Visual analytics techniques have been widely used in many areas, covering a variety of data types and analytical objects. For example, the Jigsaw (Stasko, Görg, and Liu 2008) and IN-SPIRE™ (PNNL 2014) systems allow users to analyze and investigate large document clusters. These systems allow users to extract latent information and semantic abstractions (e.g., relations among people, time, places) from a large corpus of documents. Malik et al. (2012) provide a spatio-temporal visual analytics system to allow the exploration of crime data. This system provides a suite of linked spatial and temporal visualizations to assist users in the exploration of their data. Other examples of visual analytics systems that enable users to explore their data in order to discover patterns can be found in Chang et al. (2007); Lam et al. (2007); Ziegler et al. (2010); Ko et al. (2012; 2014); Sun et al. (2013)).

In the cyber domain, more and more systems are increasingly inter-connected through networks hierarchies (e.g., internet services). This high level of connectivity can trigger latent cyber threats from different adversaries, who can make use of existing vulnerabilities in the systems to conduct illicit activities (e.g., network intrusions, malware attacks). Cyber systems can also span across a diverse network of cyber infrastructures (Lavigne and Gouin 2014). Accordingly, manual detection of these cyber attacks can become cumbersome for security experts. In order to alleviate these challenges, researchers apply automatic machine intelligence methods to detect cyber threats. However, these methods can prove to be inadequate due to the lack of a comprehensive cyber attack database for training the different models. Accordingly, visual analytics techniques can be utilized to augment automatic machine intelligence methods to efficiently collect forensic information, and further allow domain experts to apply their domain knowledge to enhance these processes (Marty 2013).

Researchers have provided several visual analytic systems that support analysts in various types of cyber analysis tasks. In order to allow monitoring of network vulnerability, Harrison et al. (2012) provide a visual analytics system designed to support the discovery of adverse activities in a network, while Roveta et al. (2011) present an approach of detecting a rogue network among a group of autonomous systems. Fowler et al. (2014) designed a system for internet security by considering the internet as a network in their visual analytics system. Their system enables security threat detection by presenting visualizations for dynamic evolution of hierarchical network activities in the internet topology. In order to protect a network, Kintzel, Fuchs, and Mansmann (2011) used IP addresses in their visualizations that help monitor daily activities of a cyber network. Other visual analytics systems include the work done by Boschetti et al. (2011) where visual-based querying methods combined with different levels-of-detail techniques are used for tracing networks and the work done by Matuszak et al. whose focus is on providing situational awareness in cyber space. In order to evaluate visualizations in the realm of cyber security, Fischer et al. (2014) present performance evaluation with common visualizations. We note that Staheli et al. (2014) provide an overview of the current visual analytics techniques and approaches to identify the threats and vulnerabilities in cyber systems.

In order to help surveil cyber spaces in real time, various systems have been developed. Fischer and Keim (2014) provide a real-time visual analysis system for analyzing workflow of data streams from the cyber infrastructures. In the work done by Chen et al. (2014), analysts are allowed to collaborate online during analysis. In addition to network intrusion, malware attacks are also a common form of cyber attacks. Malware threats are usually recognized based on behaviors of malicious software. Wagner et al. (2014) explore the theoretical description in visual analytics to discover behavior-based malware patterns. On the other hand, Long, Saxe, and Gove (2014) propose a visual tool for detecting malware samples based on similar image sets. In the same manner, Gove et al. (2014) also use and compare attributes in malicious samples in order to detect malicious activity.

Cyber logs have been considered as useful resources for detecting security breaches in various domains and have also been utilized in the cyber security domain. Legg et al. (2014) provide a tool that analyzes latent user behaviors among embedded emails of sociolinguistics, while Alsaleh et al. (2013) present an interactive visual system helping to find server attacks. The Visual Filter (Stange et al. 2014) and EVis (Humphries et al. 2013) systems focus on visual exploration of log files for network security. These systems enable analysts to monitor and mitigate the risks and vulnerabilities associated with potential cyber attacks.

In this chapter, we focus on exploring the benefits of visual analytics applied in the maritime cyber security domain. We provide examples of how visual analytics can assist analysts to identify the threats and vulnerabilities in their cyber systems, and aid in establishing resilient maritime cyber systems. We also discuss the applicability of visual analytics in performing economic risk assessments in the event of cyber attacks in the maritime domain.

VISUAL ANALYTICS FOR MONITORING LARGE SCALE MARITIME CYBER SYSTEMS

A key aspect of designing resilient maritime cyber systems involves developing effective monitoring solutions. It has become increasingly difficult, if not impossible, to prevent the occurrence of cyber attacks particularly due of the nature of the adversary. Modern day adversaries have low barriers to entry, can be located anywhere on the planet, and can range from a single cybercriminal, to nation states. However, detecting cyber attacks is particularly challenging because of the size and complexity of the data sets associated with cyber systems. Furthermore, these data sets may exhibit data quality issues (e.g., missing and uncertain data) that can further decrease the effectiveness of analysts. Visual analytics facilitates the discovery of anomalies in the data by embedding the human in the loop for understanding, exploring, and validating the underlying data and algorithms. In this section, we provide examples of a visual analytics system to demonstrate the efficacy of visual analytics in monitoring for maritime cyber systems.

Visual Analytics for Large-Scale High-Dimensional Cyber Security Network Data

Network-enabled machines are important for communications and operations. Breaches in these machines could not only cease all maritime operations but also affect other industries and economic sectors. Therefore, it is important to monitor all the operational machines and to efficiently investigate any incidents caused by possible breaches. However, the task of monitoring network-enabled machines becomes non-trivial when there are a very large number of devices that generate multivariate data in multiple locations. For proper analysis, analysts need to see and compare all of these different dimensions at multiple granularities (e.g., enterprise to individual machines in individual offices). In order to address these challenges, the visual analytics system called SemanticPrism (Chen et al. 2012) has been developed for the cyber security domain in order to help users visually analyze the large high-dimensional cyber security data (e.g., different machines at multiple locations of a connected maritime network, different maritime electronic devices connected to a network) from three perspectives: spatiotemporal distribution of machines and their health, overall temporal trends, and pixel-based IP address blocks. All visualizations in the system are linked and provide 2–4 levels of semantic zooming. In this system, the analyst can not only grasp the overall situation of the enterprise network, but also drill down to more detailed information for regions, locations, and even at the level of individual machines. Next, we present a description and a use-case on visualizations in the map and pixel views in SemanticPrism. We demonstrate the system using the data set from the VAST 2012 Mini-Challenge 1 (VAST 2012) that includes logs of a million machines from an imaginary bank.

The system consists of multiple views, including the map view (Figure 1) and pixel view (Figure 2). It provides different information based on different zooming levels. The goal of the map view (Figure 1) is to allow users to keep track of activities of geographically distributed machines based on time on a map. An example anomaly that could be detected using the system is the

CHAPTER 21

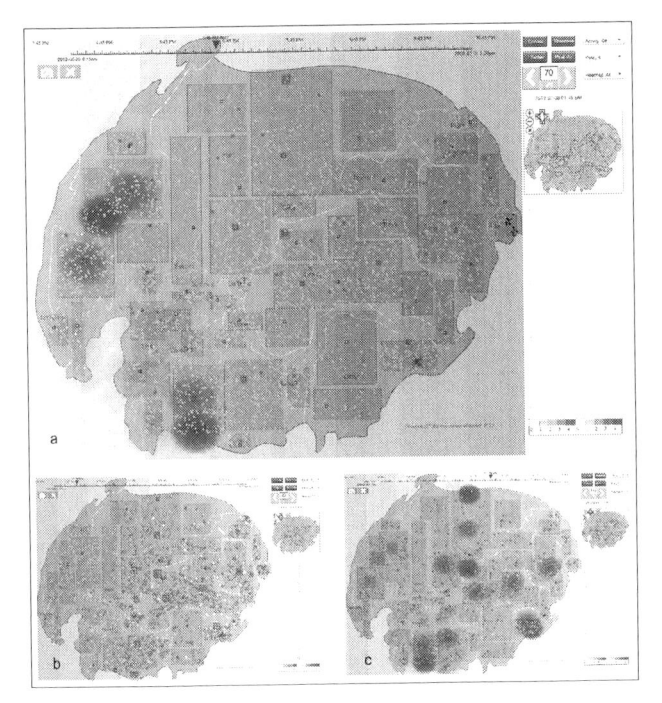

Figure 1: The map view helps users find which areas have active machines based on different time zones. (Right-top) A time slider (arrow) is used for exploring different time steps

machines that are active at off-office hours (e.g., invalid logins or intrusions). Such an anomaly can be easily seen in the map view by turning on the activity highlight layer and the time zone layer in the system. Here, as time passes, an analyst can monitor how offices become idle when their local times reach the off-working time at 6 p.m. and then become active again at 8 a.m. (Figure 1, bottom-left).

Another feature in SemanticPrism is the support of operational policies that all the machines should obey (e.g., inactive hours) and search for any violations on the policies. For example, the pixels on the map view blink when machines are violating an organization's policies. The map view can also allow users to visualize the evolution of machines that exhibit high-risk policies over time. For example, Figure 1, bottom-right shows that the number of squares in orange color gets larger as time progresses, which indicates that more and more computers are under high-risk.

Pixel-based visualizations (Keim 2000) map a value to a color-coded pixel in order to best utilize the screen space. This visualization is advantageous when a very large number of data items are analyzed. SemanticPrism utilizes pixel-based visualizations by mapping a machine to a pixel in order to allow monitoring a large number of operational machines. Here, activity of machines is judged by the color of a pixel. If a machine is violating an operational policy, the pixel representing the machine turns to red. This can be seen in Figure 2, where it is observable that many machines are violating policies with red pixels. This could be an initial symptom indicating cyber attacks that inject a virus on the network if red pixels are spreading over time.

SemanticPrism thus allows simultaneous investigations with a very large number of data items. This system can also be applied for monitoring electrical devices used for operations in the maritime realm. For example, it is possible that malwares imbedded in one of machines can spread across whole networks. These situations can be easily detected using the system; thereby, allowing maritime stakeholders to effectively monitor their cyber networks for potential security breaches and resilience analysis.

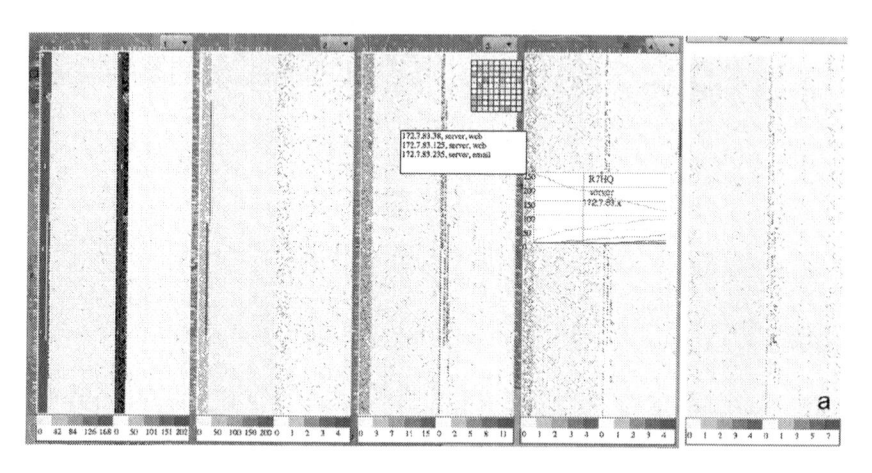

Figure 2: Pixel matrices (top) present which devices and machines are violating an organization's policies. When devices are found to be in violation, the pixel for the machine turns to red

INTERACTIVE CYBER ATTACK MODELING AND SIMULATION FOR RESILIENCY ANALYSIS

The maritime industry is comprised of several critical components that are interconnected through fragile interrelationships (e.g., communication systems, control and automation systems, vessel engine and propulsion systems, navigation and bridge systems, cargo handling systems). Disruptions (e.g., cyber attacks) in one or more of these components can cause cascading effects that can disrupt entire port operations; thereby, causing major local and national economic losses. This requires maritime analysts to perform simulations of critical maritime components in order to detect vulnerable systems. Maritime stakeholders are also required to study the economic impacts due to disruptions caused by cyber attacks in order to identify critical economic components. Visual analytic systems can assist analysts in understanding the high level risks and consequences of cyber attacks on entire maritime sectors, as well as assess the risks due to specific vulnerabilities, facilities, and systems. In the subsections that follow, we discuss how visual analytics can assist analysts and decision makers with these tasks.

Simulation of Cyberattacks on Critical Maritime Infrastructures

Simulations involve modeling real-world processes and studying them over time. Simulations have been a standard tool for analysts and policymakers to design for resilience and preparedness when perturbation in one seemingly minor aspect of a complex system can have vast and far-reaching impacts across an industry. Using complex simulations of critical maritime components, analysts can create multiple "what-if" scenarios, calculate the impact of cyber threats depending on their severity, and—last but not least—study optimal mitigation measures to address them. The simulation models can be designed to study the likelihood of the occurrence of cyber attacks on different maritime systems (e.g., critical infrastructures), and design response strategies that minimize the impact and restore full operations if a cyber attack were to occur.

Critical Infrastructure and its Application in Maritime Domain

The Visual Analytics for Simulation-based Action (VASA) system (Ko et al. 2014), shown in Figure 3, is a visual analytics platform that consists of a desktop application, a component model, and a suite of distributed simulation components for modeling the impact of societal threats (e.g., cyber, weather, food contamination, traffic) on critical infrastructure (e.g., cyber networks, supply chains, road networks, power grids). Each component encapsulates a high-fidelity simulation model that together forms an asynchronous simulation pipeline: a system of systems of individual simulations with a common data and parameter exchange format.

We demonstrate the efficacy of visual analytics by utilizing the critical infrastructure module in VASA (Figure 3) that allows analysts to identify vulnerabilities in critical infrastructures (e.g., maritime supply chain networks, power plants, management command centers) in case they are compromised by adverse elements (e.g., cyber attacks). This component, shown in Figure 4, models a hierarchical network (e.g., power grid, supply chain), where the

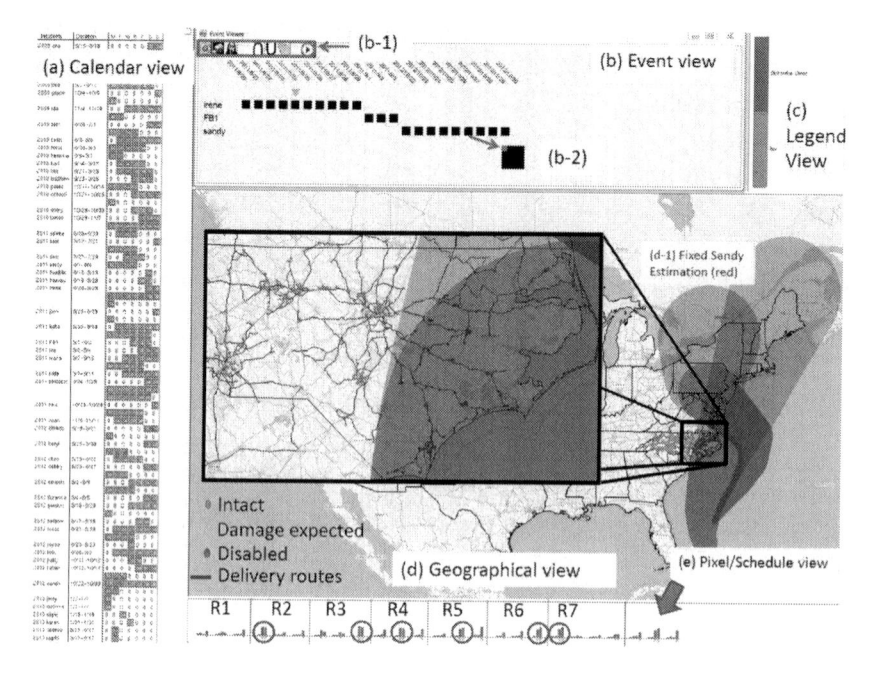

Figure 3: VASA system overview

nodes (e.g., power generators) are connected with edges (e.g., transmission lines). This model simulates the impacts of the closure of a node and provides information on the other impacted regions in the network (e.g., power outage areas). It helps analysts answer questions like when a main network node is compromised, how do the effects propagate through the network? What other nodes connected to the affected node are impacted, and thus, which critical areas are vulnerable to threats? We note that the VASA framework provides analysts with the ability to input different models, and can be used to study the effects of different cyber threat vectors on critical maritime infrastructures in order to detect vulnerabilities. Further, we note that the incorporation of multiple displays in the visual analytics environment could enhance the cyber threat monitoring capability because cyber attacks combined and cascaded with other natural (e.g., severe weather) and cyber threats (e.g., attacking systems to disable ports) could drastically exacerbate damages.

In our example, we utilize the power grid model to present a scenario where a cyber terrorist has disabled a power generation plant. This has been shown in Figure 5 (leftmost), where the analyst first disables the plant by selecting a power plant shown by a red rectangle. The model instantly estimates the affected operational facilities in the network and a full simulation is initiated for more accuracy (second-left, Figure 5). The simulation result is presented by a polygon that presents the area where all facilities are down (second-right, Figure 5). The right-most image provides magnification of the result.

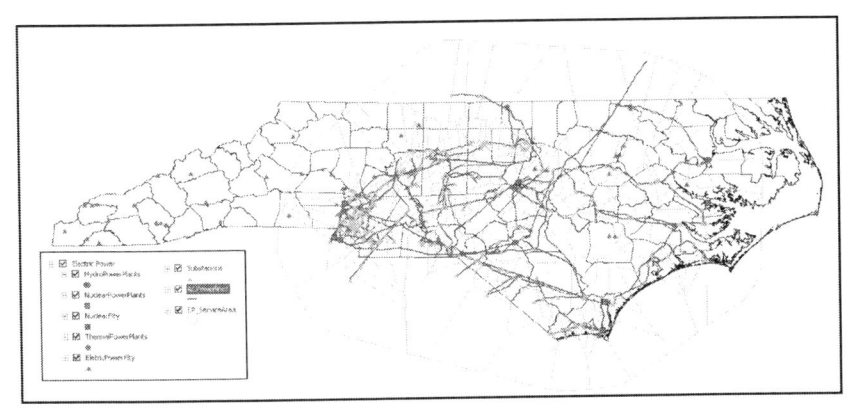

Figure 4: Power transmission grid with parts of the distribution network

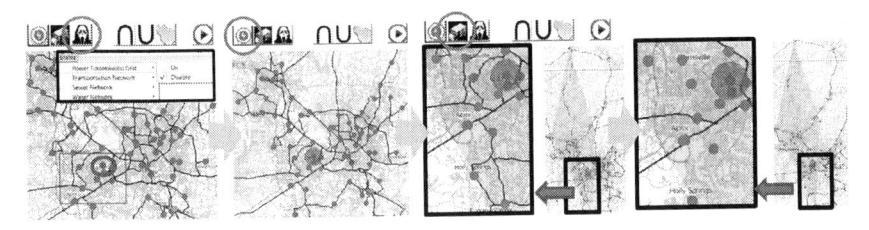

Figure 5: An example of a cyber attack simulation using the VASA system. (left) A user selects to disable a power plant by a cyber attack and selects the region shown in the red rectangle. One main server (purple dot) is included within this rectangle (shown in the dark red circle). (second-left) The model instantly estimates the affected subsidiary servers and workstations (red dots), and a full simulation is initiated. (second-right) Network-disabled regions are presented by the polygon (right)

Assessing Economic Impacts of Port Closures due to Cyber Attacks

In this section, we demonstrate the role of visual analytics in studying for the economic impacts of port closures due to adverse threat vectors (e.g., cyber attacks, weather disasters). There have been concerns of cyber attacks on ports potentially disrupting operations and causing detrimental impacts on economy (Wilshusen and Caldwell 2014). We utilize the data from the maritime economic impact study for port closures in the area of Port Arthur, Texas (DHS 2011). It should be noted that although this study was conducted to evaluate the economic impacts of port closures due to waterway issues, it can also be applied to assess the short and long term impacts of port closures due to cyber threats. This study has generated a large volume of market sector data for the impact of closures of different durations. It has also explored the impacts of different mitigation strategies. However, the large data generated from the analysis is difficult to use by decision makers in analysis consequence, response scenarios, and making decisions with trade-offs. Interactive visual analytics can alleviate these problems and provide a powerful tool since the results are more easily interpreted visually. Our visual analytics system designed for this study utilizes spatial and temporal analytics and visualizations blended with a multivariate correlative analysis to form an interactive tool that facilitates an interactive display of economic impact by choosing the type of vessel and types of inspection programs. The user can explore different strategies and assess the economic impacts over time.

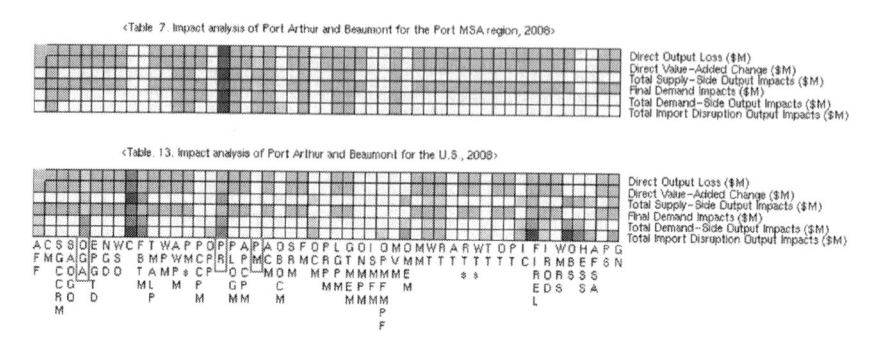

Figure 6: The system easily enables analysts to search major sectors affected by the import disruption. The PR and PM sectors are the two most affected sectors for the Port MSA region while other sectors are also much affected in a point of view for U.S.

System and Case Study: Simulation of Economic Impact of Port Arthur closed by a Cyber Attack

Figure 6 provides a snapshot of our system's visualization. In the matrices, 51 sectors are mapped horizontally while six models estimating loss (Rose and Wei 2011) are assigned vertically. Abbreviations are used for sector names to avoid visual clutter on X-axis (full industry sector names on the X-axis are presented in Appendix 1). When an impact is zero, gray color is assigned to the corresponding cell of the matrix. The square encompassed by the green outline means a field where the analyst's mouse is pointing and the system provides detailed information for the data such as a sector and model names, and impacts to the port (upper matrix) and the U.S (bottom matrix). With the

CHAPTER 21

system, analysts can easily observe which major sectors are affected and how much impact is caused by the import disruption generated by cyber attacks. For example, PR (Petroleum refineries, 17th column in Figure 6) and PM (Petrochemical mfg., 20th column in Figure 6) are the most affected sectors for the Port MSA regions (upper matrix), but not for the U.S (bottom matrix). In contrast, there are sectors that will not be affected in the Port region as a result of the cyber attacks, but have large impacts for the U.S. For example, OGA (fifth column in Figure 6) does not have a Direct Output Loss in the port region, while it can have severe impacts for the United States.

Scenarios	Scenarios	Scenarios
✓ A. Base Case(No Resilience)	A. Base Case(No Resilience)	A. Base Case(No Resilience)
B. With Re-routing	✓ B. With Re-routing	✓ B. With Re-routing
C. With SPR	✓ C. With SPR	✓ C. With SPR
D. With Use of Inventories	✓ D. With Use of Inventories	✓ D. With Use of Inventories
E. With Export Diversion	✓ E. With Export Diversion	✓ E. With Export Diversion
F. With Conservation	F. With Conservation	✓ F. With Conservation
G. With Production Reschedulig	G. With Production Reschedulig	G. With Production Reschedulig

Figure 7: Multiple scenarios can be chosen to apply. F will be selectable after B, C, D, and E scenarios are applied. Likewise, G can be applied after F is applied

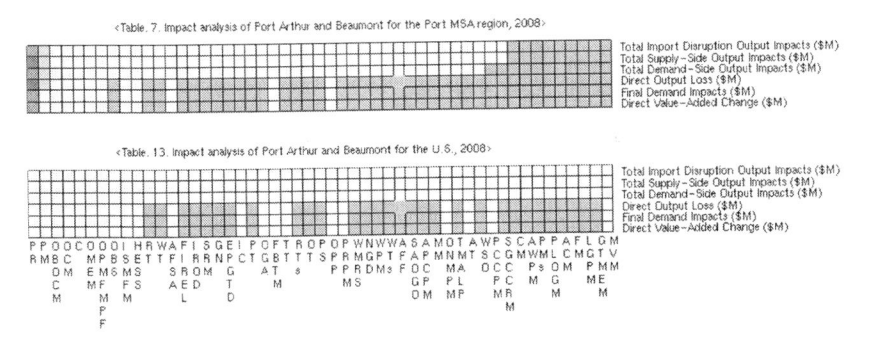

Figure 8: Re-routing has the largest effect in reducing damage to United States.

Resilience Adjustments

The system enables users to analyze the different perspectives by applying resilience adjustments (A to F in Figure 7) on-the-fly. When each adjustment is applied, a new color is assigned to the cells of the matrices. The lighter the new assigned color, the smaller the impacts. The resulting visualizations present different effects of adjustments to the Port region and the U.S. For example, the "With Re-routing" adjustment (i.e., using nearby ports in case of port closures) has two drastic effects as shown in Figure 8. It reduces the damage in "Direct Output Loss" for the Port region, and further reduces the impacts for the U.S. (when compared against Figure 6). We note that the method used in this visual analytics approach can be adapted for other systems and analyses where information needs to be weighted differently. For example, the visualization can be used for focusing on the economic impacts with various aspects and weights (e.g., entire sectors versus specific vulnerabilities of facilities, systems).

Conclusions

Cyber attacks are a threat to the modern society, and it has become increasingly difficult to prevent cyber attacks from occurring due to the nature of the adversary. Stakeholders must therefore monitor for cyber attacks, and plan, prepare, and mitigate for them in order to create resilient maritime systems. There is a need for analysts to evaluate and understand the vulnerabilities of their systems in order to minimize disruptions in case of cyber attacks. This chapter has provided an overview of the potential of visual analytics in the maritime cyber security domain. Visual analytics harnesses the cognitive abilities of analysts and decision makers in an integrated visual analysis and exploration environment. We presented a visual analytics cyber security monitoring system that can provide maritime analysts with the tools to effectively monitor their large cyber systems for malicious behaviors. We also presented interactive cyber security visual analytic modeling and simulation methods that enable analysts to apply different models in order to perform what–if analyses and simulate the effects of different cyber threat vectors on critical infrastructures. Finally, we demonstrated how visual analytics can enable the assessment of economic impacts of disruptions caused due to cyber attacks.

Although we have demonstrated the efficacy of visual analytics in the maritime cyber security domain, there are many opportunities for visual analytics that remain unexplored that can empower maritime stakeholders to help counter and respond to cyber terrorists. Visual analytics can facilitate stakeholders to identify the greatest vulnerabilities in the different components of the maritime industry by conducting importance-based risk evaluations. Analysts can also utilize visual analytics to analyze and predict the different dynamic cyber threats by allowing them to "detect the expected and discover the unexpected". Decision makers can also apply visual analytic tools to harness the data generated by the different maritime systems for improved situational awareness. Visual analytics research can also provide guidance on the vulnerabilities in the maritime transportation systems and the cascading effects to the nation and economy in case they are impacted due to cyber attacks. Finally,

CHAPTER 21

visual analytics can enable decision makers to augment their data-driven analysis processes by utilizing their domain knowledge in the different sectors of the maritime industry.

REFERENCES

Alsaleh, Mansour, Abdullah Alqahtani, Abdulrahman Alarifi, and AbdulMalik Al-Salman. 2013. "Visualizing PHPIDS Log Files for Better Understanding of Web Server Attacks." *Proceedings of the Tenth ACM Workshop on Visualization for Cyber Security*: 1–8.

Boschetti, Alberto, Luca Salgarelli, Chris Muelder, and Kwan-Liu Ma. 2011. "Tvi: A Visual Querying System for Network Monitoring and Anomaly Detection." *Proceedings of the 8th International Symposium on Visualization for Cyber Security* (VizSec '11): 1–10.

Chang, Remco, Mohammad Ghoniem, Robert Kosara, William Ribarsky, Jing Yang, Evan Suma, Caroline Ziemkiewicz, Daniel Kern, and Agus Sudjianto. 2007. "Wirevis: Visualization of Categorical, Time-varying Data from Financial Transactions." *Proceedings of IEEE Symposium on Visual Analytics Science and Technology*: 155–162.

Chen, Siming, Cong Guo, Xiaoru Yuan, Fabian Merkle, Hanna Schaefer, and Thomas Ertl. 2014. "OCEANS: Online Collaborative Explorative Analysis on Network Security." *Proceedings of the Eleventh ACM Workshop on Visualization for Cyber Security*: 1–8.

Chen, Victor Yingjie, Sungahn Ko, David S. Ebert, Cheryl Zhenyu Qian, and Ahmad M. Razip. 2012. "SemanticPrism: A Multi-Aspect View of Large High-Dimensional Data: VAST 2012 Mini Challenge 1 Award: Outstanding Integrated Analysis and Visualization." *Proceedings of IEEE Conference on Visual Analytics Science and Technology (VAST)* (VAST '12): 259–260.

Eick, Stephen G., and Graham J. Wills. 1993. "Navigating Large Networks with Hierarchies." *Proceedings of IEEE Conference on Visualization*: 204–210.

Fischer, Fabian, James Davey, Johannes Fuchs, Olivier Thonnard, Jörn Kohlhammer, and Daniel A. Keim. 2014. "A Visual Analytics Field

Experiment to Evaluate Alternative Visualizations for Cyber Security Applications." *Proceedings of EuroVA International Workshop on Visual Analytics.*

Fischer, Fabian, and Daniel A. Keim. 2014. "NStreamAware: Real-time Visual Analytics for Data Streams to Enhance Situational Awareness." *Proceedings of the Eleventh ACM Workshop on Visualization for Cyber Security (VizSec '14)*: 65–72.

Fowler, J. Joseph, Thienne Johnson, Paolo Simonetto, Michael Schneider, Carlos Acedo, Stephen Kobourov, and Loukas Lazos. 2014. "IMap: Visualizing Network Activity Over Internet Maps." *Proceedings of the Eleventh ACM Workshop on Visualization for Cyber Security (VizSec '14)*: 80–87.

Gove, Robert, Joshua Saxe, Sigfried Gold, Alex Long, and Giacomo Bergamo. 2014. "SEEM: A Scalable Visualization for Comparing Multiple Large Sets of Attributes for Malware Analysis." *Proceedings of the Eleventh ACM Workshop on Visualization for Cyber Security (VizSec '14)*: 72–79.

Hamming, Richard. 2012. *Numerical Methods for Scientists and Engineers.* New York: Courier Corporation.

Harrison, Lane, Riley Spahn, Mike Iannacone, Evan Downing, and John R. Goodall. 2012. "Nv: Nessus Vulnerability Visualization for the Web." *Proceedings of the Ninth International ACM Symposium on Visualization for Cyber Security (VizSec '12)*: 25–32.

Humphries, Christopher, Nicolas Prigent, Christophe Bidan, and Frédéric Majorczyk. 2013. "Elvis: Extensible Log Visualization." *Proceedings of the Tenth ACM Workshop on Visualization for Cyber Security*: 9–16.

Keim, Daniel A. 2000. "Designing Pixel-Oriented Visualization Techniques: Theory and Applications." *IEEE Transactions on Visualization and Computer Graphics* 6 (1): 59–78.

Keim, Daniel A., Jörn Kohlhammer, Geoffrey Ellis, and Florian Mansmann. 2010. *Mastering the Information Age-solving Problems with Visual Analytics.* Florian Mansmann.

Kim, SungYe, Ross Maciejewski, Abish Malik, Yun Jang, David S. Ebert, and Tobias Isenberg. 2013. "Bristle Maps: A multivariate Abstraction Technique for Geovisualization." *IEEE Transactions on Visualization and Computer Graphics* 19 (9): 1438–1454.

CHAPTER 21

Kintzel, Christopher, Johannes Fuchs, and Florian Mansmann. 2011. "Monitoring Large IP Spaces with Clockview." *Proceedings of the 8th International Symposium on Visualization for Cyber Security* (VizSec '11): 2.

Ko, Sungahn, Shehzad Afzal, Simon Walton, Yang Yang, Junghoon Chae, Abish Malik, Yun Jang, Min Chen, and David Ebert. 2014. "Analyzing High-dimensional Multivariate Network Links with Integrated Anomaly Detection, Highlighting and Exploration." *Proceedings of IEEE Visual Analytics Science and Technology (VAST)*: 83–92.

Ko, Sungahn, Ross Raciejewski, Yun Jang, and David Ebert. 2012. "Market Analyzer: A Visual Analytics System for Analyzing Competitive Advantage Using Point of Sales Data." *Computer Graphics Forum* 31 (3): 1245–1254.

Ko, Sungahn, Jieqiong Zhao, Jing Xia, Shehzad Afzal, Xiaoyu Wang, Greg Abram, Niklas Elmqvist Len Kne, David Van Riper, Kelly Gaither, Shaun Kennedy, William Tolone, William Ribarsky, and David S. Ebert. 2014. "VASA: Interactive Computational Steering of Large Asynchronous Simulation Pipelines for Societal Infrastructure." *IEEE Transactions on Visualization and Computer Graphics* 20 (12): 1853–1862.

Lam, Heidi, Daniel Russell, Diane Tang, and Tamara Munzner. 2007. "Session Viewer: Visual Exploratory Analysis of Web Session Logs." *Proceedings of IEEE Symposium on Visual Analytics Science and Technology*: 147–154.

Lavigne, Valérie, and Denis Gouin. 2014. "Visual Analytics for Cyber Security and Intelligence." *The Journal of Defense Modeling and Simulation: Applications, Methodology, Technology* 11 (2): 175–199.

Legg, Philip, Oliver Buckley, Michael Goldsmith, and Sadie Creese. 2014. "Visual Analytics of E-mail Sociolinguistics for User Behavioural Analysis." *Journal of Internet Services and Information Security (JISIS)* 4 (4): 1–13.

Long, Alexander, Joshua Saxe, and Robert Gove. 2014. "Detecting Malware Samples with Similar Image Sets." In *Proceedings of the Eleventh ACM Workshop on Visualization for Cyber Security VizSec '14*: 88–95.

Malik, Abish, Ross Maciejewski, Niklas Elmqvist, Yun Jang, David S. Ebert, and Whitney Huang. 2012. "A Correlative Analysis Process in a Visual Analytics Environment." *Proceedings of IEEE Conference on Visual Analytics Science and Technology (VAST)*: 33–42.

Marty, Raffael. 2013. "Cyber Security: How Visual Analytics Unlock Insight."

Proceedings of the 19th ACM SIGKDD International Conference on Knowledge Discovery and Data Mining: 1139.

Matuszak, William J., Lisa DiPippo, and Yan Lindsay Sun. 2013. "CyberSAVe: Situational Awareness Visualization for Cyber Security of Smart Grid Systems." *Proceedings of the Tenth ACM Workshop on Visualization for Cyber Security*: 25–32.

Michel, Chuck. 2015. "Coast Guard Cyber Strategy". Keynote presentation at the Maritime Cyber Security Learning Seminar and Symposium at CCICADA at Rutgers University, New Brunswick, NJ.

Pacific Northwest National Laboratory (PNNL). 2014. "IN-SPIRE™ Visual Document Analysis". http://in-spire.pnl.gov.

Rose, Adam, and Dan Wei. 2011. "Measuring Economic Risk Benefits of USCG Marine Safety Programs". In *Report No. RDC-UDI-1184*. Coast Guard New London CT Research and Development Center.

Roveta, Francesco, Giorgio Caviglia, Luca Di Mario, Stefano Zanero, Federico Maggi, and Paolo Ciuccarelli. 2011. "Burn: Baring Unknown Rogue Networks." *Proceedings of the 8th International ACM Symposium on Visualization for Cyber Security*: 6.

Simon, Herbert A. 1971. "Designing Organizations for an Information-rich World." In *Computers, Communications and the Public Interest*: 37–72. Johns Hopkins University Press.

Staheli, Diane, Tamara Yu, R. Jordan Crouser, Suresh Damodaran, Kevin Nam, David O'Gwynn, Sean McKenna, and Lane Harrison. 2014. "Visualization Evaluation for Cyber Security: Trends and Future Directions." *Proceedings of the Eleventh ACM Workshop on Visualization for Cyber Security*: 49–56.

Stange, Jan-Erik, Marian Dörk, Johannes Landstorfer, and Reto Wettach. 2014. "Visual Filter: Graphical Exploration of Network Security Log Files." *Proceedings of the Eleventh ACM Workshop on Visualization for Cyber Security (VizSec '14)*: 41–48.

Stasko, John, Carsten Görg, and Zhicheng Liu. 2008. "Jigsaw: Supporting Investigative Analysis Through Interactive Visualization." *Information Visualization* 7 (2): 118–132.

Thomas, James J., and Kristin A. Cook. 2006. "A Visual Analytics Agenda." *IEEE Computer Graphics and Applications* 26 (1): 10–13.

United States Coast Guard (USCG). 2008. "Maritime Safety Performance Plan." http://www.uscg.mil/hq/cg5/cg54/docs/MSPerformancePlan.pdf.

VAST. 2012. "IEEE VAST Challenge 2012." http://www.vacommunity.org/VAST+Challenge+2012.

Wagner, Markus, Wolfgang Aigner, Alexander Rind, Hermann Dornhackl, Konstantin Kadletz, Robert Luh, and Paul Tavolato. 2014. "Problem Characterization and Abstraction for Visual Analytics in Behavior-based Malware Pattern Analysis." *Proceedings of the Eleventh ACM Workshop on Visualization for Cyber Security* (VizSec '14): 9–16.

Wilshusen, Gregory, and Stephen Caldwell. 2014 "Maritime Critical Infrastructure Protection: DHS Needs to Better Address Port Cybersecurity." U.S Government Accountability Office. http://www.gao.gov/assets/670/663828.pdf.

Zhao, Jian, Fanny Chevalier, Emmanuel Pietriga, and Ravin Balakrishnan. 2011. "Exploratory Analysis of Time-series with Chronolenses." *IEEE Transactions on Visualization and Computer Graphics* 17 (12): 2422–2431.

Ziegler, Hartmut, Marco Jenny, Tino Gruse, and Daniel A. Keim. 2010. "Visual Market Sector Analysis for Financial Time Series Data." *Proceedings of IEEE Symposium on Visual Analytics Science and Technology (VAST)*: 83–90.

APPENDIX 1:
ABBREVIATIONS FOR MARITIME INDUSTRY SECTORS

ACM	Alkalies and chlorine mfg.
AFF	Agriculture, forestry and fishing
AFSA	Accommodations, food services, and amusements
APCPM	All other petroleum and coal products mfg.
AT	Air transportation
AWPM	All other miscellaneous wood product mfg.
C	Construction
CM	Coal mining
EPGTD	Electric power generation, transmission, and distribution
FBTM	Food, beverage, and tobacco mfg.
FIREL	Finance, insurance, real estate, and leasing
FM	Fertilizer mfg.
GN	Government and Non-NAICS
GTMEM	Ground or treated mineral and earth mfg.
HESS	Health, education and social services
IC	Information and Communication
IROD	Imputed rental for owner-occupied dwellings
ISMFM	Iron and steel mills and ferroalloy mfg.
LGPM	Lime and gypsum product mfg.
MM	Miscellaneous mfg.
MVM	Motor vehicle mfg.
NGD	Natural gas distribution
OBOCM	Other basic organic chemical mfg.
OBS	Other business services
OCM	Other chemical mfg.
OGA	Oil and gas extraction and all other mining
OMEM	Other machinery and equipment mfg.
ONMPM	Other nonmetallic mineral product mfg.
OPMFMPF	Other primary metal and fabricated metal product mfg.

CHAPTER 21

OPP	Other paper and printing
OT	Other transportation
PCCPM	Paperboard container and coated paper mfg.
PLOGM	Petroleum lubricating oil and grease mfg.
PM	Petrochemical mfg.
PMs	Pulp mills
PR	Petroleum refineries
PRPM	Plastics and rubber products mfg
PS	Personal services
PT	Pipeline transportation
RT	Retail trade
RTs	Rail transportation
SAOGO	Support activities for oil and gas operations
SGCCRM	Sand, gravel, clay and ceramic and refractory minerals
SRM	Synthetic rubber mfg.
TT	Truck transportation
TMALP	Textile and mills, apparel and leather product
WSO	Water, sewage and other systems
WPM	Wood product mfg.
WT	Wholesale trade
WTs	Water transportation
WMRS	Waste management and remediation services

CHAPTER 22:
GPS AND SHIPPING: COUNTERING THE THREAT OF INTERFERENCE

Jeffrey D. Coffed and Joe Rolli[1]
Harris Corporation

Abstract

The global navigation satellite system (GNSS), including the global positioning system (GPS), has become a ubiquitous service supporting many critical infrastructures including shipping. The U.S. Government acknowledges this importance to seafarers, "GPS provides the fastest and most accurate method for mariners to navigate, measure speed, and determine location. This enables increased levels of safety and efficiency for mariners worldwide" (United States Government 2006). GPS also plays an important role in port facilities in the tracking and moving of cargo containers. With the increased reliance on GPS there has been a rise in the number of people who want to interfere with GPS signals, because they are engaged in smuggling, theft, or simply trying to escape tracking of their daily movements. The availability and usage of low-cost GPS jamming devices has resulted in the increased threat of intentional and unintentional disruption to commercial and industrial systems that rely on precise GPS data. Recognizing this threat, technologies to strengthen, augment and protect GPS signals must be identified. This chapter will highlight the benefits of GPS in maritime applications, discuss the interference risk and touch on what is being done to protect, toughen, and augment GPS.

---

[1] This document is not subject to the controls of the International Traffic in Arms Regulations (ITAR) or the Export Administration Regulations (EAR).

ISSUES IN MARITIME CYBER SECURITY

INTRODUCTION

The global positioning system (GPS) has become a ubiquitous "utility" supporting critical infrastructure depended on by many across a multitude of industries—including maritime. Originally developed for providing precision navigation and timing to the United States military, society has rapidly developed a strong reliance on GPS information in many facets of life including transportation, communications, and financial transactions. GPS has become an indispensable source of information with significant economic benefits, making it increasingly important that GPS data be available and reliable. In fact, GPS boosts productivity in many sectors including law enforcement, civil safety response, farming, mining, surveying, and package delivery. Conservatively, it is estimated more than 3.3 million jobs rely heavily on GPS in the United States, generating approximately $122.4 billion in annual economic benefits (PR Newswire 2011).

This chapter focuses on GPS, its use in maritime applications, and its susceptibility to service outages due to both intentional and unintentional jamming. Jamming is the act of interfering with a systems capacity to transmit a stronger signal on the same frequency, blocking the GPS signal, eliminating a user's ability to receive the necessary information. The first widely reported incident happened in January 2007, where GPS services were significantly disrupted throughout San Diego, California. Air-traffic controllers, Naval Medical Center emergency pagers, and the harbor traffic-management system used were all impacted. It took 3 days to find an explanation for this mysterious event: two Navy ships in the San Diego harbor had been conducting a training exercise when technicians jammed radio signals (Hambling 2011). Unwittingly, they also blocked GPS signals across a broad swath of the city.

Future satellite designs are being prepared to improve the transmission capability and security of GPS data. However, addressing GPS outages needs to occur now as the availability and usage of low-cost GPS jamming devices has resulted in a growing threat of GPS signal disruption and increasing the likelihood of future outages to systems that rely on GPS data. Addressing the security of GPS signals and preventing denial of services should be a priority for port facilities and maritime applications.

BACKGROUND

GPS is a satellite based navigation system comprised of a network of orbiting satellites that provide location and time information, anywhere on or near the Earth. It was originally intended to be used for military applications, but was made available for civilian use in the 1980s. GPS is maintained by the U.S. government and is freely accessible to anyone with a GPS receiver. Examples of receivers include car and boat navigation systems, hand-held units for hiking and embedded receiver chip sets in smart-phones.

There are a myriad of GPS maritime applications including ocean and inshore navigation, dredging, harbor entrance, docking, automatic identification system (AIS) and cargo handling. GPS has made life easier and safer for mariners worldwide by allowing fast access to accurate position, course and speed information. Additionally, GPS has improved container management in ports around the globe.

With numerous applications growing in their dependence on GPS, the risk to its interference has risen as there are people who want to interfere with GPS signals because they are engaged in smuggling, theft, or simply trying to escape tracking of their daily movements. These individuals can employ a GPS jammer to meet their nefarious goals. For example, thieves have started to hijack trucks and using jammers so the cargo cannot be tracked. Jammers are devices that use radio frequency transmitters to intentionally block, jam or interfere with lawful communications.

GPS—Growing Use and Growing Threat

How GPS Works

GPS satellites circle the earth twice a day at an altitude of approximately 13,000 miles in a precise orbit transmitting signal information to earth. These continually broadcasted signals provide precise timing and satellite orbital positioning and trajectory information to users. A GPS receiver compares the timing of each received signal with its own internal clock. These time differences correspond to the physical distances between the receiver and each of the satellites. The receiver uses the time difference measurements and the other data received from the satellites to calculate its own precise position. A GPS receiver needs signals from a minimum of four satellites in order to calculate its own three-dimentional position.

GPS satellites contain multiple atomic clocks that contribute to the very precise timing of GPS signals. Receivers decode these signals, effectively synchronizing their time to the atomic clocks. This enables users to determine time to within 100 billionths of a second, without the cost of owning and operating an atomic clock. Precise time is crucial to a variety of economic activities around the world. Financial networks, communication systems, and electrical power grids all rely on precision timing for synchronization and operational efficiency. The free availability of GPS time has enabled cost savings for companies that depend on precise time and has led to significant advances in capabilities.

Impact of GPS

Over the past 20 years, GPS technology has transformed lifestyles and businesses with many applications across all spheres of life. In addition to creating efficiencies and reducing operating costs, the adoption of GPS technology has improved safety, emergency response times, environmental quality, and has delivered many other benefits. The market for GPS has already grown into a multibillion-dollar industry with far-reaching potential. The direct economic benefits of GPS technology on commercial users are estimated to be over $67.6 billion per year in the United States (Pham 2011). Paul Verhoef, Program Manager for the European Union (EU) Satellite Navigation Program, stated that "6 to 7 percent of the European Union gross domestic product (GDP) is directly dependent on the availability of GPS" (Cameron 2010).

While there are varying estimates of the economic impact of GNSS and GPS there is a general agreement that it is large and growing. For example, the European GNSS Agency (GSA) released its third market report ("GNSS Market Report Issue 3") and predicted "for 'GNSS core' and 'GNSS-enabled revenues will increase by 9 percent through 2016 and 5 percent through 2020, reaching a total of about €350 billion (US$478 billion) per year" (Cameron 2010).

GPS technology creates jobs and other economic activities that spur economic growth. Studies have shown sustained productivity benefits and job creation in downstream industries and the U.S. economy as a whole (GPS World Staff 2011). Additionally, GPS technology has created both positive direct and indirect spillover:

- 3.3 million jobs that rely on GPS technology, including 130,000 jobs in GPS manufacturing industries and 3.2 million in the downstream commercial GPS-intensive industries (GPS World Staff 2011).

- Improved public safety and national defense.

- Time savings, workplace health and safety gains, job creation and reduction in emissions from fuel savings.

GPS and Maritime Use

GPS is an essential element of the global information infrastructure. In fact, GPS supports nearly every facet of modern life including travel, phones, banking, construction equipment, and shipping. The shipping and maritime industry is no exception as there are many GPS maritime applications including ocean and inshore navigation, dredging, harbor entrance, docking, automatic identification system (AIS) and cargo handling.

GPS has had a great impact on marine operations and has led to advances in safety and efficiency by providing accurate position, course, and speed information. Additionally, GPS has enhanced container management in port facilities (Wales 2011). Container placement can be tracked from port entry to exit and automated when GPS is used in conjunction with Geographic Information System (GIS) software. GPS is used for underwater surveying, mapping, buoy placement and fishermen use it to navigate to fishing spots, ensure regulation compliance as well as tracking fish migrations. There are many on-board systems that rely on GPS positioning information. A ship's dynamic positioning (DP) system utilizes GPS data and automatically maintains a vessel's position and heading. GPS data are used by the Electronic Chart Display and Information Systems (ECDIS). GPS data are also embedded in AIS, a system used for vessel traffic control, information and improved port security.

GPS at Risk

Brandon Wales, Director of the Homeland Infrastructure Threat and Risk Analysis Center for the Department of Homeland Security, warned in late 2011, "U.S. critical infrastructure sectors are increasingly at risk from a growing dependency on GPS for positioning, navigation and timing services." He stated that GPS reliance in many critical systems is either not fully understood and is often taken for granted and describes GPS as a "largely invisible utility" (Wales 2011). The at risk infrastructure was highlighted by Christine Evans-Pughe (2011) in Engineering and Technology Magazine, "Power distribution networks, banking and financial trading systems, broadcasting and industrial-control networks all use GPS timing, making them equally vulnerable to unintentional or deliberate interference."

There are people who want to interfere with GPS signals because they are engaged in smuggling, theft, or simply trying to escape tracking of their daily movements. GPS jammers are being used by criminals to hide the fact they are driving stolen cars or by commercial drivers to conceal the fact that they are driving for dangerously long hours. It has been reported that a 500 m shield is created around a vehicle that uses a simple jamming device plugged into a cigarette lighter. Within that shielded area, other GPS devices are left inoperable. For example, airplane safety and other technology is unintentionally disrupted when an exhausted taxi cab driver wants to take a break and plugs in a GPS jammer to avoid detection. Some notable examples are included below.

The Risk is Real

Both the motive and the means to jam or render GPS capabilities useless are clearly evident in our society. Whether intentional or inadvertent, jamming GPS signals poses a significant risk to our daily lives, with potentially larger, long standing impacts. Some examples that highlight the real risk of GPS jamming have been recently reported.

In January 2007, GPS services were significantly disrupted throughout San Diego, California (Hambling 2011). Air-traffic controllers at the San Diego Airport noticed that their system for tracking incoming planes was malfunctioning; Naval Medical Center emergency pagers stopped working; and the harbor traffic-management system used for guiding boats failed. Additionally, cell phones users found they had no signal and bank customers trying to withdraw cash from automated teller machines (ATMs) were refused. In addition, it took 3 days to find an explanation for this mysterious event: two Navy ships in the San Diego harbor had been conducting a training exercise when technicians jammed radio signals (Hambling 2011). Unwittingly, they also blocked GPS signals across a broad swath of the city.

In August 2013, the FCC fined a Readington, New Jersey, man nearly $32,000 after concluding he interfered with Newark Liberty International Airport's satellite-based tracking system by using an illegal GPS jamming device in his pickup truck to hide from his employer (Strunsky 2013). The signals emanating from the vehicle blocked the reception of GPS signals used by an aircraft landing system and potentially put travelers in danger.

In another and widely noted incident at the Newark Airport in 2009, the GPS signal was being lost at certain times of the day, effecting navigation (The Economist 2011). This and other risk were discussed in an article by Capt. Charles A. Barton (2012),

The Federal Aviation Administration (FAA) investigated the problem and found that a local truck driver had installed an inexpensive jammer in his vehicle. On his way to work every day, he passed the airport and caused its systems to fail. The driver was using the jammer to prevent his employer from tracking his movements. The scary fact is that anyone can purchase this equipment for as little as $30 on the Internet.

In June of 2013, in international waters off the coast of Italy, a luxury yacht went off its GPS-determined course (Cameron 2014). It was under the control of hijackers, but according to the ship's GPS equipment it was where it was supposed to be. Thankfully, this was just an experiment that the University of Texas at Austin conducted. The school of engineering team was able to successfully "spoof" the private yacht. Spoofing created false GPS signals to gain control of the vessel's GPS receivers (University of Texas at Austin 2013).

Solutions and Recommendations

Most people do not realize how much of our infrastructure is dependent upon GPS signals and information. Industries around the world utilize GPS for a variety of applications that effect everyday life. However, GPS interference is an increasing worldwide problem that could seriously impact the maritime industry. The GPS industry is currently diligently working to protect, toughen and augment GPS.

Augment

Enhanced Loran (eLoran) is a positioning, navigation, and timing (PNT) service that can be used as a back-up system for GPS. It is the latest in a series of low-frequency, LOng-RAnge Navigation (LORAN) systems. In fact, it has been deployed as a back-up ship navigation system in the English Channel to tackle the ever increasing risk of disruption to vessel GPS navigation devices. Without GPS, dense traffic patterns would require additional shipboard crew. The General Lighthouse Authorities (GLA) of the U.K. and Ireland use eLoran to counter threats of jamming and GPS signal loss.

Toughen

According to the U.S. Government (2006), "the GPS modernization program [is] a multibillion-dollar effort to upgrade the GPS space and control segments with new features to improve GPS" security and performance. The next generation of GPS satellites—GPS III—will provide stronger signals with improved resistance to interference.

The improvements in GPS will impact the maritime industry. "With the modernization of GPS, mariners can look forward to even better service. In addition to the current GPS civilian service, the United States is committed to implementing two additional civilian signals. Access to the new signals will mean increased accuracy, more availability, and better integrity for all users" (GPS Tracking Application for the Maritime Industry 2014).

Protect

To protect the GPS signal, the first and best step to take is to pursue technologies that can detect and locate GPS jammers. Although, GPS jamming is growing in frequency and gaining attention, there are relatively few commercially available technologies to help combat this problem. Chronos Technology and Harris are two companies that currently offer solutions. Chronos offers several handheld and vehicle mounted products and that can detect and log an interference event. Harris, Inc. offers a product, Signal Sentry 1000, that can installed at high-risk areas, such as marine ports, to instantaneously sense and determine the location of jamming sources.

CONCLUSION

GPS is a ubiquitous "utility" that supports critical infrastructure including: law enforcement, transportation, communications, financial transactions, and maritime applications. It has become a vital source of information with significant economic benefit—millions of U.S. jobs rely heavily on GPS. It is paramount that GPS is kept reliable, available, and secure.

GPS is susceptible to service outages due to intentional and unintentional interference. To maintain critical infrastructure, proactive measures must be taken to eliminate GPS outages—it must be protected, toughened and augmented. Detection and tracking is the first step to take as the availability and usage of low-cost GPS jamming devices has resulted in an increased threat of GPS signal disruption. Addressing the security of GPS signals and preventing denial of service should be a priority.

References

Barton, Charles A. 2012. "Global Positioning System is a Single Point of Failure." *Signal*, October 1. http://www.afcea.org/content/?q=global-positioning-system%E2%80%A8-single-point-failure.

Cameron, Alan. 2010. "Galileo from the Top: Interview with the EC's Paul Verhoef." *GPS World*, November 20. http://gpsworld.com/galileo-top-interview-with-ecs-paul-verhoef/.

Cameron, Alan. 2013. "A Glowing Report Doth Not a Golden Future Make." *GPS World*, November 27. http://gpsworld.com/a-glowing-report-doth-not-a-golden-future-make/.

Cameron, Alan. 2014. "Spoofer and Detector: Battle at the Titans at Sea." *GPS World*, August 5. http://gpsworld.com/spoofer-and-detector-battle-of-the-titans-at-sea/.

Evans-Pughe, Christine. 2011. "GOS Vulnerability to Hacking." *Engineering and Technology Magazine* 6 (4). (April 15). http://eandt.theiet.org/magazine/2011/04/gps-vulnerabilities.cfm?origin=EtOtherStories.

"GPS Tracking Application for the Marine Industry" 2014. Boomerang Tracker. September 14. http://www.boomerangtracker.com/latest-news/GPS-Tracking-Applications-Marine-Industry.

GPS World Staff. 2011. "The Economics of Disruption: $96 Billion Annually at Risk." *GPS World*, July 1. http://gpsworld.com/gnss-systemthe-

CHAPTER 22

economics-disruption-96-billion-annually-risk-11825/.

Hambling, David. 2011. "GPS Chaos: How a $30 Box Can Jam your Life." *New Scientist*, March 4. https://www.newscientist.com/article/dn20202-gps-chaos-how-a-30-box-can-jam-your-life/.

Pham, Nam D. 2011. "The Economic Benefits of Commercial GPS Use in the U.S. and the Costs of Potential Disruption." NDP Consulting. June. http://www.gpsalliance.org/docs/GPS_Report_June_21_2011.pdf.

PR Newswire. 2011. "Study Shows Interference with GPS Poses Major Threat to U.S. Economy." June 22. http://www.prnewswire.com/news-releases/study-shows-interference-with-gps-poses-major-threat-to-us-economy-124352063.html.

Strunsky, Steve. 2013. "N.J. Man Fined $32K for Illegal GPS Device that Disrupted Newark Airport System." *NJ.com*, August 8. http://www.nj.com/news/index.ssf/2013/08/man_fined_32000_for_blocking_newark_airport_tracking_system.html.

The Economist. 2011. "No Jam Tomorrow." March 10. http://www.economist.com/node/18304246.

United States Government. 2006. "Marine." *GPS.gov*. http://www.gps.gov/applications/marine/.

University of Texas at Austin. 2013. "UT Austin Researchers Spook Superyacht at Sea." July 29. http://www.engr.utexas.edu/features/superyacht-gps-spoofing.

Wales, Brandon. 2011. "The Increasing Risks of GPS Systems." *Homeland Security News Wire*, November 22. http://www.homelandsecuritynewswire.com/dr20111122-the-increasing-risks-of-gps-systems.

CHAPTER 23:
GPS JAMMING AND SPOOFING:
MARITIME'S BIGGEST CYBER THREAT

Dana Goward
Resilient Navigation and Timing Foundation

Using a $35 device easily purchased on the internet, mobsters jam GPS reception and temporarily disable cargo and fleet tracking devices in a major east coast port. To all appearances legitimate truckers, they depart with two high value containers and a vehicle. Five minutes after exiting the waterfront area they pull into a non-descript warehouse, the container and vehicle tracking devices are located and removed, and the illegal GPS jammer is turned off.

A relatively simple, unattended device has been placed aboard a tanker bound from Cape Town to the Malacca Straits. It spoofs the ship's GPS into sailing a route five degrees to the right of the captain's desired course. At the end of the 10,000 km transit, the ship is 875 km off course, but is exactly where the pirates what it to be. The pirates board out of sight of land, the crew is placed in a lifeboat, and the vessel's cargo is sold on the black market. After all the identifying marks are ground off, the ship itself is sold for scrap to a breaking operation in a third world country.

Near-universal dependence on weak, easy to disrupt satellite signals as the sole source of Position, Navigation and Timing (PNT) services is undoubtedly Maritime's biggest on-going cyber vulnerability. GPS disruption[1] is a cyber vulnerability because it:

- Interferes with networks by disrupting communications pathways (GPS timing is essential to many private networks),

---

[1] "GPS" is the most complete, capable and widely adopted of all global navigation satellite systems (GNSS) and is used in this paper as a proxy for all GNSS.

- Interferes with network end-use devices, and
- Can introduce faulty and/or intentionally misleading data into IT systems

Highly precise, accurate, and free GPS signals have been incorporated into virtually every civil electronic system including cargo tracking and distribution systems, shore-side maritime domain awareness systems and vessel traffic co-ordination (e.g., Vessel Traffic Services). The same is true for virtually every electronic system aboard ships. Unfortunately, these signals are exceptionally weak, can be easily disrupted intentionally, by accident, and by natural phenomena. For users with unsophisticated equipment, the signals can also be fairly easily spoofed, or replaced by a false signal that causes the receiver to believe it is in a completely different location.

Dr. Brad Parkinson of Stanford, widely known as "the father of GPS" has called America's reliance on GPS "… a single point of failure for much of America …" and "our largest, unaddressed infrastructure vulnerability" (Gain Perspective 2015).

Disruption Scenarios

Invalid or blocked GPS signals can result from a variety of sources and for a variety of reasons.

Natural Disruptions

- Solar flares can temporarily ionize the atmosphere preventing signals from penetrating to the earth and, if strong enough, permanently damage satellites.
- GPS signals can be blocked by heavy foliage and terrain. While not normally a problem for maritime users, this could manifest in riverine areas, and in high Polar Regions where satellites are low in the sky and the geometry is poor.
- GPS does not penetrate underwater and cannot be accessed by the growing number of sub-surface mariners and vessels.

Accidental Disruption

- Many transmitters, if used improperly, can jam GPS. Instances of US Navy personnel inadvertently jamming GPS across wide areas in both San Diego and Newport News have found their way into the public record.

CHAPTER 23

- Malfunctioning electronic and electrical equipment can often interfere with GPS signals. Improperly rigged TV antennas, poorly shielded telecommunications equipment, even an elevator that was sparking as it went up and down, have all been traced as sources of GPS interference.

Intentional Jamming

- The militaries of most nations have the capability to jam GPS for their own purposes, and many have. North Korea has jammed reception in South Korea for several years running and military and paramilitary forces are currently jamming GPS reception in parts of the Ukraine (Ars Technica; "OSCE Drone Jammed Over Eastern Ukraine" 2014).
- The FBI has issued at least one alert about organized crime use of GPS jammers in cargo theft (Public Intelligence 2014).
- "Personal Privacy Devices" are legal to own, but illegal to use. Easily and inexpensively purchased on the internet, they block GPS reception over varying distances. Landing systems at Newark International Airport have periodically been disrupted by persons with such devices driving past on I-95 (Strunsky 2013). In spite of moving the system away from the highway, the airport still detects five or more drivers with jammers each day.

All of the above means that jamming GPS is well within the reach of any terrorist or other bad actor.

Spoofing

- By sending a similar, slightly stronger signal, adversaries and academics have been able to "take over" GPS controlled vehicles and direct them where they will (University of Texas at Austin 2014). Whether diverting vessels, confusing those monitoring vessel positions from shore, or misdirecting those trying to track stolen cargo, spoofing has the potential to greatly degrade maritime safety and security.
- In 2015 spoofing became available at the consumer level when step by step instructions for building a spoofing device were published at a hacker's convention in Las Vegas and kits to assemble a device were sold for about $300 (Olson 2015).

Shore-side Systems

Almost every information technology system uses some aspect of the location and/or time signal from GPS for its proper functioning. Maritime-related systems ashore that are of specific concern include:

- Cargo Tracking—The website "Public Intelligence" recently published an FBI advisory that stated "various law enforcement and private sector partners have reported that GPS tracking devices have been jammed by criminals engaged in nefarious activity including cargo theft and illicit shipping of goods" (FBI 2014). Examples cited included:

    o "Auto thieves shipping vehicles to China used GPS jammers placed in shipping containers in an attempt to thwart tracking of the containers, according to July 2014 information from the National Insurance Crime Bureau. In 46 reported incidents, the thieves placed one or more GPS jammers in cargo containers with stolen automobiles" (FBI 2014).

    o Cargo thieves in North Florida used GPS jammers with a stolen refrigerated trailer containing a temperature controlled shipment, according to a July 2014 report from the Pharmaceutical Cargo Security Coalition.

In another effort, GPS consultant Logan Scott and the company Rohde and Schwarz took a short term sample in one port city. They determined that 25%–30% of all trucks were using GPS jamming devices (Scott 2014).

- Maritime Domain Awareness/VTS—While radar and other sources of information play important roles, the data rich, highly automated, high volume data stream from the automatic identification system (AIS) is the lynchpin for creating maritime situational awareness at sea and ashore. While the AIS system's design allows for integration of any location and timing source, GPS is the only system currently used (Grant et al. 2011).

- Aids to Navigation—Transitioning from industrial age buoys and lighthouses to information age electronic aids to navigation has the potential for huge economies and efficiencies. We are in the early stages of this transition with "virtual aids" growing in popularity. All of this depends upon AIS, which depends upon GPS, and makes maritime even more vulnerable to GPS disruption.

CHAPTER 23

VESSEL SYSTEMS

Experiments by the General Lighthouse Authorities (GLAs) of the UK and Ireland have shown that, when GPS signals are not available or jammed, these systems either fail completely, or, even worse, provide false and misleading information (Grant et al. 2011).

These systems include:

- Electronic Chart Displays
- Ship's autopilot
- Heli-deck stabilization system
- Satellite voice and data communications
- DSC-GMDSS: Maritime Distress Safety System
- Radar
- Gyro-compass
- Automatic Identification System

Reports from the GLAs' first trial, which was held with the full knowledge of the ship's bridge team showed that:

- At low levels of interference, all the ship's systems received and used the bad time and location information. This resulted in erroneous information, including improper positions, tracks and speeds.
- At higher levels of interference there were so many alarms on the bridge that the experienced Captain and crew had difficulty determining what was wrong.

Spoofing maritime receivers and "taking over" vessels' automatic course keeping systems has also been demonstrated (GPS Flaw Could Let Terrorists Hijack Ships, Planes 2013).

These jamming and spoofing demonstrations highlight a major at-sea/underway cyber- vulnerability that could lead or contribute to an Incident of National Significance. Possible scenarios include:

- Misdirection of a vessel during an open sea transit due to spoofing of GPS by a device concealed on-board.
- A mishap in pilotage waters due to GPS spoofing or jamming and

confusion among a vessel's bridge crew at a critical point in the transit. Other malicious activities could include spoofing or jamming GPS to:

- Misdirect or hinder first responder/law enforcement operations
- Hinder Vessel Traffic Services, and halt or slow commerce in a busy port area.
- Avoid vessel tracking. This would enable numerous nefarious activities including:
    o Lightering and import of petroleum and other products without paying taxes and duties
    o Commerce in violation of the Jones Act
    o Smuggling human, narcotic, and other illegal cargoes

MITIGATION

Satellite signals are, by their very nature, exceptionally weak and easy to disrupt. At the same time, modern transportation systems require wireless, precise, wide area, time, and location data to properly function. Maritime safety, security, and efficiency, especially as we seek to reap the benefits promised by eNavigation, require that we not depend upon satellite signals alone. Unfortunately, at present most of the mitigation measures available to maritime stakeholders address only portions of the risk and/or vulnerabilities.

Multiple GNSS—In addition to the United States' Global Positioning Systems, Russia, China, and the European Union have or are establishing independent navigation satellite systems of their own. These are all interoperable and accessing multiple systems will increase users' resilience over relying upon one system alone. Additionally, a satellite communications company has recently established a fee-based service that provides navigation information from its satellites with a higher power signal than is available from other GNSS. All of these signals and services operate in generally the same frequency band and are vulnerable to disruption by solar events, and terrestrial intentional or unintentional jamming.

Improved Receivers and Antenna—Generally, the less expensive the equipment, the more susceptible it is to jamming and spoofing. Antennae should be structured and placed so as to be unlikely to receive signals that are not clearly from space. Receivers should have integrity monitoring and be able to distinguish between genuine GNSS signals and unsophisticated attempts to mimic them. Professional users of GPS/GNSS should thoroughly understand available equipment and its limitations and make their purchasing decisions deliberately.

Other Sources of Information—The prudent mariner uses every means at their disposal to determine position. Visual aids such as buoys and lights, depth soundings, radar ranging, radio beacons, celestial sightings ... all must remain in the mariners' tool kit and be used. Automated inclusion of some or all of these sources of navigation information is possible and efforts are on-going. If and when they are economically available, they will undoubtedly contribute to safer electronic navigation. Ashore, independent local positioning systems can provide location redundancy for container terminals. None of these systems, though, will be able to provide the precise and synchronized timing required by telecommunications and information technology systems for proper operation.

eLoran—In 2004 the problem of broad dependence upon GPS and its vulnerability was identified in National Security Presidential Directive 39. It directed the Secretary of Transportation, working with the Secretary of Homeland Security, to procure a backup navigation and timing system for the nation to use when GPS was not available.

After 3 years of effort, and with the endorsement of the National Space-based Position, Navigation, and Timing Executive Committee (PNT ExComm), the departments concluded that converting the old terrestrial Loran-C system to a modern, more precise, and less expensive eLoran system would be far and away the most effective and least costly alternative. While eLoran would provide the same kind of time and location information as GPS, its vastly different signal (terrestrial, high power, low frequency signal, etc.) made it an ideal complement for GPS as a phenomena that might disrupt one of the systems would be very unlikely to disrupt the other.

Another year passed consumed with inter-departmental consultation about which government organization would own and operate the system. Finally, in 2008, DHS announced that it would transform the old Loran infrastructure into eLoran as a part of its infrastructure protection responsibilities. Despite these decisions and announcements, and numerous studies and recommendations by high-level panels and boards since then, over the last seven years, the eLoran project has not been executed.

However, December of 2014 saw evidence of renewed interest in establishing eLoran as a complement/ backup to GPS. The President signed into law legislation requiring preservation of the old Loran infrastructure and authorizing DHS to enter into a cooperative agreement to build an eLoran system. Additionally, the National PNT Executive Committee, led by the deputy secretaries of Defense and Transportation received a report from its "Tiger Team" which took a look at systems to complement GPS, including eLoran, and considered ways forward to reduce the risk to the nation. eLoran was again determined to be the only practical method of providing a navigation and timing complementary and backup capability for GPS.

A year later, in December of 2015, the deputy secretaries of Defense and Transportation informed concerned members of Congress that the administration would partner with industry to build an eLoran system that would protect the nation's critical infrastructure and services. As of this writing in early 2-17, U.S. government representatives publicaly state that the effort to establish a national complementary and backup system for GPS is underway.

References

Ars Technica. "Technology Lab / Information Technology." http://arstechnica.com/information-technology/2012/05/north-korea-pumps-up-the-gps-jamming-in- week-long-attack/.

Federal Bureau of Investigation. 2014. "FBI Cyber Division: Private Industry Notification." 141002-001. https://info.publicintelligence.net/FBI-CargoThievesGPS.pdf.

"Gain Perspective." 2015. Resilient Navigation and Timing Foundation. http://rntfnd.org/get-involved/perspective/.

"GPS Flaw Could Let Terrorists Hijack Ships, Planes." 2013. Resilient Navigation and Timing Foundation. http://rntfnd.org/2013/07/?cat=66.

Grant, Alan, Paul Williams, George Shaw, Michelle De Voy, and Nick Ward. 2011. "Understanding GNSS Availability and How it Impacts Maritime Safety." *Proceedings of the 2011 International Technical Meeting of the Institute of Navigation*. San Diego, CA. http://rntfnd.org/wp-content/uploads/GNSS-Maritime-GLA.pdf.

Olson, Parmy. 2015. "Hacking A Phone's GPS May Have Just Got Easier." *Forbes.com* http://www.forbes.com/sites/parmyolson/2015/08/07/

gps-spoofing-hackers-defcon/#617457eb7a48.

"OSCE Drone Jammed Over Eastern Ukraine." 2014. *UT Ukraine Today.* (November6).http://uatoday.tv/geopolitics/osce-drone-jammed-over-eastern-ukraine-390227.html.

Public Intelligence. 2014. "FBI Cyber Division Bulletin: Cargo Thieves Use GPS Jammers to Mask GPS Trackers." https://publicintelligence.net/fbi-cargo-thieves-gps-jammers/.

Scott, Logan. 2014. "Strategies for Limiting Civil Interference Effects Inspired by Field Observations: And Why Civil Receivers Need to Have Jamming Meters." http://www.gps.gov/governance/advisory/meetings/2014-06/scott.pdf.

Strunsky, Steve. 2013. "N.J. Man Fined $32K for Illegal GPS Device that Disrupted Newark Airport System." (August 8). http://www.nj.com/news/index.ssf/2013/08/man_fined_32000_for_blocking_newark_airport_tracking_system.html.

University of Texas at Austin. 2014. "Professor Todd Humphreys' Research Team Demonstrates First Successful GPS Spoofing of UAV." *Aerospace Engineering and Engineering Mechanics.* http://www.ae.utexas.edu/news/504-todd-humphreys-research-team-demonstrates-first-successful-uav-spoofing.

CHAPTER 24:
WINDOWS ON SUBMARINES: CYBER VULNERABILITIES AND OPPORTUNITIES IN THE MARITIME DOMAIN

Erik Gartzke
University of California, San Diego

Jon Lindsay
University of Toronto

Abstract

Cyber war appears on its face to be different from any form of conflict experienced previously, but in many ways it is similar to naval conflict today and in the past. Warships are endowed with great firepower and mobility, especially since the industrial age, but they are also quite vulnerable to increasingly lethal and long range munitions. Together these qualities put a premium on maneuver and stealth. Similarly, the cyber environment is highly dynamic, and critical infrastructure dependent on digital technology is vulnerable to remote attack. In cyber space, too, the key to success is deception, which can be used to enhance both offense and defense. Networking expands the scope and importance of naval power and compounds the importance of deception, long a key feature of both naval affairs and information operations. Navies still have much to learn about how to fight in the internet age, but perhaps less to learn than they did in converting from the age of sail.

The *Astute* Class, Britain's newest nuclear attack submarine, uses the Microsoft *Windows* operating system to control many onboard systems. Submariners have generally avoided physical windows on their boats for good reason. The vast ocean outside might inundate this small, mobile bubble of air and kill the crew. Now that Britain has introduced the *Windows* operating system onboard, many fear that this is no wiser than penetrating the hull of a submersible warship with physical portals.

Utilization of *Windows* on the *Astute* Class and other, less colorful acts of growing military reliance on cyber-based systems for command, control, communications, and intelligence has concentrated attention on the mischief that could conceivably result, should enemy operatives or errant hackers gain access to electronic systems (Majumdar 2014). Sophisticated weapons of war might well be rendered ineffective, or even turned against the home nation. Nuclear attack submarines, among the most sophisticated and lethal military platforms ever devised, are heavily dependent upon stealth to remain effective and to survive. An enemy that can locate, damage or destroy a sub through cyber space can alter the balance of power.

There are genuine reasons that the most advanced and thus digitally dependent navies of the world should be concerned that cyber war may change the game of global influence, but the same technological developments can have contradictory political implications. Two factors deserve consideration. First, the dangers posed by the internet are balanced both by its advantages for commerce and war and by the democratic reality that threats can work both ways. If strong nations must be concerned about cyber war at sea, they may also find that it suits their purposes in maintaining dominance against lesser adversaries. If their adversaries fear that this might be the case and are themselves dependent on the internet, then cyber space may reinforce the status quo (Lindsay 2014).

A second factor reinforces the first. While cyber space is widely believed to be heavily offense dominant, this perception is at least partially mistaken. The deceptive methods on which cyber attacks rely are as usable in the defense as in the offense (Gartzke and Lindsay 2015). If this is true in general, then we should expect it to be especially meaningful at sea, where there is an even greater premium on stealth and maneuver. Reliance on outmoded concepts of deterrence and defense may have led observers to conclude that cyber aggression in the maritime domain is more of a game changing threat than is in fact the case. The very factors that make it difficult to deter or defend against cyber attacks increase the ability of naval platforms to exploit vulnerabilities or pathologies present in enemy computer systems for defensive deception.

Deception at Sea

Deception has long been a pivotal determinant of success in naval operations (Winton 1981). Even more than in land engagements, where mass and terrain can blunt an attack, not being where an enemy expects is a huge advantage at sea. Conversely, being where an enemy expects can be lethal. Mobility over the oceans, the lack of cover and concealment, the fragility of a ship or submarine against modern weapons, and the ability to rapidly concentrate assets means that deception has long been a key component of the operational art in Naval warfare. At sea, deception wins battles and allows an inferior force to survive unscathed. Sir Francis Drake successfully raided up and down the Americas, absurdly outnumbered (Corbett 1890; Bawlf 2004). In World War I, the German light cruiser *Emden* managed to destroy two Entente warships and over 70,000 tons of allied merchant shipping, destabilizing Asian trade, while evading action with superior British forces in the Indian Ocean (Hoyt 1968; Lochner 1988). In the Second World War, Japanese commanders lured Admiral Halsey's battle group away from Leyte Gulf at a decisive moment using a force of their own carriers devoid of aircraft and crews (Woodward 1947; Morison 2004 [1956]). The Battle of the Atlantic was a contest of deception on an epic scale, the Germans hiding mortal threats to Allied shipping under the vast ocean, and the Allies hiding codebreaking successes from German code users. Allied "Beach Jumpers" in World War II successfully deceived the enemy about the location and timing of amphibious operations (Dryer 1992). It is of course possible to produce many similar examples.

Victory at sea (as in combined arms maneuver warfare) depends heavily on differential concentration, or getting to a location at a given moment with preponderant power, even if your side has an overall disadvantage in power (Brodie 1944; House 1984). This can only be achieved through tactical surprise. Without surprise, counter-concentration will nullify just about every attack. An example of this can be seen in Copeland's (2010) discussion of the role of Tunny decryption in forewarning the Soviet leadership of the German intended assault on Kursk in July 1943.

Deception creates surprise by showing things that are false and hiding things that are true, creating opportunities for differential concentration while persuading the adversary to remain dispersed. This is easiest with concealment, operating at night or in the terrain, weather, or radar or sonar shadows to screen movement. Active deception to create false or misleading signals is somewhat harder because extra context must be generated to make the simulation convincing and to hide disconfirming evidence. When deception is in play, whether as simulation or dissimulation, there is always more relevant information available in the environment than an opponent is able to utilize or exploit. The deceiver depends upon these extra degrees of freedom to bait and spring the trap.

Deception can be used to draw an opponent out, turning the apparent advantages of the first mover against the enemy and buying more time and safety for friendly forces by letting the other side make the first mistake. The most decisive naval engagements often involve striking second, not first. Midway was a huge victory for the United States because it allowed the Japanese Imperial Navy to execute a flawed plan of battle, compromised in advance by a creative American intelligence coup. Deception makes it possible to persuade an opponent to attempt differential concentration, creating a ready target for friendly forces. The enemy's maneuver is thus converted into vulnerability when it reveals its intentions and allows defensive counter-concentration. Much as the Greek leader Themistocles persuaded the Persians to deploy at Salamis by ensuring that informants told King Xerxes what he wanted to hear, so too cyber deception can work to ensure that an enemy commits in a time and place of one's own choosing.

The increasing reliance on computers and other electronic systems to manage complex human affairs means that there is always a potential spy in one's camp. Learning how to manage the ubiquity and interdependence of this new kind of vulnerability is critical to success in modern warfare. Fortunately, the manipulation of enemy intelligence collection is something that navies already have practiced to some degree. The Soviet "trawlers" that shadowed U.S. carrier battle groups and naval exercises during the Cold War were ubiquitous, but they were also obvious. Their operations and procedures were as much the subject of western attention as were Soviet attempts to eavesdrop on the West. Eventually, the United States Navy and other western naval powers developed procedures that minimized the effectiveness of these electronic spies and even on occasion turned them into conduits for disinformation. Today, there are trawlers in cyber space that are intent on compromising naval operations, but they too can be tamed and even turned.

Deception in Cyber Space

Parallels between cyber war and maritime conflict abound. Skillful naval commanders appear from over the horizon without warning to strike and then disappear quickly so as to avoid retaliatory fire. Conversely, commanders who telegraph their intentions too obviously will be met with a deadly reception, even from an inferior adversary. Analogous processes exist within computer networks. Remote operators can obfuscate their identity and prowl undetected, lying in wait to cause vital machines to malfunction at the most inopportune moment. Yet intruders can be tossed out of a network and can also potentially be punished if they are detected. Counterintelligence ploys can lure complacent cyber attackers into a trap that puts their broader attack infrastructure at risk.

Cyber space enhances the role of naval deception by opening up new dimensions of maneuver and vulnerability in a domain where warships are already highly maneuverable and vulnerable. Deceive or be deceived.

Offensive cyber operations are intrinsically deceptive because they either exploit false confidence in technical designs or user gullibility to get a machine to do the right thing at the wrong time (or vice versa). The attacker uses the actions of designers and users against their own interests, rather than actually forcing members of the targeted organization to do something they do not want to. Malware only works if a logical door is left open. Open doors abound in cyber space for two reasons. First, it takes nuance and skill to dupe a skeptical human being, but a deterministic machine lacks access to the rich context of face-to-face interaction. Second, programming abstractions are too complex to audit in real time. No one truly understands how all of their software tools work and must thus take it on faith, to some extent, than software components do what they are supposed to, and nothing else. The many layers and players in modern computing infrastructure expand the scope for deception, and the operational art of cyber operations relies on tailoring this deceptive potential to the target and objective.

Strangely, however, the temptation to date has been to conceptualize cyber conflict in land-based, and often archaic terms. Discussion of deterrence and defense in cyber space often seems intent on a positional concept of battle, designed to protect critical infrastructure as virtual "territory," even as modern military tactics have increasingly abandoned positional warfare. In a world of long-range precision strike, it is no longer safe to stay in one place, or worse to outline one's position from the air with linear trench systems. Targets under the gaze of the "persistent stare" of modern ISR will use decoys, camouflage, or simply melt away into the civilian population. Many of these changes have been folded into the concept of a revolution in military affairs (RMA) (Cohen 1996; Biddle 1998, 2006; Krepinevich 2002; Vickers and Martinage 2004). This is all the more true of the maritime domain, where warfare has long emphasized maneuver and deception to both deliver firepower and to avoid it. Rather than encircling key assets in cyber space with metaphorical moats and trenches, it may be wiser to accept that systems are inherently permeable and that communication will often be intercepted. There is thus the need to plan and conduct naval operations accordingly. It is in this dynamic environment that Western navies may find that their mode of war is an advantage rather than a deficiency.

Deception works by taking the enemy's strength and turning it into a liability. Navies have long played complex games of hide and seek. With the advent of electronic communications, navies gained an ability to fool one another even more thoroughly. A component of the Japanese raid on Pearl Harbor was the disinformation campaign, which relied heavily on radio telegraphy signals, fooling American officials into believing that the Imperial Fleet was still anchored in Japanese home waters. U.S. naval intelligence repaid the stratagem by feeding disinformation to Japanese intelligence and monitoring the response, fooling Japan into tipping its hand about Midway. In many ways, cyber deception operations are simply an extension of these practices. Yet, deception in cyber space can perhaps be more comprehensive and convincing.

Even more than an elaborate façade of radio traffic, cyber deception offers the promise of bringing an adversary into the fold. If enemy hackers believe they have "penetrated" a target network through their own skill, the eavesdroppers are more likely to trust the information they unearth. In this case, however, the targets of a cyber attack become the counterintelligence attackers, baiting the adversary to accept information that is false but believable. Defenders might even sow enemy systems with disruptive malware, primed to disable critical enemy communication, command and intelligence networks at the decisive moment in a multidomain engagement. An adversary with asymmetric advantages in a cyber conflict would suddenly find that their comparative advantage in cyber space had betrayed them and that the decisive domain had suddenly shifted to naval operations, or combined operations, where the enemy was less prepared to prevail, or at least persist.

Information provided to an enemy can even be factually correct. Sometimes it is critical that an adversary be informed in order to be more pliant. Deterrence and compellence require that an enemy understand that they are being deterred or compelled. When the U.S. carrier *Nimitz* and her battle group sailed to the Taiwan Strait in December 1995, followed in March 1996 by the deployment of *Independence* near Taiwan, efforts were apparently made to apprise Chinese officials of the proximity of U.S. Navy warships (*CNN*, March 10, 1996). Navies have sought to broadcast their intentions in order to maximize influence, often with little or no actual kinetic action. Today this type of communication is more feasible than ever, even as careful management of information has become more complex and difficult. While a natural reaction to vulnerability is to wish it were not so, the strategic response is to figure out how to turn vulnerability into advantage. Being exposed to enemy monitoring and cyber espionage is also a conduit to provide the information we want them to receive.

DECEPTION ACROSS DOMAINS

Winning at cyber war is still about winning more kinetic contests. Unless one's enemy agrees to lose in cyber space, one must expect that cyber war will envelop other domains. If various technical advantages are reduced by cyber attacks, it still remains the case that a technologically advanced power can use its advantages, albeit less efficiently, to win. A key component of U.S. operational effectiveness early in the Pacific campaign in World War II was superior damage control. Despite being outnumbered and in some cases outclassed in naval platforms, U.S. forces were able to keep their ships at sea. Japanese commanders at Midway famously refused to believe sightings of the *Yorktown*, as the ship was believed to have been sunk at the Battle of the Coral Sea, yet American damage control efforts saved her. The impact of cyber operations on modern warfare will similarly depend heavily on the ability of adversaries to compensate for the effects of deception in their networks. Victory will not go to the side that prevails in cyber space, but to the side that combines cyber operations with more traditional operations at sea, or to the side that performs the best damage control (physical and virtual) as operations create damage on both sides.

Most cyber threat narratives focus on the use of computer network attack to catastrophically paralyze critical infrastructure of command and control. The threat of a "cyber Pearl Harbor" is typically illustrated with technical narratives about how remote disruption is possible with little attention to why it might be politically or militarily useful (Peterson 2013). Threats of surprise attack are self-defeating since the threat invites nullifying countermoves, and actual surprise attack is only decisive in a strategic sense if the temporary window thus opened can be exploited through additional, typically kinetic, actions. The issue is not whether vulnerabilities exist, but how they are used (Gartzke 2013). The Japanese attack in December 1941 was a costly gamble that failed in two important respects: Japan's carrier force was able to find and sink the U.S. carrier fleet, while the larger goal of intimidating the United States into a negotiated compromise in the Pacific was also not achieved. On the contrary, the attack stiffened American resolve to mobilize resources to defeat Japan. Japan took the gamble in hopes that the attack on Pearl Harbor would delay U.S. counteraction long enough for Japan to set up defense in depth in the South Pacific. Any military act must find its justification in its relation to a broader military, and ultimately political objective, not simply the effects of the attack itself.

Cyber war, by itself, is extremely unlikely to deal a knockout blow. To the contrary, to be effective, a cyber attack must be followed up with physical acts that exploit vulnerabilities created or increased in the virtual domain. Indeed, because of this interdependence between virtual and physical, the decisive use of cyber at sea may be in convincing an enemy to take actions that make physical intervention easier or more effective for the attacker. The admiral that is deceived into committing his forces at the wrong time or place is doing for the enemy what the enemy would otherwise have to do for itself. Conversely, fear of being duped in this manner might lead a commander to be cautious, even unwilling to commit his forces, as happened with the Japanese carrier force at Pearl Harbor. Taken to the strategic or grand strategic level, cyber war could lead countries into mutual risk aversion and peace.

Actors who expect to be decisive in cyber space will be surprised if they are not also ready to act in other domains. Just as threats from cyber space have become a source of concern to operators in other domains, so too victory over a cyber threat need not occur in cyber space. The trick for cross-domain operations is to ensure that decisive engagements occur when and where one's own capabilities are most robust, where one's costs are lowest or one's resolve highest. This can be accomplished in the cyber domain in particular by deploying deception to convert enemy advantages (penetration of friendly systems) into liabilities (vulnerability to deception campaigns). As greater access to computer networks increases enemy exposure to information, it increases exposure to disinformation.

Strategy in multiple domains is inherently more complex (Lindsay and Gartzke 2016). While it is easier to get things wrong, this is true for both sides in a crisis, contest, or in general deterrence in peacetime. Indeed, success in conflict requires only that one be better than the adversary, not perfect. The inherent uncertainty of maritime cyber operations also magnifies the effects of maneuver and vulnerability at sea. Some see great potential for crisis instability as both sides have strong incentives to move first to achieve a cyber coup before the other (Gompert and Libicki 2014). However, the inherent reliance of cyber operations on deception should lead both sides to become more risk averse, fearing to risk their fleets on a possible trap in anything short of total war (and maybe not even then).

While cyber is typically seen as highly offense dominant, when combined with the naval domain, cyber space may actually become a much more conservative environment. Naval commanders on both sides of the North Sea in the First World War sought a decisive engagement where they could concentrate dreadnaughts in superior numbers against an inferior concentration of the enemy fleet. Beatty at first appeared to fall into Scheer's trap at Jutland, only for Jellicoe to turn the tables on the German admiral. Fear of a reprise of this harrowing experience kept the respective fleets from putting to sea in force for the rest of the war. The prospect of disastrous consequences brought about by deception in the cyber age could cause opposing navies to show considerable hesitancy to commit.

Paranoia about deception might even lead enemy commanders to forego opportunities for decisive attack where one is vulnerable. There is no doubt that elaborate deception is difficult. An array of falsehoods and partial truths must be presented consistently and backstopped with persuasive context. The prominence of deception online may simply lead attackers to be even more suspicious about traps. Yet this is also valuable (for the status quo defender) in that attackers are forced to expend more resources on operational security and move more deliberately. Combined arms offensives, especially those up against a maneuvering defense in depth, must similarly slow their momentum in order to coordinate covering and supporting movements.

The point is not that deception is a silver bullet for the offense or the defense, but simply that it is an increasingly essential facet of warfare in any domain. It is possible that with the rise of cyber space and cyber-enabled naval platforms, especially between peer or near peer competitors, that deception will be more often attempted but less often believed, resulting in hesitancy, caution, and a bias for inaction. Ironically, increased potential for deception may reinforce the status quo. Rather than a destructive and destabilizing force, cyber war may discourage challengers and prop up existing hierarchies of naval power. Strong navies are vulnerable but also more capable.

The vulnerability of warships makes deception necessary. Maritime mobility makes deception possible. Ships move. Naval vessels are as notable for their ability to disperse and to maneuver as for their ability to concentrate. It is not knowing where the enemy is, not knowing whether he is running away or setting a trap, that makes deception so potent. The crew of the deceptive vessel, a tight-knit human network, can all be in on the secret. Cyber space, by contrast, seems to be a more connected domain, where it might be harder to keep secrets. Interestingly, in combination the two domains might offer new potential for deception not available in either one. The comparative advantages of different domains and the results of their combination for warfighting or deterrence is still poorly understood in strategic theory, though commanders perhaps have better intuitions in practice.

Capitalizing on Cyber Deception in the Naval Domain

Cyber threats are best assessed in terms of their functional role. Some potential attacks are meant to obstruct communications or interfere with intelligence gathering. Other attacks may be designed to disable a ship's onboard systems (power plant, weapons, etc.) or to reveal the location of platforms in order to accurately target a vessel with kinetic attacks. In each case, electronic systems are used as a means and target of operations. Commands or queries are given over the internet or airwaves that interact with either shipboard systems or with shore-based systems that in turn have knowledge of, or a link to, electronic systems at sea.

Simple things can prove important in parameterizing the risk of penetration and in limiting negative effects. For example, there is no requirement that cyber systems be truthful. A ship's navigational software does not have to accurately report latitude and longitude in mid-ocean. A ship's crew can create their own daily corrections to azimuth and distance, something that does not need to be shared through any electronic means. Naval tradition has encouraged maintenance of paper charts long after they were no longer necessary for day-to-day navigation. Paper cannot be hacked. Similarly, weapons systems can be protected, not just with passwords and keys, but with multiple duplicate false command and control sequences that the crew on board recognizes as bogus, but which will appear genuine to those lacking basic situational awareness of the material, non-cyber environment on the ship. The tight-knit nature and routines of shipboard life ensure that information can be disseminated through non-cyber means. The interface between cyber and human reality must be carefully managed and kept secret from others outside this environment.

Pundits have reasonably, but incorrectly, inferred that societies and organizations that are the most dependent on cyber systems are in turn the most vulnerable to attack. Western navies clearly fit this bill. However, whether reliance translates to weakness depends on the ease with which an adversary can make use of vulnerabilities to coerce or otherwise undermine the capabilities of its target.

Western powers have for some time been aware that adversaries conduct espionage using cyber systems. At the same time, open societies have proven so economically productive as a result of global interconnectivity and so organizationally flexible in adapting to these new challenges—e.g., by leveraging a burgeoning cyber security industry—that in the end they have generally prevailed in maintaining commercial and military advantages, even against opponents that have achieved important tactical victories. While it is difficult to obstruct espionage completely, the suspicion with which adversaries treat the spoils of their spying, or the difficulties they encounter in attempting to absorb illicitly gained data into their internal organization, undermine the opportunities for advantage via cyber espionage.

Similarly, the side that is already good at complex distributed deception possesses important organizational advantages, which may become more prominent and consequential with greater dependence on information technology. Cyber operations in the naval domain may not be so much about adjusting naval operations to cyber space as adjusting the use of the cyber domain to insights from naval tactics and strategy. Western navies, with a wealth of experience in integrating deception into complex naval operations, may find that war in the context of cyber space is a fairly natural extension of their existing ways of warfare. Even if a weaker adversary has better hackers, their cyber operations might not be sufficiently well integrated with the broader campaign so that gaining surprise becomes a pivotal advantage. Cyber attacks could even become counterproductive if they alert an enemy of an attack or betray aspects of strategy, for instance if an overly ambitious stratagem lacking in careful tradecraft leaves too many digital clues (Lindsay 2015).

More problematic, potentially, is the ability to use cyber to disable or degrade the kinetic capacity of friendly forces. Cyber attacks like the Stuxnet worm, used to degrade Iranian enrichment, suggest that mere code might undermine the effectiveness of naval platforms in numerous ways. Ships could be idled by disabling or damaging electronic control systems, or by instructing those systems to damage or destroy the ship's power plant, electrical supply, environmental systems, water and sewage, etc. Platforms may be rendered mission ineffective if they cannot fire weapons or lack fire control information, or if their navigational systems are subverted to prevent maneuver or cause dangerous maneuvers when steaming in formation, such as during underway replenishment.

Disruptive cyber attacks will most likely involve "soft kills" on naval platforms where the adversary intends to exploit the temporary vulnerability through other means, typically kinetic. Timing of a soft kill is critical because the window for exploiting a vulnerability can fail to line up with other operational, strategic or grand strategic concerns. The Stuxnet attack, a covert operation to reduce enrichment efficiency over the long run, was not constrained by these tactical temporal considerations (Lindsay 2013). Suppose, for example, that Admiral Yamamoto and the Japanese fleet had executed the Pearl Harbor attack earlier than December 1941. The lives lost and equipment destroyed at Pearl Harbor would still have weakened U.S. capabilities in the Pacific, but other Japanese forces were not yet ready to project power southward. The delay in execution would have allowed the United States to move from defense to offense much more quickly, further weakening the Japanese strategy of defense in depth in the Western Pacific. Moving the Pearl Harbor raid back even earlier would have increased its ineffectiveness, as the United States had not yet moved the fleet forward and the Dutch and British still retained strong contingents in their possessions in South and Southeast Asia.

Any enemy that moves too soon to disable friendly capabilities is either wasting an advantage or warning an adversary of impending attack. If instead enemy forces hold damaging cyber attacks in reserve, then friendly forces may be able to strike quickly disabling an opponent's ability to exploit temporary weakness, before the enemy can mobilize. Timing is everything in military operations, not least in cyber war, especially given the complementarity/interdependence of cyber operations with other domains.

The decisive nature of naval engagements means that surprise and concentration are only needed temporarily. Having an enemy that *thinks* they know the location of friendly forces is even better than having an adversary who knows that they do not know this information. Conversely, operational security involves knowing how to prioritize one's secrets, and how to parse the desirable from the impractical. It may be more value to a commander to know for how long communications or command and control are likely to be in his grasp than to believe, falsely, that they can be maintained this way indefinitely.

Conclusion

Like any brave new world, the cyber era appears daunting, not least because it is new and unfamiliar. Yet as Bernard Brodie pointed out in 1944, when many observers believed that airpower had rendered sea power obsolete, "the war of today is being fought with new weapons, but so was the war of yesterday and the day before. Drastic change in weapons has been so persistent in the last hundred years that the presence of that factor might be considered one of the constants of strategy" (Brodie 1944, ix).

While it is tempting to give way to fear, the novelty of the cyber environment is actually an ally to those who are concerned about security, even or especially for organizations that are increasingly dependent on automation to function in the air, on land, in space or at sea.

Just as a century ago, radio telegraphy exposed modern navies to the risk of discovery in the largest oceans, so too today the fear is that the ubiquity of cyber will deny to modern navies the historic advantage of stealth. Fears that telecommunications would undermine naval power were discovered to be overstated. More to the point, the security implications of telecommunications were found to be a two-way street. Electronic deception—for gaining intelligence and sowing disinformation—could as often produce advantages from eavesdropping as disaster. An apparent vulnerability was turned on its head, making the capable more capable still.

Today we face new challenges as electronics find their way into every aspect of shipboard operations. While the desire to protect these systems from attack is natural, it must be recognized that protection may neither be practical nor catastrophic if it fails. Navies can often hide in the vastness of the ocean. Cyber space is a still vaster territory where again it is possible to prove inconspicuous. Navies proved a critical asset to evolving nation states as mobile platforms capable of concentration and with considerable firepower. Modernity is mobile and dynamic; cyber will be no different. Given familiar precepts of naval strategy, protection in the cyber era will have more to do with stealth and deception than with moats and palisades, and with preventing an adversary from discovering vulnerabilities as with preventing vulnerabilities from occurring or being exploited at all.

While concerns about cyber attacks on maritime operations are real, and in many respects may be more important at the strategic level than equivalent efforts against shore based assets, we wish to emphasize here the considerable opportunities that deception provides to defenders in the cyber domain. The very intelligence advantages that cyber penetration seems at first to offer for offense become the Achilles heel of the adversary's strategic and tactical awareness.

References

Bawlf, Samuel R. 2004. *The Secret Voyage of Sir Francis Drake: 1577–1580*. New York: Penguin.

Biddle, Stephen. 1998. "Assessing Theories of Future Warfare." *Security Studies* 88 (1): 1–74.

Biddle, Stephen. 2006. *Military Power: Explaining Victory and Defeat in Modern Battle*. Princeton, NJ: Princeton University Press.

Brodie, Bernard. 1944. *A Guide to Naval Strategy*. Princeton, NJ: Princeton University Press.

Cohen, Eliot A. 1996. "A Revolution in Warfare." *Foreign Affairs* 75 (2): 37–54.

Copeland, B. Jack. 2010. "Colossus: Breaking the German 'Tunny' Code at Bletchley Park." *The Rutherford Journal*. Volume 3.

Corbett, Julian. 1890. *Sir Francis Drake*. London and New York: Macmillan.

Dryer, John B. 1992. *Seaborne Deception: The History of U.S. Navy Beach Jumpers*. New York: Praeger.

Gartzke, Erik. 2013. "The Myth of Cyberwar: Bringing War in Cyberspace Back Down to Earth." *International Security* 38 (2): 41–73.

Gartzke, Erik, and Jon Lindsay. 2015. "Weaving Tangled Webs: Offense, Defense, and Deception in Cyberspace." *Security Studies* 24 (2): 316–348.

Gompert, David C., and Martin Libicki. 2014. "Cyber Warfare and Sino-American Crisis Instability." *Survival* 56 (4): 7–22.

House, Jonathan M. 1984. *Toward Combined Arms Warfare: A Survey of 20th-Century Tactics, Doctrine, and Organization*. Ft. Leavenworth, KS: US Army Command and General Staff College. Combat Studies Institute, Research Survey No. 2.

Hoyt, Edwin Palmer. 1968. *Swan of the East: The Life and Death of the German Cruiser Emden in World War I*. New York: Macmillan.

Krepinevich, Andrew F. 2002. *The Military-Technical Revolution: A Preliminary Assessment*. Washington, DC: Center for Strategic and Budgetary Assessments.

Lindsay, Jon R. 2013. "Stuxnet and the Limits of Cyber Warfare." *Security Studies* 22 (3): 365–404.

Lindsay, Jon R. 2014. "The Impact of China on Cybersecurity: Fiction and Friction." *International Security* 39 (3): 7–47.

Lindsay, Jon R. 2015. "Tipping the Scales: The Attribution Problem and the Feasibility of Deterrence Against Cyber Attack." *Journal of Cybersecurity* 1 (1): 53–67.

Lindsay, Jon R., and Erik Gartzke. 2016. "Cross-Domain Deterrence as a Practical Problem and a Theoretical Concept." In *Cross-Domain Deterrence: Strategy in an Era of Complexity*, eds. Erik Gartzke and Jon R. Lindsay. La Jolla, CA: Book Manuscript.

Lochner, R.K. 1988. *The Last Gentleman of War: The Raider Exploits of the Cruiser Emden*. Annapolis, MD: Naval Institute Press.

Majumdar, Dave. 2014. "NAVSEA: Submarines Control Systems Are at Risk for Cyber Attack." *USNI News*, December 8. http://news.usni.org/2014/10/22/navsea-submarines-control-systems-risk-cyber-attack.

Morison, Samuel Eliot. 2004[1956]. *Leyte, June 1944 – January 1945*, vol. 12 of *History of United States Naval Operations in World War II*. Champaign, IL: University of Illinois Press.

Peterson, Dale. 2013. "Offensive Cyber Weapons: Construction, Development, and Employment." *Journal of Strategic Studies* 36 (1): 120–124.

"U.S. Navy Ships to Sail Near Taiwan: Officials call Chinese Moves 'Reckless.'" *CNN*. March 10, 1996. http://edition.cnn.com/US/9603/us_china/.

Vickers, Michael G., and Robert C. Martinage. 2004. *The Revolution in War*. Washington, DC: Center for Strategic and Budgetary Assessments.

Winton, John. 1981. *Below the Belt: Novelty, Subterfuge and Surprise in Naval Warfare*. New York: HarperCollins.

Woodward, C. Vann. 1947. *The Battle for Leyte Gulf*. New York: Macmillan.

CHAPTER 25:
CYBER SECURITY RESILIENCY IN THE MARITIME SECTOR: A SYSTEMS APPROACH TO ANALYZING GAPS IN THE NIST FRAMEWORK

Kimberly Young-McLear
U.S. Coast Guard Academy

John Fossaceca
Ultra-Electronics 3eTI

Abstract

The world is becoming increasingly complex and dynamic, yet more interconnected than ever before through the advances of technology. Technology has played an instrumental role in facilitating commerce in the maritime transportation system (MTS). Systems engineering principles and modeling approaches offer new insights into the examination of the MTS and the National Institute of Standards and Technology (NIST) Cybersecurity Framework. This paper presents findings of actual mini-cases of cyber related vulnerabilities in the MTS. This chapter is targeted for students and professionals interested in systems thinking and risk management modeling. Mini-cases can be used for lecture-based discussions and modeling assignments.

Keywords: Systems Engineering, Maritime Transportation System, NIST Framework

A Systems View of the MTS

The Maritime Transportation System (MTS) is a complex sociotechnical system (and can also be viewed as a system of systems) which is vast, diverse, and geographically distributed around the world. From a systems theory perspective, systems are a "phenomenon seen as a whole and not simply the sum of elementary parts ... the focus is on the interactions and relationships between parts in order to understand" the entire system (U.S. Department of Homeland Security 2004). The "MTS is a network of maritime operations that interface with shore-side operations as part of the overall global supply chains ..." (U.S. Department of Homeland Security 2004). The MTS consists of four sub-systems: components, interfaces, information systems, and overarching networks. Components are the physical entities including facilities, vessels, and navigational aids. When the components interact with other critical elements in the MTS, it creates critical interfaces. These critical interfaces include port operations at intermodal connections. Key information systems are used by shipping lines and port operations, for example, to manage the flow of goods and commerce. Networks include the overarching systems that facilitate the sustainability of the MTS as a whole. Using a systems perspective, vulnerabilities can exist within the MTS through any of these four interrelated sub-systems. From Figure 1, these components, interfaces, information systems, and networks must be secure in order to prevent terrorists, hackers, or other actors from causing major disruptions in the MTS.

There are many disciplines that have contributed to research advancing operations in the Maritime Transportation System, however, since the MTS is a complex system and critical infrastructure, it is valuable to analyze the vulnerabilities, resiliency and existing standards through a systems perspective. Systems engineering is multidisciplinary and "integrates all the disciplines and specialty groups into a team effort forming a structured development process that proceeds from concept to production to operation" (What is Systems Engineering). Systems engineering also considers the business and technical factors with the goal of developing a viable system. Systems engineers have also contributed to supply chain, information systems, disaster response, and cyber security, and resiliency body of knowledge. Systems theories, modeling and simulation, and systems engineering principles can help analyze and evaluate MTS resiliency (Mele, Pels, and Polese 2010). PPD-21 "defines resilience as the ability to prepare for and adapt to changing conditions and withstand and recover rapidly from disruptions" (U.S. Department of Homeland Security 2015). INCOSE, the International Council on Systems Engineering, which is a professional society for the systems engineering discipline, chartered a working group to examine resilience of critical infrastructures (MITRE). From this perspective, MTS resilience is desirable, especially with respect to both man-made disasters as well as cyber attacks. These disruptions can affect the system's ability to recovery which may have devastating impacts to the economy.

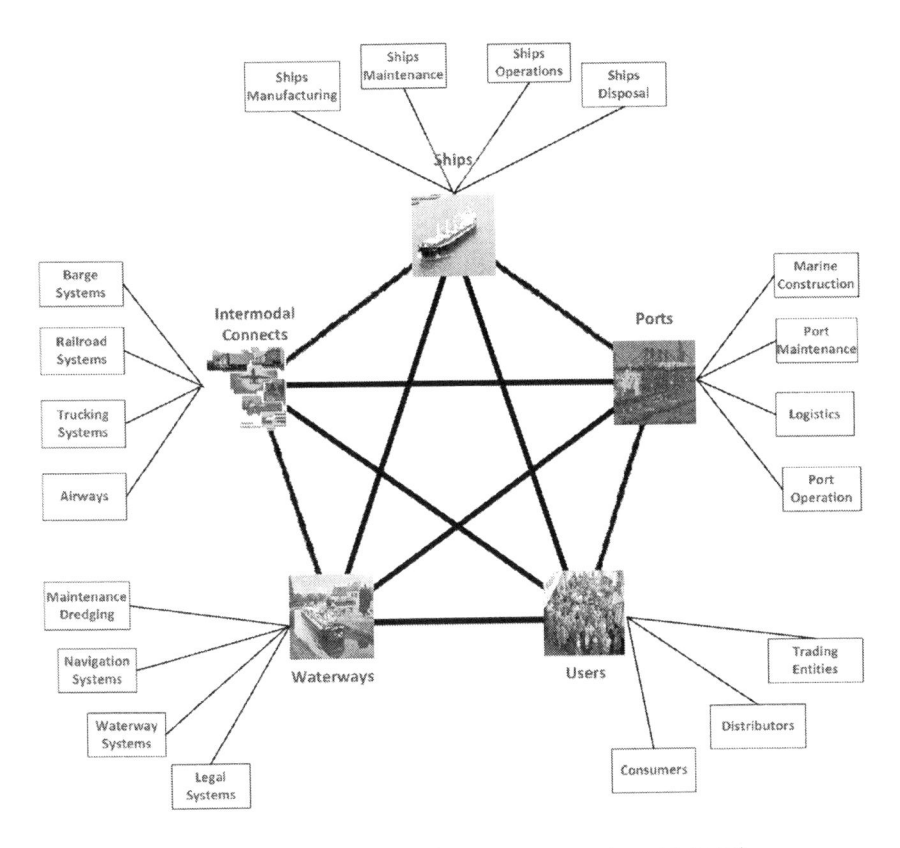

Figure 1: MTS as a System (Mansouri, Gorod, et. Al. 2009)

RESILIENCE CHALLENGES IN MTS

The landscape of the maritime industry has evolved drastically over the past decades with the complexity of computers, increase in cyber-physical systems in vessels and ports, reliance on information systems, rise in terrorism and smuggling operations, lack of emphasis on cyber security, and an economic necessity for faster delivery of products to market. Maritime businesses are interconnected, complex, but tend to have a low profit margins. Businesses must continually evaluate the cost–benefits of implementing new requirements for their internal operations, protocols, and training through methodologies such as business continuity planning (BCP) to minimize disruptions and increase the likelihood of their businesses recovering from an attack or incident. In the United States alone, the nation's ports facilitate more than $1.3 trillion in cargo annually (U.S. Department of Homeland Security). Maritime ports are a major component to facilitating supply chain efficiencies and commerce. According to the Government Accountability Office (GAO), "by exploiting vulnerabilities in information and communications technologies supporting port operations, cyber attacks can potentially disrupt the flow of commerce, endanger public safety, and facilitate the theft of valuable cargo" (U.S. GAO 2015). Some of the major disruptions of information and communications technologies (ICT) can include the network becoming unavailable, corrupted, or sensitive information is compromised. Actors within the system that may threaten these maritime ICTs include hackers, company insiders, competitors, terrorists, or employees who caused unintended harm through human error. Some of these will be illustrated in the mini-cases later in this paper.

From a policy perspective some of the domestic and international maritime challenges to resilience stem from fragmented governance and lack of a systems or holistic approach to risk management. In the wake of the September 11, 2001 attacks, the maritime sector has seen new requirements to address physical resiliency through port security. However, the Maritime Transportation Security Act (MTSA) of 2002 did not include cyber security. The Federal Emergency Management Agency (FEMA) and the United States Coast Guard (USCG) are responsible for cyber related threats in critical infrastructure. The Coast Guard is the lead agency for enforcing maritime security in domestic ports, however, it had not addressed cyber security in its port and area facility security plans. The GAO in a 2015 study concluded that "until the Coast Guard improves these mechanisms, maritime stakeholders in different locations are at greater risk of not being aware of, and thus not mitigating, cyber-based threats" (U.S. GAO 2014). Admiral Thomas, the Coast Guard's Assistant Commandant for Prevention Policy stated in a 2015 testimony that "as port facilities and vessels continue to incorporate information technology systems into their operations, the Coast Guard must adapt its regulatory regime accordingly" (U.S. Coast Guard 2015). The European Network of Information Security Agency (ENISA) also concluded in a 2011 report that there was an absence of a holistic approach to addressing cyber risks in the maritime sector (Maritime Cyber Security).

An Evaluation of NIST Cyber Security Framework Using the SIMLAR Method

President Barrack Obama issued Executive Order 13636, "Improving Critical Infrastructure Cybersecurity," on February 12, 2013 which established that

CHAPTER 25

"[i]t is the Policy of the United States to enhance the security and resilience of the Nation's critical infrastructure and to maintain a cyber environment that encourages efficiency, innovation, and economic prosperity while promoting safety, security, business confidentiality, privacy, and civil liberties" (National Institute of Standards and Technology). To address vulnerabilities in the critical infrastructures, the National Institute for Standards and Technology (NIST) developed a Framework for Improving Critical Infrastructure Cybersecurity (Version 1.0) (National Institute of Standards and Technology). This is a comprehensive and voluntary framework for businesses and organizations to evaluate and manage cyber related risks. The MTS is a critical infrastructure and since 2014, businesses and organizations within maritime sector have been using NIST to evaluate risk management approaches. The core of the NIST Framework uses business drivers to inform cyber security activities. The core consists of five high-level functions, which are Identify, Protect, Detect, Respond, Recover, shown in Figure 2. This is a concurrent and continuous systems view of how an organization, over its lifecycle, can manage cyber-related risks.

Function Unique Identifier	Function	Category Unique Identifier	Category
ID	Identify	AM	Asset Management
		BE	Business Environment
		GV	Governance
		RA	Risk Assessment
		RM	Risk Management
PR	Protect	AC	Access Control
		AT	Awareness and Training
		DS	Data Security
		IP	Information Protection Processes and Procedures
		PT	Protective Technology
DE	Detect	AE	Anomalies and Events
		CM	Security Continuous Monitoring
		DP	Detection Processes
RS	Respond	CO	Communications
		AN	Analysis
		MI	Mitigation
		IM	Improvements
RC	Recover	RP	Recovery Planning
		IM	Improvements
		CO	Communications

Figure 2: Five Phases of NIST Cybersecurity Framework (National Institute of Standards and Technology)

The NIST Cybersecurity Framework is currently voluntary, but if the Federal government were to implement risk management standards in the MTS, the NIST Cybersecurity Framework would more than likely be used. The NIST Cybersecurity Framework presents a common taxonomy to compliment an organization's risk management policies and procedures to include:

- "Describe their current cyber security posture;
- Describe their target state for cyber security;
- Identify and prioritize opportunities for improvement within the context of a continuous and repeatable process;
- Assess progress toward the target state;
- Communicate among internal and external stakeholders about cybersecurity risk" (National Institute of Standards and Technology 2014).

The SIMLAR method is the systems engineering process for designing products or services to meet customer needs. In this iterative method, designers must $\underline{S}$tate the problem, $\underline{I}$nvestigate alternatives, $\underline{M}$odel the system, $\underline{I}$ntegrate, $\underline{L}$aunch the system, $\underline{A}$ssess performance, and continually $\underline{R}$e-evaluate as shown in Figure 3 below. In this section, it is used as a methodology for analyzing how effective the NIST Cybersecurity Framework provides a "service" of risk management to customers within the MTS. Although the NIST Cybersecurity Framework is not meant to replace a business or organization's risk management process, it does outline steps for examining how to reduce and manage risk.

The Systems Engineering Process

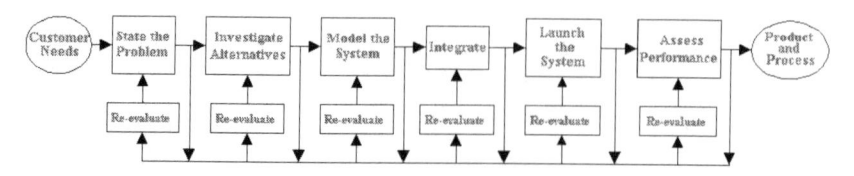

Figure 3: The Systems Engineering Process (NASA)

Although there are many similarities with between the taxonomy of the NIST Cybersecurity Framework and the SIMLAR Method as applied to risk management, there are some distinct differences. Current cybersecurity posture can be mapped to the problem statement. The target state for cyber security can be linked to the investigation of alternatives. The identification and prioritization of opportunities for improvement can be expressed as the re-evaluation of the entire repeatable process. The assessment of progress toward the target state can be mapped to the performance assessment. The communication among internal and external stakeholders can be linked to the customer needs. A distinction of the SIMLAR Method as compared to the NIST Cybersecurity taxonomy is the application of modeling and simulation. Another major distinction is that the NIST Cybersecurity Framework is for individual businesses or organizations. The NIST Cybersecurity Framework can be used to optimize components of the MTS, including vessels or facilities, however, it does not necessarily address the system-of-systems nature of the MTS. Optimizing the components or businesses within the MTS omits the risk management of the interfaces, information systems, and overarching networks. The SIMLAR Method can be used as a complimentary process to examining risks in the MTS by modeling the how components are interconnected with other sub-systems in the MTS.

Modeling Applications for the MTS

System dynamics modeling (SDM) and agent-based modeling (ABM) have been used to model complex systems, understand policy analysis, and design systems. System dynamics modeling "applies to dynamic problems arising in complex social, managerial, economic, or ecological systems ... characterized by interdependence, mutual interaction, information feedback, and circular causality" (What is System Dynamics) SDM is ideal for capturing the behavior, management, policy and technology of a system (Siegel and Houghton 2015). SDM is also closely related to the field of "Systems Thinking" which leverages "systemigram" diagrams whereas SDM uses stock flow diagrams and provides more quantitative data on the level of impact of certain scenario. Agent-based modeling "consists of a system of agents and the relationships between them" where emergent phenomena may develop from flows, markets, organizations, and diffusion (Bonabeau 2002). Although there are other types of modeling that can be applicable to the MTS, both SDM and ABM can offer a depth to analysis beyond other modeling techniques, such as game theory which may be limited to self-imposed constraints. ABM is also a helpful approach to studying operational risk in organizations which may arise "from the potential that inadequate information systems, operational problems, breaches in internal controls, fraud, or unforeseen catastrophes will result in unexpected losses" (Franzetti 2011). Popular SDM simulation packages include Vensim, STELLA, PowerSim, and AnyLogic.

In the literature, there have been examples of using case studies to model system interdependencies to uncover policy causal relationships and effects. An example of one such application is the hypothetical case study of a cyber attack in the oil and gas sector that created a shortage in the Gulf Coast crude oil supply" (Santos, Haimes, and Lian 2007). Using stocks and flows in Figure 4, MIT modeled how cyber security and threat processes are managed (Siegel and Houghton 2015). Another example makes use of agent-based modeling is the simulation of a new policy to mitigate the risks of pirates to vessel traffic management (Vaněk et al. 2013).

Mansouri, Sauser, and Boardman (2009) have already applied a "Systems Thinking approach" to the MTS as an "integration of interdependent constituent systems" where they apply a tool called a Systemigram to study critical properties of the MTS system including resilience and security. Leveraging the diagrams such as the systemigram in Figure 4, systemic interrelationships can be analyzed in a structured manner providing a clearer view of interdependencies and interconnections among the components of the MTS which is considered as a system of systems in the prescribed framework.

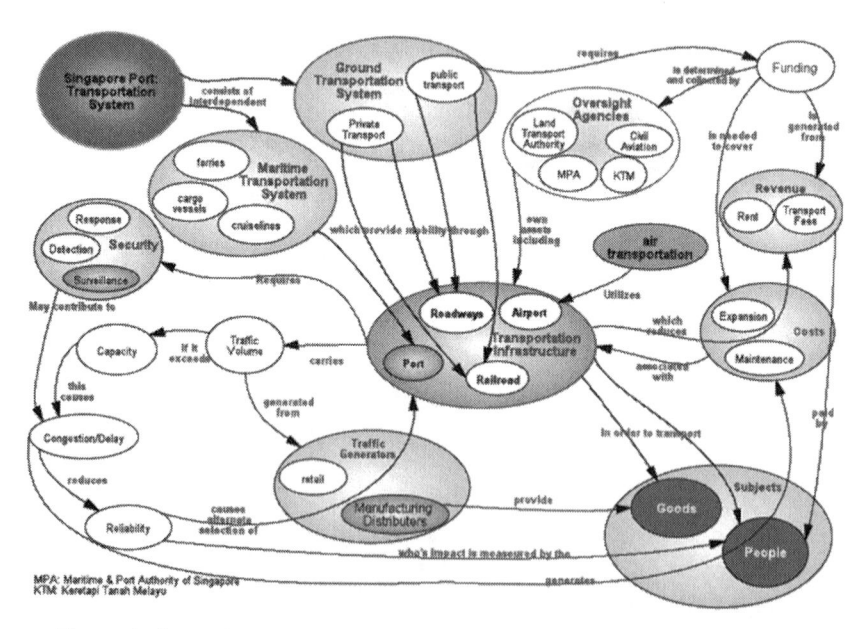

Figure 4: Systemigram representation of the transportation systems at the Port of Singapore (Mansouri, Sauser, and Boardman 2009)

In Behmer et al. (2016) ABM is leveraged to analyze and develop resilient combat architectures for littoral operations (i.e., naval engagements near the shoreline supporting amphibious operations). In this paper, the authors emphasize the importance of resilience in a system of systems and found that by distributing managerial control in the architecture improved overall system level performance as measure in terms of in terms of "time to threat elimination." The authors of this paper expressed that ABMs can be rapidly developed and use to assess the ability for prototype architectures to defend against a variety of threats and disruptions and continue to function to some minimal target level during network failures and attacks. A similar approach if applied to the MTS could prove beneficial.

MTS Cyber Risks Illustrated in Mini-Cases

By examining actual documented threats and vulnerabilities from the lenses of the hacker, the following mini-cases explore risks in the sub-systems of the Maritime Transportation System including the components, interfaces, and information systems. This section of the paper illustrates how each of these sub-systems can be penetrated by hackers or potential hackers. Specifically, the mini-cases show how vessels, ports, and supply chains are especially vulnerable.

Case 1: Disrupting Vessel Navigation Systems (Component Security)

Vessels are one of the most important components in the Maritime Transportation System. Vessels are needed to move cargo throughout the world's waterways and ports. To facilitate safer and more cost efficient transits, vessels have become more reliant on technology. Many domestic and international regulations require the use of electronic navigational equipment. Researchers and government organizations, however, have discovered methods for disrupting the safe navigation of vessels by exploiting vulnerabilities in Global Positioning Systems (GPS), Electronic Chart Display & Information Systems (ECDIS), and other critical integrated situational awareness systems.

Global positioning systems have significantly increased the ease of navigation of both small vessels and larger vessels, such as cruise ships. Students from the University of Texas at Austin, with the assistance of their faculty advisor, successfully spoofed the navigation equipment on board an $80 million superyacht (University of Texas at Austin 2013). In June of 2013, the students and faculty advisor were invited on board the White Rose of Drachs. As it sailed off the coast of Italy toward its destination in Greece, the students created false GPS signals using equipment which costs about $3,000 and was about the size of a briefcase. The students covertly gained control of the yacht's GPS receivers with ease. Unlike GPS jamming, spoofing did not alert navigation watchstanders because the false GPS signals were indistinguishable from the real GPS signals.

In 2008, British authorities and scholars conducted research on GPS jamming effects on maritime navigation equipment (Grant et al. 2009). Over 3 days of sea trials, the NLV Pole Star and its crew were subjected to GPS jamming experiments. The results of the experiments revealed that the jamming

ISSUES IN MARITIME CYBER SECURITY

equipment did successfully disrupt the situational awareness of the crew. Because the bridge also had integrated navigational equipment, jamming the GPS also disrupted the Electronic Chart Display & Information System (ECDIS). The crew reported that the screen was not providing updates and became still. The crew responded by turning the ECDIS completely off. The British authorities also revealed that the eLoran system was unaffected by the denial of GPS by the jamming equipment. By manipulating information or denying navigational information service, this case illustrates how other aspects of the MTS could be affected. Not only would the safe navigation of vessels be compromised, but Vessel Traffic Management Systems may not have situational awareness, and also aids to navigation which rely on the same GPS technology may not work properly.

Case 2: Insiders and Hackers Smuggle Drugs through a Port Facility (Interface Security)

The Port of Antwerp is vital to the industrial and consumer markets throughout Europe. From a global perspective, the Port of Antwerp is the 14th largest container port and the 10th busiest international freight shipping port in the world (WPS—Port of Antwerp Port Commerce). Located in Belgium, the port handles more than 170 million tons of cargo by 14 thousand vessels annually (WPS—Port of Antwerp Port Commerce). The container terminals were modernized after significant infrastructure investments, including streamlined use of technology to facilitate the flow of their 24 × 7 port operations. More than 20,000 containers pass through Antwerp in a typical day of operations (Antwerp's Success). Because of the high shipping volume, reliance in technology, and location, the Port of Antwerp was an ideal target for a drug smuggling plot in 2011 (Bateman 2013).

For 2 years, criminals were exploiting cyber and physical vulnerabilities at the Port of Antwerp unbeknownst to authorities. Drugs were being physically smuggled into containers carrying bananas and timber from South America and then being transported to the Port of Antwerp undetected by authorities and the shipping line. Drug smugglers leveraged the dark web and internet to recruit talented hackers to execute criminal activities. Ultimately in 2013, authorities discovered the nefarious activities and seized more than a ton of heroin and cocaine from the smugglers. To accomplish their large-scale smuggling operations, these persistent criminals launched an attack on the port in two phases.

The first phase was gaining remote access to port terminal systems through human and cyber vulnerabilities. The port terminal computer systems track secure information, such as, the containers, its contents, location, and delivery schedule. Hackers first used social engineering to send phishing e-mails to port and shipping line employees (Williams 2014). Some employees downloaded the malware containing key stoke logging software which allowed the hackers to access the terminal systems. Penetrating this system allowed the hackers to identify the planned location in the port of their drug filled containers from South America, and they could also very easily virtually change the location and planned delivery schedule. Port operators viewing the computer system were completely deceived and unaware of this cyber vulnerability. Operators were oblivious to the fact that the physical location of the containers did not match the output of the computer system's information. The criminals orchestrating the smuggling operations recruited insiders within the port who were truck drivers. Once the port terminal system information was manipulated, the smugglers relied on these truck drivers to pick up the containers from within the port and deliver it to their handlers. Additionally, hackers were able to remotely remove information associated with the container after the smuggling operations, essentially, covering their tracks.

The second phase relied on physical vulnerabilities. The remote access was eventually discovered and removed by shipping line when containers were "vanishing" and clients were not receiving their goods. The port installed a firewall, but the criminals persisted by physically breaching the facility and installed wireless bridges on the computers, logging devices in keyboards, and screen capture tools, to resume monitoring their robust smuggling operations. The criminals no longer had remote access to manipulate the data, but could still monitor the transportation information. Not all of their drivers were able to intercept the containers within the port. This resulted in an incident where a driver not affiliated with the smuggling operation, had no awareness that he was transporting a container full of cocaine. The criminals tracked his movement and shot at the driver to commandeer the truck miles away from Antwerp (Bateman 2013).

Since it took 2 years for the port to realize their systems were breached, it is speculated that the ton of cocaine and heroin are only a fraction of the amount of drugs the organized criminals and hackers were able to transport through the Port of Antwerp (Bateman 2013).

Case 3: Malware from Supply Chain Penetrate Critical Systems (Information Systems)

Supply chain management is an integral part of the Maritime Transportation System. Beyond the supply of the goods being transported on vessels through ports and waterways, supply chain of products and equipment to support the movement of goods is also critical to the MTS. Logistics and shipping companies look to enterprise software and hardware solutions to effectively and efficiently manage goods within the MTS. Many companies have found the use of enterprise software and hardware to manage logistical and financial operations are a necessity to doing business in a global economy. Enterprise resource planning (ERP) systems "represent the world's business-critical infrastructure," integrating many core business functions including finance, human resources, and operations management (Nunez 2012). However, these ERP systems can be susceptible to advanced persistent cyber attacks (Nunez 2012).

In 2014, TrapX, a U.S. cyber security solutions firm, discovered malicious software that was pre-installed on barcode scanners designed to be used with popular ERP systems (TrapX 2014). The Chinese hackers had advanced knowledge of zero day vulnerabilities in ERP systems and installed sophisticated malware to penetrate logistics companies. Hackers installed polymorphic malware on the embedded Windows XP operating system of the scanner between the manufacturing and the delivery to the company. Once the malware-infected barcode scanner was plugged into the company's network, it attacked the company in a series of stages to gain access to its logistics and financial systems. During the first phase, the malware identifies remote administration protocol (RAP) and server message block (SMB) protocol with finance related terms in the hostname. Even though most companies secure their SMB ports with firewalls, the administration ports may be more vulnerable to allow administrators flexibility in server management. This allowed hackers to establish a remote connection into the "crown jewels" of the company (TrapX 2014). Even though the company installed authentication security certificates on the barcode scanners, it was compromised because the malware was installed (Marko 2014). During the second phase of the attack, sensitive data from device scans were then sent to a Chinese botnet, later determined to be connected to the Lanxiang Vocational School, a school previously affiliated with other sophisticated attacks such as those on Google (Google Shuts Down China Site). Once the botnets had this data, it could be used to access customer data, financial data, and disrupt shipping data, causing a major disruption to the company and ability to provide products accurately and timely to market.

This type of advanced attack exploits zero-day vulnerabilities, arrives infected with adaptive malware from the manufacturer, and can bypass or deceive common network security protocols (Google Shuts Down China Site). Carl Wright, General Manager at TrapX, says these factors are significant and make this type of supply chain cyber attack a major contributing factor to potential disruptions in the MTS (TrapX 2014). Since there are only a few different reputable ERP solutions available on the market, such as Oracle and SAP, hackers can find methods to exploit vulnerabilities over a large number of logistics companies at once.

CHAPTER 25

DISCUSSION QUESTIONS AND CONCLUSION

The following five discussion questions are meant to complement existing literature in a cyber security course or professional engagement. By examining actual vulnerabilities in the MTS, students and professionals have an opportunity to have meaningful dialogue on the NIST Cybersecurity Framework and other policies or dynamics that are relevant to advancing the security of vessels, ports, and information systems to facilitate commerce.

Question 1: How are cyber vulnerabilities interconnected throughout the MTS?

Question 2: How might each of the mini-case scenarios presented be modeled and/or simulated in order to gain a deeper understanding of the attack surface? How might these models and simulations then be applied to MTS scenarios?

Question 3: How can businesses better protect themselves against vulnerabilities in their supply chain management processes?

Question 4: Using the SIMLAR Method, how can the NIST Cybersecurity Framework be expanded to address risk management of the MTS as a whole?

Question 5: What is the future role of government and how can it leverage private sector relationships to address risk management of the MTS?

The maritime sector is integral to the functioning of our society facilitating the reliable transport of domestic and international goods. Because the MTS is an interconnected and complex network, it is highly vulnerable to single point of failures and hackers breaching systems with a lack of visibility from port authorities and operators. Lack of resilience across the MTS would be devastating to businesses, consumers, and commerce. Further research is needed to examine how redundancy versus integrated systems can mitigate risks in the MTS. Systems engineering and risk management disciplines offer this perspective to solve emerging issues threatening the resilience of the MTS. Whether the disruptions in the MTS are caused inadvertently or deliberately, they have real consequences that can result in loss of life, property, or constrain of the flow of goods required to maintain the health of the global economy.

REFERENCES

"Antwerp's Success." Port of Antwerp. http://www.portofantwerp.com/en/antwerps-success (accessed January 15, 2016).

Bateman, Tom. 2013. "Police Warning After Drug Traffickers' Cyber-attack." *BBC News*, October 16. http://www.bbc.com/news/world-europe-24539417 (accessed December 18, 2015).

Behmer, James, Kolawole Ogunsina, Parth Shah, and Suhas Srinivasan. 2016. "An Agent-Based Modeling Approach to Creating More Resilient Littoral Combat Architectures." *2016 IEEE Aerospace Conference*. IEEE.

Bonabeau, Eric. 2002. "Agent-Based Modeling: Methods and Techniques for Simulating Human Systems." *Proceedings of the National Academy of Sciences* 99 (Supplement 3): 7280–7287. doi:10.1073/pnas.082080899.

Franzetti, Claudio. 2011. *Operational Risk Modelling and Management*. Boca Raton: CRC Press.

Grant, Alan, Paul Williams, Nick Ward, and Sally Basker. 2009. "GPS Jamming and the Impact on Maritime Navigation." *Journal of Navigation* 62 (2): 173–187. doi:10.1017/s0373463308005213.

Helft, Miguel, and David Barboza. 2010. "Google Shuts Down China Site in Dispute Over Censorship." *New York Times*, March 22. http://www.nytimes.com/2010/03/23/technology/23google.html?_r=0.

Mansouri, Mo, Brian Sauser, and John Boardman. 2009. "Applications of Systems Thinking for Resilience Study in Maritime Transportation System of Systems." *Systems Conference, 2009 3rd Annual IEEE*. IEEE.

Mansouri, A. Gorod, T. H. Wakeman and B. Sauser. 2009. "A systems approach to governance in Maritime Transportation System of Systems," *2009 IEEE International Conference on System of Systems Engineering (SoSE)*, Albuquerque, NM.

"Maritime Cybersecurity: A Growing, Unanswered Threat." *The Maritime Executive*. http://www.maritime-executive.com/article/Maritime-Cybersecurity-A-Growing-Unanswered-Threat-2014-10-24 (accessed October/November 24, 2014).

Marko, Kurt. 2014. "Trojan Hardware Spreads APTs." *Forbes*. http://www.forbes.com/sites/kurtmarko/2014/07/10/trojan-hardware-spreads-apts/#2715e4857a0b450e02bb4342 (accessed January 23, 2016).

Mele, Cristina, Jacqueline Pels, and Francesco Polese. 2010. "A Brief Review of Systems Theories and Their Managerial Applications." *Service Science* 2 (1-2): 126–135. doi:10.1287/serv.2.1_2.126.

MITRE. "Cyber Resiliency Engineering Framework." The MITRE Corporation. https://www.mitre.org/sites/default/files/pdf/11_4436.pdf (accessed January 7, 2016).

NASA. n.d. "The Systems Engineering Process." *NASA—Academy of Aerospace Quality*. http://aaq.auburn.edu/node/126.

National Institute of Standards and Technology. 2014. *Framework for Improving Critical Infrastructure Cybersecurity (Version 1.0)*. Washington, DC. http://www.nist.gov/cyberframework/.

Nunez, Mariano. 2012. "Cyber-Attacks on ERP Systems." *Datenschutz Datensich* 36 (9): 653–656. doi:10.1007/s11623-012-0220-5.

Santos, Joost R., Yacov Y. Haimes, and Chenyang Lian. 2007. "A Framework for Linking Cybersecurity Metrics to the Modeling of Macroeconomic Interdependencies." *Risk Analysis* 27 (5): 1283–1297. doi:10.1111/j.1539-6924.2007.00957.x.

Siegel, Michael, and James Houghton. 2015. "Interdisciplinary Consortium for Improving Critical Infrastructure Cybersecurity (IC)3." Technology Research for Industry | ARC Advisory Group. http://www.arcweb.com/events/arc-industry-forum-orlando/arcindustryforum2015presentations/MSiegel-MIT-presentation.pdf (accessed January 13, 2016).

TrapX. 2014. "TrapX Discovers 'Zombie Zero' Advanced Persistent Malware." (July 10). http://trapx.com/07-09-14-press-release-trapx-discovers-zombie-zero-advanced-persistent-malware/ (accessed January 15, 2016.)

University of Texas at Austin. 2013. "UT Austin Researchers Successfully Spoof an $80 Million Yacht at Sea." *UT News*, July 29. http://news.utexas.edu/2013/07/29/ut-austin-researchers-successfully-spoof-an-80-million-yacht-at-sea (accessed January 11, 2016).

"USCG Says Port Cyber Security Efforts Ongoing, but GAO Report Expresses Skepticism | Government Security News." *Government Security News | The News Leader in Physical, IT and Homeland Security*. http://gsnmagazine.com/article/45491/uscg_says_port_cyber_security_efforts_ongoing_gao (accessed January 15, 2016).

U.S. Coast Guard. 2015. "Written testimony of USCG Assistant Commandant for Prevention Policy RDML Paul Thomas for a House Committee on Homeland Security, Subcommittee on Border and Maritime Security hearing titled 'Protecting Maritime Facilities in the 21st Century: Are Our Nation's Ports at Risk for a Cyber-Attack?.'" Reading, 311 Cannon House Office Building, Washington, DC, October 8, 2015.

U.S. Department of Homeland Security. 2004. *Maritime Transportation System Security Recommendations for the National Strategy for Maritime Security*. Washington, DC. http://oai.dtic.mil/oai/oai?&verb=getRecord&metadataPrefix=html&identifier=ADA474574.

U.S. Department of Homeland Security. 2015. "What Is Security and Resilience?" http://www.dhs.gov/what-security-and-resilience (accessed January 15, 2016).

U.S. Government Accountability Office. 2014. *Maritime Critical Infrastructure Protection: DHS Needs to Better Address Port Cybersecurity*. GAO-14-459. http://www.gao.gov/products/GAO-14-459.

U.S. Government Accountability Office. 2015. *Maritime Critical Infrastructure Protection: DHS Needs to Enhance Efforts to Address Port Cybersecurity*. GAO-16-116T. http://www.gao.gov/products/GAO-16-116T.

Vaněk, Ondřej, Michal Jakob, Ondřej Hrstka, and Michal Pěchouček. 2013. "Agent-Based Model of Maritime Traffic in Piracy-Affected Waters." *Transportation Research Part C: Emerging Technologies* 36: 157–176. doi:10.1016/j.trc.2013.08.009.

"What is System Dynamics." System Dynamics Society. http://www.systemdynamics.org/what-is-s/ (accessed January 15, 2016).

"What is Systems Engineering." International Council on Systems Engineering Website. http://www.incose.org/AboutSE/WhatIsSE (accessed January 10, 2016).

Williams, Colin. 2014. "Security in the Cyber Supply Chain: Is it Achievable in a Complex, Interconnected World?" *Technovation* 34 (7): 382–384. doi:10.1016/j.technovation.2014.02.003.

"WPS - Port of Antwerp Port Commerce." World Port Source. http://www.worldportsource.com/ports/commerce/BEL_Port_of_Antwerp_25.php (accessed January 8, 2016).

CHAPTER 26:
GAME THEORETIC DEFENSE FOR MARITIME SECURITY

Sara McCarthy, Arunesh Sinha, and Milind Tambe
University of Southern California, Los Angeles, USA

Abstract

Security is a critical concern around the world. In many domains from cyber-security to sustainability, limited security resources prevent complete security coverage at all times. Instead, these limited resources must be scheduled (or allocated or deployed), while simultaneously taking into account the importance of different targets, the responses of the adversaries to the security posture, and the potential uncertainties in adversary payoffs and observations, etc. Computational game theory can help generate such security schedules. Indeed, casting the problem as a Stackelberg game, we have developed new algorithms that are now deployed over multiple years in multiple applications for scheduling of security resources. These applications are leading to real-world use-inspired research in the emerging research area of "security games." The research challenges posed by these applications include scaling up security games to real-world sized problems, handling multiple types of uncertainty, and dealing with bounded rationality of human adversaries. In cyber-security domain, the interaction between the defender and adversary is quite complicated with high degree of incomplete in- formation and uncertainty. While solutions have been proposed for parts of the problem space in cyber-security, the need of the hour is a comprehensive understanding of the whole space including the interaction with the adversary and the human in the loop. We highlight the innovations in security games that could be used to tackle the game problem in cyber-security.

ISSUES IN MARITIME CYBER SECURITY

Introduction

Security is a critical concern around the world that manifests in problems such as protecting our cyber infrastructure from attacks by criminals and other nation states, protecting our ports, airports, public transportation, and other critical national infrastructure from terrorists, in protecting our wildlife and forests from poachers and smugglers, and in curtailing the illegal flow of weapons, drugs, and money across international borders. In all of these problems, there are limited security resources which prevents security coverage of all the targets at all times; instead, security resources must be deployed intelligently taking into account differences in the importance of targets, the responses of the attackers to the security posture, and potential uncertainty over the types, capabilities, knowledge and priorities of attackers faced.

Game theory, which models interactions among multiple self-interested agents, is well-suited to the adversarial reasoning required for the security resource allocation and scheduling problem. Casting the physical problem as a Stackelberg game, we have developed new algorithms for efficiently solving such games that provide randomized patrolling or inspection strategies in physical security. These algorithms have led to successes and advances over previous human-designed approaches in security scheduling and allocation by addressing the key weakness of predictability in human-designed schedules. These algorithms are now deployed in multiple applications. The first application was ARMOR, which was deployed at the Los Angeles International Airport (LAX) in 2007 to randomize checkpoints on the roadways entering the airport and ca- nine patrol routes within the airport terminals (Jain et al. 2010). Following that, came several other applications: IRIS, a game-theoretic scheduler for randomized deployment of the US Federal Air Marshals (FAMS), has been in use since 2009 (Jain et al. 2010); PROTECT, which schedules the US Coast Guard's randomized patrolling of ports, has been deployed in the port of Boston since April 2011 and is in use at the port of New York since February 2012 (Shieh et al. 2012), and has spread to other ports such as Los Angeles/Long Beach, Houston, and others; another application for deploying escort boats to protect ferries has been deployed by the US Coast Guard since April 2013 (Fang, Jiang, and Tambe 2013); and TRUSTS (Yin et al. 2012) which has been evaluated in field trials by the Los Angeles Sheriffs Department (LASD) in the LA Metro system. These initial successes point the way to major future applications in a wide range of security domains. Indeed, researchers have started to explore the use of security games models in tackling security issues in the cyber world, such as deep packet inspection (Vanek et al. 2012), optimal use of honey pots (Durkota et al. 2015) and enforcement of privacy policies (Blocki et al. 2013; Blocki et al. 2015).

CHAPTER 26

Given the many game-theoretic applications for solving real-world security problems, this paper provides an overview of the models and algorithms, key research challenges and a brief description of our successful deployments. We also provide an overview of applying Stackelberg game based models to cyber-security and compare with other existing approaches to model defender-adversary interaction in cyber-security. Overall, the work in security games has produced numerous game-theoretic decision aids that are in daily use by security agencies to optimize their limited security resources. The implementation of these applications required addressing fundamental research challenges and has led to an emerging *science of security games* consisting of a general framework for modeling and solving security resource allocation problems. We categorize the research challenges associated with security games into four broad categories: (1) addressing scalability across a number of dimensions of the game, (2) tackling different forms of uncertainty that are present in the game and (3) addressing human adversaries' bounded rationality.

The issues in cyber security provide an even richer set of challenges. These include (1) risk and economic analysis: game theoretic approaches to cyber security relies on risk assessment of various cyber assets. Risk assessment is a fundamental challenge in the cyber domain, given the dynamic and complex nature of cyber world. For example, values of assets can change over time depending on what data they store. (2) Scalability: as stated the cyber world is more intricate than the physical world with very complex interactions among the different components. A detailed model of cyber security would present greater scalability challenges than in the physical world. (3) Uncertainty: apart from the various uncertainties present in the game parameters itself, a unique uncertainty in the cyber world arises due to the ability of the attacker to conceal its attack resulting in partial observability of the defender. This also raises problems of detecting an attack itself. Finally, (4) Human in the loop: human security analyst have limited cognitive abilities and have often been unable to examine the large number of possible scenarios that could exist in the context of cyber security. Further, social engineering attacks have exploited human limitations to enable attacks on cyber systems. Modeling the human in the loop is a critical challenge in cyber security.

The rest of the paper is organized as follows: Section 2 introduces the general security games model, Section 3 describes the approaches used to tackle scalability is- sues, Section 4 provides details of field evaluation of the science of security games and Section 5 describes some approaches of applying security game models to cyber-security and privacy and also other game theoretic approaches to cyber-security.

Stackelberg Security Games

Stackelberg games were first introduced to model leadership and commitment (von Stackelberg 1934). The term Stackelberg Security Games (SSG) was first introduced by Kiekintveld et al (2009) to describe specializations of a particular type of Stackelberg game for security as discussed below. This section provides details on this use of Stackelberg games for modeling security domains. We first give a generic description of security domains followed by security games, the model by which security domains are formulated in the Stackelberg game framework.

Stackelberg Security Game

In Stackelberg Security Games, a defender must perpetually defend a set of targets T using a limited number of resources, whereas the attacker is able to surveil and learn the defender's strategy and attack after careful planning. An action, or *pure strategy*, for the defender represents deploying a set of resources R on patrols or checkpoints, e.g., scheduling checkpoints at the LAX airport or assigning federal air marshals to protect flight tours. The pure strategy for an attacker represents an attack at a target, e.g., a flight. The *mixed strategy* of the defender is a probability distribution over the pure strategies. Additionally, with each target are also associated a set of payoff values that define the utilities for both the defender and the attacker in case of a successful or a failed attack.

A key assumption of Stackelberg Security Games (we will sometimes refer to them as simply security games) is that the payoff of an outcome depends only on the target attacked, and whether or not it is *covered* (protected) by the defender (Kiekintveld et al. 2009). The payoffs do *not* depend on the remaining aspects of the defender allocation. For example, if an adversary succeeds in attacking target t_1, the penalty for the defender is the same whether the defender was guarding target t_2 or not.

This allows us to compactly represent the payoffs of a security game. Specifically, a set of four payoffs is associated with each target. These four payoffs are the rewards and penalties to both the defender and the attacker in case of a successful or an unsuccessful attack, and are sufficient to define the utilities for both players for all possible outcomes in the security domain. More formally, if target t is attacked, the defender's utility is $Udc(t)$ if t is covered, or $Udu(t)$ if t is not covered. The attacker's utility is $Uac(t)$ if t is covered, or $Uau(t)$ if t is not covered. Table 1 shows an example security game with two targets, t_1 and t_2. In this example game, if the defender was covering target t_1 and the attacker attacked t_1, the defender would get 10 units of reward whereas the attacker would receive −1 units. We make the assumption that in a security game it is always better for the defender to cover a target as compared to leaving it uncovered, whereas it is always better for the attacker to attack an uncovered target. This assumption is consistent with the payoff trends in the real-world. A special case is *zero-sum games*, in which for each outcome the sum of utilities for the defender and attacker is zero, although general security games are not necessarily zero-sum.

	Defender		Attacker	
Target	Covered	Uncovered	Covered	Uncovered
t_1	10	0	-1	1
t_2	0	-10	-1	1

Table 1. Example of a security game with two targets.

Solution Concept: Strong Stackelberg Equilibrium

The solution to a security game is a *mixed* strategy for the defender that maximizes the expected utility of the defender, given that the attacker learns the mixed strategy of the defender and chooses a best-response for himself. The defender's mixed strategy is a probability distribution over all pure strategies, where a pure strategy is an assignment of the defender's limited security resources to targets. This solution concept is known as a Stackelberg equilibrium (Leitmann 1978).

The most commonly adopted version of this concept in related literature is called Strong Stackelberg Equilibrium (SSE) (Breton, Alg, and Haurie 1988; Conitzer and Sandholm 2006; Paruchuri et al. 2008; von Stengel and Zamir 2004). In security games, the mixed strategy of the defender is equivalent to the probabilities that each target t is covered by the defender, denoted by $C = \{c_t\}$ (Korzhyk, Conitzer, and Parr 2010). Furthermore, it is enough to consider a pure strategy of the rational adversary (Conitzer and Sandholm 2006), which is to attack a target t. The expected utility for defender for a strategy profile (C,t) is defined as $Ud(t,C) = c_t Udc(t) + (1 - c_t)Udu(t)$, and a similar form for the adversary. A SSE for the basic security games (non-Bayesian, rational adversary) is defined as follows:

Definition 1. *A pair of strategies* (C^*, t^*) *form a* Strong Stackelberg Equilibrium (SSE) *if they satisfy the following:*

1. *The defender plays a best-response:* $Ud(t^*, C^*) \geq Ud(t(C), C)$ *for all defender's strategy C where $t(C)$ is the attacker's response against the defender strategy C.*

2. *The attacker plays a best-response:* $Ua(t^*, C^*) \geq Ua(t, C^*)$ *for all target t.*

3. The attacker breaks ties in favor of the defender: $Ud(t^*,C^*) \geq Ud(t,C^*)$ for all target t such that $t = \text{argmax}_t \, Ua(t,C^*)$

Challenge	Scalability				Uncertainty		Attacker Bounded Rationality
	Large defender strategy space	Large defender & attacker strategy spaces	Mobile resources & moving targets	Multiple boundedly rational attackers	Unifications of uncertainties	Dynamic execution uncertainty	Learning attacker behaviors
Domain Example	Federal Air Marshals Service	Road Network Security	Ferry Protection	Fishery Protection	Security in LAX Airport	Security in Transit System	Green Security Domains: wildlife/fishery protection
Algorithmic Solution	ASPEN: strategy generation approach	RUGGED: double oracle approach	CASS: compact representation of strategy	MIDAS: cutting plane approach	URAC: multi-dimensional reduction & divide-and-conquer approach	Markov Decision Processes approach	Behavioral models & Human subject experiments

Figure 1. Summary of Real-world Security Challenges

The assumption that the follower will always break ties in favor of the leader in cases of indifference is reasonable because in most cases the leader can induce the favorable strong equilibrium by selecting a strategy arbitrarily close to the equilibrium that causes the follower to strictly prefer the desired strategy (von Stengel and Zamir 2004). Furthermore an SSE exists in all Stackelberg games, which makes it an attractive solution concept compared to versions of Stackelberg equilibrium with other tie-breaking rules. Finally, although initial applications relied on the SSE solution concept, we have since proposed new solution concepts that are more robust against various uncertainties in the model (Yin et al. 2011; An et al. 2011; Pita et al. 2012) and have used these robust solution concepts in some of the later applications.

In the following sections, we present four key scalability challenges in solving real-world security problems which are summarized in Figure 1. While Figure 1 does not provide an exhaustive overview of all research in SSG, it provides a general overview of the areas of research. In each case, we will use a domain example to motivate the specific challenge and then outline the key algorithmic innovation needed to address the challenge.

Addressing Scalability in Real-world Problems

For simple examples of security games, such as the one shown in the previous section, the Strong Stackelberg Equilibrium can be calculated by hand. However, as the size of the game increases, hand calculation is no longer feasible and an algorithmic approach for generating the SSE becomes necessary. Conitzer and Sandholm (2006) provided the first complexity results and algorithms for computing optimal commitment strategies in Bayesian Stackelberg games, including both pure and mixed-strategy commitments. An improved algorithm for solving Bayesian Stackelberg games, DOBSS (Paruchuri et al. 2008), is central to the fielded application ARMOR in use at the Los Angeles International Airport (Jain et al. 2010). These early works required that the full set of pure strategies for both players be considered when modeling and solving Stackelberg security games. However, many real world problems feature billions of pure strategies for either the defender or the attacker.

In addition to large strategy spaces, there are other scalability challenges presented by different real world security domains. There are domains where, rather than being static, the targets are moving and thus the security resources need to be mobile and move in a continuous space to provide protection. There are also domains where the attacker may not conduct the careful surveillance and planning that is assumed for a SSE and thus it is important to model the bounded rationality of the attacker in order to predict their behavior. In the former case, both the defender and attacker's strategy spaces are infinite. In the latter case, computing the optimal strategy for the defender given attacker behavioral (bounded rationality) model is computationally expensive. In this section, we thus highlight the critical scalability challenges faced to bring Stackelberg security games to the real world and the research contributions that served to address these challenges.

Scale Up with Large Defender Strategy Spaces

This section provides an example of a research challenge in security games where the number of defender strategies is too enormous to be enumerated in computer memory.

Domain Example—IRIS for US Federal Air Marshals Service. The US Federal Air Marshals Service (FAMS) allocates air marshals to flights departing from and arriving in the United States to dissuade potential aggressors and prevent an attack should one occur. Flights are of different importance based on a variety of factors such as the numbers of passengers, the population of source and destination cities, and international flights from different countries. Security resource allocation in this domain is significantly more challenging than for ARMOR: a limited number of air marshals need to be scheduled to cover thousands of commercial flights each day. Furthermore, these air marshals must be scheduled on tours of flights that obey various constraints (e.g., the time required to board, fly, and disembark). Simply finding schedules for the marshals that meet all of these constraints is a computational challenge.

For an example scenario with 1000 flights and 20 marshals, there are over 1041 possible schedules that could be considered. Yet there are currently tens of thousands of commercial flights flying each day, and public estimates state that there are thousands of air marshals that are scheduled daily by the FAMS (Keteyian 2010). Air marshals must be scheduled on tours of flights that obey logistical constraints (e.g., the time required to board, fly, and disembark). An example of a schedule is an air marshal assigned to a round trip from New York to London and back.

Against this background, the IRIS system (Intelligent Randomization In Scheduling) has been developed and deployed by FAMS since 2009 to randomize schedules of air marshals on international flights. In IRIS, the targets are the set of n flights and the attacker could potentially choose to attack one of these flights. The FAMS can assign $m < n$ air marshals that may be assigned to protect these flights.

Since the number of possible schedules exponentially increases with the number of flights and resources, DOBSS is no longer applicable to the FAMS domain. Instead, IRIS uses the much faster ASPEN algorithm (Jain et al. 2010) to generate the schedule for thousands of commercial flights per day.

Algorithmic Solution—Incremental Strategy Generation (ASPEN). We describe one particular algorithm ASPEN, that computes strong Stackelberg equilibria (SSE) in domains with a *very large* number of pure strategies (up to billions of actions) for the defender (Jain et al. 2010). These pure strategies can be represented as integral points in a high dimensional space. ASPEN builds on the insight that there exist solutions with small support sizes, which are mixed strategies in which only a small set of pure strategies are played with positive probability (applying Carathéodory theorem (Carathodory 1911) to the convex hull of pure strategies). ASPEN exploits this by using a *column generation* (Barnhart et al. 1998) based approach for the defender, in which defender pure strategies are iteratively generated and added to the optimization formulation. The novel contribution of ASPEN is to provide a linear formulation for the master and a minimum- cost integer flow formulation for the slave, which enables the application of strategy generation techniques.

Scale Up with Large Defender & Attacker Strategy Spaces

Whereas the previous section focused on domains where only the defender's strategy was difficult to enumerate, we now turn to domains where both defender and attacker strategies are difficult to enumerate.

Domain Example—Road Network Security. One area of great importance is securing urban city networks, transportation networks, computer networks and other net- work centric security domains. For example, after the terrorist attacks in Mumbai of 2008 (Chandran and Beitchman 2008), the Mumbai police started setting up vehicular checkpoints on roads. We can model the problem faced by the Mumbai police as a security game between the Mumbai police and an attacker. In this urban security game, the pure strategies of the defender correspond to allocations of resources to edges in the network—for example, an allocation of police checkpoints to roads in the city. The pure strategies of the attacker correspond to paths from any *source* node to any *target* node—for example, a path from a landing spot on the coast to the airport.

The strategy space of the defender grows exponentially with the number of available resources, whereas the strategy space of the attacker grows exponentially with the size of the network. For example, in a fully connected graph with 20 nodes and 190 edges, the number of defender pure strategies for only 5 defender resources is $\binom{190}{5}$ or almost 2 billion, while the number of attacker pure strategies (i.e., paths without cycles) is on the order of 10^{18}. Real-world networks are significantly larger, e.g., the entire road network of the city of Mumbai has 9,503 nodes (intersections) and 20,416 edges (streets), and the security forces can deploy dozens (but not as many as number of edges) of resources.

Algorithmic Solution–Double Oracle Incremental Strategy Generation (RUGGED). We describe the RUGGED algorithm (Jain et al. 2011), which generates pure strategies for both the defender and the attacker. This algorithm is inspired by the double oracle algorithm of solving large scale games (McMahan, Gordon, and Blum 2003). RUGGED models the domain as a zero-sum game, and computes the minimax equilibrium, since the minimax strategy is equivalent to the SSE in zero-sum games. Starting with an initial small set of strategies, RUGGED iterates over two oracles: the defender best response and the attacker best response oracles, and adds the best response to the set of strategies. The algorithm stops when neither of the generated best responses improve on the current minimax strategies.

The contribution of RUGGED is to provide the mixed integer formulations for the best response modules which enable the application of such a strategy generation approach. The key once again is that RUGGED is able to converge to the optimal solution without enumerating the entire space of defender and attacker strategies. However, originally RUGGED could only compute the optimal solution for deploying up to 4 resources in real-city network with 250 nodes within a time frame of 10 hours (the complexity of this problem can be estimated by observing that both the best response problems are NP-hard themselves (Jain et al. 2011). More recent work (Jain, Tambe, and Conitzer

2013) builds on RUGGED and proposes SNARES, which allows scale-up to the entire city of Mumbai, with 10–15 checkpoints.

Scale Up with Mobile Resources & Moving Targets

Whereas the previous two sections focused on incremental strategy generation as an approach for scale-up this section introduces another approach: use of compact marginal probability representations. This alternative approach is shown in use in the context of a new application of protecting ferries.

Domain Example—Ferry Protection for the US Coast Guard. The United States Coast Guard is responsible for protecting domestic ferries, including the Staten Island Ferry in New York, from potential terrorist attacks. here are a number of ferries carrying hundreds of passengers in many waterside cities. These ferries are attractive targets for an attacker who can approach the ferries with a small boat packed with explosives at any time; this attacker's boat may only be detected when it comes close to the ferries. Small, fast, and well-armed patrol boats can provide protection to such ferries by detecting the attacker within a certain distance and stop him from attacking with the armed weapons. However, the numbers of patrol boats are often limited, thus the defender cannot protect the ferries at all times and locations. We thus developed a game-theoretic system for scheduling escort boat patrols to protect ferries, and this has been deployed at the Staten Island Ferry since 2013 (Fang, Jiang, and Tambe 2013).

The key research challenge is the fact that the ferries are continuously moving in a continuous domain, and the attacker could attack at any moment in time. This type of moving targets domain leads to game-theoretic models with continuous strategy spaces, which presents computational challenges. Our theoretical work showed that while it is "safe" to discretize the defender's strategy space (in the sense that the solution quality provided by our work provides a lower bound), discretizing the attacker's strategy space would result in loss of utility (in the sense that this would provide only an upper bound, and thus an unreliable guarantee of true solution quality). We developed a novel algorithm that uses a compact representation for the defender's mixed strategy space while being able to exactly model the attacker's continuous strategy space. The implemented algorithm, running on a laptop, is able to generate daily schedules for escort boats with guaranteed expected utility values.

Algorithmic Solution—Compact Strategy Representation (CASS). In this section, we describe the CASS (Solver for Continuous Attacker Strategy) algorithm (Fang, Jiang, and Tambe 2013) for solving security problems where the defender has mobile patrollers to protect a set of mobile targets against the attacker who can attack these moving targets at any time during their movement. In these security problems, the sets of pure strategies for both the defender and attacker are continuous w.r.t the continuous spatial and time components of the problem domain. The CASS algorithm attempts to compute the optimal mixed strategy for the defender without discretizing the attacker's continuous strategy set; it exactly models this set using sub-interval analysis which exploits the piecewise-linear structure of the attacker's expected utility function. The insight of CASS is to compactly represent the defender's mixed strategies as a *marginal* probability distribution, overcoming the short-coming of an exponential number of pure strategies for the defender. CASS shows that *any strategy in full representation can be mapped into a compact representation* as well as *compact representation does not lead to any loss in solution quality*. This compact representation allows CASS to reformulate the resource allocation problem as computing the optimal *marginal* coverage of the defender.

Scale Up with Boundedly Rational Attackers

One key challenge of real-world security problems is that the attacker is boundedly rational; the attacker's target choice is non-optimal. In SSGs, attacker bounded rationality is often modeled via behavior models such as Quantal Response (QR) (McFadden1972; McKelvey and Palfrey 1995). In general, QR attempts to predict the probability the at- tacker will choose each target with the intuition is that the higher the expected utility at a target is, the more likely that the adversary will attack that target. Another behavioral model that was recently shown to provide higher prediction accuracy in predicting the attacker's behavior than QR is Subjective Utility Quantal Response (SUQR) (Nguyen et al. 2013). SUQR is motivated by the lens model which suggested that evaluation of adversaries over targets is based on a linear combination of multiple observable features (Brunswik 1952). However, handling multiple attackers with these behavioral models in the context of large defender's strategy space is computationally challenging. In this section, we mainly focus on handling the scalability problem given behavioral models of the attacker.

To handle the problem of large defender's strategy space given behavioral models of attackers, we introduce yet another technique of scaling up: we use the compact marginal representation, discussed earlier, but refine that space incrementally if the solution produced violates the necessary constraints.

Domain Example—Fishery Protection for US Coast Guard. Fisheries are a vital natural resource from both an ecological and economic standpoint. However, fish stocks around the world are threatened with collapse due to illegal, unreported, and unregulated (IUU) fishing. The United States Coast Guard

(USCG) is tasked with the responsibility of protecting and maintaining the nation's fisheries. To this end, the USCG deploys resources (both air and surface assets) to conduct patrols over fishery areas in order to deter and mitigate IUU fishing. Due to the large size of these patrol areas and the limited patrolling resources available, it is impossible to protect an entire fishery from IUU fishing at all times. Thus, an intelligent allocation of patrolling resources is critical for security agencies like the USCG.

Natural resource conservation domains such as fishery protection raise a number of new research challenges. In stark contrast to counter-terrorism settings, there is frequent interaction between the defender and attacker in these resource conservation domains. This distinction is important for three reasons. First, due to the comparatively low stakes of the interactions, rather than a handful of persons or groups, the defender must protect against numerous adversaries (potentially hundreds or even more), each of which may behave differently. Second, frequent interactions make it possible to collect data on the actions of the adversaries actions over time. Third, the adversaries are less strategic given the short planning windows between actions.

Algorithmic Solution—Incremental Constraint Generation (MIDAS). Generating effective strategies for domains such as fishery protection requires an algorithmic approach which is both *scalable* and *robust*. For scalability, the defender is responsible for protecting a large patrol area and therefore must consider a large strategy space. Even if the patrol area is discretized into a grid or graph structure, the defender must still reason over an exponential number of patrol strategies. For robustness, the defender must protect against *multiple* boundedly rational adversaries. Bounded rationality models, such as the quantal response (QR) model (McKelvey and Palfrey 1995) and the subjective utility quantal response (SUQR) model (Nguyen et al. 2013), introduce stochastic actions, relaxing the strong assumption in classical game theory that all players are perfectly rational and utility maximizing. These models are able to better predict the actions of human adversaries and thus lead the defender to choose strategies that perform better in practice. However, both QR and SUQR are non-linear models resulting in a computationally difficult optimization problem for the defender. Combining these factors, MIDAS models a population of boundedly rational adversaries and utilizes available data to learn the behavior models of the adversaries using the SUQR model in order to improve the way the defender allocates its patrolling resources.

Figure 2 provides a visual overview of how MIDAS operates as an iterative process. Similar to the ASPEN algorithm described earlier, given the sheer complexity of the game being solved, the problem is decomposed using a column generation based master-slave formulation. The master utilizes multiple simplifications to create a relaxed version of the original problem which is more efficient to solve.

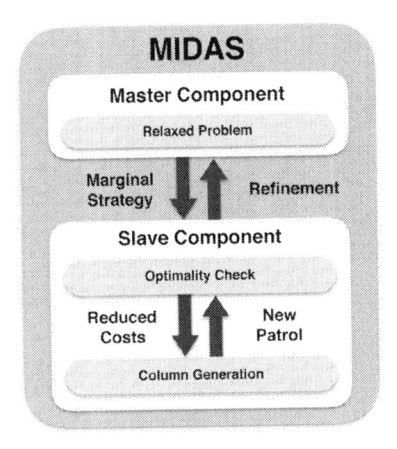

Figure 2. Overview of the multiple iterative process within the MIDAS algorithm

Due to the relaxations, solving the master produces a marginal strategy x which is a probability distribution over targets. However, the defender ultimately needs a probability distribution over patrols. Additionally, since not all of the spatio-temporal constraints are considered in the master, the relaxed solution x may not be a feasible solution to the original problem. Therefore, the slave checks if the marginal strategy x is a feasible solution. However, given the exponential number of patrol strategies, even performing this optimality check is intractable. Thus, column generation is used *within* the slave where only a small set of patrols is considered initially in the optimality check and the set is expanded over time. If the optimality check fails, then the slave generates a cut which is returned to refine and constrain the master, incrementally bringing it closer to the original problem. The entire process is repeated until an optimal solution is found. Finally, MIDAS has been successfully deployed and evaluated by the USCG in the Gulf of Mexico.

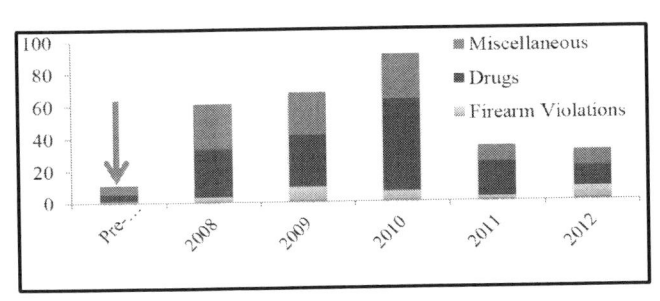

Figure 3. ARMOR evaluation results.

ISSUES IN MARITIME CYBER SECURITY

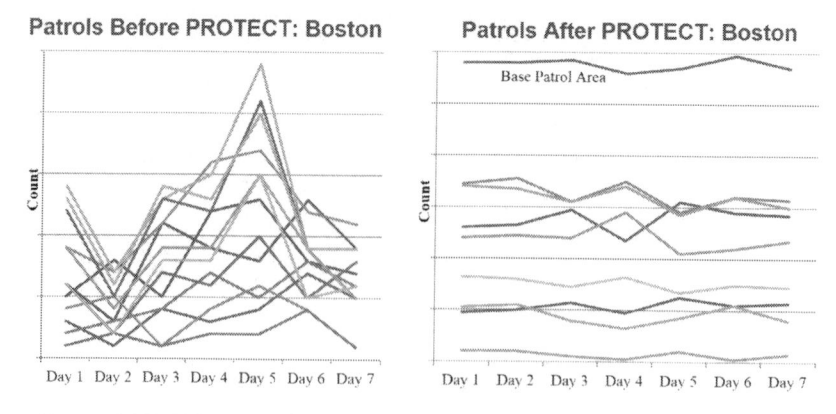

Figure 4. PROTECT evaluation results: pre deployment (left) and post deployment patrols (right).

ADDRESSING FIELD EVALUATION IN REAL-WORLD PROBLEMS

Evidence showing the benefits of the algorithms discussed in the previous sections is definitely an important issue that is necessary for us to answer. Unlike conceptual ideas, where we can run thousands of careful simulations under controlled conditions, we cannot conduct such experiments in the real world with our deployed applications. Nor can we provide a proof of 100% security —there is no such thing.

Instead, we focus on the specific question of: are our game-theoretic algorithm better at security resource optimization or security allocation than how they were allocated previously, which was typically relying on human schedulers or a simple dice roll for security scheduling (simple dice roll is often the other automation that is used or offered as an alternative to our methods). We have used the following methods to illustrate these ideas. These methods range from simulations to actual field tests.

1) **Simulations (including using a machine learning attacker):** We provide simulations of security schedules, e.g., randomized patrols, assignments, comparing our approach to earlier approaches based on techniques used by human schedulers. We have a machine learning based attacker who learns any patterns and then chooses to attack the facility being protected. Game-theoretic schedulers are seen to perform significantly better in providing higher levels of protections (Pita et al. 2008; Jain et al. 2010). This is also shown in Figure 3.

2) **Human adversaries in the lab:** We have worked with a large num-

CHAPTER 26

ber of human subjects and security experts (security officials) to have them get through randomized security schedules, where some are schedules generated by our algorithms, and some are baseline approaches for comparison. Human subjects are paid money based on the reward they collect by successfully intruding through our security schedules; again our game-theoretic schedulers perform significantly better (Pita et al. 2009).

3) **Actual security schedules before and after:** For some security applications, we have data on how scheduling was done by humans (before our algorithms were deployed) and how schedules are generated after deployment of our algorithms. For measures of interest to security agencies, e.g., predictability in schedules, we can compare the actual human-generated schedules vs our algorithmic schedules. Again, game-theoretic schedulers are seen to perform significantly better by avoiding predictability and yet ensuring that more important targets are covered with higher frequency of patrols. Some of this data is published (Shieh et al. 2012) and is also shown in Figure 4.

4) **"Adversary" teams simulate attack:** In some cases, security agencies have deployed adversary perspective teams or mock attacker teams that will attempt to conduct surveillance to plan attacks; this is done before and after our algorithms have been deployed to check which security deployments worked better. This was done by the US Coast Guard indicating that the game-theoretic scheduler provided higher levels of deterrence (Shieh et al. 2012).

5) **Real-time comparison: human vs algorithm:** This is a test we ran on the metro trains in Los Angeles. For a day of patrol scheduling, we provided head-to-head comparison of human schedulers trying to schedule 90 officers on patrols vs an automated game-theoretic scheduler. External evaluators then provided an evaluation of these patrols; the evaluators did not know who had generated each of the schedules. The results show that while human schedulers required significant effort even for generating one schedule (almost a day), and the game-theoretic scheduler ran quickly, the external evaluators rated the game theoretic schedulers higher (with statistical significance) (Fave et al. 2014).

6) **Actual data from deployment:** This is another test run on the metro trains in LA. We had a comparison of game-theoretic scheduler vs an alternative (in this case a uniform random scheduler augmented

with real time human intelligence) to check fare evaders. In 21 days of patrols, the game-theoretic scheduler led to significantly higher numbers of fare evaders captured than the alternative (Fave et al. 2014; Fave et al. 2014).

7) **Domain expert evaluation (internal and external):** There have been of course significant numbers of evaluations done by domain experts comparing their own scheduling method with game theoretic schedulers and repeatedly the game theoretic schedulers have come out ahead. The fact that our software is now in use for several years at several different important airports, ports, air-traffic, and so on, is an indicator to us that the domain experts must consider this software of some value.

Cyber-security: Challenges and Opportunities

The domain of computer security and privacy provides a rich set of challenges that requires new innovation and techniques. The application of game theory to maritime cyber security is a new and promising research field. The potential benefits of applying game theory to cyber security problems are:

1) Game theory captures the adversarial nature of cyber security interactions and provides quantitative and analytical tools that may help find the optimal defense strategies.

2) Computer implementations of those methods allow examination of a large number of threat scenarios, which human analyst can miss due to cognitive limitations and biases.

3) Game theory provides methods for predicting actor's behavior in uncertain situations and suggesting probable actions along with predicted outcomes.

Maritime security in particular stands to benefit greatly from a game theoretic approach to their cybersecurity. Maritime security is primarily concerned with protection against intelligent adversaries looking to do damage through sabotage, subversion and terrorism. In a cyber system, there are many different classes of hackers which may potential pose a threat. These range from unskilled hackers who rely on automated tools such as script kiddies, to a more elite class of hackers, such as nation states, organized criminals or cyber terrorists. It is this latter type, the resourceful, careful, adaptive and intelligent hacker which concerns maritime security. As such a game theoretic model is appropriate to properly model the dynamics of the cyber maritime environment.

Several challenges we identify in the maritime domain are those of general net- work security, protecting against sophisticated attacks such as advanced persistent threats (APT), adversary modeling, and the security of cyber physical systems. We survey some prior work on game theory approaches to cyber-security problems and discuss security games based approaches to these cyber-security problems.

Network Security

There is already some work on using the security games model to address general problems in cyber-security and privacy; in particular, the security of networks is a challenge that affects any organization with a cyber presence. Game theory in the context of network security is most often discussed as a decision module which leverages information about the network and has the responsibility of choosing or suggesting the best security strategy. It has been proposed to model the interaction between security specialists and attackers well as it captures the cat and mouse dynamic present in the ongoing cycle of intrusion detection and response in network systems.

Most network security solutions involve some form of intrusion detection system, however these IDS lack any sort of decision framework, and rely mainly on a library of known signatures, issuing an often overwhelming amount of alerts while leaving the decision of what problems to address up to the network administrator. The use of game theory can leverage the information presented by an IDS system and provide valuable suggestions on optimal policies and strategies to use, most of which may not be immediately obvious to a network administrator.

A key challenge we identify here is being able to filter through a large volume of suspicious activity alerts and determine what is the relevant information necessary to make a decision on the state of the network as well as what measures to take in response. There is also a clear need to find the balance between safety of the network and performance; the defender must expend enough resources in order to be able catch an attack, all while minimizing the impact on network performance of these actions. There is a dynamic resource allocation problem here which is not present in past work in physical security. With physical security, the defender must commit to an investment in security resources prior to deployment; there is no room for further optimization once the interaction between defender and attack takes place. Because the cyber environment is so volatile, resources an be quickly deployed, making it necessary to consider a dynamic set of resources rather than a static one. This is challenging as it adds further complexities to the problem, but potentially allows for a richer set of solutions which can better defend against potential threats. We now discuss some examples of how how game theory can be used for this analysis.

In (Vanek et al. 2012), the authors study the problem of optimal resource allocation for packet selection and inspection to detect potential threats in large computer networks with multiple computers of differing importance. A number of intrusion detection and monitoring systems are deployed in real world computer networks with the goal of detecting and preventing attacks. One countermeasure employed is to conduct *deep packet inspections*, a method that periodically selects a subset of packets in a computer network for analysis, but is costly in terms of throughput of the network. The security problem is formulated as a Stackelberg security game between two players: the attacker (or the intruder), and the defender (the detection system), which is played on a computer network modeled as a graph. The intruder wants to gain control over (or to disable) a valuable computer in the network by scanning the network, compromising a more vulnerable system, and/or gaining access to further devices on the computer network. The actions of the attacker can therefore be seen as sending malicious packets from a controlled computer (termed source) to a single or multiple vulnerable computers (termed targets). The objective of the defender is to prevent the intruder from succeeding by selecting the packets for inspection, identifying the attacker, and subsequently thwarting the attack. However, packet inspections cause unwanted latency and hence the defender has to decide where and how frequently to inspect network traffic in order to maximize the probability of a successful malicious packet detection. The authors provide polynomial time approximation algorithm that benefits from the sub-modularity property of the discretized zero-sum variant of the game and finds solutions with bounded error in polynomial time.

In another recent paper (Durkota et al. 2015), the authors study the problem of optimal number of *honeypots* to be placed in a network. Honeypots are fake copies of electronic resources (servers, computers, routers, etc.) that aim to confuse the attacker so that the attacker attacks these honeypots. Attacks on honeypots also enable the defender to study the attacker and possibly catch them. The use of honeypots as a deceptive defense mechanism seems promising, but has an associated cost in setting up the fake electronic assets. Thus, a central question in this defense mechanism is how many and which types of honeypots should be used? The authors use attack graphs to model possible attack trajectories that the attacker may use. The nodes in attack graphs are annotated with costs of attacks and benefits of successful attack, and also the probability of success of attack. In particular, the number and types of honeypots deployed influence the probability of success of attacks. The authors model the game as a Stackelberg security game with the defender choosing the number and type of honeypots to deploy. The attacker chooses an attack path with the best utility. The authors provide heuristic algorithm for the NP-Hard problem of finding the optimal attack by converting the problem to a Markov Decision Process.

Much of the work here focuses on how to best use a single type of defense mechanism, whether it be deep-packet inspection, honeypots or any other type of security resource. This raises a important question in the economics of security; given that there are so many security solutions available, each with the ability to protect against different kinds of threats, what software should we invest in order to maximize our overall security? While there has been a large body of work in portfolio optimization, this work fails to consider the actions of the attacker. Combining portfolio optimization with a more game theoretic analysis would be a challenge, due to the increased complexity of the problem, but could greatly benefit sectors where the adversary plays a very active role, such with maritime security.

Due to the large space of vulnerabilities and attacks that need to be defended against, addressing the problem of general network security is extremely challenging. We next look at a sub-class of attacks more pertinent to maritime security, known as active persistent threats, and some existing work on modeling and solving these threats.

Advanced Persistent Threats

Advanced persistent threats can be one of the most harmful attacks for any organization with a cyber presence, as well as one of the most difficult attacks to defend against. These attacks are sophisticated in nature and often targeted to the vulnerabilities of a particular system. They exploit vulnerabilities in the system using exploits like zero day attacks as well as leveraging the human as a weak link, employing social engineering tactics to gain access to a system. They operate quietly, over long periods of time and actively attempt to cover their tracks and remain undetected.

A popular model for the interaction between a defender and attacker in cyber- security, meant to address the problem of advanced persistent threats, is the FlipIt game (van Dijk et al. 2013). The Flip-it game is an abstract model originally developed by the RSA community (van Dijk et al. 2012). Their threat model involves a very intelligent and careful adversary who is well funded and organized, having very specific objectives. Because there are many possible ways an APT can occur, the attack surface for APT is very large. The Flip-it game takes a very high level view of the APT threat, in an attempt to generalize to all kinds of APTs and to model many kinds of attack vectors.

Flip-it is a continuous time game that models the fact that any cyber-system will ultimately be compromised and the defender will have to expend effort to detect and recover. In the game model, each player is fighting for the control of a shared resource. The resource can be an entire network, single user's computer, or even a password or digital signature key. At any given time a player can choose to forcibly take control of the resource. For the defender this action models any steps the defender can take which allows them to either assert that their control of the system or remove the attacker from the system though some form of resetting, recovery or disinfection of the system. For the attacker this action allows them to compromise the system if it is currently under the control of the defender, or install new malware, steal a new password or corrupt data. The identity of the player controlling the system is not known to either player until they make a move to control it; this means that the only time a player can be 100% certain they control the resource, is during the time step that they choose to take control. It is for this reason that the game is also dubbed the "stealthy takeover." Players can choose to act at any time; however each player pays a cost to do so. The goal of each player is then to maximize the time that they control the resource while minimize the cost of their actions. An example is given in Figure 5.

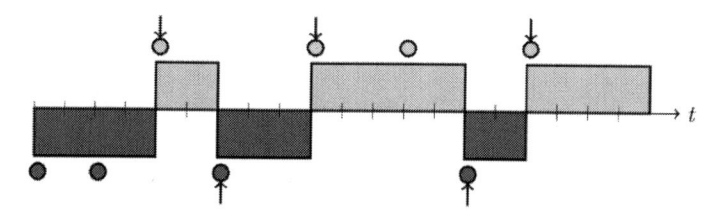

Figure 5. An instance of the Flipit game with the state of control of the system being shown as the bottom for the defender as and on top for the attacker. Actions are indicated by circles, while actions which resulted in a flip in the resource are indicated with arrows.

Many variants of the game have been studied, such as playing FlipIt with actual human subject experiments (Nochenson et al. 2013), the three player FlipIt game (Xiaotao Feng and Mohapatra year) used to model insider threats, as well as the FlipThem game (Laszka et al. 2013), a generalization of FlipIt to multiple resource. The game has been discussed in the context of credential management, particularly for password resets and key management where the game can be used to evaluate common password reset and key rotation policies. It can be shown that the best reset policies should take place at random intervals; (Bowers 2012) provides an analysis of several case studies and are able to compute non-adaptive strategies playing against an adaptive attacker. It has also been applied to cloud-enabled cyber-physical systems in order to influence when a device should trust commands from a potentially compromised cloud (Pawlick, Farhang, and Zhu 2015).

There are two main strategies that can be employed when playing the Flip-it game. These are classified as *adaptive* and *non-adaptive* play. Most of the work in solving this model focuses on non-adaptive play, where each player does not consider the reactions of other player in the game. Non-adaptive strategies could correspond to periodic strategies, where the player issues a take-over move with some fixed rate, or perhaps following some probability distribution (like a poisson process). Because current solutions use non-adaptive play, these strategies do not well model the details of a defender-adversary interaction in a cyber-settings and cannot be used directly in any real world network to provide guidance about how to use and deploy defense resources at a fine grained level.

Addressing APT's remains one of the most challenging problems for cyber security. The main reasons that these types of attacks are so difficult to defend against is because they are so difficult to detect. Recently there has been some work which borrows from the machine learning community, and models the problem as an active sensing problem. For machine learning, the problem amounts to deciding amongst a set of expensive la- bels, or test, which test are the most informative. Such a problem can be easily mapped to the cyber domain, if we consider sets of detectors which may be dispersed throughout our network, or sets of diagnostics which may be run. The idea behind using something like active sensing, is that we may be able to dynamically respond to suspicious activity. A particular series of alerts may raise suspicion about some malicious activity which might be occurring. However, due to the noisy nature of the detectors, we cannot be certain if it as actual attack; rather than being overly conservative and blocking that network activity, we can strategically deploy additional detectors to gather more information in order to make a more informed decision.

There has been some work that looks at active sensing in the cyber domain, but treats it in a decision theoretic framework rather than a game theoretic frame work. One line of work looks at using decision making and planning models like Partially Observable Markov Decision Processes (POMDPs) in order to address the problem of data exfiltration. While the end goal of advanced persistent threats may be diverse, it is often the case that intent of an attack is the theft of sensitive data. In the maritime sector, this can be disastrous, resulting in not only loss of competitive advantage and trade secrets of any companies which are part of the maritime system, but the leaking of confidential documents, and endangerment of national. Attackers will use covert channels to slowly and quietly transfer information, hiding themselves in legitimate network traffic to avoid detection. A recent trend in these attacks has relied on exploiting Domain Name System (DNS) queries in order to provide channels through which exfiltration can occur.

This work looks at leveraging information from noisy detectors in a network in order to correctly label channels and domains as malicious or legitimate. The model balances the cost of deploying these detectors with the benefit of additional information on the channels they are sensing on. Because this work does not explicitly use an adversary model, they are able to scale up to model realistic network sizes. These decision theoretic models do well at modeling the uncertainty in the network state. A big challenge that still remains in this domain is being able to solve models that can capture both the uncertain as well as the adversarial nature of these problems.

Threat Modelling

One of the challenges of applying game theory to cyber security is choosing the appropriate game model for a given security problem, in particular, building a good threat model is of critical importance. This has been shown in past work, such as in poaching and wildlife protection domains, where using a good adversary model is very important to performance of game theoretic algorithms. Depending on the scenario, a hacker's motives can be varied, ranging from simple trouble makers, motivated by the simple challenge of hacking into a system to criminals who are looking to cause as much dam- age as possible. While in the maritime sector, it may be more appropriate to consider a more elite class of hacker, more thought needs to be put in on who exactly we are trying to defend against.

While in reality, a network may be attacked by many hackers, (Lye and Wing 2005) argues that these can all be treated as a single entity performing multiple actions. This assumption only holds if we are considering a single behavioral model for our adversary, which may not always be the case. In the event that we are only attempting to protect against a single adversary type, then the assumption is valid. However, it may be the case that we need to consider multiple adversary models. Perhaps there are several types of assets that need to be protected, where each might have a different motivation for being attacked. There is then a benefit to being able to capturing multiple adversary types in a game theoretic model. Moayedi and Azgomi (2012) models the effects of different classes of adversaries and how this impacts the security policies in place. They classify the adversaries, or hackers in this context, based on two main characteristics; motivation and skill and model the system as a game with k-players, with each player belonging to a different class of hacker.

Human in the Loop

Human error is also an extremely large security threat. IBM's security services reports that over 95% of all security incidences have human error as a contributing factor. Even with the strongest system security available, the best detectors and firewalls in place, if users of a network fail to adhere to proper security protocol, bypassing the security systems becomes extremely easy. Dealing with the human aspect of cybersecurity is a huge challenge, as there will always be a human element present in the cyber ecosystem. While educating users on safe practices and protocols, is vital to minimizing this problem, we can also use game theory to help ensure that these proper practices are being adhered to.

One interesting work, called audit games (Blocki et al. 2013; Blocki et al. 2015), enhances the security games model with choice of punishments in order to capture scenarios of security and privacy policy enforcement in large organizations. Large organizations (such as Google, Facebook, hospitals) hold enormous amounts of privacy sensitive data. These organizations mandate their employees to adhere to certain privacy policies when accessing data. Auditing of access logs is used by organizations to check for policy violating accesses and then the violators are punished. Auditing often requires human help to investigate suspicious cases, and thereby arises the problem of allocating few resources to the huge number of cases to investigate. Another relevant question in this domain is how much should the organization punish in case of a violation? The audit game models the adversary as an agent that performs certain tasks (e.g., accesses to private data), and a subset of these tasks are policy violations. The auditor inspects a subset of the tasks, and detects violations from the inspected set. As punishments do affect the behavior of the adversary, it is critical for the auditor to choose the right level of punishment. As a consequence, the choice of a punishment level is added to the action space of the auditor. However, punishment is not free for the auditor; the intuition being that a high punishment level creates a hostile work environment, leading to lack in productivity of employees that results in loss for the organization (auditor). As a consequence, the auditor cannot impose infinite punishment and deter any adversary. The auditor's cost for a punishment level is modeled as a loss proportional to the choice of the punishment level. The auditor moves first by committing to an inspection and punishment strategy, followed by the best response of the adversary. The resultant Stackleberg equilibrium optimization turns out to be non-convex due to the punishment variable. The authors present efficient algorithms for various types of scheduling constraints.

Cyber Physical Systems

As game-theoretic methods have been applied with great success in securing physical systems such as airports, ports and cities, a natural extension would be to consider applications in cyber-physical systems. Cyber-physical systems (CPS) are typically sensor based communication-enabled autonomous systems, and are an important aspect to consider for maritime security. They include wireless censor networks, autonomous auto- motive systems, medical monitoring, process control systems, distributed robotics, and smart grid systems.

We focus on the examples of the power grid and smart grid system along with a more general example of industrial control systems to illustrate the current state of work and challenges faced when considering cyber physical systems. While the smart grid system is expected to revolutionize the future of the power industry, security is still a major challenge. Smart grids, or the Advanced Metering infrastructure (AMI) replaces old power meters with smart meters, which can enable two way communication between the user and the power utility. AMI's allow utilities to monitor electricity consumption without needing to physically access the site. This real time feed back allows them to perform load balancing and curtailment when power demand is high, diagnose causes of power outages, as well as remotely disconnecting or connecting meters. (Amin et al. 2015). However, these meters are often installed in physically insecure areas, making them vulnerable to attack (U.S Department of Energ 2010). This is a feature common to many cyber-physical systems, where physical sensors or components are vulnerable to physical attacks.

Although there are many possible attack vectors in an smart grid system, a large portion of research has been focused on the problems of *electricity theft*, as well as disruptions in *energy distribution*. Electricity theft in smart grid systems can be the result of either cyber or physical attacks on the smart meters, where an attacker can tamper with the meter's low-level components, alter the meter's long-term storage, or spoof transmissions to the power utility (McLaughlin et al. 2010). While some solutions to this problem are tamper-evident seals and balance meters, these can be easily defeated. This problem can be formulated as a theft-detection game' Cardenas et al. (2012) proposes a multiplayer game, played among a set θ of customers and a single power utility. A customer may be either a malicious user of the system or not. Protecting against theft can be done by investing in meter data management and anomaly detection, increasing redundancy/balance meters, or investing in tamper resistant solutions; however this paper focuses only on the use of anomaly detection. The defender (power utility) is at- tempting to minimize their operation cost, defined to be a function of both the cost of meeting the power consumption demands of consumers as well as the cost of incurred by employing security resources to detect and protect against theft. Factors which go into determining the operational cost of managing the anomaly detector are the cost of resources used when dealing with alarms (which in turn is a function of the amount of effort e the defender puts in their response), the probability of detecting fraud (which increases with the effort and with the frequency with which malicious users will attempt to steal power) as well as the probability density function determining the scale of theft occurring (the amount of electricity stolen by the attacker). They compute a Nash equilibrium which allows for the determination of the optimal false alarm rate (a measure of the ratio of utility effort to stolen power). The model is extremely detailed in terms of accurately modeling the physical control system, however the defender's actions are left quite abstract and amount to an effort parameter in the anomaly detector.

Disruptions in energy distribution are another serious problem, as they can result in loss of voltage regulation, incurring large costs for the power utility as they are forced to perform load control or face the threat of brownouts or blackouts in the grid. One area that this problem has been studied in, is smart grids involving distributed energy systems, which generate or store energy in a decentralized fashion, relying on a grid of small-scale distributed energy resources (DER) (Consortium on Energy Restructuring 2007). Shelar and Amin (2015) proposes a sequential Stackelberg game to address vulnerabilities in the DER networks. Their threat model is of an attacker who's main objective is to cause loss of voltage regulation and force the power utility to perform load curtailment in response. The actions of the attacker and defender are as follows; the attacker can disconnect DER nodes from the system by manipulating the set-points of their inverters. The defender can perform load balancing on the connected nodes. Computing the equilibrium strategies allows for the determination of the optimal load control response for the defender. While the model itself is not that complex, a key contribution of this work is that it offers insight into optimal strategies for the attacker, and they show that when considering an intelligent adversary, the nodes downstream to the control center are much more vulnerable the upstream nodes. This result shows that even if game theoretic models are not able to provide complete security solutions they can often provide valuable insight which can be leveraged to develop other solutions.

Another critical piece of infrastructure in cyber physical systems, perhaps the most relevant to maritime security, are **industrial control systems (ICS)**, which are charged with the control, monitoring and distribution of physical and chemical processes. In (Zhu and Basar 2012), the authors establishes a game theoretic framework in order to facilitate the design of resilient control systems. A resilient control systems is one with the ability to maintain state awareness and operational normalcy in response to disturbances, and threats. (Rieger, Gertman, and McQueen 2009). The interaction between the cyber and physical system is modeled as a set of coupling equations (Zhu and Basar 2011). Each system, both the cyber and physical, are modeled as a separate game; the physical system is a feedback controller modeled using a zero-sum game, where the defender is minimizing a cost function, following state feedback control law, state- feedback disturbance law (parameters like the control input, physical state dynamics and disturbances) as functions of time, all taken as an expectation value over the state of the (cyber system) controller θ which can be operational or in various states of failure.

The cyber system is modeled using a stochastic, zero sum, dynamic game (Zhu and Basar 2009), where the defender is minimizing a cost function determined by the state of Markov decision process which models the cyber system. The objective functions defined for each system are coupled using the sets of coupled optimality equations are solved in order to find the optimal control policies.

Models in this field generally combine game theory and complex control frame- works in order to properly model both the adversary and defender decision making as well as the responses from the physical components. Much of the complexity is introduced by the control equations which influence both the defender and attacker's utility. This is both a strength and a weakness for these kinds of models; since much of the structure relies on the control equations, the models do not generalize very easily and are highly specialized to the problem they are trying to solve. However, this allows them to create models with much more realism and fidelity with respect to real world situations.

Conclusion

Security is recognized as a world-wide challenge and game theory is an increasingly important paradigm for reasoning about complex security resource allocation. We have shown that the general model of security games is applicable (with appropriate variations) to varied security scenarios. There are applications deployed in the real world that have led to a measurable improvement in security. We presented approaches to address four significant challenges: scalability, uncertainty, bounded rationality and field evaluation in security games. Cyber-security provides additional challenges that include limited observability and deception.

In short, we introduced specific techniques to handle each of these challenges. For scalability, we introduced three approaches: (i) incremental strategy generation for ad- dressing the problem of large defender strategy spaces; (ii) double oracle incremental strategy generation w.r.t large defender & attacker strategy spaces; (iii) compact representation of strategies for the case of mobile resources and moving targets; and (iv) cutting plane (incremental constraint generation) for handling multiple boundedly rational attacker. For handling uncertainty we introduced two approaches: (i) dimensionality reduction in uncertainty space for addressing a unification of uncertainties; and (ii) Markov Decision Process with marginal strategy representation w.r.t dynamic execution uncertainty. In terms of handling attacker bounded rationality, we propose different behavioral models to capture the attackers' behaviors and introduce human subject experiments with game simulation to learn such behavioral models. Finally, for addressing field evaluation in real-world problems, we discussed two approaches: (i) data from deployment; and (ii) mock attacker team.

While the deployed game theoretic applications have provided a promising start, significant amount of research remains to be done. In particular, cyber security provides challenges and opportunities in modeling multiple agents interacting in the extremely complicated cyber world. In particular, we identified several important challenges that are related to the maritime sector. With regards to general network security we dis- cussed (i) dealing large volumes of traffic and suspicious activity, (ii) the economics of security, and (iii) advanced persistent threats. We also discussed the importance of modeling the human players in the game, looking at both the (iii) threat model as well as dealing with the (iiii) human in the loop. Lastly we looked at some of the challenges in (iiiii) cyber-physical systems.

These are large-scale interdisciplinary research challenges that call upon multi- agent researchers to work with researchers in other disciplines, be "on the ground" with domain experts and examine real-world constraints and challenges that cannot be abstracted away.

Acknowledgment

This research was supported by MURI Grant W911NF-11-1-0332 and by the United States Department of Homeland Security through the Center for Risk and Economic Analysis of Terrorism Events (CREATE) under grant number 2010-ST-061-RE0001. We wish to acknowledge the contribution of all the rangers and wardens in Queen Elizabeth National Park to the collection of law enforcement monitoring data in MIST and the support of Uganda Wildlife Authority, Wildlife Conservation Society and Mac Arthur Foundation, US State Department and USAID in supporting these data collection financially.

References

Amin, Saurabh, Galina A. Schwartz, Alvaro A. Cardenas, and S. Shankar Sastry. 2015. "Game-Theoretic Models of Electricity Theft Detection in Smart Utility Networks: Providing New Capabilities With Advanced Metering Infrastructure." *Control Systems, IEEE* 35(1): 66–81.

An, Bo, Milind Tambe, Fernando Ordóñez, Eric Shieh, and Christopher Kiekintveld. 2011. "Refinement of Strong Stackelberg Equilibria in Security Games." In *Proceedings of the 25th AAAI Conference on Artificial Intelligence, San Francisco, CA*, 587–593. AAAI Digital Library.

Barnhart, Cynthia, Ellis L. Johnson, George L. Nemhauser, Martin W. P. Savelsbergh, and Pamela H. Vance. 1998. "Branch-and-Price: Column Generation for Solving Huge Integer Programs." *Operations Research* 46(3): 316–329.

Blocki, Jeremiah, Nicholas Christin, Anupam Datta, Ariel D. Procaccia, and Arunesh Sinha. 2013. "Audit Games." In *Proceedings of the 23rd International Joint Conference on Artificial Intelligence*, Beijing, China.

Blocki, Jeremiah, Nicholas Christin, Anupam Datta, Ariel D. Procaccia, and Arunesh Sinha 2015. "Audit Games With Multiple Defender Resources." In *Proceedings of the 29th AAAI Conference on Artificial Intelligence, Austin, TX*, 791-797. AAAI Digital Library.

Breton, M., A. Alg, and Alain Haurie. 1988. "Sequential Stackelberg Equilibria in Two-Person Games." *Optimization Theory and Applications* 59(1): 71–97.

Brunswik, Egon. 1952. *The Conceptual Framework of Psychology*. Volume 1. University of Chicago Press.

Carathédory, C. 1911. "Uber Den Variabilitäsbereich Der Fourier'schen Konstanten von Positiven Harmonischen Funktionen." *RendicontidelCircolo Matematico di Palermo (1884-1940)* 32(1): 193–217.

Cárdenas, Alvaro, Saurabh Amin, Galina Schwartz, Roy Dong, and Shankar Sastry. 2012. "A Game Theory Model for Electricity Theft Detection and Privacy-Aware Control in Ami Systems." In *Proceedings of the 50th Annual Allerton Conference on Communication, Control, and Computing (Allerton), Monticello, IL*, 1830–1837. IEEEXplore Digital Library.

Chandran, R. and G. Beitchman. 2008. "Battle for Mumbai Ends, Death Toll Rises to 195." *Times of India*. November 29. http://articles.timesofindia.indiatimes.com/2008-1129/india/27930171_1_taj-hotel-three-terrorists-nariman-house.

Conitzer, Vincent and Tuomas Sandholm. 2006. "Computing the Optimal Strategy to Commit Too." In *Proceedings of the 7th ACM Conference on Electronic Commerce (ACM-EC), Ann Arbor, MI*, 82–90. ACM: New York.

Consortium on Energy Restructuring. 2007. "Distributed Generation—Educational Modules." Virginia Tech. http://www.dg.history.vt.edu/.

Delle Fave, Francesco M., Matthew Brown, Chao Zhang, Eric Shieh, Albert X. Jiang, Heather Rosoff, Milind Tambe, and John P. Sullivan. 2014. "Security Games in the Field: An Initial Study on a Transit System." In *Proceedings of the 2014 International Conference on Autonomous Agents and Multiagent Systems, Istanbul, Turkey*, 1363-1364. International Foundation for Autonomous Agents and Multiagent Systems: Richland, SC.

Delle Fave, Francesco M., Albert Xin Jiang, Zhengyu Yin, Chao Zhang, Milind Tambe, Sarit Kraus, and John P. Sullivan. 2014. "Game-Theoretic Security Patrolling with Dynamic Execution Uncertainty and a Case Study on a Real Transit System." *Journal of Artificial Intelligence Research* 50(1): 321-367.

Durkota, Karel, Viliam Lisý, Branislav Bošanský, Christopher Kiekintveld. 2015. "Game Theoretic Algorithms for Optimal Network Security Hardening Using Attack Graphs." In *Proceedings of the 2015 International Conference on Autonomous Agents and Multiagent Systems, Istanbul, Turkey*,1773-1774. International Foundation for Autonomous Agents and Multiagent Systems: Richland, SC.

Fang, Fei, Albert Xin Jiang, and Milind Tambe. 2013. "Optimal Patrol Strategy for Protecting Moving Targets with Multiple Mobile Resources." In *Proceedings of the 2013 International Conference on Autonomous Agents and Multiagent Systems, Saint Paul, MN*, 957-964. International Foundation for Autonomous Agents and Multiagent Systems: Richland, SC.

Feng, Xiaotao, Zizhan Zheng, Pengfei Hu, Derya Cansever, and Prasant Mohapatra. "Stealthy Attacks Meets Insider Threats: A Three-Player Game Model." Technical Report, University of California, Davis, USA. http://spirit.cs.ucdavis.edu/pubs/tr/TechnicalReportMilcom15.pdf

Jain, Manish, Erim Kardes, Christopher Kiekintveld, Fernando Ordóñez, and Milind Tambe. 2010. "Security Games with Arbitrary Schedules: A Branch and Price Approach." In *Proceedings of the 24th AAAI Conference on Artificial Intelligence*, 792–797. AAAI Digital Library.

Jain, Manish, Dmytro Korzhyk, Ondřej Vaněk, Vincent Conitzer, Michal Pěchouček, and Milind Tambe. 2011. "A Double Oracle Algorithm for Zero-Sum Security games on Graphs." In *Proceedings of the 10th International Conference on Autonomous Agents and Multiagent Systems, Taipei, Taiwan*, 327-334. International Foundation for Autonomous Agents and Multiagent Systems: Richland, SC.

Jain, Manish, Milind Tambe, and Vincent Conitzer. 2013. "Security Scheduling for Real-World Networks." In *Proceedings of the 2013 International Conference on Autonomous Agents and Multiagent Systems, St. Paul, MN*, 215-222. International Foundation for Autonomous Agents and Multiagent Systems: Richland, SC.

Jain, Manish, Jason Tsai, James Pita, Christopher Kiekintveld, Shyamsunder Rathi, Milind Tambe, and Fernando Ordóñez. 2010. "Software Assistants for Randomized Patrol Planning for the LAX Airport Police and the Federal Air Marshal Service." *Interfaces* 40(4): 267–290.

Keteyian, Armen. 2010. "TSA: Federal Air Marshals." Accessed February 1, 2011. http://www.cbsnews.com/stories/2010/02/01/earlyshow/main6162291.shtml.

Kevin D. Bowers, Marten van Dijk, Robert Griffin, Ari Juels, Alina Oprea, Ronald L. Rivest, Nikos Triandopoulos. 2012. "Defending Against the Unknown Enemy: Applying Flipit to System Security." http://www.emc.com/emc-plus/rsa-labs/staff-associates/defending-against-the-unknown-enemy.htm.

Kiekintveld, Christopher, Manish Jain, Jason Tsai, James Pita, Milind Tambe, and Fernando Ordóñez. 2009. "Computing Optimal Randomized Resource Allocations for Massive Security Games." In *Proceedings of the 8th International Conference on Autonomous Agents and Multiagent Systems, Budapest, Hungary*, 689–696. International Foundation for Autonomous Agents and Multiagent Systems: Richland, SC.

Korzhyk, Dmytro, Vincent Conitzer, and Ronald Parr. 2010. "Complexity of Computing Optimal Stackelberg Strategies in Security Resource Allocation Games." In *Proceedings of the 24th AAAI Conference on Artificial*

CHAPTER 26

Intelligence, Atlanta, Georgia, 805–810. AAAI Digital Library.

Laszka, Aron, Gabor Horvath, Mark Felegyhazi, and Levente Buttyan. 2013. "Flipthem: Modeling Targeted Attacks with Flipit for Multiple Resources." Technical report. http://aronlaszka.com/papers/laszka2014flipthem.pdf

Leitmann, G. 1978. "On Generalized Stackelberg Strategies." *Optimization Theory and Applications* 26(4): 637–643.

Lye, Kong-wei, and Jeannette Wing. 2005. "Game Strategies in Network Security." *International Journal of Information Security* 4(1-2): 71-86.

McFadden, Daniel. 1972. "Conditional Logit Analysis of Qualitative Choice Behavior." 105-142. https://eml.berkeley.edu/reprints/mcfadden/zarembka.pdf

McKelvey, Richard D. and Thomas R. Palfrey. 1995. "Quantal Response Equilibria for Normal Form Games." *Games and Economic Behavior* 10(1): 6–38.

McLaughlin, S., D. Podkuiko, S. Miadzvezhanka, A. Delozier, and P. McDaniel. 2010. "Multi-Vendor Penetration Testing in the Advanced Metering Infrastructure." In *Proceedings of the 26th Annual Computer Security Applications Conference, New York, NY*, 107–116. ACM.

McMahan, H. Brendan, Geoffrey J. Gordon, and Avrim Blum. 2003. "Planning in the Presence of Cost Functions Controlled by an Adversary." In *Proceedings of the Twentieth International Conference on Machine Learning, Washington, D.C.*, 536-543. AAAI Press: Menlo Park, CA.

Moayedi, Behzad Zare and Mohammad Abdollahi Azgomi. 2012. "A Game Theoretic Framework for Evaluation of the Impacts of Hackers Diversity on Security Measures." *Reliability Engineering & System Safety* 99: 45-54.

Nguyen, Than Hong, Rong Yang, Amos Azaria, Sarit Kraus, and Milind Tambe. 2013. "Analyzing the Effectiveness of Adversary Modeling in Security Games." In *Proceedings of the 25th AAAI Conference on Artificial Intelligence, Bellevue, WA*, 718-724. AAAI Digital Library

Nochenson, Alan, Jens Grossklags. 2013. "A Behavioral Investigation of the Flipit Game." In *12th Workshop on the Economics of Information Security, Washington, D.C.* http://www.econinfosec.org/archive/weis2013/papers/NochensonGrossklagsWEIS2013.pdf

Paruchuri, Praveen, Jonathan P. Pearce, Janusz Marecki, Milind Tambe,

Fernando Ordóñez, and Sarit Kraus. 2008. "Playing Games with Security: An Efficient Exact Algorithm for Bayesian Stackelberg Games." In *Proceedings of the 7th Annual Joint Conference on Autonomous Agents and Multiagent Systems, Estoril, Portugal*, 895-902. International Foundation for Autonomous Agents and Multiagent Systems: Richland, SC.

Pawlick, J., S. Farhang, and Q. Zhu. 2015. "Flip the Cloud: Cyber-Physical Signaling Games in the Presence of Advanced Persistent Threats." CoRR abs/1507.00576.

Pita, James, Harish Bellamane, Manish Jain, Chris Kiekintveld, Jason Tsai, Fernando Ordóñez, and Milind Tambe. 2009. "Security Applications: Lessons of Real-World Deployment." ACM SIGecom Exchanges 8(2). http://www.sigecom.org/exchanges/volume_8/2/pita.pdf

Pita, James, Manish Jain, Janusz Marecki, Fernando Ordóñez, Christopher Portway, Milind Tambe, Craig Western, Praveen Paruchuri, and Sarit Kraus. 2008. "Deployed ARMOR protection: The Application of a Game-Theoretic Model for Security at the Los Angeles International Airport." In *Proceedings of the 7th Annual Joint Conference on Autonomous Agents and Multiagent Systems, Estoril, Portugal*, 125-132. International Foundation for Autonomous Agents and Multiagent Systems: Richland, SC.

Pita, James, Richard John, Rajiv Maheswaran, Milind Tambe, and Sarit Kraus. 2012. "A Robust Approach to Addressing Human Adversaries in Security Games." In *Proceedings of the 11th International Conference on Autonomous and Multiagent Systems, Valencia, Spain*, 1297-1298. International Foundation for Autonomous Agents and Multiagent Systems: Richland, SC.

Rieger, Craig G., David I. Gertman, and Miles A. McQueen. 2009. "Resilient Control Systems: Next Generation Design Research." In *Proceedings on the 2nd Conference on Human System Interactions, Catania, Italy*, 632–636. Institute of Electrical and Electronics Engineers (IEEE).

U.S. Department of Energy. 2010. "NISTR 7628 - Guidelines for Smart Grid Cyber Security: Vol. 2, Privacy and the Smart Grid." (July). The Smart Grid Interoperability Panel Cyber Security Working Group. https://www.smartgrid.gov/document/nistr_7628_guidelines_smart_grid_cyber_security_vol_2_privacy_and_smart_grid.html

Shelar, Devendra and Saurabh Amin. 2015. "Analyzing Vulnerability of Electricity Distribution Networks to DER Disruptions." In *Proceedings of the*

2015 American Control Conference (ACC), Chicago, IL, 2461–2468. IEEE Explore.

Shieh, Eric, Bo An, Rong Yang, Milind Tambe, Craig Baldwin, Joseph DiRenzo, Ben Maule, and Garrett Meyer. 2012. "PROTECT: A Deployed Game Theoretic System to Protect the Ports of the United States." In *Proceedings of the 11th International Conference on Autonomous Agents and Multiagent Systems, Valencia, Spain*, 13-20. International Foundation for Autonomous Agents and Multiagent Systems: Richland, SC.

van Dijk, Marten, Ari Juels, Alina Oprea, and Ronald L. Rivest. 2012. "Flipit: The Game of 'Stealthy Takeover.'" ePrint Archive. Report 2012/103. http://eprint.iacr.org/.

van Dijk, Marten, Ari Juels, Alina Oprea, and Ronald L. Rivest. 2013. "Flipit: The Game of Stealthy Takeover." *Journal of Cryptology* 26(4): 655–713.

Vaněk, Ondřej, Zhengyu Yin, Manish Jain, Branislav Bošanský, Milind Tambe, and Michal Pěchouček. 2012. "Game-Theoretic Resource Allocation for Malicious Packet Detection in Computer Networks." In *Proceedings of the 11th International Conference on Autonomous Agents and Multiagent Systems, Valencia, Spain*, 905-912. International Foundation for Autonomous Agents and Multiagent Systems: Richland, SC.

von Stackelberg, Heinrich. 1934. *Marktform und Gleichgewicht*. Vienna: Springer.

von Stengel, Bernhard and Shmuel Zamir. 2004. "Leadership with Commitment to Mixed Strategies." Technical Report. CDAM Research Report, LSE-CDAM-2004-01. Centre for Discrete and Applicable Mathematics. http://www.cdam.lse.ac.uk/Reports/Abstracts/cdam-2004-01.html

Yin, Zhengyu, Manish Jain, Milind Tambe, and Fernando Ordóñez. 2011. "Risk-Averse Strategies for Security Games with Execution and Observational Uncertainty." In *Proceedings of the 25th AAAI Conference on Artificial Intelligence, San Francisco, CA*, 758–763. AAAI Digital Library.

Yin, Zhenyu, Albert Xin Jiang, Matthew Johnson, Milind Tambe, Christopher Kiekintveld, Kevin Leyton-Brown, Tuomas Sandholm, and John P. Sullivan. 2012. "TRUSTS: Scheduling Randomized Patrols for Fare Inspection in Transit Systems." In *Proceedings of the 24th Conference on*

Innovative Applications of Artificial Intelligence, Toronto, Canada, 2348-2355.

Zhu, Q. and T. Basar. 2012. "A Dynamic Game-Theoretic Approach to Resilient Control System Design for Cascading Failures." In Proceedings of the 1st International Conference on High Confidence Networked Systems, HiCoNS '12, New York, NY, USA, pp. 41–46. ACM.

Zhu, Q. and T. Basar. 2009. "Dynamic Policy-Based Ids Configuration." In Decision and Control, 2009 held jointly with the 2009 28th Chinese Control Conference. CDC/CCC 2009. Proceedings of the 48th IEEE Conference on, pp. 8600–8605.

Zhu, Q. and T. Basar. 2011. "Robust and Resilient Control Design for Cyber-Physical Systems with an Application to Power Systems." In Decision and Control and European Control Conference (CDC-ECC), 2011, Orlando, FL, December 12-15.

CHAPTER 27:
MARITIME CYBER SECURITY:
WHAT ABOUT DIGITAL FORENSICS?

Scott Blough and Gordan Crews
Tiffin University

INTRODUCTION

The concept of security has exponentially evolved over the millennia. A central component of this evolution has always revolved around recognizing threat vectors (e.g., a path or means by which an attacker can gain access to a potential victim) and posturing against them (Bowen, Hash, and Wilson 2006). Part of the recognition of a threat vector often involved the success of that threat vector, or one closely associated. After a successful attack from a ground-based carnivore (e.g., tiger or lion), early human beings discovered that retreating to the trees was an effective defense. Although, this security tactic was not effective against carnivores that had the ability to climb trees. The response to that threat was to utilize enclosures for effective defense. Human use of enclosures evolved from simply finding caves to building structures as the threat vectors evolved.

The evolution of the threat vectors encompassed not only physical threats but also threats to property. To protect both, the structures became more sophisticated and resulted in the concept of defense-in-depth (i.e., the coordinated use of multiple security countermeasures to protect the integrity of the information assets in an enterprise), which utilized multiple walls, each stronger than the first, to mitigate a variety of threats (Easttom 2014). Castles are an excellent example of the defense-in-depth strategy as they were designed for multiple threat vectors, including siege weapons and siege warfare. Often, the evolution of security is viewed as a linear process brought about by innovation. This popular view fails to recognize the importance of one of the main components of security: forensics.

Forensics has many definitions, but it is simply the analysis of an incident (Digital Evidence and Forensics 2016). Early humans could use analysis to discover that tree-climbing carnivores remained a threat even though they had mitigated the ground-dwelling carnivore threat. Using that analysis, they moved toward using a structure as security. Further use of analysis of an incident can be seen in the evolution of defense-in-depth construction of castles. Thus, forensics has and continues to be a very important component of security.

Incidents are time constrained. They do not last forever. Thus, for purposes of this article, an incident can be divided into three separate, but very much overlapping, parts:

- Pre-Incident—What types of security steps are taken before the incident;
- During-Incident—What incident response steps are taken during the incident;
- Post-Incident—What analysis or reconstruction steps are taken after the incident.

The below is a graphical representation (see Figure 1) of this view of the incident. This process can be best represented as a cycle diagram. This suggests that information from the parts of the incident should be considered when designing the evolution of a security posture.

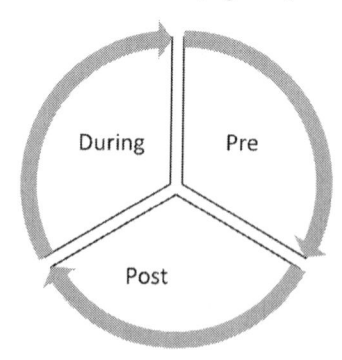

An Incident (Figure 1)

A Scenario: An Unknown Disgruntled Employee

To better illustrate how this applies to maritime cyber security, consider the following scenario involving the Maritime Transportation System.

Pre Incident: The Last Voyage

Shipping company *Shipping Are Us, Inc.* (SAU) has been contracted by logis-

CHAPTER 27

tics company *Logistics Are Us, Inc.* (LRU) to move 10,000 TEU containers (i.e., 20-food equivalency unit) from New York, NY to Houston, TX. There are 1,000 TEU containers from 10 different companies that comprise the total cargo. After off-loading at Houston, the intermodal company *Pretty Darn Quick, Inc.* (PDQ) will ship the containers to 50 different vendors throughout the Western United States.

The Captain of the US M/V OOPS prepares the ship to leave New York and conducts the normal pre-voyage inspections. The ship has a crew of 15, each of which has many years of experience. All systems are functioning properly, and the US M/V OOPS begins the voyage down the East Coast of the United States.

DURING INCIDENT: INTRODUCTION OF MALWARE

As the ship steams along, one of the senior crew members, Ima Crook, is relieved from watch at 2400 and heads toward the galley. Ima stops in the ships communications room to check company email and validate TEU container manifest, as he has done for 15 years with SAU. This time, however, he inserts a thumb drive into the USB port. After about 3 minutes, he removes it and continues his work on the computer.[1] Ima closes his email at 0020 and makes his way to the galley.

Unfortunately for the Captain of the US M/V OOPS, SAU Human Resources did not tell him that Ima has been told that they are downsizing and that he will be let go in 30 days. Thus, Ima still performs his normal duties on the OOPS. In addition to being let go, Ima has a wife and four children, one of which has a genetic condition that requires extensive medical care. In addition to losing his job, Ima will be losing the insurance that allows him to provide care for his child. Ima also has connections to several competing shipping companies that have targeted some of SAU's longtime customers.

The rest of the voyage is mundane, and the US M/V OOPS arrives in Houston where the containers are off-loaded to LRU for transport. It takes about 2 days for the first deliveries of the OOPS's cargo to arrive at the purchasing vendor facilities. It is then that the complaints begin to hit LRU and SAU. As the containers are traced to the OOPS, the Captain is called by SAU's corporate headquarters. The Captain is informed that they are conducting an internal investigation into his management of the OOPS based on multiple vendor complaints.

---

1 For more information on the handling of cyber threats posed by employees, see https://insights.sei.cmu.edu/insider-threat/2015/07/handling-threats-from-disgruntled-employees.html.

Post Incident: The Impact on an Intermodal System

As with most ships transporting cargo, the OOPS uses Intermodal freight transport (Muller 1999). This involves the transportation of freight in an intermodal container or vehicle, using multiple modes of transportation (rail, ship, plane, and truck), without any handling of the freight itself when changing modes. This method reduces cargo handling, and in doing so improves security, reduces damage and loss, and allows freight to be transported faster (Muller 1999). Reduced costs over road trucking is the key benefit for inter-continental use. This may be offset by reduced timings for road transport over shorter distances.

The malware introduced by Ima Crook was a simple program which would affect the tracking numbers of all cargo being transported by this ship. The malware simply accessed the cargo tracking database and switches the cargo tracking number on containers.

Off-Load

As stated above, it may take several days or weeks for the final off-loading of a piece of cargo occurs at its "final" destination. This off-loading may involve extensive costs and/or machinery at a particular location. If a piece of cargo arrives at the wrong location, that location may not be equipped to deal with that cargo—thus additional effort may be incurred to off-load the cargo.

Inspection

Obviously, once a piece of cargo is off-loaded at a site, it will be inspected. Such off-loading and inspection will require manpower, time, and expense. If the correct piece of cargo is being inspected, the requirements of said inspection will be expected. If a highly paid inspector is there to inspect extremely expensive pieces of technology, but opens a crate full of bananas instead, their time and expertise will be wasted. In turn, the produce manager who is expecting a crate of bananas will have no idea what to do with 100 laptop motherboards.

Destination

Thus, the cargo destined for San Diego, California ends up in Tiffin, Ohio and the cargo destined for Toronto, Canada ends up in Tijuana, Mexico. The benefits of the intermodal system (e.g., no direct handling, no direct inspection, and speed) have resulted in very costly mistakes which must be absorbed by the shipping partner.

The cargo in Tiffin, Ohio will need to get to San Diego, California and the cargo from Tijuana, Mexico will need to get to Toronto, Canada as soon as possible. The cost of this delivery is only the beginning of a very expensive mistake if the cargo is perishable or time specific. The nature of the cargo and additional length of delivery time may bring about many lawsuits as well.

Incident Response

No type of response can occur until an incident becomes known or is detected. As discussed, it may be days or weeks after an attack occurs that anyone becomes aware of its occurrence, much less its impact.

Detection

Before any positive actions can be taken in regards to a negative event, the negative event must be known. Given this scenario, the detection of something being wrong may be on a port dock one month after an item has been shipped and 2 months after its original ordering.

Reconstruction of Events

As with any negative incident, people want to know "why" it has occurred. When this involves extremely expensive perishable items, they will certainly want to know who is responsible. As with any type of investigation, the best way to determined "how" and "why" something has happened is to trace the steps leading to the negative incident. This will most often involve reconstructing past events of the movement of the cargo. The first place to start in any type of reconstruction is to examine who has been affected by the incident.

Affected Parties

Initially, the number of affected parties involved in this incident may not be obvious. Depending on the situation or type of incident, there could be a cascading effect to this individual act.

Original Company

First, there is the original company which accepted the order for a product and trusted the shipping company to deliver the product to the buyer. Ultimately, the original company will be the entity responsible to the customer in that they guaranteed proper delivery. In addition to financial impact of this situation, their reputation will also be damaged through no fault of their own.

Purchasing Company

Secondly, the purchasing company will be negatively impacted. The product they purchased may be items that they desperately need to do their own business. They may also be an "in-between" entity who has contracts with others to get ordered products moved to the next buyer. They will be impacted financially, and their reputation with others will be negatively impacted.

Shipping Company

The shipping company will inevitably have to respond to this incident in that they have obligations to the Seller Company and Buyer Company. Their reputation will also suffer in that others may question their reliability in the future.

Port Management Company

Those managing the movement of a shipment on the ground at a port will also suffer in such an incident. Such an incident will inevitably cost time, money, and manpower trying to determine exactly what has occurred. It will make them question their own processes and that of all associated with the movement of cargo used on a daily basis.

Intermodal Transport Company

As stated earlier, a piece of cargo may travel by land, sea, and air. The company or companies involved in the transportation of just one piece of cargo will also be affected negatively by an incident such as the above scenario. As with all involved, there will be a loss in time and money which illustrates the cascading effects of the incident.

Collection of Evidence

The concept of evidence includes everything that is used to determine or demonstrate the truth of an assertion (Reiber 2016). Giving or procuring evidence is the process of using those things that are either (a) presumed to be true, or (b) were themselves proven via evidence, to demonstrate an assertion's truth. Evidence, ultimately, is how the burden of proof is fulfilled in civil or criminal court.

Evidence collection techniques vary by the evidence type and location (Reiber 2016). The main purpose of evidence collection is to retrieve the evidence "as-is" and without damaging the evidence. In a criminal investigation, rather than attempting to prove an abstract or hypothetical point, the evidence gatherers attempt to determine who is responsible for a criminal act. The focus of criminal evidence is to connect physical evidence and reports of witnesses to a specific person(s).

In the scenario presented previously, this incident would mostly result in the criminal prosecution of Ima Crook for his actions introducing the malware. This scenario would also probably result in several civil lawsuits which

CHAPTER 27

will be handled in national or international courts. In a criminal trial, the evidence will need to be such that proves guilt "beyond a reasonable doubt" and "preponderance of the evidence" in civil court (Reiber 2016).

ORIGINAL DOCUMENTS EVIDENCE

As with any type of criminal or civil investigation, evidence will be crucial in determining who is responsible for any event. The most important type of evidence is tangible evidence. This is an evidence which can be treated as fact; real or concrete (Reiber 2016). It is capable of being touched or felt and has a real substance, a tangible object. The following is a brief overview of the various types of tangible evidence which may be pertinent in such an event.

Contracts

The very first piece of evidence which needs to be examined is the original contract which initiated the entire sale and, thus, the shipment of a piece of cargo. Civilly, this will determine who is *financially responsible* for the event. Criminally, this will begin to expose who is *legally responsible* for this incident.

Bill of Sale

A very important part of this contract will be the bill of sale. A bill of sale is a certificate of transfer of personal property from one individual to another (Ferrell, Hirt, and Ferrell 2014). The bill of sale in this scenario will determine those affected the most in this incident. These entities will be the most concerned as to who is responsible for the incident.

Bill of Lading

The bill of lading, a detailed list of a shipment of goods in the form of a receipt given by the carrier to the person consigning the goods, will also be paramount in the ensuing investigation (Ferrell, Hirt, and Ferrell 2014). This will be used in determining who, at a minimum, the victims of this incident are.

Ship Manifest

The ship manifest will also be of extreme importance in the initial investigation of an incident such as this presented. The ship manifest or customs manifest or "cargo document" is a document listing the cargo, passengers, and crew of a ship, aircraft, or vehicle, for the use of customs and other officials (Muller 1999). In this scenario, this manifest will allow the investigators to begin to determine the extent of the incident and all the affected parties.

Port Company Manifest

As with the ship manifest, the port company manifest will allow investigators to determine where the cargo they have received originated. It will also allow the determination of the actual correct destination of the cargo which has been received.

Intermodal Transport Company Manifest

Ultimately, it will be determined that somewhere along the transportation of a particular piece of cargo, something went wrong. This manifest is where investigators will ultimately determine a breach of security has occurred. This is where it will be determined that the malware was introduced which derailed the shipment(s). This is the piece of evidence which will lead investigators to determine those who are criminally and financially responsible for this incident.

Forensic Digital Evidence

Per the National Institute of Justice, "digital evidence is information stored or transmitted in binary form that may be relied on in court. It can be found on a computer hard drive, a mobile phone, a personal digital assistant (PDA), a CD, and a flash card in a digital camera, among other places" (Digital Evidence and Forensics 2016). Digital evidence can be easily contaminated if the device is accessed before it is properly imaged. Thus, digital forensics investigators must use a write-blocking device while imaging the suspect hard drive to prevent evidence corruption. This is because any time a digital device is accessed, the file structure of the device changes. Thus, accessing the device would alter the original evidence. To prove that the image of the device is an exact copy of the original, digital forensics investigators must use a method known as hashing. Hashing is described by Rothstein, Hedges, and Wiggins (2007) as:

> A unique numerical identifier that can be assigned to a file, a group of files, or a portion of a file, based on a standard mathematical algorithm applied to the characteristics of the data set. The most commonly used algorithms, known as MD5 and SHA, will generate numerical values so distinctive that the chance that any two data sets will have the same hash value, no matter how similar they appear, is less than one in one billion. "Hashing" is used to guarantee the authenticity of an original data set and can be used as a digital equivalent of the Bates stamp used in paper document production.

This method ensures that the evidence is copied in a bit for bit manner and allows the forensic investigator to perform the investigative analysis on the copy, thus preserving the original evidence.

Network Logs

One of the main objectives of a digital forensic investigation is to identify who was responsible for the incident. This can be most effectively accomplished in the digital arena by using logs. According to Easttom (2015), "a device's log files contain the primary records of a person's activities on a system or network." In the scenario presented above, Ima logged into the shipboard network using his own credentials. Thus, it would be relatively easy to identify the time and network access point by checking the system logs. System logs provide records that indicate access by the user account and what actions were performed on the system during the event (Kim and Solomon 2016).

In the above scenario, Ima Crook introduced malware onto the system that randomly changed the TEU Container Manifest. Since this was an action taken on the network, it would have been recorded in the application log, which records the date and time that the user accessed the data and the changes made to the data (Kim and Solomon 2016). By reviewing the application logs, a digital forensic investigator would be able to identify the user account, time, and track the changes made during the incident.

Workstation Evidence

Since Ima accessed a shipboard workstation, there are also evidentiary artifacts contained on that system as well. One of the problems that digital forensic investigators encounter during an investigation is the workstation that has multiple users. There are two distinct ways to identify a unique user in this type of circumstance (Reiber 2016). The first way is to use timelines to determine who had access to a workstation. In the above scenario, a ship schedule would be an effective way to begin to eliminate suspects. The ship schedule would enable the forensics investigator to determine who could not have accessed the machine due to physical positioning and post orders on the ship. Once that list is determined, it is relatively easy to narrow the suspect list.

The actual workstation will provide a significant amount of evidence. The forensic investigator must first use proper digital forensics evidence preservation procedures before accessing the files on the suspect workstation (Reiber 2016). The digital evidence will yield information that is helpful in identifying activities that took place during the time of the incident that would link a certain person to the workstation. These include such things as email transactions and online purchases. In addition to utilizing time stamping as evidence, each user on a workstation has a unique identifier that is known as the Security Identifier or SID. The SID is a unique number given to each user on the workstation (Reiber 2016). That number is associated with each file and event transaction that the user makes during the time they are logged on to the workstation. This is the easiest and most effective way to identify and place Ima at the keyboard during the investigation.

In addition to the aforementioned methods, there is also a way to identify the USB drive that Ima placed into the workstation in the above scenario. This would be easily identified by using the workstation's system file, which is contained in the registry. This file enumerates the information on drive letter mappings that would identify the USB drive that Ima attached to the workstation. This would be helpful in proving that Ima was responsible for inserting the USB drive from which the malware originated. Identifying what software Ima used during the malicious act is another important aspect of the investigation. This could be accomplished by utilizing another registry file, NTUSER.DAT, which would provide a list of Ima's most recently used files and would likely indicate the use and change of the TEU Container Manifest database (dependent upon the type of database utilized). Additionally, it would be important in this scenario to prove that Ima had sufficient access to the affected system. This could be done by utilizing the security file contained within the registry. This file contains all of Ima's user and group assigned policies, which would allow for such things as accessing the TEU Container Manifest.

The digital forensics investigator, having used the information described above, would be able to build a very solid case against Ima. This evidence, coupled with the physical evidence described earlier, would be provided to both Human Resources and, if SAU desired, to law enforcement officials.

Mitigation

It is vital that SAU review the evidence of the incident for disciplinary and possible criminal reasons; however, a more important reason to review and reconstruct the incident is for mitigation. SAU does not want this type of incident to happen again, as it has negatively impacted their brand. Much like the earlier discussion about defense in depth, SAU must learn from their mistakes.

Training

One of the key issues in mitigating these types of issues is employee training (Reiber 2016). The human factor has been an issue since the inception of security. According to IBM's 2014 Cyber Security Intelligence Index, approximately 95% of all attacks involved human error. Interestingly, Human Resources, the division of the organization usually responsible for training, views anything related to information technology as an issue for the Information Technology division (Muller 1999). This view often includes employee training. Although the above scenario would not have been stopped by training Ima, there are clearly issues that could have been addressed through training.

The major mistake that SAU made involved Human Resources communication with the Captain of the OOPS. Since the Captain was Ima's supervisor, he should have been notified that Ima was being terminated. In addition to this notification, the IT division should have been consulted so that they could have removed or lessened Ima's user privileges from the network perspective. Thus, a simple written policy and training on that policies execution could have removed Ima's ability to execute this incident and saved SAU an enormous amount of money and brand goodwill.

Access Policies

In addition to something as simple as policy and training, there are other steps that could be taken to prevent such an incident from reoccurring (United States 2012). As SAU reconstructs the incident, they would be wise to examine all the parts of their organization that were effectively compromised. There are several key areas in which SAU could improve their security posture. The first of these is access policies. Access policies govern the way in which organizations manage information system accounts. The National Institute of Standards and Technology (NIST) promulgated useful standards for ABC to consider when revising its current policy. A summary of those standards is provided below:

NIST 800-100 and NIST 800-53r4:

- Identifying account types
- Establishing policies for group membership;
- Identifying authorized users and specifying levels of access privilege;
- Requiring approvals to establish, modify, disable, or remove accounts;
- Authorizing and monitoring the use of guest/anonymous and temporary accounts;
- Notifying managers when temporary accounts are no longer required and when users are terminated, transferred, or access privileges are modified;
- Deactivating: accounts of terminated or transferred users;
- Granting access to the system based on: valid access authorization; intended system usage; and other attributes required by the organization's missions/business functions; and
- Reviewing accounts.

By complying with the aforementioned NIST standard, SAU would have been able to mitigate this incident. There are four specific sections of NIST 800-100 and 800-53r4 that were vital to this incident. They pertain to the establishment of conditions for group membership, review, modification, and deactivation of accounts. A timely completion of account review would have determined that Ima should not have maintained some of the privileges in his user or group access profile. The review would also have indicated that another person be assigned to manage the OOPS Container Manifest. As discussed previously in relation to training, Human Resources should have notified both the Information Technology division and the Captain, in the role of Ima's supervisor, of the impending termination.

Configuration of Hardware and Software

Another area that SAU should focus on is the configuration of hardware and software. One of the easiest and most effective ways to prevent the introduction of malware into a network environment is disable the USB ports on workstations. This is also effective when the focus is on preventing data leakage of sensitive data. By utilizing this simple configuration tool, SAU could have prevented Ima from downloading the malware from his USB drive onto the system.

Another important aspect of configuration in security is software configuration. In SAU's case, their database containing the TEU Container Manifest was compromised. Proper configuration of the database could have prevented the incident from occurring. Databases are utilized for storing and organizing records. In the case of the TEU Container Manifest database, it was used to store information related to the vendor, contents, off-loading, intermodal transport, and final destination. Ima was able to introduce malware into the database that randomized the final destination field. To prevent this, database configuration could have prevented changes to any or all fields unless approved by another user. This would have rendered the malware useless, unless it was specifically written to spoof the other user's approval. If that were the case, a second configuration management tool could have notified a supervisor that the database had been changed. Although this would not have prevented the change, it would have allowed SAU the ability to restore the proper database, thus mitigating the incident.

Auditing

Auditing is an effective security tool for the discovery and, more importantly, the prevention of incidents (Digital Evidence and Forensics 2016). Conducting regular audits allow companies to more effectively understand their security posture. Auditing also provides information on patch management, which is a term used to describe updates to existing software systems. In the scenario mentioned above, Ima was able to introduce malware that infected the database. SAU had not properly patched their database over the past few years, resulting in vulnerabilities to several known database attacks. Since Ima was authorized to access the database, he was able to get information on the database version and simple search of the National Vulnerabilities Database enabled him to identify and download the malware that was utilized during the incident (NVD 2016).

In addition to patch management, auditing would have perhaps identified anomalous behavior from Ima's account. An audit of his account could have raised red flags if he was attempting to gather information before about the database and system before his attack. An analysis of Ima's network traffic could have indicated a violation of the access control issues that were previously discussed. Once this traffic was flagged, the resulting inquiry could have allowed the disparate divisions previously discussed to share information on Ima's status, which could have ultimately prevented the issue.

Physical Security

The physical room in which Ima initiated the attack could have had some physical security restrictions (Kim and Solomon 2016). A key-card access control device would have been useful in identifying Ima as a suspect. A key-card access control device can also limit the times in which employees have access to a certain location, which can be an effective security measure. When employees access the room during restricted hours, a notification or alarm is sent to the appropriate business function. This can trigger further investigation into the activities of that employee. Utilizing time-based access controls is the most effective way to mitigate the insider threat.

Conclusions and Recommendations

Although the incident took place on the US M/V OOPS, which was under the control of the Captain, the vessel's staff had little opportunity to prevent Ima Crook from exploiting SAU's weaknesses. As this scenario illustrates, there were multiple critical errors in SAU's overall security posture. These errors ranged in scope from organizational communication to policy and training to the more technical aspects of network security. The other major mistake that SAU made was not adopting a security posture that utilized threat intelligence and vulnerability analysis. As with a vast majority of companies, SAU put the vast majority of its policy and training efforts toward fulfilling requirements related to its core business. SAU is likely to learn from its mistakes and begin to develop a security posture within the entire organization. To do this, they should initially focus on the following critical areas.

The Information Technology division should develop a patch management program. This program should ensure that both software and hardware updates are identified and installed in a timely manner. This will reduce SAU's vulnerability footprint for both the inside and outside threat.

Although developing a patch management program will reduce the vulnerability footprint, it is not a security panacea. To manage other threats, SAU should establish a protocol for using intrusion detective systems and intrusion prevention systems. These will enable SAU to detect or prevent malicious traffic on their network. These systems are very effective when used in conjunction with a robust organizational security plan.

In support of the intrusion detection and intrusion prevention systems, SAU should implement a ship-board monitoring program to analyze network traffic. This program should monitor both internal and external (coming and going) network traffic. The key to the development of a success program in this area lay in establishing a baseline for the internal and external network traffic. This allows the monitoring function to determine anomalous traffic and would alert identified roles within SAU's organization.

The above scenario would have created a host of issues that had nothing to do with the recovery of evidence. Issues such as vendor notification, press coverage, crisis communication, and continuity of operations, disaster recovery, and financial obligations are but a few. Creating an effective incident response protocol would integrate and coordinate the response to the issues that arise from a scenario such as this. Another important aspect of the incident response protocol involves roles and responsibilities for the investigation. It enables the team members to be properly trained and have the appropriate equipment to successfully complete their assigned roles. This allows for a much more thorough response from an investigative standpoint. It also minimizes the chances of people making critical mistakes that could jeopardize the investigation.

A major part of the incident response plan is the forensics recovery plan. This is an important part for several reasons. The forensic recovery plan will outline roles and responsibilities for team members. It provides for training and equipment to properly conduct the forensic investigation. Training in the forensics recovery plan is not limited to those designated as recovery specialists, it must be organization-wide. This is due to the sensitive and volatile nature of digital evidence. Supervisors must be trained to properly secure devices that may contain evidence. This includes not allowing the employee to access the machine before it is properly processed.

CHAPTER 27

In addition to the technical aspects of the forensics recovery plan, a more mundane aspect is very important. The after-action review of the incident should include a reconstruction of the events that led to the incident. It should also include a review of the steps taken during the response and investigation. This after-action review is vital to the organization's overall security program. Determining how and why and the incident took place allows the organization to mitigate future incidents of the same or similar nature. It also identifies policy and training deficiencies throughout the organization.

Another important part of having an effective security program is to develop a threat intelligence strategy. By completing the aforementioned after-action report, SAU would be able to see with some clarity what the threat was, how it entered, what it did, and how it was mitigated. Perhaps not as clear may be why it happened. SAU should develop a program designed to identify threats to their specific business. That program should include actors, motivations, and desired goals. By doing this, SAU would have better understanding of threat landscape and would be better positioned to integrate that knowledge into their security program.

Finally, the most important part of the security program is organizational communication. The above scenario highlighted the lack of communication between divisions that resulted in Ima's ability to execute the attack. Aside from ensuring that policies exist and training is provided on access control issues, it is vital to include employees in the overall security posture of the organization. Employees need to know what types of threats exist, why they exist, and what happens to SAU if a threat becomes an incident. Training employees on security policies should consist of not only what not to do (such as clicking on a link in a suspicious email), but what the consequences of that click are for themselves and the organization. This tactic raises awareness on the part of the employee, which ultimately raises the security posture of the entire organization.

Perhaps the most important lesson that SAU can learn from this scenario is that there is never a magic bullet for security. Security is an organizational issue. Developing an inclusive security program is the best way to ensure that the organization has defense in depth, which is the multi-layered security platform best illustrated by the medieval castle.

REFERENCES

Bowen, Pauline, Joan Hash, and Mark Wilson. 2006. *Information Security Handbook a Guide for Managers*. Gaithersburg, MD: U.S. Dept. of Commerce, Technology Administration, National Institute of Standards and Technology.

"Digital Evidence and Forensics." 2016. National Institute of Justice. http://www.nij.gov/topics/forensics/evidence/digital/Pages/welcome.aspx (accessed May 10, 2016).

Easttom, Chuck. 2014. *System Forensics, Investigation, and Response*. Burlington, MA: Jones & Bartlett Learning.

Ferrell, O.C., Geoffrey Hirt, and Linda Ferrell. 2014. *Business: A Changing World*. 9th edition. New York, NY: McGraw-Hill.

Kim, David, and Michael G. Solomon. 2016. *Fundamentals of Information Systems Security*. Jones & Bartlett Learning.

Muller, Gerhardt. 1999. *Intermodal Freight Transportation*. 4th edition. New York, NY: Eno Transportation Foundation, Inc./Intermodal Association of North America.

Reiber, Lee. 2016. *Mobile Forensic Investigations: A Guide to Evidence Collection, Analysis, and Presentation*. New York, NY: McGraw-Hill.

Rothstein, Barbara J., Ronald J. Hedges, and Elizabeth Corinne Wiggins. 2007. *Managing Discovery of Electronic Information: A Pocket Guide for Judges*. 1st edition. Federal Judicial Center. http://www.fjc.gov/public/pdf.nsf/lookup/eldscpkt.pdf/$file/eldscpkt.pdf (accessed December 4, 2016).

United States. 2012. "Joint Task Force Transformation Initiative." *Security and Privacy Controls for Federal Information Systems and Organizations*. Gaithersburg, MD: U.S. Dept. of Commerce, National Institute of Standards and Technology.

CHAPTER 28:
MARITIME CYBER THREAT INTELLIGENCE

Tom Gresham
San Diego Unified Port District

Cyber threat intelligence (CTI) is a critical component in a maritime organization's overall risk mitigation strategy. In recent years, the use of threat intelligence has gained popularity as the value of threat prediction and proactive defense building has proven itself effective at mitigating threat activities. CTI provides information that links the probability and impact of a cyber attack by providing a framework for timely analysis and prioritization of potential threats and vulnerabilities given the maritime industry's unique threat landscape. The maritime industry in particular is decentralized and lacks common standards and mechanisms to create and exchange threat intelligence. While basic information can be assessed through monitoring tools and log review, it is not actionable intelligence until it has been analyzed by competent individuals with the appropriate training and education. CTI can assist in providing maritime organizations with the information needed to direct resources to address threats to the industry.

Many industry standard methods exist for the generation, interpretation and application of CTI. CTI is a type of event or profile data that contributes to security monitoring and response. Most maritime organizations with a basic cyber security program already collect similar information such as event logs, firewalls and intrusion detection platforms. Those boasting a more advanced cyber security program fully integrate CTI into response procedures and systems. The need to rapidly exchange CTI is recognized in the development of defined standard data formats that are now growing in popularity with technology providers. These standards will aid in the exchange, analysis and consumption of CTI data. This in turn, will enhance the quality and value of the information making it clear and actionable.

When leveraged correctly, CTI can help maritime organizations at multiple levels. Operational-level personnel may be alerted to attacks at early stages by providing indicators of activity during the various threat attack phases. Cyber security front-line managers are able to analyze and leverage CTI to detect and respond to adversaries already operating within the organization's network. Executive management is also able to leverage CTI in developing strategy-based plans and policies to protect the organization against potential adversaries. Incorporating CTI into a risk management framework reduces risk to an organization's mission and assets through meaningful and actionable intelligence that allows for informed choices that bolster security. The benefits of CTI may be recognized on a larger scale when maritime organizations exchange information through product sharing or direct system-to-system data feeds. It is through the sharing of CTI that ports and other facilities can increase their threat awareness and preempt attacks by taking proactive steps to bolster cyber defenses. The sharing of CTI has become a priority within the U.S. federal, state and local governments as well as the private sector.

What is Cyber Threat Intelligence?

The Carnegie Mellon Software Engineering Institute (SEI) defines Cyber Intelligence as "the acquisition and analysis of information to identify, track, and predict cyber capabilities, intentions, and activities to offer courses of action that enhance decision making" (Townsend and McAllister 2013, 6). More specifically, Cyber Threat Intelligence (CTI) can be thought of as applying intelligence techniques to the aggregation and analysis of contextual and situational risks and that are tailored to the organization's specific threat landscape. The goal of CTI is to provide the ability to recognize and take corrective action upon indicators of attack and exploit scenarios in a timely manner to avoid impact to assets or services. CTI takes into account many sources of information including: the organization's threat landscape, threat actor capabilities, attack surface, as well as data provided by security systems.

The first stage to developing meaningful CTI is defining an organization's unique threat landscape and the identification of the threat actors most likely to attack and their capabilities. This is particularly important for properly matching an integration strategy with the right sources of intelligence. The Intelligence and National Security Alliance (INSA) (2015) describes the threat landscape as the "Operating Environment" with the following definition:

> the Operating Environment helps identify potential adversaries and the level of risk each adversary represents, identify an organization's vulnerability to malicious behavior in a dynamic global environment, and bridge strategic and operational level analyses. (6).

For example, a defense contractor may be a likely target for a nation state while a retail company may be a more attractive target for a cyber-criminal threat due to the nature of the information assets. The threat landscape can be determined through examination of the organization's mission, the value and type of information the organization processes and an examination of incident history. Who has targeted the organization in the past or similar organizations? The maritime industry has been the target of cyber criminals, particularly those assisting narcotics smugglers. Directly attacking cyber systems is often a tactic used to facilitate drug smuggling through the disabling of detection systems. Cyber criminals are also actively engaged in the hijacking of ship and shore inventory systems to surreptitiously move cargo containers that may contain narcotics. The same tactic may also be of interest to terrorist organizations for the smuggling of weapons.

Threat attributes are another source of information for CTI. Threats are commonly measured through two elements or aspects. These aspects of a "threat attribute" are defined as "a discrete characteristic or distinguishing property of a threat. The combined characteristics of a threat describe the threat's willingness and ability to pursue its goal" (Mateski et al. 2012, 14). Specific nation states may be considered threats that possess a high commitment and high capability. Crippling a nation's maritime shipping system can have detrimental economic effects. According to the U.S. Department of Transportation, "in 2011, 53% of all U.S. imported goods arrived via maritime shipping" (Chambers and Liu 2011, 1). In the larger context, "around 90% of world trade is carried by the international shipping industry" (International Chamber of Shipping 2013, 3). Delays in the shipment of goods can lead to consumer inventory, fuel, and medical shortages. Other threats such as narcotics traffickers may have a high commitment but a low degree of capability. These threat attributes feed into CTI in terms of the degree of preparedness and necessary countermeasures to mitigate threat activity.

With increasing automation and interconnectivity, the attack landscape has become more complex and sophisticated. Thus, understanding the vectors with which a threat actor can attack a maritime organization is vital. This is often referred to as the attack surface. Identifying the attack surface for an organization will drive the level of data required to accurately develop and maintain situational awareness. For example, an organization with a mature security architecture will be centralized physically or logically and would therefor require a smaller intelligence framework to protect its resources. Even when taking into account Cloud Services, the same security architecture may be logically extended to the cloud through single sign-on credential control, event reporting and cloud network monitoring. Conversely, a decentralized organization with an immature security architecture, and perhaps covers a larger geographic area, would need more resources to effectively manage its situational awareness. Ports and other maritime facilities are particularly sensitive in this area as most ports operate independently with little control standardization. Many supervisory control and data acquisition (SCADA) systems are often antiquated as they were designed without a robust security framework, leaving them unprotected from cyber attacks. And while many SCADA systems are physically air-gapped from an organization's business network, often times those systems require maintenance through the connection of a modern computing system that may be compromised. Another area of concern is the increased automation of Port operations such as crane loaders. These newer SCADA systems are built at least with the capability to cross-connect with business systems for monitoring and remote control. Elements of an organization's infrastructure including points of entry for an attacker such as perimeter Internet connections, mobile devices, and remote access, should be given special attention when evaluating the attack surface as they may provide easy access to SCADA and other maritime systems.

Along with the evaluation of a maritime organization's environment and attack surface, the ability to measure, process, and provide meaningful data is necessary to provide actionable CTI. Raw data provided by security systems such as firewalls and filters in and of itself is simply feedback on specific events that occur in the moment. These data may be of value to system administrators or others for specific decision actions, but does not provide a comprehensive analysis of the entire infrastructure or any historic trending. Intelligence on the other hand is produced through the meaningful collection and examination of patterns, correlation and tracking of that raw data over time from multiple sources. This intelligence can be leveraged by educated, trained, and experienced analysts and decision makers to have a significant and positive impact on the organization's ability to anticipate breaches before they occur. These decisions will allow an organization to respond rapidly, decisively and effectively to confirmed breaches by implementing countermeasures before and during an attack.

CTI products must provide meaningful and actionable information, modeling the same standards of traditional intelligence analysis. The Office of the Director of National Intelligence (ODNI), Intelligence Community Directive (ICD) 203 (2015) states:

> All intelligence products should follow five analytic standards: Objectivity, Independent of Political Considerations, Timeliness, Based on all available sources of intelligence and exhibits traditional tradecraft standards such as source quality and reliability, logical argumentation, consistency and accuracy (2).

For example, a single CTI product produced for a Cyber Security Manager in the maritime domain may include a report containing a specific observed threat, the sources that provided detection, impact level, timeline of needed action and recommended actions. Other CTI products may be targeted at the operational level such as automated system feeds on observed threats that in turn automatically implement technical countermeasures. CTI products may also be leveraged at the executive level. Report information may be fed into more strategic documents such as a Business Impact Analysis (BIA), Enterprise Risk Strategy or other organizational strategic plan.[1] Moreover, CTI can be used to share threat information between maritime organizations through system-to-system data feeds. While automatic in nature, these feeds require active management by security experts. Protecting feed confidentiality is of critical importance as its accidental disclosure may provide a threat actor with exploitable information.

Cyber Threat Intelligence Sources and Formats

CTI is leveraged by maritime organizations from both internal and external sources in various formats. Many CTI reports are generated by government and vendor organizations such as the Department of Homeland Security's (DHS), Homeland Security Information Network (HSIN). HSIN access is granted to personnel based on enrollment and approval of requests. HSIN provides both open source and government generated CTI in the form of intelligence announcements that are either distributed to subscribing agencies or posted on HSIN interest group websites. This material is often broken down by critical infrastructure sector and access to information is based on the enrollment, vetting, and approval of personnel such as maritime operators. The Maritime Sector is reported on through this portal. HSIN CTI material is often marked UNCLASSIFIED for open source and FOR OFFICIAL USE ONLY (FOUO) for more sensitive information. These products provide information for cyber security stakeholders in terms of the observed threat, dates, recommended actions and all other aspects of an intelligence product. It should be noted that many HSIN products as well as those generated by other government affiliated organizations such as Infragard, are limited. General users such as SCADA system operators will require dissemination from within the organization. This may be accomplished through regular safety briefings and/or other channels such as email.

Organizations may also generate CTI from within. These products are often created by a cyber security team to provide intelligence to senior manage-

---

1 See Appendix A, Figure 1 for a chart identifying the relationship levels of cyber intelligence.

ment, the general user community or to other partner organizations in a human-readable format. Organizations with limited resources, that may not have dedicated cyber security teams, may leverage intelligence products produced through contracted services such as the Multi-State Information Sharing & Analysis Center (MS-ISAC) or through a private security firm.

CTI may also be used to automate defensive mechanisms within a security infrastructure. Many industry technologies provide proprietary report information. Operational security systems such as Intrusion Detection Systems (IDS), Security Information and Event Management (SIEM) systems do generate CTI at a basic level. According to a 2015 SANS survey of organizations, "to get to that visibility, 55% of organizations are currently using SIEM, and 54% are using intrusion monitoring platforms to aggregate, analyze and present CTI" (Shackleford 2015, 12). Automated instructions flow from the SIEM or IDS to the operational security systems such as instructions for blocking a particular threat via network firewalls or host countermeasures.

SIEM systems, for example, ingest and correlate data from disparate systems into a single collection point for pattern analysis. The term "SIEM" was first introduced into the industry by Gartner in 2005 (Williams and Nicolett 2005). It has since gained popularity among organizations hosting cyber security programs. This evolution in security log analysis removed the labor intensive process of manually inspecting systems for event logs. Instead, events and other information types are managed through the automatic collection of log data among multiple systems in real-time. SIEM systems can monitor SCADA and other Industrial Control Systems (ICS), providing information in terms of configuration changes or operation of equipment beyond set thresholds including temperature, load, etc. Such information can assist the maritime industry in the tracking of industrial assets and the assurance that the systems are operating within nominal levels.

While these CTI products and feeds benefit a single organization, the sharing of CTI data becomes a force multiplier as threat information can be quickly disseminated to interested parties. As more organizations share CTI, the intelligence pool becomes richer and the aggregate of information can be leveraged across industry verticals, government institutions, critical infrastructure sectors, etc., in real-time. Maritime organizations seeking to implement a threat intelligence subscription should ensure that the feed addresses the types of threats that are likely to be active in their environment such as the hostile nation states, narcotic smugglers or terrorism. Considerable research has been spent by the U.S. government in attempting to develop universal formats for exchanging CTI across organizations of varying types, sectors, and technology. DHS, the National Institute of Standards and Technology (NIST), and the U.S. Computer Emergency Readiness Team (US-CERT) have aided in the development CTI standards through the MITRE Corporation.

In 2010, the MITRE Corporation was the first organization to introduce a CTI data format standard known as Cyber Observable eXpression (CybOX). These data format provided a common structure for representation cyber observable elements such as malware detection, intrusions and threat attack patterns. By 2013, MITRE introduced a second standard, built upon CybOX, known as Structured Threat Information eXpression (STIX).[2] The STIX language articulates the full range of potential cyber threat information and provides data that is fully expressive, flexible, extensible, automatable, and human-readable. This format is recognized by the industry as many vendors have adopted connectors to interpret and process STIX feeds. Transporting STIX feeds across organizations requires another protocol created by the MITRE Corporation, known as Trusted Automated eXchange of Indicator Information (TAXII). TAXII messages carry the payload of cyber threat data in STIX format. TAXII is defined as, "the protocols and data formats for securely exchanging cyber threat information for the detection, prevention, and mitigation of cyber threats in real time" (Connolly, Davidson, Richard and Skorupka 2012, 3).[3] When these standards are combined and put into practice, one can think of STIX as a language that can use CybOX words, and the communication is possible with TAXII. STIX characterizes what is being told, while TAXII defines how the STIX language is shared. By adopting these CTI data standards, organizations can enhance the ease and speed with which to share information. It allows an entire community of interest to add to and extend the context of threat information and CTI. Different sharing models exist for organizations to tap into CTI feeds. According to Bret Jordan of Blue Coat Systems (2015),

> there are the three basic types of sharing models: Open Source Intelligence (OSI) provider, similar what we have today with URL blacklists; Subscription based private intelligence, usually found in the vendor space and may be tied to products and subscription fees and Open sharing within a restricted ecosystems such as financial services, industrial control systems, governments, vendor alliances. (47).

DHS published support for the STIX and TAXII formats for CTI exchange among organizations (DHS 2015). The STIX and TAXII standards have matured well beyond their initial drafts and first release in 2013. Major security

---

2   See Appendix A, Figure 2 for an illustration of the STIX 1.1.1 architecture.
3   See Appendix A, Figure 3 for an illustration of the TAXII concept.

vendors such as Bluecoat and FireEye have announced support for the formats. Governments, incident responders and CERTs, the Financial Services Information Sharing and Analysis Center (FS-ISAC), and the Industrial Control Systems Information Sharing and Analysis Center (ICS-ISAC) are leveraging STIX and TAXII in their production environments.

In July of 2015, DHS announced that it will transition ownership and maintenance of the CybOX, STIX and TAXII languages to the Organization for the Advancement of Structured Information Standards (OASIS). OASIS is a private global non-profit collaborative with ties to the United Nations and the International Standards Organization. The newly formed OASIS Cyber Threat Intelligence (CTI) Technical Committee will oversee further development of the standards and will promote their adoption to allow threat intelligence to be more readily analyzed and shared among trusted partners and communities. OASIS is an American National Standards Institute (ANSI) accredited developer as well as an authorized Publicly Available Specification (PAS) submitter to the International Standards Organization (ISO). This will allow STIX and TAXII to reach a very broad stakeholder community, paving the way forward for certified standardization.

There exist other CTI formats as well such as Open Threat Exchange (OTX) and Collective Intelligence Framework (CIF). OTX is a publicly available service created by Alien Vault for sharing threat data. CIF is client/server system for sharing threat intelligence data. CIF was developed out of the Research and Education Network Information Sharing and Analysis Center (REN-ISAC). The Managed Incident Lightweight Exchange (MILE) Working Group is working on standards for exchanging incident data. The group works on the data format to define indicators and incidents. This group has defined a package of standards for CTI which includes Incident Object Description and Exchange Format (IODEF), IODEF for Structured Cyber Security Information (IODEFSCI) and Real-time Inter-network Defense (RID). Advantages and disadvantages exist with each format, but two formats have risen to industry standards. According to SANS author Greg Farnham (2013), "if an organization is looking for a package of industry standards then the MILE package (IODEF, IODEF-SCI, RID) or the MITRE package (CybOX, STIX, TAXII) would be suitable. Both have the capability to represent a broad array of data and support sharing of that data" (22).

Cyber Threat Intelligence Benefits to Maritime Organizations

CTI has a direct and powerful benefit for maritime organizations in terms of risk management. CTI products provide the information necessary for decision makers to direct resources to address well-defined threats to a maritime organization and better understand the potential impact of a successful breach through knowledge of their own threat landscape and attack surface. Meaningful intelligence can be produced through the correlation and analysis of massive amounts of data that would otherwise be problematic if not impossible for cyber security specialists to examine manually. Similarly, managers at the executive level benefit from CTI products as they impart concise knowledge that may be acted upon to steer strategic investments in organizational policies, processes and technology. "CTI allows an organization to strategically prioritize countermeasures, focusing efforts on what may cause the most damage. It helps to ensure business continuity and the overall success of the organization" (Ernst and Young 2014, 6). By integrating CTI into various security operations, the resulting intelligence can be used to map out the threat landscape and put historical data into context that is easy to understand and take action upon.

One specific benefit of CTI is the ability to break the Cyber Attack Lifecycle. Most of the major breeches, allegedly perpetrated by nation states, occurred via a threat concept known as advanced persistent threat (APT). Organizations that are highly motivated and have the technical resources to carry out cyber-attacks will often follow a similar pattern. For instance, a nation state may perform a scan on a port's network then develop malicious tools for specifically targeting assets found during the reconnaissance scan. The attacker will then exploit a target's weakness through zero-day attacks which will then allow entry for more sophisticated tools to be delivered, gaining further access within an organization's infrastructure to the eventual valued information such as a sensitive database. NIST is currently drafting special publication (SP) 800-150, "Guide to Cyber Threat Information Sharing (Draft)." The publication illustrates how threat intelligence can be applied to thwart an APT attack. The key to developing a strong defense is to understand the threat's Tactics Techniques and Procedures (TTPs). "To mount an active defense, an organization should understand an adversary's TTP within the cyber attack lifecycle and make use of detailed threat intelligence that is relevant, timely, and accurate" (Johnson, Badger, and Waltermire 2014, 17). Through the use of advanced threat modeling, maritime organizations can characterize their attack surface and use benign threat penetration testing to expose vulnerabilities in likely attack scenarios. Modeling results will provide the organization with instructions on where to deploy countermeasures to block APT attack channels, remediate system vulnerabilities and detect abnormal system behavior. For instance, port SCADA systems may be vulnerable to a TTP similar to Stuxnet. Other attacks are likely to occur on container scanning systems or inventory applications.

Managing risk through the application of CTI at the organization level is beneficial. However, the value of CTI is enhanced greatly through the sharing of information among maritime organizations and through partnerships between the government and private sector shipping companies/tenants. Smaller ports in particular benefit greatly from the ingestion of CTI material from larger more prominent entities that have larger attack surface areas. This collaboration empowers smaller organizations that may lack the budget or expertise to internally generate CTI. To make a comparison, the effectiveness of anti-virus technology relies on the ability to identify and deliver known attack signatures of malicious code such as the famous "BLASTER" and "NIMDA" worms. In prior years, if anti-virus signature databases were up-to-date, organizations had little to fear. With the advent of zero-day attacks, the effectiveness of anti-virus signatures is almost neutralized. Attacks are more coordinated and are occurring across multiple systems before anti-virus vendors can identify and deploy signatures. This is where the sharing of CTI can help mitigate that issue. CTI examines the attack patterns, heuristics, and TTPs of threats. The sharing of this information can position maritime organizations in a protective manner prior to a zero-day attack commencing. Taking a step further, CTI sharing among friendly nations dramatically increases the ability to defend against hostile nation states through the immediate sharing of observed threat TTPs.

The needs and benefits of CTI sharing are recognized in the federal government with the establishment of the Cyber Threat Intelligence Integration Center (CTIIC) under the Director of National Intelligence (DNI). The CTIIC's function, among others, will be to "provide integrated all-source analysis of intelligence related to foreign cyber threats or related to cyber incidents affecting U.S. national interests" (80 FR 11317, 2015, 1). The CTIIC is chartered with the oversight, development and implementation of intelligence sharing capabilities (including systems, programs, policies, and standards) and will share CTI contained in intelligence channels that are downgraded to the lowest classification possible for distribution to both the United States government and U.S. private sector entities including the critical infrastructure sectors such as maritime. The CTIIC will be the cyber intelligence hub when a cyber crisis occurs, managing the sharing of CTI among not only the intelligence community, but partner organizations in the private sector, state and local governments. DHS is the authoritative agency for the dissemination of CTI to maritime stakeholders. It is through DHS that maritime sector specific intelligence will be provided through DHS sponsored organizations such as US-CERT and MS-ISAC. Distribution of declassified material to the private sector maritime industry can occur through outreach efforts by the United States Coast Guard

CHAPTER 28

(USCG) Area Maritime Security Committees (AMSC) as well as Port partners and Port tenant associations.

CONCLUSION

CTI is an integral part of a maritime organization's overall risk management strategy. CTI may be viewed as the next evolution of cyber risk management in that the information provided is based on solid principles of analysis, correlation and objectivity. Incorporating the identification, tracking and prediction of threat activities is accomplished not only by cataloging the likely threat actors, but also the study of the maritime industry's threat environment and attack surface. This knowledge can be used to produce meaningful intelligence that may be leveraged by all levels of a maritime organization to manage risk through the incorporation of CTI. Most importantly, CTI is being leveraged in maritime organizations to offer courses of action that enhance informed decision-making. This will allow for ports and other maritime entities to develop a robust threat landscape through the exchange of information occurring at each organization.

The U.S. federal government recognizes the value of CTI through the establishment and adoption of standards for information exchange. Based on the same tenants of traditional intelligence products, CTI is presented as objective, non-biased, timely, and actionable information that can readily be used in a decision-making process. These standards are applied in both human readable intelligence products such as intelligence digests as well as machine-to-machine information exchanges. Automated exchanges of CTI follow industry recognized data formats such as STIX and TAXII. By adopting these formats, technology providers can ensure a seamless exchange of CTI that is applied to a variety of security systems within the maritime environment in real-time.

The benefit of CTI for maritime organizations is that decision makers are able to direct resources to address well-defined threats and better understand the potential impact of a successful breach through knowledge of their own threat landscape and attack surface. CTI is an excellent tool for defending against complex attacks such as APTs from determined and capable threat actors. Through the analysis of known attack TTPs, maritime organizations can apply anomaly detection and heuristics to quickly assess and protect against APTs at all stages of the attack. The benefits of CTI are dramatically increased from the sharing of information between organizations. Already, the U.S. government is taking steps to increase the exchange of CTI not only among the intelligence agencies, but also state and local governments as well as private sector industries, particularly those in the maritime critical infrastructure sector.

There is much to gain from adopting a CTI program as the resulting intelligence provides insight to the specific threats that pose a risk to the organization. In turn, this leads to the establishment of better policies and processes that can be used to strategically allocate resources to safeguards and focus efforts on what threats may have the greatest impact. CTI helps to ensure mission readiness, resiliency, and ultimately the success of the maritime industry.

References

Chambers, Matthew, and Mindy Liu. 2011. "Maritime Trade and Transportation by the Numbers." The U.S. Department of Transportation. http://www.rita.dot.gov/bts/sites/rita.dot.gov.bts/files/publications/by_the_numbers/maritime_trade_and_transportation/pdf/entire.pdf.

Connolly, Julie, Mark Davidson, Matt Richard, and Clem Skorupka. 2012. "The Trusted Automated eXchange of Indicator Information (TAXII)." The Mitre Corporation. https://taxii.mitre.org/about/documents/Introduction_to_TAXII_White_Paper_November_2012.pdf.

Department of Homeland Security (DHS). 2015. "Information Sharing Specifications for Cybersecurity." US-CERT United States Computer Emergency Readiness Team. https://www.us-cert.gov/Information-Sharing-Specifications-Cybersecurity.

Ernst & Young Global Limited. 2014. Nov. *Cyber Threat Intelligence – How to Get Ahead of Cybercrime*. http://www.ey.com/Publication/vwLUAssets/EY-cyber-threat-intelligence-how-to-get-ahead-of-cybercrime/$FILE/EY-cyber-threat-intelligence-how-to-get-ahead-of-cybercrime.pdf.

Farnham, Greg. 2013. "Tools and Standards for Cyber Threat Intelligence Projects." SANS Institute. https://www.sans.org/reading-room/whitepapers/warfare/tools-standards-cyber-threat-intelligence-projects-34375.

Intelligence and National Security Alliance (INSA). 2014. "Operational Cyber Intelligence." INSA Cyber Intelligence Task Force. https://www.insaonline.org/CMDownload.aspx?ContentKey=fd1a4a3e-cbe7-4afb-8c98-5a54ec878fb1&ContentItemKey=5527e2f9-2236-40bc-8012-08fef290bbc4.

International Chamber of Shipping (ICS). 2013. *Sustainable Development, IMO World Maritime Day 2013.* http://www.ics-shipping.org/docs/default-source/resources/policy-tools/sustainable-development-imo-world-maritime-day-2013.pdf.

Johnson, Chris, Lee Badger, and David Waltermire. 2014. "Guide to Cyber Threat Information Sharing (Draft)." National Institute of Standardsand Technology (NIST) Special Publication 800-150 (Draft). Gaithersburg, MD. http://csrc.nist.gov/publications/drafts/800-150/sp800_150_draft.pdf.

Jordan, Bret. 2015. *Evolution of Cyber Threat Intelligence.* European Union Agency for Network and Information Security (ENISA).

https://www.enisa.europa.eu/activities/risk-management/events/enisa-workshop-on-eu-threat-landscape/07PresentationBretJordan.

Mateski, Mark, Cassandra M. Trevino, Cynthia K. Veitch, John Michalski, J. Mark Harris, Scott Maruoka, and Jason Frye. 2012. "Cyber Threat Metrics." Sandia National Laboratories Report. SAND2012-2427. Albuquerque, New Mexico. http://nsarchive.gwu.edu/NSAEBB/NSAEBB424/docs/Cyber-065.pdf.

Office of the Director of National Intelligence (ODNI). 2015. "Analytic Standards." Intelligence Community Directive 203. http://www.dni.gov/files/documents/ICD/ICD%20203%20Analytic%20Standards.pdf.

Presidential Documents. "Establishment of the Cyber Threat Intelligence Integration Center." 80 Fed. Reg. 11317. (March 3, 2015).

Shackleford, Dave. 2015. "Who's Using Cyberthreat Intelligence and How?" SANS Institute. http://www.sans.org/reading-room/whitepapers/analyst/cyberthreat-intelligence-how-35767.

Townsend, Troy, and Jay McAllister. 2013. *Implementation Framework—Cyber Threat Prioritization.* Carnegie Mellon University. http://www.sei.cmu.edu/about/organization/etc/upload/framework-cyber.pdf.

Williams, Amrit T., and Mark Nicolett. 2005. "Improve IT Security with Vulnerability Management." Gartner. https://www.gartner.com/doc/480703/improve-it-security-vulnerability-management.

Appendix A

Figure 1: Relationships among Levels of Cyber Intelligence

RELATIONSHIPS AMONG LEVELS OF CYBER INTELLIGENCE

	STRATEGIC	OPERATIONAL	TACTICAL
Scope	General; "Art of the possible"	Industry/Sector Partners, Suppliers, Competitors, Customers, other Trust Relationships	Company "Inside the wire"
Focus	Political, Social, Behavioral	Adversarial Campaign Planning	On-the-network
Consumer	C-suite	Executive Management CIO/CISO	Incident Response Teams
Purpose	Maintain Competitive Advantage	Avoid Distribution	Remediate and Return to Normal Operations
Posture	Proactive	Proactive	Reactive
Interrogatives	Who, Why, Where	When, Where, How	How, What
Time Horizon	Far	Near	Immediate
Kill Chain	Motivation/Decision to Act Determine Objectives	Avenues of Approach Acquire Capabilities Develop Access	Implement Actions Assess Status Restrike
Attack Surface	Geographic Physical	Persona Logical	Logical Devices
Adversary		Opportunistic Targeted (President?) Threat Shifting: timing, resources, target, methods	
Types of Intelligence	Estimative intelligence General intelligence Scientific and technical intelligence Identity intelligence	Warning intelligence Counterintelligence	Current intelligence
Nature of	Non-technical, Contextual indicators Arguments traditional technology-centric Defense-in-Depth/Layered Security approaches		traditional technologies (e.g., IDS)
Sharing	Public, Private partnerships; ISACs; Private security reports		Automated means (e.g., IOC, STIX, TAXII)
Decisions	Driven by organizational Strategy	Driven by risk-based resource allocation	Driven by operational restoral or LE evidence collection
Relevant Artifacts	Organizational Strategy Plans of Action & Milestones Business Impact Analysis Enterprise Risk Strategy	Plans of Action & Milestones Business Impact Analysis Business Continuity Disaster Recovery	

Adapted from Operational Cyber Intelligence, p. 12, 2014, Retrieved from the Intelligence and National Security Alliance (INSA)

CHAPTER 28

Figure 2: STIX Architecture 1.1.1

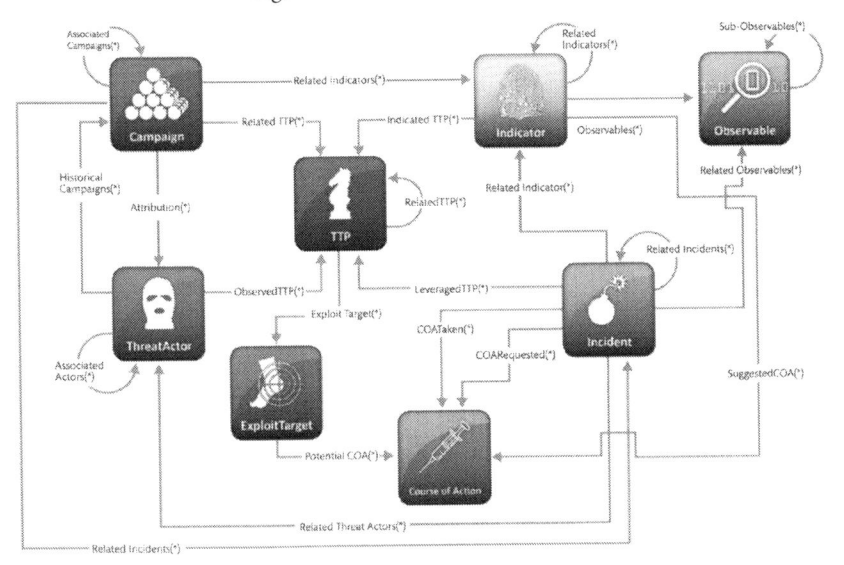

Adapted from Evolution of Cyber Threat Intelligence, p. 30, by B. Jordan, 2015, Retrieved from the European Union Agency for Network and Information Security (ENISA)

Figure 3: TAXII High-Level Concept

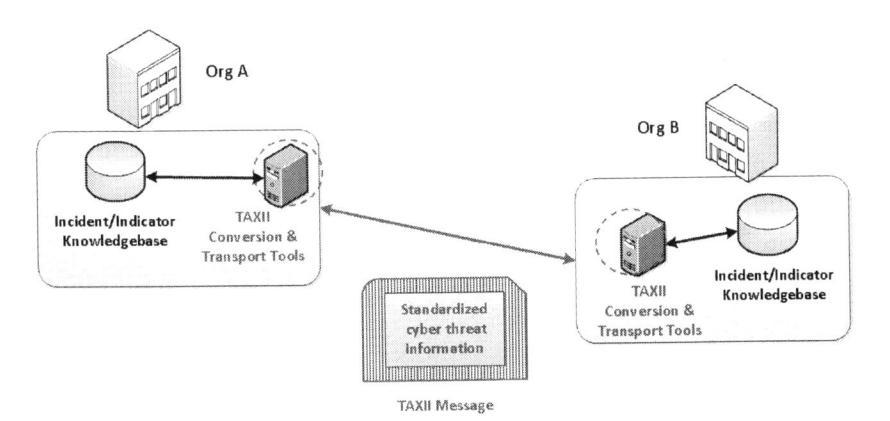

Adapted from The Trusted Automated eXchange of Indicator Information (TAXII™), p. 12, by Connolly, J., Davidson, M., Richard, M., and Skorupka, C., 2012, Retrieved from MITRE

513

CHAPTER 29:
THE NEED FOR A USER-FRIENDLY CYBER SECURITY VULNERABILITY TOOL FOR COAST GUARD MARINE INSPECTORS

Mark Behne, Benjamin Chapman, Michael Clancy, Koachar Mohammad, and Kimberly Young-McLear
U.S. Coast Guard Academy

Abstract

As the Coast Guard's requirements for mission capability continue to increase, so does its dependence on technology and computer systems that are used to perform everyday duties. Considering the fact that our service is the largest component of the Department of Homeland Security, the Coast Guard is at the forefront of developing effective and efficient means of protecting our nation's critical infrastructure within the cyber realm. However, our own reliance on cyber systems has put our asset's security atop the priority list since we have become so reliant on technology. Therefore, it is paramount that a proper risk analysis is constructed for our assets in order to determine where the vulnerabilities with the greatest potential consequences might be found in our networks and instruments. Specifically, this paper proposes a cyber security risk analysis tool for Coast Guard marine inspectors to use for vessel examinations.

Keywords: Integrated Bridge Systems, Risk Assessment, Marine Inspector

Background

The U.S. Coast Guard's 2015 Cyber Strategy was released by its Commandant, Admiral Zukunft, which tasks the Coast Guard to identify a tool that can be used to assess the cyber security risk for the Maritime Transportation

System (MTS). Although the Coast Guard currently has a risk assessment tool for physical security, called the Maritime Security Risk Analysis Model (MSRAM), it does not include vulnerabilities and risks associated with cyber. The Coast Guard currently does not have plans to modify MSRAM; however, efforts to develop a separate cyber risk tool are ongoing.

Risk can be estimated by assessing the threat, vulnerability, and consequence. The Coast Guard multiplies these all together (Total Risk = Threats × Vulnerabilities × Consequences) to determine the total risk of an event. While that may appear to be an easy approach to determining risk, each variable in the equation is complex to figure out. Threat is likely the simplest one, as it is comprised of elements such as the capability of an adversary and their intent to initiate an attack. However, assessing the vulnerabilities, or the weaknesses, of an asset is typically subjective because there are many things to consider. Additionally, consequences can also vary in both total numbers and magnitude. In an article titled, "Risk Analysis and Port Security: Some Contextual Observations and Considerations," consequences can affect things such as physical injury, property damage, environmental aspects, economic effects, or result in human deaths, to name a few (Greenberg 2009). A simple scorecard model, therefore, can be useful to determine the risk.

Within the MTS, scenarios that are likely to occur and may have serious consequences include vulnerabilities in in automatic identification systems (AIS) and global positioning systems (GPS). With AIS, it is connected to a network on the internet and is constantly submitting data to all vessels within a certain region. These networks are not entirely impenetrable, and are susceptible to denial of service (DoS) attacks as well as false identification hacks (Middleton 2014). DoS is simple to achieve if an adversary has the equipment to send AIS signals at a very high rate, which could overload a local AIS network in an area with high traffic and essentially shut it down, erasing one the Coast Guard's main vessel identification tools. Furthermore, jamming and spoofing are a commonly seen approach from targets who wish to disrupt another vessel's primary navigation equipment, which can interrupt operations and even force a vessel to become lost at sea (Thompson 2014).

The maritime industry has only recently released guidelines for risk management of vessels, offshore platforms, and ports. BIMCO, ABS, and Lloyd's Register have all released cyber security guidelines between January 2016 and February 2016. This is beneficial for the U.S. Coast Guard because it is an excellent source of information and to evaluate the self-guiding priorities from industry. These guidelines, however, have not yet been used to develop a simple tool that owners and operators can use as a self-audit. The tool presented later in this paper describes a tool such that a Coast Guard marine inspector without any previous technical knowledge can accurately assess the risk of a vessel (of a likely scenario) with the philosophy that if every vessel maintained the minimum level of risk needed, the MTS would be improved. Over time, the tool can be expanded to include more robust requirements.

KEY FINDINGS FROM CYBER SECURITY VULNERABILITIES

Safe navigation is a concern of vessels because ships rely heavily on GPS for navigation. GPS spoofing is a form of attack on a vessel that can feed the navigators onboard faulty information or take down the navigation system altogether. The focus is usually on the actual navigation system, ECDIS/ECS as well as AIS which we use for collision avoidance. These systems can be compromised by GPS spoofing. However, sometimes GPS jamming could result in all navigation systems being lost, including steering, radar, and radios if all these systems are connected on one network, GPS spoofing or hacking could render all these systems ineffective (Moskoff 2014). Moskoff (2014) states that cyber security awareness in the maritime industry needs to increase; that identification and assessment of vessels own risk needs to occur; that vessels need to identify possible threats; and lastly that there needs to be a mechanism to ensure maritime cyber security response plans are in place and are being practiced through drills.

Control systems have become increasingly more vulnerable to cyber-attacks because of the maritime industry's increasing dependence on computerized remote operations. A Norwegian company, who owns many oil and gas rigs, who has also been attacked numerous times by terrorists in the cyber domain, has developed one practice to reduce their vulnerability with regard to cyber attacks. That method is to practice information security, cyber security auditing, and vulnerability and robustness testing. The Norwegian company has an information security policy where they treat each individual computer on their oil rigs essentially as a weapon system, which instills in the minds of workers the dangers of compromising the system. With cyber security auditing, they use an approach based on crisis intervention and operability analysis. This evaluates the control center personnel's ability to handle all modes of operation safely and efficiently by using checklists and step diagrammed scenarios. How the company tests their vulnerabilities and robustness is through scanning of communications networks. Communications networks include human interaction with the systems and the systems interacting and communicating with other systems. These scans probe for unknown vulnerabilities by using a method called "fuzzing" which recognizes more than just two basic responses that include true or false. It essentially triggers unexpected behavior in the software being tested. Another form of testing includes the flooding of their systems with network traffic to test their robustness and capacity of the systems (Csorba, Husteli, and Johnson 2014).

The Navy has developed their own method to combat potential cyber-attacks as well. The Navy has fears that hackers might also try to disable or take control of machines in our physical world—from large systems like electric power grids and industrial plants, to transportation assets like cars, trains, planes, and even ships. It has been discussed that anything computer related is susceptible to being hacked. The Navy has developed a method called the Resilient Hull, Mechanical, and Electrical Security (RHIMES) system which enables its vessels to defend against a cyber-attack and protect shipboard physical systems, such as, weapon systems, navigation systems, and other mechanical systems (Office of Naval Research RHIMES). GPS spoofing and control system infiltration are primary concerns of vessels because attacks like these hinder the safety of the vessel, crew, and any vessels or people around the affected vessel. GPS spoofing hinders the safe navigation of the vessel while control system hacks hinder the safe control of the vessel.

As society continues to progress, the dependence on technology is also becoming overwhelming in certain regards. With an increase in the use of technology individuals and entities open up the possibility to cyber attacks that could cause a great deal of harm. Critical Infrastructures (CIs) are systems or programs that are interconnected between physical and cyber based systems which are used for essential operations on a daily basis in both the government and economy (Ralston, Graham, and Heib 2007). Some examples that we can relate to in the Coast Guard of critical infrastructures are many of our operating systems which are supervisory control and data acquisition (SCADA) systems, used in valves to transfer fuels, sewage, propulsion, and even fire suppression systems (Radgowski 2014). In fact, most modern ships use SCADA systems which can be vulnerable to a cyber-attack without the proper preventive measures.

Some of the main vulnerabilities of SCADA systems are that most systems used today are much older and considered legacy systems. A lot of these SCADA systems were created in an era before cyber security was a concern, and with increased demand organizations have been able to create hybrid networks with newer corporate IT (Radgowski 2014). Now, most SCADA systems can connect to the open internet however this convenience increases the risk of cyber attacks.

Major threats to SCADA systems include the introduction of viruses or malware, such as a Trojan horse, which potentially could manipulate the system and make it more vulnerable and even send false messages causing undesirable outcomes (Radgowski 2014). All consequences that can have grave impacts especially in our service that we so readily rely on our operational ability.

A cyber risk assessment tool is advantageous especially on the Coast Guard's newest assets such as the National Security Cutters which depend on several SCADA systems.

From an economic impacts perspective, it is noteworthy to draw attention to the reality that seaborne trade represents about 90% of the world's trade (Burns 2013). For this reason any policy or event that would impact maritime industries is of great concern for the United States government and more specifically the Coast Guard who is the lead agency for securing the Maritime Transportation System. Burns (2013) discusses the burdens, in terms of reduced efficiency and capital spent on security measures which the maritime industries experience. As a countries economy grows, so too does the value of items transported through its supply chain and the likelihood of an attack to a part of its supply chain (Burns 2013). According to this model the United States, a country with one of the strongest economies in the world, is a likely target for some form of maritime attack.

There are two general forms of attacks on the maritime industry including terrorism and piracy. Terrorist attacks serve a primary goal of destruction and attempt to create fear and intimidation. For these attacks, targets are carefully chosen for sentimental value or to inflict a large number of human casualties (Burns 2013). However, attacks by pirates are more generally motivated by financial gains. The goal of a pirate is to attack a target carrying valuable cargo that has weak security (Burns 2013). Both forms of attack would have a significant financial impact along with whatever physical or sentimental impact corresponds with the target. Generally speaking, governments are most concerned with terrorist attacks and spend significantly more trying to prevent and mitigate the risks and impact that those attack could have. Maritime business's and ship-owners allocate more of their resources toward preventing piracy (Burns 2013).

Port security costs are estimated to reach 12 billion dollars annually (Burns 2013). The International Maritime Organization after the attacks of September 11th established the International Ship and Port Facility Security (ISPS) Code. This establishes mandatory minimum security standards that need to be met by ships and port facilities (Burns 2013). The standards require significant investments by ports and ships. Increased security personnel, biometric technology, and radiation scanning equipment are just some examples of these investments. In order to satisfy these standards initially, U.S. Ports were required to spend between $3,000 and $35,500,000. Since then the annual cost to maintain these standards are between $1,000 and $19,000,000 per facility (Burns 2013).

Shipping companies also experience many security costs including the hiring of armed escorts for vessels traveling in high risk areas. An armed escort can cost around $50,000 (Burns 2013). Companies also experience increased fuel expenses by traveling around, or at faster speeds, through high risk areas. Also, delays and forced deviations caused by security backups and concerns cost shipping companies a significant amount of time and money (Burns 2013).

While the price of increased maritime security is very high there are also significant financial implications of successful attacks. An attack on a major port that could impede operations would have a significant impact on the economy both locally and nationally. It is estimated that a disruption at the country's largest port in Los Angeles and Long Beach, which impeded operations for a week, would prevent half a billion dollars in trade (Burns 2013). This type of interruption could be from a physical attack or a cyber attack that disrupts the ports automated systems. Events in the past have shown the significant financial impacts of attacks. The attack on the World Trade center resulted in over $30 billion in losses.

With all of the increased expenses associated with increased port security there are questions about who should be responsible for covering the cost. Many port facilities operators have chosen to pass the security costs on to their customers. However, many shipping lines believe that many of these security costs are the responsibility of the ports who are responsible for providing a safe shipping environment (Khalid 2006). In addition to the ports and shipping lines supporting the costs of increased security measures, the government should as well. Ports represent significant and strategic national and regional assets. As such, the government has some responsibility to fund and ensure the safety and security of those assets. In 2005 the U.S. government, through the Department of Homeland Security, spent $1.6 billion on port security which was over 700 times as much as before the September 11th attacks (Khalid 2006). In the future, the Coast Guard, as the federal agency in charge of port security, will have a significant role in determining how and where funds should be allocated for port security. Furthermore, it is likely that based on new and developing threats, that cyber security will see increased funding from the Coast Guard.

Overview of Proposed Cyber Security Vulnerability Tool

The chapter proposes a tool (shown at the end this chapter) to be used to audit integrated bridge systems onboard vessels. The tool has three specific attributes: quantitative, inclusive, and scalable. For the quantitative attribute, the tool has a scoring system which possesses a group of cyber-related elements to be assessed and ultimately given an overall grade. For the inclusive attribute, the tool is applicable to multiple systems on-board the vessel (i.e., the tool can-

CHAPTER 29

not be aimed solely toward GPS, for instance). For the scalable attribute, the tool is broad enough so that it can be considered adaptable for advances in cyber technology.

The three primary dimensions of the scorecard are Detection, Accessibility, and Prevention. These three aspects of a cyber system are important to evaluate. Detection capabilities of a system are necessary because crew members would otherwise not have awareness of an intrusion into one of the bridge systems. The lack of detection and alert of attacks is problematic. Accessibility is another key aspect the tool considers for vessels. The vessel operators should be aware of who is able to gain access or control of certain systems, which inventory of items are interconnected on a network, and the number of points of network connection. Prevention is the last dimension of the tool. The auditor can assess if there has been any preventative measures to mitigate cyber risks.

The tool uses three functional categories to evaluate the performance of the vessel. The three scored categories are Information Systems, Applications, and Networks. For each of the functional categories, a score is tallied of either 0 or 1 to the answer of a Yes/No question, with a score of "1" being assigned to the less preferable Yes/No answer. For example, within the first element under the Information Systems category in the "Ability to Detect Attack," the question asks, "Can system detect attack/intrusion?" If the answer is "No" then the evaluator would add one point to this category, since it is preferred that the system actually be able to detect an intrusion as previously mentioned.

The totals from each category are tallied up out of 8 points, and then combined for an overall score out of 24 points, which sends the evaluator to the last step of the scorecard—the General Assessment of Risk (GAR) score. In order to remain aligned with the traditional General Assessment of Risk model that is used throughout the U.S. Coast Guard in many types of evolutions, we decided to emulate the same model. Total scores that make the "Green" range insinuate that the evolution has little risk and the crew shall proceed as planned; "Amber" scores mean that the evolution has its downsides, but the crew can proceed with caution and remain cognizant of dangerous factors; "Red" scores are almost unacceptable, and typically mean that the evolution should not be carried out unless enough mitigating factors are established to get the score within the Amber range. Since scores of "0" were assigned to the more favorable answers to each element, a potential perfect assessment of a system would receive an overall score of "0". Conversely, a potential colossal failing assessment of a system would receive an overall score of "18." Scores from 0 to 6 are labeled as Green, 7 to 12 are Amber, and 13 to 18 are Red. The reasoning for making the range of Red scores larger than the other two is because a score of "13" would reveal that over sixty percent of the elements were evaluated as unfavorable.

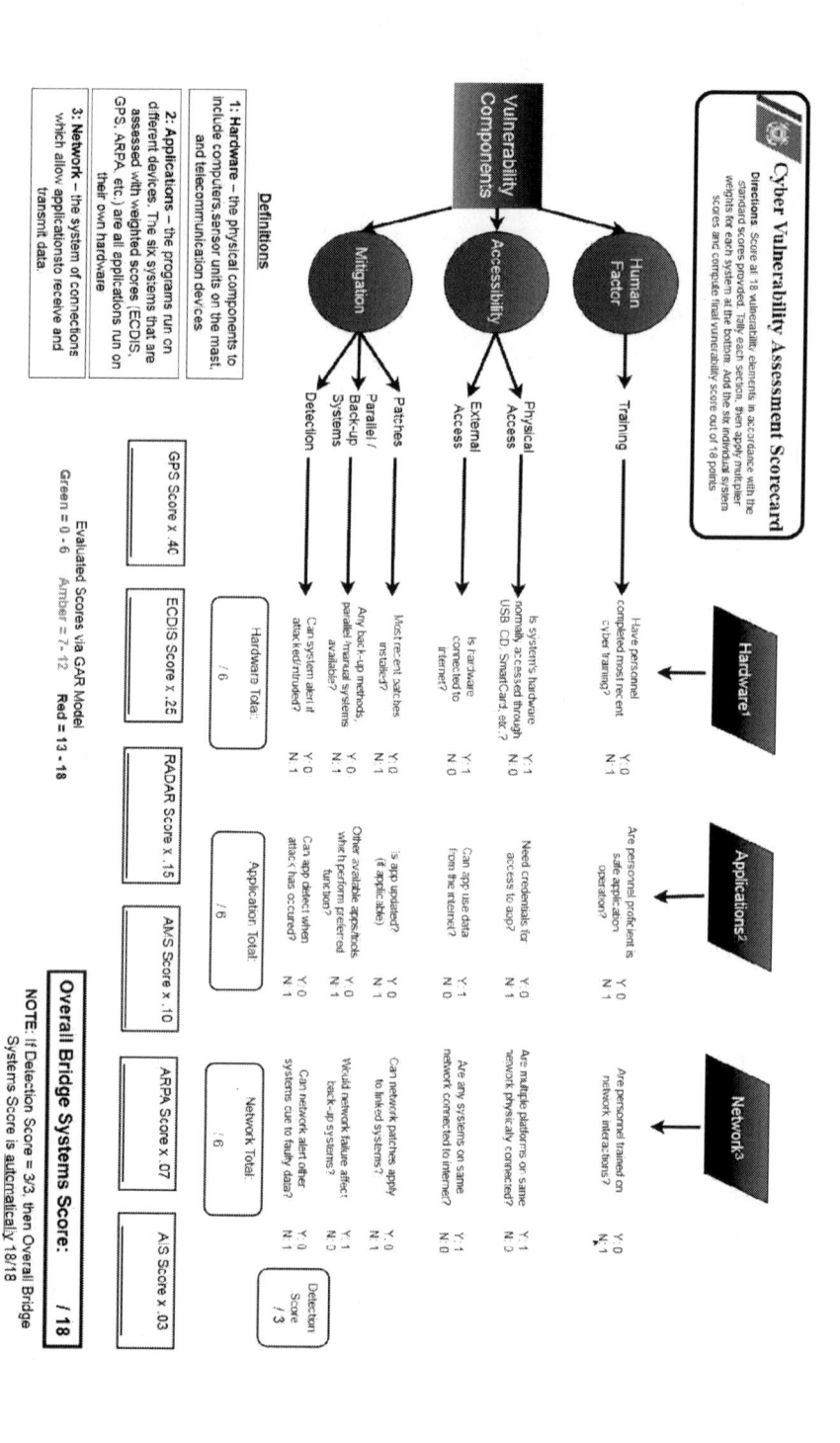

ISSUES IN MARITIME CYBER SECURITY

Cyber Vulnerability Assessment Scorecard

Directions: Score all 18 vulnerability elements in accordance with the standard scores provided. Tally each section, then apply multiplier weights for each system at the bottom. Add the six individual system scores and compute final vulnerability score out of 18 points.

Vulnerability Components

Human Factor
- Training → Have personnel completed most recent cyber training? Y: 0 N: 1

Accessibility
- Physical Access → Is system's hardware normally accessed through USB, CD, SmartCard, etc.? Y: 1 N: 0
- External Access → Is hardware connected to internet? Y: 1 N: 0

Mitigation
- Patches → Most recent patches installed? (if applicable) Y: 0 N: 1
- Parallel / Back-up Systems → Any back-up methods, parallel 'manual' systems available? Y: 0 N: 1
- Detection → Can system alert if attacked/intruded? Y: 0 N: 1

Hardware[1]
- Are personnel proficient in safe application operation? Y: 0 N: 1

Applications[2]
- Need credentials for access to app? Y: 0 N: 1
- Can app use data from the internet? Y: 1 N: 0
- Is app updated? (if applicable) Y: 0 N: 1
- Other available apps/tools which perform preferred function? Y: 0 N: 1
- Can app detect when attack has occurred? Y: 0 N: 1

Network[3]
- Are personnel trained on network physically interactions? Y: 0 N: 1
- Are multiple platforms on same network connected to internet? Y: 1 N: 0
- Are any systems on same network connected to internet? Y: 1 N: 0
- Can network patches apply to linked systems? Y: 0 N: 1
- Would network failure affect back-up systems? Y: 1 N: 0
- Can network alert other systems due to faulty data? Y: 0 N: 1

Hardware Total: / 6
Application Total: / 6
Network Total: / 6

Definitions

1: Hardware – the physical components to include computers, sensor units on the mast, and telecommunication devices

2: Applications – the programs run on different devices. The six systems that are assessed with weighted scores (ECDIS, GPS, ARPA, etc.) are all applications run on their own hardware

3: Network – the system of connections which allow applications to receive and transmit data

Evaluated Scores via GAR Model:
Green = 0 - 6   Amber = 7 - 12   Red = 13 - 18

GPS Score x .40
ECDIS Score x .25
RADAR Score x .15
AMS Score x .10
ARPA Score x .07
AIS Score x .03

Detection Score / 3

Overall Bridge Systems Score: / 18

NOTE: If Detection Score = 3/3, then Overall Bridge Systems Score is automatically 18/18

522

CHAPTER 29

Conclusion

This scorecard tool was designed to be used in conjunction with existing standards and guidelines. The tool is scalable, intuitive, and quantitative. As the Coast Guard matures its knowledge and understanding of cyber best practices, the tool can be updated. It is appropriate to have a tool that is at the same level of the knowledge of the marine inspector. Other training would need to exist, such as, in marine inspector school to provide awareness of how to use the tool for maximum effectiveness. Findings from the assessment of the cyber security risk levels of the vessels should be captured in the Coast Guard's Marine Information for Safety and Law Enforcement (MISLE) database along with other deficiencies noted during an examination.

References

Burns, Maria G. 2013. "Estimating the Impact of Maritime Security: Financial Tradeoffs Between Security and Efficiency." *Journal of Transportation Security* 6 (4): 329–338.

Csorba, Mate J., Nicolai Husteli, and Stig O. Johnson. 2014. "Information Systems: Securing Your Control Systems." *Proceedings of the Marine Safety & Security Council* 71 (4): 34–37.

Greenberg, Michael R. 2011. "Risk Analysis and Port Security: Some Contextual Observations and Considerations." *Annals of Operations Research* 187 (1): 121–136.

Khalid, Nazery. 2006. "Too Much of a Good Thing? Some Reflections on Increased Port Security and its Costs." *Defense & Security Analysis* 22 (3): 261–273.

Middleton, Allison. 2014. "Hide and Seek: Managing Automatic Identification System Vulnerabilities." *Proceedings of the Marine Safety & Security Council* 71 (4): 48–49.

Moskoff, David B. 2014. "GPS Jammers a Top Concern in Maritime Cyber Readiness." *Professional Mariner* (June 3). www.professionalmariner.com/June-July-2014/GPS-jammers/ (accessed December 22, 2016).

Radgowski, Jeff. 2014. "Cyberspace—the Imminent Operational Domain: A Construct to Tackle the Coast Guard's Toughest Challenges." *Proceedings of the Marine Safety & Security Council* 71 (4): 18–21.

Ralston, Patricia A.S., James H. Graham, and Jeffrey Hieb. 2007. "Cyber Security Risk Assessment for SCADA and DCS Networks." *ISA Transactions* 46 (4): 583–594.

Thompson, Brittany. 2014. "GPS Spoofing and Jamming: A Global Concern for All Vessels." *Proceedings of the Marine Safety & Security Counc*

PART IV:
The Way Forward

CHAPTER 30:
THE NEXT GREAT BATTLEFRONT

Gerald Feltman
American Military University

Abstract

The cyber realm is an integral part of everyday modern life, used by many Americans for everything from banking to keeping in contact with friends and relatives; however, the general public often, disregards the threat posed by cyber-attacks. With respect to maritime security, attention to the possibility of a cyber attack specific to the critical maritime infrastructure is lacking. Such an occurrence has the potential to incur both economic and loss of life consequences. However, the impending chance of a cyber attack against the assets within the maritime environment is more a question of "when" rather than "if". This paper examines the factors that put the cyber-domain at risk for an attack and further reviews the specific vulnerabilities within the maritime sector such as denial of service attacks (DSA), and hacking naval vessel GPS hacking. The chapter concludes with a discussion of possible ways to combat these threats.

Introduction

This chapter asserts that the next great threat vector the United States will encounter will be on the cyber-front. As is evident from recent news stories such as the Apple iCloud hacking (Cellan-Jones 2014), or the 2014 hack on the Sony Corporation servers before the release of its controversial movie *The Interview* ("The Interview: Sony's North Korea film to be screened in US" 2014), and increasing amount of cyber-attacks have made headlines. A 2007 case involving China emphasized the critical situation that the U.S. faces, demonstrating that even the Pentagon is vulnerable to a system attack (Ward,

Sevastopulo, and Fidler 2007). Such a high vulnerability demonstrates that the cyber front presents clear challenges which should not be taken lightly. The Chinese case demonstrates a very different kind of threat than attacks posed by non-state actors (Clapper 2013). Unlike states, non-state actors do not have the same political and economic constraints. This gives these actors greater freedom when selecting targets and also places less restraints on them when it comes to carrying out an attack which causes harm and induces panic.

Background

Despite the potentially devastating ramifications posed by cyber attacks, such attacks do not appear to receive much attention from the general public. Even though government agencies work to address cyber-attacks, these efforts go largely unnoticed by the general public, by comparison to the attention paid to kinetic operations that result in the death or capture of a terrorist leader. For example, greater public attention was paid to Operation Geronimo where Navy SEALs infiltrated the complex where Osama Bin Laden had been hiding in Pakistan and killed him almost ten years after the attacks on 9/11 than to any recent cyber attack (Schmidle 2011). One of the key reasons cyber attacks are so harmful to a nation is the way in which they are conducted and can take place. In an article entitled "Cyber Attack" by Brad Gilmer, the author explored the potentially disastrous effects a denial of service attack poses to a network's security and accessibility. Gilmer (2013) described a denial of service attack as an instance where a specific website is put to near 100% capacity of the network servers, thereby preventing even the administrator from accessing key functions. In this example, the attack is a mere inconvenience; however, if the same sort of attack is perpetrated against key infrastructure, the possible damage increases greatly.

One such example of a potentially devastating cyber attack would be one executed against the U.S. power grid. The U.S. power system has already been infiltrated by hackers who implanted software capable of remotely disabling the grid at the whim of the hacker; if such an attack were to take place against the Western U.S. power grid, it could lead to cascading power failures across the nation's power system (Wei et al. 2011). Another example of a possible attack includes the use of multiple computers and networks to overwhelm a targeted server (Gilmer 2013). This is accomplished through a computer virus that is spread to thousands of computers which then directs each computer to access a website's server simultaneously, thus overwhelming and crashing the site (Gilmer 2013). When considering that many Americans download viruses without even realizing it, it is not hard to see the potential for a large-scale attack on a national defense server without the owner of the "hijacked computers" even knowing it (Gilmer 2013). Examples such as this highlight serious vulnerabilities the United States has within the cyber-domain, vulnerabilities that could have devastating effects on millions of Americans.

The main problem the United State must attempt to tackle, is to determine the best course of action in aiding civilian companies that run critical national infrastructure in preventing such an attack. The U.S. is among the most ill prepared countries in the world when it comes to cyber-security in the private sector and civil/government communication ("ANALYSIS: Under Cyber Attack" 2011). One-third of organizations polled in a study published in *Process Engineering*, stated they had not received contact from the government relating to cyber security. In contrast, of the Chinese executives surveyed, they stated that the Chinese government initiated both formal and informal communication with their companies on the matter of cyber security ("Analysis: Under Cyber Attack" 2011). This is cause for concern because while the U.S. government has a multitude of agencies that are dedicated to counterintelligence and counteracting cyber-attacks, the private sector does not have the same kind of resources needed to thwart an attack. The vulnerability of private sector companies to cyber attacks becomes even more concerning when considering that systems such as the U.S. power grid are becoming increasingly automated. These systems now rely on computers to complete the daily operations and communication between control points and to the grid itself, making the systems much more susceptible to a cyber attack (Wei et al. 2011).

Types of Cyber Security Threats

"Main Stream" Cyber-Attacks

The threat cyber attacks pose to the United States is argued to rival a conventional military strike from an enemy state. Saini and Singh (2012) make the case that so much vital information is stored digitally from every sector of life, from health care to the finances to homeland security and defense; the exploitation of the information passing through this openly accessible realm could forever damage the economy and society itself. The previous example given specific to an attack against the U.S. electric grid further demonstrates this point. The systems within the grid were designed for ease of communication and information maximization, at the time of it original development, little thought went to consideration of future cyber security issues that could impact the electric gird (Wei et al. 2011). Another example of the impending cyber threat posed to the electrical grid: if the same method for a denial of service attack that used to disrupt service for a website (Gilmer 2013) was applied to taking down a critical server for the distribution of electricity across the nation, the results could have far reaching consequences and dire consequences if the effects lasted for an extended period of time (Wei et al. 2011).

Another possible cyber attack target which could prove catastrophic is the airline industry. As far back as 2002, potential vulnerabilities were identified within U.S. based airline companies' cyber security. It was determined that a hacker could obtain information on employees for use in the extortion of information/manipulation on where and when aircraft are scheduled to fly or where they are kept (Morris 2002). Access to the systems described above could grant an attacker the ability to eliminate an aircraft without requiring an individual on board during flight to execute the plan. These capabilities could potentially expand the reach of terrorist groups by using information gathered, by hacking into airline records, to coerce employees into planting an explosive charge on an aircraft set to go off at a predetermined point, thus inflicting the greatest possible damage. All of this could potentially be done without a single member of the terrorist cell being physically present, or needing to gain access to the plane or even airport terminal.

With so much of people's everyday lives intertwined with technology, and the increase of society's critical framework run by computers, an attack on the networks that control these functions would be devastating, to say the least. Because of the potentially crippling effects they would cause, one popular example often used in the literature involved attacks on the financial sector; which if carried out, would be devastating to the U.S. economy. In the financial arena, the cost of cyber crimes and the countermeasures that banks take to safeguard their funds is estimated between 375 and 575 billion dollars annually ("Cyber Attacks" 2014). Traditionally, banks attempt to lower their expenditures on things like security systems, but now have no choice other than to spend more and more money to keep their customers money safe, and prevent from having to pay the cost of security and reimbursement due to lost funds ("Cyber Attacks" 2014). Even popular retail companies such as Target have had breaches in cyber security, with information from over 40 million credit cards put at risk after a cyber breach in to the company's credit card machines ("Target Card Heist Hits 40 Million" 2013).

The Maritime Environment

In 2011, the United States received 53% of all imported goods by maritime vessels. Additionally, the maritime transpiration system (MTS) was responsible for 38% of all exported goods (Chambers and Liu 2012). During this time, there were over 62,000 vessels arriving at U.S. seaports, equating to around 172 per day, per calendar year (Chambers and Liu 2012). While these numbers seem staggering, it is important to recognize that this only reflects vessels carrying goods—it does not include vessels within the tourism industry. The amount of trade conducted via the MTS accounted for 1.7 billion dollars' worth of trade to the U.S. through both import and exports in 2011 (Chambers and Liu 2012). In addition to the massive amount of commerce occurring within the MTS, in 2011 over 2.8 million passengers aboard 1,075 cruise liners departed from over 2,887 ports worldwide (Office of Policy and Plans 2012). Due to the staggering commerce and travel occurring within the MTS, one might expect a large amount of effort to be placed into the cyber security of maritime vessels and ports; however, the opposite is occurring.

While the cyber domain is rife with examples of exploitable vulnerabilities, many of the main-stream cyber defenses and infrastructures designed to combat attacks have developed in kind. Unfortunately, however, the cyber defenses of the maritime sector, which as previously mentioned, sees a great deal of goods and personnel moved through its systems, has not kept pace as efficiently as other sectors in this rapidly advancing field. Greenwald (2014) argues the disastrously low level of maritime cyber security is summed up best by a statement from the U.S. Coast Guard stating that, "'Cyber-related vulnerabilities are a growing portion of the total risk exposure facing the marine transpiration system,' ... Cyber threats 'continue to grow and represent one of the most serious national security challenges we must confront'"(30). Additionally, in October 2013, a company based out of Tokyo examining the field of cyber vulnerabilities, discovered a critical flaw in the automated identification software installed on over 400,000 ships, this vulnerability created a situation where a person could hijack the communications between vessels or even create a fake vessel (Greenwald 2014). The creation of a ghost vessel is particularly troubling as this could allow attackers to simulate a vessel in distress, to lure in vessels for aid, and then either commandeer it, hold it for ransom, or steal cargo onboard.

Another issue specific to maritime cyber is the susceptibility of navigational features such as global positioning systems (GPS) to be hacked and disrupted. A group of researchers from Texas A&M University demonstrated this vulnerability when the researchers caused an $80 million yacht off the coast of Italy to believe it was in a completely different location than it actually was (Greenwald 2014). Imagine the implications of a similar attack on a cruise ship carrying thousands of people—of which there are many traversing the high seas at any given time—leading it to run aground, perhaps injuring or even killing innocent people. Robert Parisi, a senior vice president, private practice leader, and a network security administrator, at Marshal L.L.C. in New York, stated the problem eloquently saying that, "... when you're dealing with a ship that weighs hundreds of tons, if that goes sideways on you, it's not quite the same as a point-of-sale terminal not working" (Greenwald 2014, 30). Part of the reason the consequences stemming from a cyber attack on the MTS are especially poignant is because of the potentially fatal effects a simple GPS miscalculation can have on a cruise liner filled with passengers.

Cyber security at ports is no further developed than the security aboard ships being used to carry the people and goods across the oceans of the world. In March 2013, the U.S. Coast Guard released data that revealed that, of all the ports reviewed, there was only one that even conducted an assessment of its cyber security, and there were no ports that created a response to a cyber attack (Greenwald 2014). Captain David B. Moskoff, a recognized expert in maritime cyber security and Master Mariner, outlined how dangerous even simple disruptions to port operations through denial of service attacks or GPS hacking can be stating, "GPS jamming could close a major port for days, resulting in economic losses as a high as a billion dollars or more" ("Cyber Attacks Could Paralyze U.S. Ports, Captain Moskoff Tells CCICADA" 2015, 1). There are over 360 commercial ports that serve the United States and approximately 3,200 passenger/cargo handling centers (U.S. Public Port Facts 2013). The damage that a GPS jamming attack could yield from taking down even a handful of these ports would have far reaching impacts on the U.S. economy. An article in the *Navy Times* (2013) stated that a recent report done by an U.S. Coast Guard JAG advocates U.S. ports are at a dangerous risk for a cyber-attack, one that could halt commerce in the country on everything from crude oil to food goods (Rico 2013). This article reinforced that it is crucial to foster better communication between the private sector and government with regards to cyber security (Rico 2013). Standardized procedures or regulations for cyber security among the private sector are limited, additionally there has seemingly been little thought on what government agency, specifically, would enforce these regulations should they exist (Rico 2013). The decentralized control among the private ports around the U.S. creates a vulnerable spot within the cyber realm for terrorists to take advantage of, a vulnerability with crippling consequences.

With so many vulnerabilities looming in the cyber world, the next logical step is to ascertain an effective defense for the various strategies that cyber-attackers can utilize. Fink et al. (2014) discuss a possible defense mechanism against the increasingly aggressive cyber attacks in their research "Defense on the Move: Ant-Based Cyber Defense." They explain a possible way to prevent millions of dollars from being wasted in revitalizing the infrastructure is to "mobilize" the network defense measures already in use, rather than re-build the entire system infrastructure (Fink et al. 2014). It is vital that cyber security measures be kept from becoming stagnant like traditional firewalls, measures that the attacker can probe and test, eventually eliminating the incorrect ways to access the system and identifying the proper procedures necessary to violate it (Fink et al. 2014). The overarching idea behind new security measures is that, often, the actual hardware and original systems were not designed with cyber security in mind (Wei et al. 2011).

Another interesting perspective postulated in cyber security stems from the use of artificial intelligence (AI) as an aid for cyber-defense (Saini and Singh 2012). In this example, AI is utilized to identify common patterns found in cyber-attacks—e.g., repeated failed attempts for a password, the same username and password being accessed from different locations at the same time, etc. (Saini and Singh 2012). The AI would be able to aid human administrators in the removal of some uncertainty inherent when dealing with cyber attacks through taking into account the probability a certain cyber domain has in coming under attack at a given moment (Saini and Singh 2012). This kind of outside the box thinking is essential in combating the incredible threat posed cyber-attacks against the U.S., and the ever-increasing technological society of today's world.

Conclusion

The dynamic cyber threats facing the maritime sector, coupled with woefully inadequate and outdated cyber security technology, make a cyber attack against the maritime system not a question of "if" but "when" (Greenwald 2014). Even as the maritime sector has evolved, now heavily utilizing new technology, cyber security within this realm has not kept pace (Greenwald 2014), meaning the number of possible holes within these systems are staggering. Combating these threats will require the use of programs such as the Anti-Based Cyber Defense method, and the integration of Artificial Intelligence to aid human administrators and cyber security specialists (Saini and Singh 2012; Fink et al. 2014). Lastly, maritime cyber security will require ensuring ports and ships alike are prepared for the possibility of a crippling cyber attack and have contingency plans in place. Further research needs to be done into what the possible risks are in the commercial and military fleet alike of maritime vessels and ports having new technology being placed into older systems that have never run on such a complicated framework before.

REFERENCES

"Analysis: Under Cyber Attack." 2011. *Process Engineering* (May/June): 10. http://search.proquest.com/docview/866018763?accountid=8289.

Cellan-Jones, Rory. 2014. "A Cloud of Uncertainty." *BBC News* (September). http://www.bbc.com/news/technology-29030229.

Chambers, Matthew, and Mindy Liu. 2012. "Maritime Trade and Transportation by the Numbers." United States Department of Transportation: Bureau of Transportation Statistics (May). http://www.rita.dot.gov/bts/sites/rita.dot.gov.bts/files/publications/by_the_numbers/maritime_trade_and_transportation/index.html.

Clapper, James R. 2013. "Worldwide Threat Assessment of the US Intelligence Community." Senate Select Committee on Intelligence (March). https://edge.apus.edu/access/content/group/security-and-global-studies-common/Intelligence%20Studies/INTL634/634-wk8-worldwise%20threat%20assessment.pdf.

"Cyber Attacks." 2014. *Financial Times* (August). http://search.proquest.com/docview/1564797061?accountid=8289.

"Cyber Attacks Could Paralyze US Ports, Captain Moskoff Tells CCICADA." 2015. *Command, Control, and Interoperability for Advanced Data Analysis* (January). http://www.ccicada.org/2015/01/29/cyber-attacks-could-paralyze-us-ports-captain-moskoff-tells-ccicada/.

"Cyber Definitions." 2013. NATO Cooperative Cyber Defense Center of Excellence. https://ccdcoe.org/cyber-definitions.html.

Fink, G.A., J.N. Haack, A.D. McKinnon, and E.W. Fulp. 2014. "Defense on the Move: Ant-Based Cyber Defense." *Security & Privacy, IEEE* 12 (2): 36–43. http://ieeexplore.ieee.org.ezproxy1.apus.edu/xpls/abs_all.jsp?arnumber=6798536&tag=1.

Gilmer, Brad. 2013. "Cyber Attack." *Broadcast Engineering* 55 (5): 16–19. http://search.proquest.com/docview/1428252205?accountid=8289.

Greenwald, Judy. 2014. "Marine Sector Struggles With Cyber Risks." *Business insurance* 48 (10): 1–1, 30. http://search.proquest.com/docview/1524697289?accountid=8289.

Morris, Jefferson. 2002. "Despite Increased Security, Airlines Still Vulnerable to Cyberattack, Analysts Say." *Aerospace Daily* 201 (16). http://search.proquest.com.ezproxy2.apus.edu/docview/231514389/fulltext/684DB787DE3A41AEPQ/1?accountid=8289.

Office of Policy and Plans. 2012. "North American Cruise Statistical Snapshot, 2011." U.S. Department of Transportation Maritime Administration (March). http://www.marad.dot.gov/documents/North_American_Cruise_Statistics_Quarterly_Snapshot.pdf.

Rico, Antonieta. 2013. "Report: Lax Cybersecurity At U.S. Ports." *Navy Times* (July). http://search.proquest.com/docview/1418399133?accountid=8289.

Saini, Dinesh K., and Vikas Singh. 2012. "Soft Computing Techniques in Cyber Defense." *International Journal of Computer Applications* 50 (20) http://www.ijcaonline.org/archives/volume50/number20/7919-1217.

Schmidle, Nicolas. 2011. "A Reporter at Large: Getting Bin Laden: What Happened That Night In Abbottabd." *The New Yorker* (August). http://www.thomasweibel.ch/artikel/110808_new_yorker_getting_bin_laden.pdf.

"Signals Intelligence." 2015. National Security Agency/Central Security Service. https://www.nsa.gov/sigint/.

Tafoya, William L. 2011. "Cyber Terror." The Federal Bureau of Investigation (November). https://leb.fbi.gov/2011/november/cyber-terror.

"Target Card Heist Hits 40 Million." 2013. *BBC News* (December). http://www.bbc.com/news/technology-25447077.

"The Interview: Sony's North Korea Film to be Screened in US." 2014. *BBC News* (December). http://www.bbc.com/news/entertainment-arts-30589472.

"Understanding Denial-of-Service Attacks." 2009. United States Computer Emergency Readiness Team (November). https://www.us-cert.gov/ncas/tips/ST04-015.

U.S. Department of Defense. 2013. *Joint Publication 3-27: Homeland Defense*, July. http://www.dtic.mil/doctrine/new_pubs/jp3_27.pdf.

U.S. Department of Defense. 2014. *Joint Publication 3-26: Counterterrorism*, October. http://www.dtic.mil/doctrine/new_pubs/jp3_26.pdf.

Ward, Andrew, Demetri Sevastopulo, and Stephen Fidler. 2007. "US Concedes Danger of Cyber-Attack." *FT.Com* (September): 1. http://search.proquest.com/docview/228973691?accountid=8289.

Wei, Dong, Yan Lu, Mohsen Jafari, Paul M. Skare, and Kenneth Rohde. 2011. "Protecting Smart Grid Automation Systems Against Cyberattacks." *IEEE Transactions on Smart Grid* 2 (4): 782–795. doi:10.1109/TSG.2011.2159999.

Zhen, Lu, Zhen, Xiong, and Tu Keqin. 2014. "Research of Computer Network Information Security and Protection Strategy." *Applied Mechanics and Materials*, 496–500 (January): 2162–2165. http://search.proquest.com.ezproxy2.apus.edu/docview/1675252027?pq-origsite=summon.

CHAPTER 31:
THE COMBINED JOINT OPERATIONS FROM THE SEA CENTRE OF EXCELLENCE IS WAY AHEAD FOR MARITIME CYBER SECURITY

Ovidiu Marius Portase[1]
Romanian Naval Forces

Cyber space is a new frontier triggering research across all realms, especially in the maritime domain. New digital technologies have taken automation to a new level. Processing speeds have increased productivity, while decreasing staffing requirements. This presents new cyber security challenges that can negatively impact systems, activities, and lives. A cyber attack directed at targets in the maritime domain is a real and growing concern that needs further attention from key leaders in the maritime community.

The Combined Joint Operations from the Sea (CJOS) Centre of Excellence (COE), one of the North Atlantic Treaty Organization (NATO) accredited COEs, was one of the first organizations to recognize the need for maritime cyber security. CJOS is a recognized expert in the field of maritime situational awareness—an essential element of maritime security. The CJOS mission is to improve sponsoring nations' and NATO's ability to conduct operations in a dynamic maritime environment with emerging global cyber security challenges.

In 2012, CJOS identified maritime cyber security as a definite danger to overall maritime security and initiated a project to solve this growing threat. The objectives of the first phase of the project were to raise awareness that the cyber threat and vulnerabilities require more action, enhanced information sharing, and close cooperation amongst public and private stakeholders from the maritime community. The objectives of the second phase of the project are to identify the cyber vulnerabilities, recognize potential consequences, influ-

---

1   Disclaimer: The following article does not reflect the United States Fleet Forces Command, Combined Joint Operations from the Sea Centre of Excellence or NATO intent or policy.

ence the priorities and methods safeguarding critical infrastructure, and foster a maritime industry culture of increased awareness. The following are some of CJOS initiatives.

First, in 2013 CJOS worked closely with two other NATO accredited COEs and Old Dominion University (ODU) to identify maritime cyber security gaps that were presented in the Maritime Cyber Security Study (MCSS). CJOS launched the MCSS as part of an internally generated project, study was designed to explore the cyber, energy, and maritime security vulnerabilities in the maritime cyber security domain.

Second, in partnership with the United States Coast Guard (USCG) commands, Department of Homeland Security (DHS) Command, Control, and Interoperability Center for Advanced Data Analysis COE (CCICADA) and American Military University (AMU), CJOS co-sponsored the first major maritime cyber security event in 2015, with the aim of providing maritime cyber security expertise to the broad maritime community of interest.

Lastly, CJOS is currently developing a cyber security risk matrix that will provide a toolset addressing maritime cyber security threats. Additionally, the Romanian Navy has requested CJOS to conduct a maritime cyber security study identifying Romanian cyber vulnerabilities in port operation systems and ship systems in the Black Sea, an area of vital strategic importance.

Maritime Cyber Security Study

In November 2011, European Network and Information Security Agency (ENISA) released a report on cyber security aspects in the maritime sector (European Union Agency for Network and Information Security 2011). In July 2013, Commander Joseph Kramek (2013) published a paper on cyber vulnerabilities in U.S. port facilities sponsored by the Brookings Institute. Despite the conclusions outlined in these two publications, there are still some who are slow to see the connection between maritime cyber security and maritime security.

During the first phase of the project, CJOS collaborated extensively with its partners on research in the maritime cyber security domain. CJOS provided subject matter expertise to the Virginia Modeling, Analysis and Simulation Center (VMASC) on a DHS-sponsored project; assigned student research projects on maritime cyber security in port operations, containerized cargo management, energy domain and ship systems; co-hosted a round-table discussion; and, lastly, co-sponsored a webinar organized with the Maritime Institute. ODU made available its research resources and enriched its expertise on maritime cyber security.

CJOS findings from its collaborative efforts demonstrate the definite need for a more comprehensive approach to maritime cyber security. The results from this phase of the project determined the following:

- A lack of understanding exists between maritime and land domain cyber security challenges
- The current cyber security threat is driven mainly by economic, opportunistic, and criminal motivations, but could in the future be influenced by political aims or even as part of a hybrid warfare strategy
- The energy sector is cognizant of cyber security issues in the maritime context, but focus remains primarily on the financial aspects of their business models
- There is a limited understanding of the vulnerabilities of ship systems to cyber interference
- There is a heavy reliance on GPS navigation and automated cargo management systems used onboard ships, in ports and in the energy sector, which are increasingly susceptible to cyber-attack or interference

CCICADA AND AMU

In 2014, the United States Government Accounting Office (U.S. GAO) released a report recognizing the need to better address port cyber security. The U.S. GAO (2014) mentioned insufficient actions and deficiencies in federal agencies that limit the ability to properly mitigate cyber incidents emerging within the maritime domain. CJOS believes more attention and resources are needed to mitigate issues impacting maritime cyber security. The maritime domain is far more complex due to sustainment of undersea communication cables; protecting vital onshore connection points; defending information management systems used in maritime port operations; tracking and de-conflicting maritime vessel traffic systems; and safe-guarding global positioning systems (GPS) used for maritime navigation to name a few. None of these areas has been studied thoroughly enough to identify cyber vulnerabilities to date. To increase the desired level of cyber security in the maritime domain, the shortfalls and vulnerabilities in the maritime cyber arena need to be adequately addressed. Through teamwork, partnerships and research on common systems like Resilient Hull, Mechanical, and Electrical Security System (RHIMES) developed by the U.S. Navy could prove beneficial to the wider maritime community of interest (Freeman 2015).

Notable steps have been made to address maritime cyber security for port operations, particularly in the U.S. and Europe.[2] The energy sector has taken significant strides to protect their systems as well (Moschner 2015). Despite these achievements, an enduring lack of understanding for ship-based systems continues to plague the maritime community. This presented another opportunity for CJOS to take the lead and identify issues and provide solid recommendations. The research on cyber security of ship systems (navigation, ship control systems, at-sea cargo handling systems) was conducted in cooperation with industry, nations, and academia. In March 2015, the results were presented by CJOS representatives at the Maritime Cyber Security Symposium organized by AMU and CCICADA.

The International Maritime Organization's (IMO) existing regulations and policies focus mainly on security and safety without directly addressing the electronic threat. The International Convention for the Safety of Life at Sea (SOLAS) and the International Ship and Port Facility Security (ISPS) code demands that security measures be taken aboard ships and in port infrastructure; however, neither of the documents incorporate appropriate maritime cyber security measures. The initiative taken by Canada to identify and develop guidelines on maritime cyber security issues is commendable; an initiative followed up by the United States and other representatives from within the maritime community (International Maritime Organization 2016).

Maritime Cyber Security Risk Matrix and Romanian Naval Forces Cyber Study

Looking toward the future, CJOS has recognized another area of research that holistically identifies a risk-based approach for maritime cyber security. While cyber security has become a critical concern worldwide, more leaders and managers from the maritime domain realize there has been limited movement identifying and understanding the consequences of a series of cyber-re-

---

2   For the United States are notable USCG activities to address cyber risks in the Marine Transportation System, activities presented by Rear Admiral Paul F. Thomas during his testimony before the House Committee on Homeland Security Border & Maritime Security Subcommittee on October 8, 2015. For Europe the most notable is the European Commission-funded project called CYSM, project which addresses potential security gaps in ports related to the cyber-physical nature of their infrastructure. (See "Testimony of Rear Admiral Paul F. Thomas Assistant Commandant for Prevention Policy on Cybersecurity in U.S. Ports Before the House Committee on Homeland Security Border & Maritime Security Subcommittee, 8 October 2015," Unites States House of Representatives, http://docs.house.gov/meetings/HM/HM11/20151008/104007/HHRG-114-HM11-Wstate-ThomasP-20151008.pdf (accessed October 8, 2015); "Collaborative Cyber/Physical Security Management System," CYSM: An Innovative Physical/Cyber Security Management System for Ports, http://www.cysm.eu/index.php/en/ (accessed January 30, 2016)).

lated risks. A common toolset for assessing the maritime cyber security risk is necessary for sharing information between various elements in the maritime domain. Cooperation will be required in the future to take action and manage incidents generated by a cyber attack.

In the second phase of its initial project, CJOS is focusing its efforts on the development of a maritime cyber security risk matrix to enable maritime domain operators to assess maritime cyber-related risks. Once cyber risks are identified, defensive measures can be prepared to fight-through cyber-based attacks. In April 2015, the Romanian Navy submitted a request for CJOS support. Project objectives were adjusted to identify the cyber vulnerabilities within the maritime domain, especially those impacting ship and port operator systems; to recognize potential consequences of exploited maritime cyber security vulnerabilities in the Black Sea and in Romanian ports; to examine the priorities and methods to safeguard maritime critical infrastructure; to foster a culture of increased cyber security awareness in the maritime community; and to share information, cooperate and collaborate among government and non-government agencies.

The other aspect of this project is to develop a cyber security risk matrix that will provide a toolset addressing possible cyber threats within the maritime domain; identifying vulnerabilities of common shipboard and shore-based applications and systems; and finally, score the impact to all elements in the maritime domain. This will provide increased awareness and understanding of cyber threats that can jeopardize vital maritime interests, navigation safety, maritime security and critical infrastructure. Maritime domain users will need to re-evaluate their cyber risk in order to safeguard the maritime infrastructure and improve their cyber resilience. A risk matrix could prove useful in further concept development and experimentation.

Conclusion

Maritime cyber security has been recognized as a growing field worldwide. During his first address to the Assembly in November 2015, Mr. Kitack Lim, the new Secretary-General of the IMO, mentioned cyber security as one of the IMO's challenges and issues (International Maritime Organization 2015). The maritime domain presents some very unique and challenging cyber vulnerabilities that require unique approaches to reduce the associated risks. The work ODU has completed to date has been critical to identifying maritime cyber security issues that have assisted CJOS in developing a way ahead to tackle cyber security issues in the maritime domain. CJOS looks forward to the continued partnership with other stakeholders to work cyber security issues in the future.

REFERENCES

European Union Agency for Network and Information Security. 2011. "Cyber Security Aspects in the Maritime Sector." (December 19) https://www.enisa.europa.eu/activities/Resilience-and-CIIP/critical-infrastructure-and-services/dependencies-of-maritime-transport-to-icts/cyber-security-aspects-in-the-maritime-sector-1.

Freeman, Bob. 2015. "A New Defense for Navy Ships: Protection from Cyber Attacks." America's Navy. (September 18). http://www.navy.mil/submit/display.asp?story_id=91131.

International Maritime Organization. 2015. "IMO Assembly Confirms Mr. Kitack Lim as Secretary-General." (November 26). http://www.imo.org/en/MediaCentre/PressBriefings/Pages/50-kitack-lim-SG-.aspx.

International Maritime Organization. 2016. "MSC 96/25, Report of the Maritime Safety Committee on Its Ninety-Sixth Session." (May 31). https://docs.imo.org/Search.aspx?keywords=msc%2096.

Kramek, Joseph. 2013. "The Critical Infrastructure Gap: U.S. Port Facilities and Cyber Vulnerabilities." Brookings (July 3). http://www.brookings.edu/research/papers/2013/07/03-cyber-ports-security-kramek.

Moschner, Anne. 2015. "CMA Shipping 2015: DNV GL Addresses Cyber-security Risks." DNV GL (March 25). https://www.dnvgl.com/news/cma-shipping-2015-dnv-gl-addresses-cybersecurity-risks-21348.

U.S. Government Accountability Office. 2014. "Maritime Critical Infrastructure Protection: DHS Needs to Better Address Port Cybersecurity." Report Number GAO-14-459. (June 5). http://www.gao.gov/products/GAO-14-459.

CHAPTER 32:
TOWARD A MARITIME CYBER SECURITY COMPLIANCE REGIME

Mark R. Heckman, John McCready, David Mayhew, and Winnie L. Callahan
University of San Diego

Abstract

This whitepaper is an attempt at understanding, bounding, and providing a framework for answering the following questions posed by the United States Coast Guard: *What actions should vessel owners and port entity operators perform to evaluate their cyber safety and security postures (perhaps within the context of MTSA)? How can these steps be validated and who, if anyone, should do so and how often? How can we measure the effectiveness of a compliance regime that requires these types of performance-based protection measures?* The paper discusses cyber security standards and compliance regimes in general, risks specific to the maritime transportation system, current efforts by the USCG and other entities, the feasibility of a USCG-led maritime compliance regime, and directions for future research.

Introduction

The 2013 Presidential Policy Directive 21 (PPD-21), *Critical Infrastructure Security and Resilience,* promotes a national policy to "strengthen and maintain secure, functioning, and resilient critical infrastructure" (White House 2013b). Critical infrastructure is defined in Executive Order (EO) 1363, *Im-*

proving Critical Infrastructure Cybersecurity, as "systems and assets, whether physical or virtual, so vital to the United States that the incapacity or destruction of such systems and assets would have a debilitating impact on security, national economic security, national public health or safety, or any combination of those matters" (White House 2013a). One of the 16 critical infrastructure sectors identified in PPD-21 is Transportation Systems. The sector specific agencies (SSA) tasked with overseeing the Transportation Systems sector effort, the Department of Homeland Security (DHS) and the Department of Transportation (DOT), identified 7 key subsectors, one of which is the Maritime Transportation System (MTS) (U.S. Department of Homeland Security 2016). Under the DHS and DOT's *Transportation Systems Sector-Specific Plan* (TS SSP), DHS delegated its co-SSA responsibilities to the Transportation Security Administration (TSA) and the United States Coast Guard (USCG) (U.S. Department of Homeland Security 2015c).

Cyber systems in the MTS are intricately linked to physical systems and provide attackers an increased attack surface beyond the primarily physical vectors addressed by existing regulations such as the International Ship and Port Facility (ISPS) Code and Federal regulations that implement the Maritime Transportation Security Act (MTSA) (Office of the Federal Register 2003; International Maritime Organization 2016). One of the goals of the TS SSP is to further the implementation of EO 16363, in recognition of the increasing reliance by transportation services on networked and often remotely-accessible cyber-based systems for "positioning, navigation, tracking, shipment routing, industrial system controls, access controls, signaling, communications, and data and business management." Frequent reports of the ease with which attackers have been able to penetrate and exploit similar types of systems in other government and commercial sectors increase the urgency of addressing these new threats.

A major objective in the 2015 *United States Coast Guard Cyber Strategy* is to reduce cyber vulnerability for vessels and facilities. (U.S. Coast Guard 2015). Achieving this objective will require the USCG to "Develop guidance for commercial vessel and waterfront facility operators on how to identify and evaluate their cyber security-related vulnerabilities [and] incorporate this risk information into existing vessel and facility security assessments, or other appropriate management regimes, conducted by private industry and port authorities." The USCG asked the University of San Diego (USD) to explore the form that an organized program that provided such guidance and assessments might take. Specifically, the questions posed by the USCG to USD were:

What actions should vessel owners and port entity operators perform

CHAPTER 32

to evaluate their cyber safety and security postures (perhaps within the context of MTSA)? How can these steps be validated and who, if anyone, should do so and how often? How can we measure the effectiveness of a compliance regime that requires these types of performance-based protection measures?

An evaluation of the cyber safety and security posture of a vessel or port facility, however, is likely to always return "not secure" unless the vessel or port facility operator takes specific actions to improve the posture. A cyber security compliance regime evaluates the security posture of an entity by comparing the security practices of the entity against a security standard. So a more pertinent formulation of the questions might be this:

> What practices should vessel owners and port owner operators adopt in order to raise their security postures to a minimally acceptable level? How can their efforts be validated and by whom, and how frequently should evaluations be performed? How can the effectiveness of the compliance regime that validates the security efforts be measured?

This whitepaper is an attempt to understand, bound, and provide a framework for addressing these questions. It incorporates feedback from a November 16, 2016 working meeting of the Maritime Cyber Security USCG/University Research Initiative on an earlier draft. The following sections discuss cyber security standards and compliance regimes in general; risks specific to the MTS, current efforts by the USCG and other entities, the feasibility of a USCG-led maritime compliance regime, and directions for future research.

COMPLIANCE REGIMES

A compliance regime is a set of processes that ensure an organization and its systems are compliant with a set of obligations, usually regulatory obligations. The regime must include a standard or set of standards and processes for evaluating compliance against the standards, for resolving deficiencies found during an evaluation, and for maintaining compliance between evaluations. Additional processes are usually required for record-keeping and reporting results to the compliance enforcement agency. The USCG, for example, under Federal regulations to enact provisions of the MTSA (Office of the Federal Register 2003), administers a compliance regime for U.S. enforcement of the International Convention for the Safety of Life at Sea (SOLAS) and the International Ship and Port Facility (ISPS) Code (International Maritime Organization 2016). Many aspects of this regime may also serve as a model for a maritime cyber security compliance regime.

Compliance regimes are widely used for evaluating the security of cyber systems in many different sectors. Examples include the Payment Card Industry Data Security Standard (PCI-DSS) (Payment Card Industry 2016), required of all businesses that handle credit-card payments; the private regulatory authority North American Electric Reliability Corporation (NERC) Critical Infrastructure Protection (CIP) standards used to assure the reliability of the bulk power system in North America (North American Electric Reliability Corporation 2016b), which have been adopted by the Federal Energy Regulatory Commission (FERC) of the U.S. Department of Energy (DOE) as a mandatory standard; and the Department of Health and Human Services (HHS) Health Information Portability and Accountability Act (HIPAA) security and privacy rules, which apply to all health plans, health care clearinghouses, and any health care provider who transmits health information in electronic form (HIPAA Security Rule 2016).

Compliance Regime Attributes

In addition to the standards and processes used to assess compliance, a compliance regime is characterized by several attributes:

- Is compliance mandatory or voluntary?
- Responsibility for enforcement
- Responsibility for auditing and reporting
- Penalties for non-compliance
- Period between audits

Mandatory versus Voluntary Compliance

A cyber security compliance regime may be mandatory, required by government statute or regulation, or it may be voluntary, where entities in a particular sector voluntarily impose a compliance regime on themselves because the entities recognize and seek to mitigate risk (and, perhaps, to forestall a government-mandated regime). HIPAA is an example of a mandatory compliance regime required by laws and regulations. PCI-DSS is a voluntary commercial sector initiative mandated by credit card companies, not by a government entity. NERC CIP was originally a voluntary regime that has become a mandatory regime under FERC.

Because the MTS is critical infrastructure, it is logical that a maritime cyber security compliance regime will be mandatory, just as SOLAS and the ISPS Code are.

CHAPTER 32

Responsibility for Enforcement

A compliance regime that has any teeth at all requires enforcement. An enforcement entity is responsible for carrying out control and compliance review measures and levying sanctions. The enforcement entity could be private, as in the PCI-DSS, or a government entity, as with HIPAA, or a private entity with the backing of a government entity, as with NERC and FERC.

The NERC/FERC paradigm may be a useful model for the USCG and the MTS. A private critical sector umbrella organization, NERC, developed a cyber security compliance standard and administers that standard under regulatory review and oversight of FERC. NERC has been certified by FERC for purposes of establishing standards, enforcement, and imposing sanctions for violations, and is required to submit an assessment of its performance to FERC every 5 years.

It is not clear, however, if there is any existing non- or quasi-governmental organization that currently has the authority, widespread acceptance, and technical knowledge to serve the same role in the MTS as NERC does in the electrical power sector. Perhaps the International Maritime Organization (IMO), which provides support and guidance for the SOLAS and the ISPS Code, comes closest. But the IMO is a treaty organization that does not enforce compliance. Enforcement in the U.S. is the responsibility of the USCG.

Under the MTSA, a USCG Captain of the Port (COTP) (also the District Commander, Area Commander, and the Commandant) has the authority to enforce provisions of SOLAS and the ISPS Code. Vessel and facility owners or operators must submit proof of compliance to the COTP, request a waiver, or face sanctions. Owners and operators may carry out their own security assessments and develop their own security plans, or they may hire and supervise third parties with "appropriate skills". The role of the USCG is to check the quality of the assessments and plans and to ensure that the facilities and vessels are adhering to the plans (Office of the Federal Register 2003).

Conceivably, the USCG could enforce cyber security compliance using a similar model. Owners or operators, or an independent third-party hired by them, would be responsible for conducting compliance evaluations and reporting results to the USCG.

Responsibility for Auditing and Reporting

Auditing and reporting of audit results may be performed by the audited entity or by an independent entity. Self-auditing and reporting suffers from the obvious weakness that organizations cannot always be trusted to accurately evaluate and report their security postures, especially if there are sanctions for non-compliance. (Self-reporting can be useful, however, to gain experience when a new regime is starting up.) Independent third parties, presumed to be unbiased and trained, generally give more reliable results.

Some compliance regimes, such as HIPAA, use self-auditing and reporting. Organizations subject to the HIPAA security rule must carry out periodic technical and non-technical evaluations to make sure that their security policies and procedures meet the security requirements. HHS does not care if the evaluation is made by the organization itself or by a third-party and there are no standards for certifying auditors. The PCI-DSS, on the other hand, requires formal compliance assessments by trained "Qualified Security Assessors" (QSAs) in certain cases, such as for merchants (or their service providers) who handle large numbers of transactions or that have been breached in the past. The PCI Security Standards Council provides training and certifies QSAs, but each organization can choose its own QSA to carry out an audit. Each of the eight NERC regional entities keeps a list of contractor auditors who are deemed to have suitable training and experience and chooses which auditors are assigned to carry out an audit. NERC's approach of choosing the auditors ensures auditor independence and prevents organizations from "auditor shopping" in an attempt to more easily pass audits.

Because cyber security compliance is much more technical than the primarily physical compliance standards in the ISPS Code, adequate training and certification of evaluators will be critical for ensuring quality and consistency of evaluations. Creating a training and certification standard for evaluators will be an essential part of the creation of a maritime cyber security compliance regime. It is likely that, as with other cyber security compliance regimes, a commercial infrastructure of training courses and certifications will spring up as the regime matures, but certified evaluators should nevertheless be accredited by the enforcement agency (presumably the USCG).

Whether self-reported or reported by a third-party, results of an evaluation should be presented in a standard report format. And the compliance regime must provide some way for monitoring an organization's progress in addressing whatever deficiencies were found in an audit.

Penalties for Non-Compliance

Failure to comply with a mandatory regime may result in significant penalties. A business that is found not in compliance with the PCI-DSS, for example, may be forbidden from handling credit card transactions, which could severely hamper their operations. Sanctions for violations of NERC CIP include large monetary penalties and, in extreme cases, limitations on operations (North American Electric Reliability Corporation 2016b). And unauthorized disclosure of protected health information can lead to major fines by HHS (HIPAA Security Rule 2016).

Possible penalties under the MTSA for violations of SOLAS and the ISPS Code are as follows (Office of the Federal Register 2003):

1) Inspection of the vessel;
2) Delay of the vessel;
3) Detention of the vessel;
4) Restriction of vessel operations;
5) Denial of port entry;
6) Expulsion from port;
7) Lesser administrative and corrective measures; or
8) For U.S. vessels, suspension or revocation of security plan approval, thereby making that vessel ineligible to operate in, on, or under waters subject to the jurisdiction of the U.S. in accordance with 46 U.S.C. 70103(c)(5).

As cyber security is part of the overall security posture, these penalties may also be appropriate for violations of a maritime cyber security compliance regime.

Period between Audits

Allowing too much time between security evaluations may permit entities to grow lax with respect to compliance requirements, but audits are usually expensive and time-consuming efforts. PCI requires annual audits or security assessments, plus quarterly vulnerability scans (Payment Card Industry 2016). NERC Regional Entities typically conduct on-site audits every three years, but the results of "Inherent Risk Assessments" (IRAs) are used to scope the level of effort, so that some entities may only need to provide self-certifications or submit to spot-checks (North American Electric Reliability Corporation 2016a). Under certain conditions, however, based on the risk calculation, audits may be carried out more frequently (and FERC has begun to conduct its own random audits that are independent of NERC audits), so the period between audits can vary. The frequency of evaluations under a maritime cyber security compliance regime may be similarly variable, but that will depend on the nature of the regime and the standards used, and is still an open research question.

Cyber security Standards

The security of a system can only be evaluated for compliance against a standard or set of standards. Standards define the minimum set of security practices that an entity must adopt in order to be considered compliant. The level of compliance is a rough indicator of the entity's overall security posture. An essential first step in the creation of a maritime cyber security compliance regime, therefore, is creation of a set of standards against which compliance can be measured.

But standards only make sense in terms of helping entities to meet their organizational objectives (e.g., satisfy all relevant laws and regulations, be profitable, etc.) and enforcing a well-defined security policy against specific threats. Without a firm basis in policy and threats, a standard will likely have gaps that an adversary can exploit, or might contain unnecessary requirements that waste resources.

Security Policies

A security policy is the definition of security for a system. A system that correctly enforces the policy can be said to be "secure with respect to the policy". A security policy identifies the information that is to be protected, who is authorized to access the information, and the type of access that is authorized. Cyber security policies typically are concerned with the confidentiality of information (controlling who is allowed to read information), the integrity of information (maintaining the trustworthiness of information), and the availability of information (ensuring that information can be accessed and used when needed). The HIPAA security rule, for example, is primarily concerned with maintaining the confidentiality of protected health information. NERC CIP standards address confidentiality of sensitive data belonging to electricity sector infrastructure owners and operators, but focus primarily on the integrity of power grid components in order to maintain the reliability (availability) of the power grid.

Entities in the MTS have requirements for all three of these types of security policies. Development of a sound security policy is essential for creating a security program, and the security posture of a vessel or port facility can only be evaluated with respect to the specific policy that the vessel or facility is trying to enforce. So an essential first step in creating a maritime cyber security compliance regime will be to identify the types of information that must be protected, who is authorized to access that information, and what types of access are authorized.

CHAPTER 32

Threats, Vulnerabilities, and Risks

Threats are events or actions that can lead to a violation of a security policy. For example, a threat could be that an attacker will try to install malware on a system and thereby gain access to sensitive information, violating a confidentiality policy, or be able to tamper with the system and its data, violating an integrity policy. A vulnerability is a weakness in a system that could allow a threat to actualize. An unpatched operating system, for example, contains vulnerabilities that could be exploited to install malware.

An attack can be successful only when a threat meets a vulnerability. If there is no threat then a vulnerability does not matter and there can be no attack. If there is no vulnerability for a threat to exploit, there also can be no successful attack. Standards are written based on a threat model. The security controls in a standard are intended to eliminate or mitigate vulnerabilities so that threats cannot exploit them. That is why, for example, timely installation of operating system patches is always a part of cyber security standards. The patches eliminate vulnerabilities that could be exploited by threats.

A risk is an estimate of the potential impact of a successful attack. Risk estimates can be used to prioritize attention to protecting resources. If the cost of installing a particular security control exceeds the risk, it is not cost effective to install the control. The NIST Risk Management Framework (RMF) is a standard for protecting Federal information systems. Under the RMF, entities must determine the relative criticality of computing resources—low, medium, or high—with respect to confidentiality, integrity, and availability. The criticality level and the policy determine the number, type, and relative strength of security controls that must be used to protect each resource (National Institute of Standards Technology 2010).

Good cyber security standards are based on policies, but also must consider threats, potential vulnerabilities, and risks to ensure that the security controls mandated by the standard are necessary and sufficient, and cost effective.

Descriptive and Prescriptive Standards

A standard may be descriptive or prescriptive. A descriptive standard specifies activities, but does not provide interpretation and specific implementation requirements. Examples are the "security rule" of HIPAA and the NERC standards. The HIPAA standards are intentionally high level and generic, so that different approaches can be used as technology evolves (HIPAA Security Rule 2016). The NERC CIP standards are also high-level and open to different interpretations, but NERC has found it necessary to provide detailed implementation guidance documents (North American Electric Reliability Corporation 2016c). At the other extreme, prescriptive standards are highly specific implementation rules. Examples of prescriptive standards are Defense Information Systems Agency (DISA) Security Technical Implementation Guides (STIGs), which are exhaustive lists of system configuration settings for Department of Defense (DOD) information assurance (IA) and IA-enabled devices (Information Assurance Support Environment 2016). STIGs are part of the complete set of DOD compliance standards, which include both high-level, generic standards and the highly-specific STIGs.

As with the DOD, systems in the MTS will likely require both high-level, generic standards written so that they can still be effective as technology changes, but also highly detailed configuration standards for specific components.

Pros and Cons of Compliance Regimes

Compliance regimes, as with any human-designed system, have both advantages and disadvantages. Proper attention to processes and sanctions, however, can go a long way toward mitigating the disadvantages.

Advantages of Compliance Regimes

A compliance regime has many advantages as a defined program for focusing effort on a particular problem, such as cyber security. Some of these advantages are listed here:

- **Creates a definitive baseline for protection**—Compliance must always be measured against a standard. Creation of a compliance regime first requires creation of a clear and definitive security standard that is based on security policy, threats, and risk.

- **Improves cyber operations**—A well-designed cyber security standard focuses an entity's attention on secure configuration and management of cyber assets, which helps to normalize and streamline operations. For example, implementing a compliance regime encourages an organization to detect and identify previously unknown (and unapproved) cyber assets, to clean up tangled firewall rules that have accreted over time, and to implement monitoring that can detect changes to database schema, permissions, and dependencies.

- **Sets reasonable expectations**—Without a recognized and accepted standard, entities have no guidance about what a reasonable level of effort is.

- **Secures upper management buy-in**—Security is a very difficult property to measure. Organizations will often underestimate their risk and allocate too few resources to security in the absence of a mea-

CHAPTER 32

surable need. A compliance regime makes it possible to measure "success" (compliance with the standard) and to account for the cost of failure (the sanctions).

- **Drives budget into security departments**—Management will ensure that security departments have adequate resources to accomplish compliance. A compliance regime makes it possible for a security department to show, for example, how an expenditure of $100,000 can prevent a fine of $1,000,000, even in the absence of a detected breach.

- **Enhances interoperability and maintains security in networked entities**—Common, compatible security processes must be uniformly applied across distributed enterprises. Security is only as strong as the weakest link.

- **Improves communication and intelligence sharing**—Enterprises can share knowledge about breaches with partners who will be able to leverage their security controls to block attacks.

- **Creates efficiencies and economies of scale**—A security program based on a compliance regime requires a certain level of uniformity across the enterprise. Use of repeated, controlled processes create the opportunity for economies of scale.

- **Levels the playing field**—Entities that skimp on security may reap (temporary) benefits in terms of reduced expenditures, but a breach may have serious downstream effects and losses extend well beyond that one entity. A compliance regime sets a minimum level of effort and forces everyone to shoulder part of the burden.

- **Increases customer confidence**—With major cyber security breaches being publicized almost daily, adherence to a strong compliance standard can increase customer confidence that the entity is taking appropriate steps to protect customer and entity assets.

- **Provides defensible legal position in case of a breach**—Entities that are evaluated and found to be in compliance with an accepted cyber security standard may demonstrate that they are meeting a legal standard of due care with respect to cyber assets.

- **Provides a basis for insurers**—Cyber incident insurance is needed to help transfer risk. But insurers generally lack sufficient actuarial data to determine insurance rates. A compliance regime provides a standard that can be used by insurers as well as regulators.

Disadvantages of Compliance Regimes

While the advantages of compliance are marked enough to spark the creation of many such regimes, compliance regimes also have several weaknesses. It is imperative that the standards and processes that drive a regime mitigate the weaknesses in order to realize the advantages. Here are some of the weaknesses:

- **Compliance regimes can be expensive to implement and manage**—Organizations must maintain records, generate reports and, except when self-reporting is tolerated, support the work of an external auditor. The overhead of running a compliance regime may, in some cases, exceed the cost of the security controls themselves.

- **Regulations and standards only reflect current industry "best practices"**—"Best practices" are usually widespread and intuitive conventions, but historically have seldom had a scientific basis for assessing effectiveness. Once enshrined in a standard, they are very difficult to change.

- **Standards may lag well behind the latest attack vectors**—Yesterday's best practices may not reflect today's attacks, so standards require constant updating.

- **An organization can be totally compliant, yet completely insecure**—As the saying goes, "compliance is to security what a dance-step diagram is to dancing." The result of a compliance audit is a checklist that reflects the conditions of an entity at a particular moment in time. Checklists measure the existence of controls, but it is very difficult to measure the quality of the controls. And the controls may not be vigorously implemented between audits.

- **Instead of being the floor, the standard becomes the ceiling**—Organizations may be tempted to do just the minimum necessary to pass a compliance audit. Security is no longer the goal; passing the audit becomes the goal.

- **Inconsistent auditing**—Training and experience of auditors can vary and compliance standards are often subjective, so auditors may disagree after evaluating the same system. If third-party auditors are paid by the organization under audit, the organization may "audit shop" until they find an easy auditor.

CHAPTER 32

Measuring the Effectiveness of a Compliance Regime

The overall objective of a cyber security compliance regime is to improve the security posture of compliant organizations. But directly measuring "security" is not possible. If no successful breaches have been detected during some period of time, for example, is that because the security measures called for in the standards are effective against all attacks, because there just happened to have been no attacks of sufficient sophistication during that period of time that could get past the security measures, or because mechanisms for detecting successful attacks are too weak?

While "security" is impossible to measure, compliance with the security practices called for in the standards is intended to correlate with an improved security posture. And the reason for having a cyber security compliance regime, rather than simply having just standards, is to encourage or coerce organizations to adopt the security practices called for in the standards. A reasonable metric for the effectiveness of a compliance regime, therefore, is how well organizations comply with the standards over time. Effectiveness can be measured by counting the number of compliance deficiencies found during audits. Assuming a certain level of quality and consistency in evaluations, if the number of deficiencies found in each audit falls over time and reaches a steady, but low, level, the compliance regime can be counted a success because organizations are paying attention and making sure they are compliant. If the number of deficiencies continues to be high, however, the effectiveness of the regime may be called into question.

Another approach is to measure attributes of the security controls called for in the standards. Measurements may include, for example, counts of identified security vulnerabilities (which should drop over time) or the time it takes for an organization to respond to detected incidents. Metrics based on these measurements can be shown to be correlated with the effectiveness of the regime.

MTS-Specific Cyber Threats and Vulnerabilities

The primary cyber threats to the MTS are that attackers will attempt to (1) cripple ports; (2) track, divert, and disable vessels; and (3) disrupt cargo handling and tracking through attacks on information technology (IT) or computer-reliant systems. This risk is increasing along with "the growing reliance on cyber-based control, navigation, tracking, positioning, and communications systems" (U.S. Department of Homeland Security 2015c). The attack surface of maritime activities, formerly existing primarily in the physical realm, increasingly extends into the digital realm (and can be detected as failures across the digital to physical barrier). The USCG Office of Port & Facility

Compliance suggests that cyber attacks could "kill or injure workers, damage equipment, expose the public to harmful pollutants, substantially slow cargo operations," and even "facilitate the smuggling of people, weapons of mass destruction, or other contraband" (U.S. Coast Guard 2016b).

Port Facility Cyber Threats and Vulnerabilities

Different vessels and port and terminal facilities, of different ages and with different levels of resources, may have widely varying dependence on cyber resources. It is, therefore, difficult to generalize what threats and vulnerabilities broadly apply throughout the sector. Cyber Guard attack/defend exercise simulations make it apparent, however, that all ports face many of the same type of threats:

- **General IT threats**—Similar to threats faced by other government and commercial offices, such as attacks on office and server systems, networks, and web sites. Attackers may, for example, steal, tamper with, or destroy key databases.

- **Threats to operational technology (industrial) networks and systems**—Port facilities and terminals are also industrial sites that include remote equipment control networks—i.e., industrial control systems (ICS) and supervisory control and data acquisition (SCADA) systems, and equipment connected to them such as cranes, fuel dispensing or recovery components, and even a local power grid.

- **Threats to physical security systems**—Physical security systems, such as closed-circuit camera (CCTV) systems, are increasingly cyber-based and networked.

Vessel Cyber Threats and Vulnerabilities

Ship-builders are already demonstrating completely automated and autonomous vessels; the cyber threat to such vessels is obvious, as are the risks if command and control were lost. In many computer-based products, however, developers focus on features, time-to-market, and cost—security is usually an afterthought and serious vulnerabilities are detected long after the system is put into use. Given the risks, no automated, autonomous vessel should be permitted to operate unless the controls on the vessel have been demonstrated to comply with a rigorous cyber security standard. In the absence of clear compliance requirements, however, developers have even less incentive to build security into the product.

CHAPTER 32

Contemporary vessels already have a variety of devices and systems that are becoming increasingly cyber-dependent. As with most cyber-based devices and systems, these were undoubtedly developed with little or no awareness of cyber threats. Here are some examples of these devices and systems, and some of the threats to them:

- **Navigation**—automatic identification system (AIS), electronic chart display and information system (ECDIS), global positioning system (GPS), Radar—Information provided by the AIS, such as unique identification, position, course, and speed, may be blocked or spoofed; ECDIS navigation charts may be destroyed or tampered with, or inputs from position, heading, speed and other navigation sensors may be blocked or altered; GPS satellite signals maybe spoofed or blocked, or the system may be tampered with and provide incorrect location data.

- **Propulsion and Steering**—Digital controls are at risk of subversion. Malicious software could interfere with propulsion causing improper commands at critical times (e.g., vessels in turning basins) or could cause incorrect reporting of critical engine/propulsion indicators such as low oil pressure or high inter-cylinder temperatures, which may not immediately destroy an engine but could result in increased maintenance needs and causing reliability and cost issues.

- **Vessel balance of control**—Fire and damage control, Environmental controls, Waste controls—Intruders can disable systems or commandeer them, potentially, for example, dumping fuel, bilge, or ballast.

- **Communications**—Communications systems are ubiquitous within the entire vessel, interacting with routine message traffic, personal communications between crew members and their families, as well as providing continual reporting on position, system performance, navigation, etc. Communications are now largely automated and digital and can be blocked, eavesdropped on, or spoofed.

Cargo Handling Threats and Vulnerabilities

Disruptions to cargo management systems could lead to destruction, hiding, and theft of cargo, or allow the entrance of dangerous or illegal cargo. Refrigerated cargo monitoring ("Reefer Tracking") readings, for example, may be blocked or falsified, leading to destruction of the cargo. Fluid cargo management systems, intended to prevent overfilling and accidental product discharge of ballast or productive tanks by monitoring parameters as filling speed and temperature, may be crashed, or false readings may be presented on the console, resulting in damage to the crew, vessel, port, environment, or even loss of life. Management systems implemented using radio-frequency identification (RFID) tags are subject to attacks that block, clone, or spoof the tags, which would make it impossible to reliably identify and track shipments.

Existing, Non-secure Technology

Ideally, each of the computer-reliant devices and systems used in the MTS would be securable (and have their own STIG for secure configuration), but because they were developed without cyber security in mind, that may not be possible with the current generation of these devices. Additional development standards will need to be created that specify requirements for security that can be built into new systems. The long active service life-span of equipment and the need for backward compatibility, however, may delay full implementation of development standards for a long time, but is a necessary goal. As newer, and more securable, equipment is adopted in vessels and ports, older, less secure systems will co-exist with the newer systems, creating vulnerabilities. External devices may have to be used to provide some protection, but because the base protocol themselves in most cases do not have a security component, redundant systems may need to be used to provide some assurance that the devices or systems have not been subverted.

Multiple Spheres of Regulation

The 2013 attack on Target stores began with an attack on a vendor that provided HVAC and refrigeration systems to Target (Krebs 2014). This demonstrates that the security of a system may only be as good as the security of the weakest system to which it connects.

The MTS has many stakeholders and multiple entities may control different facilities in a port—even within the same cargo terminal. The June 2015 United States Coast Guard Cyber Strategy mentions that Coast Guard systems and networks are subject to compliance with DOD IA and Intelligence Community policies and regulations, but that most of the MTS critical cyber infrastructure is owned and operated by local governments and private companies that are not subject to the same compliance requirements (U.S. Coast Guard 2015). These different entities may have vastly different perceptions of cyber risk, different levels of resources to implement security controls, and different levels of cyber experience and training. Vessels may be flagged in many different countries, which also have the same disparities. This could complicate the creation of a broadly applicable maritime cyber security regime.

CHAPTER 32

Existing MTS Cyber Security Programs

Both the DHS and the USCG have current active efforts to address cyber security risks. USCG efforts to create a maritime cyber security compliance regime, not surprisingly given the request to create this whitepaper, are still immature. The USCG has an office of port and facility compliance whose mission is to "provide clear and timely regulations, policy and direction to Coast Guard Operational Commanders and other maritime stakeholders to achieve maritime safety, maritime security and environmental stewardship" (U.S. Coast Guard 2016a). The Domestic Ports Division (CG-FAC-1) has a web site dedicated to cyber security (U.S. Coast Guard 2016b), but this site is still underdeveloped and, absent a true standard, only makes very general suggestions to, e.g., "conduct a risk assessment" and "identify and adopt best practices". A January 2015 public meeting titled "Guidance on Maritime Cybersecurity Standards" was an attempt to grapple with some of the same questions posed in this whitepaper, including "How can vessel and facility operators reliably demonstrate to the Coast Guard that critical cyber systems meet appropriate technical or procedural standards" (U.S. Department of Homeland Security 2014)? The resulting February 2015 "Guidance on Maritime Cybersecurity Standards" document, which seems to be an attempt to answer the questions posed in the meeting, contained answers limited to procedures and tools to prevent malware infection through USB drives (U.S. Department of Homeland Security 2015a).

The DHS has created a "Transportation Systems Sector Cybersecurity Framework Implementation Guide" that purports to "provide an approach for Transportation Systems Sector owners and operators to apply the tenets of the National Institute of Standards and Technology [NIST] Cybersecurity Framework to help reduce cyber risks" but the guideline is an extremely high-level document that will require significant interpretation to turn it into something actionable by port facility and vessel operators (U.S. Department of Homeland Security 2015b). It provides no specific controls and little more direction than the NIST framework itself.

The NIST cyber security framework is considerably more developed and NIST frameworks have been used in other government agencies, but a framework is not a worksheet. It is non-trivial to interpret and implement a framework, requiring significant resources and trained staff to translate the abstract activities into concrete tasks (National Institute of Standards Technology 2014). We found no record of a successful application of the NIST framework to any MTS entity, although that could simply reflect a lack of publicity for such efforts.

The American Bureau of Shipping (ABS) has recently (September 2016) published a series of documents on Maritime cyber security, such as *Guidance Notes on the Application of Cybersecurity Principles to Marine and Offshore Operations* and a *Guide for Cybersecurity Implementation For The Marine And Offshore Industries*. (American Bureau of Shipping 2016a; 2016b). While these documents draw heavily from NIST guides, they also reference documents from other sources, such as the European Union Agency for Network and Information Security (ENISA) and standards organizations such as the International Standards Organization (ISO), International Electrotechnical Commission (IEC), IEEE, ANSI, and others. These documents are an attempt to tailor NIST and other standards to maritime cyber security. ABS intends this framework to serve as a certification standard, for which ABS will be the evaluator and certification-provider (American Bureau of Shipping 2016b).

Lacking a standard and compliance regime, additional maritime cyber security controls are appearing to satisfy a need that is increasingly evident to stakeholders. An example is the Norman SCADA Protection (NSP) system being marketed by Kongsberg Maritime to protect Kongsberg Maritime systems (used for positioning and navigation, marine automation, cargo management, and other maritime functions) from malware (Kongsberg 2012). The Kongsberg NSP was cited in the February 2015 USCG "Guidance on Maritime Cybersecurity Standards" document (U.S. Department of Homeland Security 2015a).

Additional informal and ad-hoc guidelines are developing as concern for maritime cyber security grows without a definitive standard. The results of the most recent Cyber Guard exercise, for example, identified a few key take-aways (Parsons 2016):

- Network segmentation—can apply more stringent controls to critical resources

- Intrusion detection and correlation—to identify attack vectors and compromised systems

- Credential protection (e.g., passwords)—Use two-factor

- Federal partnership—Work with partnering agencies (expertise, intelligence) and have regular exercises

These findings, however, hardly constitute a standard.

A March 2016 presentation to a large international shipping container group could only advise that, until maritime cyber security standards are developed, the group should take the following actions (K & L Gates 2016):

- Know Your Systems

- Assess Their Vulnerabilities
- Design/Install Cyber security Protections
- Educate Your Workforce
- Reassess/Strengthen Defenses Regularly
- Establish/Practice Response/Remediation Team
- Stay Abreast of USCG/IMO Standards
- Choose Insurance That Fits Your Needs/Risks

Once again, it is clear that a demand exists among sector entities for a MTS standard, but there is little in the way of progress toward a widely accepted and actionable standard.

Feasibility of a MTS Cyber Security Compliance Regime

It seems logical that the USCG has the authority to mandate a maritime cyber security compliance regime. Furthermore, it is clear that there is already a recognized need for a maritime cyber security standard, not only by the USCG, but also by port, terminal, and vessel operators, and likely by insurers. In the absence of a standard, operators are grasping at different informal and ad-hoc guidelines, while at least one private organization has created its own maritime standard. Given the demonstrated need, an official and well-designed standard and compliance regime could be an easy "sell" to operators and owners.

The span of authority of the USCG may not extend far enough to mandate full compliance with its standard to all terminal and vessel operators, however. Terminal operators have a certain amount of latitude with respect to how they run their operations, and foreign-flagged vessels have a great deal more. Compliance may start out mandatory for U.S. port facilities and vessels, but only recommended for others. Once a well-designed maritime cyber security standard becomes the U.S. standard, it could serve as a basis for international agreements under the IMO that would make compliance mandatory worldwide. A question for further research is what efforts other countries are making in this area and how those efforts could be combined with efforts in the U.S.

A comprehensive set of maritime compliance standards could be created as an interpretation of the NIST RMF tailored to protecting the MTS. Given the nature of the safety-critical devices and systems identified earlier, the standards would likely need to be a combination of high-level standards and STIGs. The feasibility to adopt or adapt previously created standards documents from other compliance regimes or from the ABS standard is still an open question, as is the full scope and shape of necessary standards.

The role of the USCG in enforcing the ISPS Code under the MTSA is oversight, approving standards, and reviewing and approving compliance. This is the same role the USCG would conceivably have in a new maritime cyber security compliance regime, and would probably not require a significant change in staffing or operations to manage the new regime. But managing a cyber security compliance regime requires different knowledge and skills than the primarily physical security regime of the ISPS Code, and will require new training for USCG personnel.

The issue of training applies just as much to port facility and vessel operators as it does to USCG personnel. A Vessel Security Officer training guide from a leading maritime training center that we reviewed, for example, contained no cyber security training (Vessel Security Officer Student Guide 2016). Cyber security practitioners are in great demand everywhere; it is highly unlikely that port facility staff and vessel crews would have these skills (large shipping companies, which have large IT departments, might be an exception). Other cyber security compliance regimes, however, have given rise to a vast infrastructure of training courses and certifications, as well as consultants, seeking to cash in on a lucrative market. It is certain that the same type of infrastructure will spring into existence around a maritime cyber security compliance regime as the regime matures.

Penalties for non-compliance with a maritime cyber security compliance regime could be very much like existing penalties for non-compliance under the MTSA. It is likely that any penalties, however, would be phased in gradually over time as standards are developed, tested, and adopted. This is the approach taken by, for example, the HHS under HIPAA, where penalties have gradually stiffened as organizations have had time to adapt to the new regulations.

As the preceding discussion shows, the form of a USCG-managed maritime cyber security compliance regime could be very similar to the way the USCG currently manages the current ISPS Code compliance regime, in terms of the USCG role, enforcement, and penalties. This suggests that a USCG-managed regime would be highly feasible. The specific standards that are to be enforced, however, are still an open research question.

RECOMMENDATIONS

A maritime cyber security compliance regime effort must begin with the creation of well-designed and actionable standards. The need for such standards is widely recognized because of increasing awareness about the risk of a cyber incident. Port facility and vessel operators are floundering without guidance.

But the MTS consists of a wide variety of cyber systems used for both

CHAPTER 32

business IT and industrial operations at ports and on vessels. There are existing cyber security compliance standards for business IT systems and even for SCADA systems that can be leveraged to create standards for maritime systems, but there are also unique aspects of the MTS that may not be addressed elsewhere. And there is no prior model for a vessel cyber security standard. Creating a comprehensive standard, even using a framework such as the NIST framework, will be a major undertaking.

Based on our findings, we recommend the following steps, in approximate order, for creating a comprehensive maritime cyber security compliance standard. Note that each of these steps is itself a research topic (or set of topics).

1) **Identify "organizational objectives" for cyber-based systems in the MTS**—This includes all relevant laws and regulations, as well as functional and performance goals.

2) **Create a complete inventory of cyber-based components (devices, networks, and systems) in the MTS**—No comprehensive threat model can be developed without such an inventory. The inventory should be categorized to help identify the types of applicable security policies and controls. Example categories could be, for example, "hardware", "software", "human", etc., but the ontology to be used is an open research question.

3) **Identify the security policies that must be enforced for digital information in port facilities and vessels**—This will require identifying the types of information that must be protected and the access policy for each information type.

4) **Create a detailed threat model for each component individually and as systems of components in vessels and port facilities**—Each item has its own threats and vulnerabilities, but each item is also part of a larger system. The larger system inherits the weaknesses of each of the components and the composition of components may induce weaknesses not present in the individual components.

5) **Create a risk model based on the threats and known or potential vulnerabilities**—The risk model must consider the impact on particular components and systems, vessels, or port facilities, but also to the MTS as a whole based on the downstream effect of a cyber attacks.

6) **Develop cyber security standards for base systems and protocols**—Standards for configuring and managing existing devices and

protocols in order to mitigate threats should be created and the threats that cannot be mitigated should be documented. New equipment and systems should be required to have security built-in to address the threats. (Getting acceptance for standards that all new systems must adhere to, and for new, secure protocols, is likely to be a long-term project, however.) Existing standards from other sectors should be leveraged, when possible.

7) **Develop cyber security standards for port facilities, terminals, and vessels as whole systems**—These standards address the risks specific to the composition of cyber components and extend over the entire entity. For example, insurance may be required to transfer residual risk when threats cannot be entirely mitigated.

8) **Foster the development of a cyber security curriculum that will become part of the required training of port and vessel personnel**—Tailor training to the needs of different staff positions.

9) **Require a cyber-component in all vessel and port security plans**—Security plans should identify responsible personnel and contain an incident handling plan that explains how a vessel's crew or a port's personnel would respond to a cyber-intrusion.

Conclusion

While the need for a maritime cyber security compliance standard is increasingly recognized, the decisions about how to create the standard and a regime built on the standard are still open research questions. Existing cyber security compliance standards and regimes in other critical sectors, such as NERC CIP in and the power industry, should be studied and adapted when possible.

The creation of a cyber security compliance regime for a transportation sector is not a problem unique to the MTS. The Federal Aviation Administration (FAA), for example, is attempting to address cyber security issues in aircraft and airports. Aircraft are no longer "isolated and independent systems" but now are integrated into a large, distributed, networked system (U.S. Department of Transportation 2014). Consisting as the MTS does of a combination of IT, commercial, industrial, and communications systems, a MTS cyber security compliance regime is more likely to resemble that of other transportation sectors than it does of sectors, such as the financial sector, with more homogeneous systems. It is not clear if the FAA has yet developed a solution to this problem, but study of their progress to date could be illuminating.

In addition to the open research questions that surround the creation of compliance standards, there are also research questions about the compliance regime that would enforce the standards. The USCG has a well-developed process for managing and enforcing a physical security compliance regime. It may be possible to preserve this process for cyber security, although the nature of the risk is very different and would require different training for USCG staff who would assume the additional responsibility. Furthermore, cyber assets, vulnerabilities, and threats are constantly evolving. It may be necessary to review and update the cyber security compliance standards much more frequently than the physical security compliance standards. The frequency of assessments, too, may need in general to be higher than for physical security, and as with other compliance regimes, there may be specific events, such as detected breaches, that trigger assessments. These decisions, too, call for further study.

The only decision that does not need to be further studied is the decision to push ahead toward an actionable cyber security compliance standard and regime for the MTS. There is a growing gap that other entities have also recognized and are racing to fill, but without coordination the results will be confusing and wasteful. The USCG should lead and coordinate these efforts.

References

American Bureau of Shipping. 2016a. "Guidance Notes on the Application of Cybersecurity Principles to Marine and Offshore Operations: ABS Cybersafety Volume 1." (September). http://ww2.eagle.org/content/dam/eagle/rules-and-guides/current/other/250_cybersafetyV1/CyberSafety_V1_Cybersecurity_GN_e.pdf.

American Bureau of Shipping. 2016b. "Guide for Cybersecurity Implementation for the Marine and Offshore Industries: ABS Cybersafety Volume 2." (September). http://ww2.eagle.org/content/dam/eagle/rules-and-guides/current/other/251_cybersafetyV2/CyberSafety_V2_Cybersecurity_Guide_e.pdf.

HIPAA Security Rule. 2016. 45 CFR Part 160 and Subparts A and C of Part 164. (January 7). http://www.hhs.gov/hipaa/for-professionals/security/.

Information Assurance Support Environment. 2016. "Security Technical Implementation Guides (STIGs)." (August 1). http://iase.disa.mil/stigs/Pages/index.aspx.

International Maritime Organization. 2016. "Maritime Security." http://www.imo.org/en/OurWork/Security/Guide_to_Maritime_Security/Pages/Default.aspx (accessed January 7, 2017).

K & L Gates. 2016. "Maritime Cybersecurity, CMA Shipping 2016." PowerPoint Presentation. (March 23).

Kongsberg. 2012. "Protecting Kongsberg Maritime IT Systems." (June 8). https://www.km.kongsberg.com/ks/web/nokbg0238.nsf/AllWeb/FFD6EB66CE1EC4C2C1257A170028505C?OpenDocument (accessed January 7, 2017).

Krebs, Brian. 2014. "Target Hackers Broke in Via HVAC Company." *Krebs on Security Blog*. (February 5). https://krebsonsecurity.com/2014/02/target-hackers-broke-in-via-hvac-company/ (accessed January 7, 2017).

National Institute of Standards and Technology (NIST). 2010. "Guide for Applying the Risk Management Framework to Federal Information Systems: A Security Life Cycle Approach." Special Publication 800-37, Revision 1. (February). http://dx.doi.org/10.6028/NIST.SP.800-37r1.

National Institute of Standards and Technology (NIST). 2014. "Framework for Improving Critical Infrastructure Cybersecurity, National Institute of Standards and Technology." (February 12). https://www.nist.gov/cyberframework (accessed January 7, 2017).

North American Electric Reliability Corporation. 2016a. "2016 ERO Enterprise Compliance Monitoring and Enforcement Program Implementation Plan." Version 2.5. July. http://www.nerc.com/pa/comp/Reliability%20Assurance%20Initiative/2016%20CMEP%20IP_v_2%205_071116_POSTED.pdf.

North American Electric Reliability Corporation. 2016b. "NERC Compliance & Enforcement." http://www.nerc.com/pa/comp/Pages/default.aspx.

North American Electric Reliability Corporation. 2016c. "NERC Compliance Guidance." http://www.nerc.com/pa/comp/guidance/Pages/default.aspx.

Office of the Federal Register. 2003. "Title 33 - Navigation and Navigable Waters." (July 1). https://www.gpo.gov/fdsys/pkg/CFR-2003-title33-vol1/pdf/CFR-2003-title33-vol1.pdf and https://www.gpo.gov/fdsys/pkg/CFR-2003-title33-vol1/pdf/CFR-2003-title33-vol1-part101.pdf (accessed January 7, 2017).

Parsons, Jim. 2016. "Cyber Guard 16." PowerPoint Presentation.

Payment Card Industry (PCI). 2016. "Data Security Standard: Requirements and Security Assessment Procedures." Version 3.2. April. https://www.pcisecuritystandards.org/documents/PCI_DSS_v3-2.pdf.

U.S. Coast Guard. 2015. *United States Coast Guard Cyber Strategy*. June. https://www.uscg.mil/seniorleadership/DOCS/cyber.pdf (accessed January 7, 2017).

U.S. Coast Guard. 2016a. "CG-FAC Office of Port & Facility Compliance." (October 6). https://www.uscg.mil/hq/cg5/cg544/default.asp.

U.S. Coast Guard. 2016b. "Cybersecurity." (October 6). https://www.uscg.mil/hq/cg5/cg544/cybersecurity.asp.

U.S. Department of Homeland Security. 2014. "Supplemental Documents." U.S. Coast Guard Cybersecurity Public Meeting. (December 17). https://www.regulations.gov/document?D=USCG-2014-1020-0002 (accessed January 7, 2017).

U.S. Department of Homeland Security. 2015a. "Guidance on Maritime Cybersecurity Standards." U.S. Coast Guard Cybersecurity Public Meeting. Docket No. USCG-2014-1020. (February 18). https://www.regulations.gov/document?D=USCG-2014-1020-0016 (accessed January 8, 2017).

U.S. Department of Homeland Security. 2015b. "Transportation Systems Sector Cybersecurity Framework Implementation Guide." (June 26). https://www.dhs.gov/publication/tss-cybersecurity-framework-implementation-guide (accessed January 8, 2017).

U.S. Department of Homeland Security. 2015c. "Transportation Systems Sector-Specific Plan 2015." https://www.dhs.gov/publication/nipp-ssp-transportation-systems-2015 (accessed March 21, 2016).

U.S. Department of Homeland Security. 2016. "Transportation Systems Sector." (July 8). https://www.dhs.gov/transportation-systems-sector (accessed January 7, 2017).

U.S. Department of Transportation. 2014. "A Summary of Cybersecurity Best Practices." National Highway Traffic Safety Administration, DOT HS 812 075. (October). https://www.nhtsa.gov/DOT/NHTSA/NVS/Crash%20Avoidance/Technical%20Publications/2014/812075_CybersecurityBestPractices.pdf.

Vessel Security Officer Student Guide. Training Resources, Ltd. Inc. Obtained October 2016.

White House. 2013a. "Executive Order—Improving Critical Infrastructure Cybersecurity." (February 2012). https://www.whitehouse.gov/the-press-office/2013/02/12/executive-order-improving-critical-infrastructure-cybersecurity (accessed January 7, 2017).

White House. 2013b. "Presidential Policy Directive—Critical Infrastructure Security and Resilience." PPD-21. (February 12). https://www.whitehouse.gov/the-press-office/2013/02/12/presidential-policy-directive-critical-infrastructure-security-and-resil (accessed January 7, 2017).

CHAPTER 33:
EVERGREEN CYBER PROJECT

Ventus Solutions
Eric Popiel
U.S. Coast Guard

Executive Summary

A combined team from the Evergreen Office and Ventus Solutions (VES) conducted a project that examined future cyber challenges for the U.S. Coast Guard (USCG) and identified strategic cyber needs of the U.S. Coast Guard (USCG) in the 2025 time frame. Following rigorous research and analysis using scenario-based planning methods, the team identified 12 key forces driving future uncertainty for the USCG. The Evergreen/VES team then leveraged details from the 12 variables to build 4 worlds.[1] In a subsequent workshop, four teams examined each alternative future to identify key areas of leadership focus or emphasis for Coast Guard consideration. These areas of emphasis—called key success factors (KSFs)—are designed to enable the Coast Guard to continue to perform above peer organizations. VES provided a number of methodologies and algorithms to both weight output from each team and to aid in selecting a temporal strategy for possible implementation. Thirty-seven KSFs were selected by the teams, with a wide range of recommendations across Coast Guard culture, organization, training, manning, and resourcing. Of these, 13 ranked significantly higher across multiple alternative futures. VES then examined interactions between KSFs, and along with workshop scores, provided a recommendation for seven KSFs to be examined in the near term by Coast Guard leadership including: Professional Cyber Career Field, Adaptable, Flexible Human Resources System, Tolerance for Innovation, Cy-

---

[1] Details of these worlds can be found on the website:
http://www.uscg.mil/strategy/evergreendocuments.asp.

ber Center of Excellence, Enhanced Operational Training for Cyber Units, Cyber Mission Teams, and Resilient Infrastructure with Enclaves.

We recommend Coast Guard leadership begin by pursuing the first three KSFs. They represent corporate cultural changes that are necessary for continued mission success in the future. Of note, workshop participants placed a high emphasis on protection of the Maritime Transportation System; accordingly, the "Resilient Infrastructure" KSF ranked highly. Furthermore, we recommend that the Coast Guard consider Innovation Pools (collaborative research and development activities with private sector organizations) for inclusion in the top group of 13, as it could be a high payoff collaboration with the private sector on cyber. Finally, the team conducted a review of the results and compared outputs with the 2015 USCG Cyber Strategy. We found the results consistent with the current strategy, but also noted that the vast majority of the KSFs align with the seven long term success vectors of the strategy. This team believes cyber poses a significant challenge to the USCG, but believes there are a number of actionable, near-term steps that can be considered to help the Coast Guard achieve its objectives in the coming years. Pursuit of the top KSFs is a prime place to start.

Background and Objective

The purpose of this project was to examine future cyber challenges for the U.S. Coast Guard and to identify and synthesize solutions to complex problems that could arise from those cyber challenges. A key element of the project workshop was to develop alternative future world scenarios focused on the cyber domain that would enable insight into the future Coast Guard operating environment. An additional project goal included identifying strategic needs in anticipation of these future-operating environments.

The White House 2015 National Security Strategy highlights cyber security as an escalating challenge (White House 2015). Ensuring cyber security remains a particularly difficult problem to solve for a number of reasons: attacks may come from anywhere in the world at any time, the physical world is increasingly connected to cyber space, and cyber networks are becoming progressively more complex, with technological change bringing both new vulnerabilities as well as opportunities. Threats may come from national governments, terrorists, organized crime groups, hackers, or disgruntled insiders. Understanding the magnitude of the challenge, U.S. Coast Guard (USCG) senior leadership published a new USCG Cyber Strategy in June 2015. This strategy articulated the Coast Guard vision for cyber operations and noted: "We will ensure the security of our cyber space, maintain superiority over our adversaries, and safeguard our Nation's critical maritime infrastructure." As part of this document the Coast Guard identified the rapidly evolving cyber domain and cyber security as one of the most serious economic and national security challenges of today (United States Coast Guard 2015).

The overall mission of the Coast Guard is to ensure the safety, security, and stewardship of our nation's maritime interests in the heartland, in the ports, at sea, and around the globe. As part of this broad overall mission, the Coast Guard must execute eleven statutory missions. In our view, virtually any operational mission conducted by the Coast Guard in support of its statutory mission incorporates use of technology and networks with potential cyber vulnerabilities. As noted in the Coast Guard's 2015 strategy, the service must strategically adapt to meet the challenges of the digital era. First and foremost, the Coast Guard must embrace cyber space as an operational domain. The USCG's Cyber Strategy identifies three distinct cyber priorities critical to overall mission success: defending cyber space, enabling operations, and protecting infrastructure. The Evergreen project objective is to enable this future mission success.

Evergreen Process

The Evergreen Office (Office of Emerging Policy) supports Coast Guard strategy development and is a key element of the USCG strategic foresight initiative. That initiative is to provide defined and vetted strategic needs and instill strategic intent throughout the service by engaging various levels of internal and external stakeholders through scenario-based planning methods. Evergreen uses scenario-based planning methods to identify and synthesize solutions to complex strategic problems, and to develop strategic needs in preparation for a highly dynamic and constantly evolving operating environment. In short Evergreen helps the Coast Guard prepare for an uncertain future.

In 2015, with Evergreen IV the Coast Guard started a trend toward executing shorter-term, more subject-based efforts with a time horizon of approximately 10 years. Evergreen intends to delve deeply into specific topics such as cyber, the Arctic, energy, climate change, and others as directed by Coast Guard senior leadership. Drilling down into a specific area enables Evergreen and associated stakeholders to consider in more detail how future uncertainties could impact the Coast Guard operating environment in these specific areas.

Organizations use Scenario-Based Planning to get around a fundamental problem of long term planning: no one can predict the future. In this process, a team attempts to bound the future by creating a number of plausible futures. Typically, these scenarios stretch ones' thinking to allow planners to consider new or unforeseen possibilities. By examining a number of different future worlds, planners gain a better appreciation of what capabilities an organization might need to emphasize to improve the odds of future success and they gain some understanding of the complex interactions among variables. Scenario planning avoids a common weakness of linear planning, (i.e., the future will be like the recent past).

Organizations tend to underperform when developments proceed along a different path than expected. In the renowned book, *The Innovator's Dilemma*, Harvard professor and businessman Clayton Christensen notes how even disruptive innovators can get blindsided by other disruptive technologies and technologists. Christensen (2011) demonstrated that leaders of these technology companies did not anticipate or comprehend how new technologies might impact their businesses. Scenario-based planning reduces the probability of these surprises and helps organizations prepare and "future proof" their decisions.

Schoemaker Method

There are a number of methods used to conduct scenario-based planning. Peter Schwartz and Paul Schoemaker have each advanced their own methods. This project used a modified Schoemaker method, for this method enables one to bound the future and then provides a well-organized approach to determining follow-on activities. Schoemaker, a pioneer in the field of decision sciences and Research Director of the Mack Institute for Innovation Management at the Wharton School (University of Pennsylvania), considers scenario planning to be an attempt to "capture the richness and range of possibilities, stimulating decision-makers to consider changes they would otherwise ignore" (Schoemaker and Mavaddat 2000). It is important to highlight that scenarios are used to bound the future instead of forecasting the future. The basic method is presented in Figure 1.

Overall Project Process

VES employed the Schoemaker process and merged it with the Evergreen process to provide a robust strategic planning process that supports the Coast Guard mission. The following sections detail critical elements of this strategic planning process.

Workshop and Development of KSFs

With the internal testing complete and reviewed, the workshop design was finalized. Results and observations from the two cells were reviewed with Evergreen and incorporated into the final workshop design. The primary purpose of the workshop was twofold: finalize the alternative futures and develop Key

CHAPTER 33

	Step	Explanation
1	Define Scope	Define issues to be understood by the organization in terms of time frame, scope and decision variables.
2	Identify Major Stakeholders	Identify major stakeholders, who are affected or may influence the issues, they may be both internal and external to the organization.
3	Identify Main Forces	Identify and study the main forces that shape the future within the scope looking at social, economic, technological, environmental and political domains.
4	Identify Trends	Identify which forces are trends and understand how they will affect the issues of interest.
5	Identify Key Uncertainties	Identify the main uncertainties from the list of forces, how they interrelate, and rank them on both importance and degree of uncertainty.
6	Construct Initial Scenario	Select the two uncertainties of greatest importance and greatest uncertainty, and develop a 2 × 2 matrix of plausible scenarios. Suitable outcomes from other key uncertainties and trends are added as elements to all scenarios.
7	Check for Consistency and Plausibility	Assess the consistency and plausibility of the initial scenarios; are trends compatible within the time frame, are outcomes of uncertainties combined in a logical manner, and are the presumed actions of stakeholders compatible with their interests?
8	Redefine Scenarios Themes	Reassess the ranges of uncertainty variables and retrace the steps to develop final scenarios.

Figure 1: Basic Schoemaker Method

Success Factors (KSFs) that enable superior Coast Guard performance in each of those futures as well as across the futures. The teams developed KSFs using a human centered design methodology to draw out diverse views and opinions and enable a broader understanding of what the Coast Guard could emphasize to enable superior performance and resilience in the face of an uncertain future.

Workshop Output and Conclusions

The teams developed 37 key success factors (see the complete report on the Evergreen website for a list of all KSF's and respective definitions http://www.uscg.mil/strategy/evergreen.asp). The teams discussed and verified that the KSFs were not ordinary activities that every Coast Guard or like organization must undertake, but rather high return activities likely to enable future Coast Guard success. VES collected team data on the weighting of each KSF and the standard deviation of scores. Additionally, each team developed definitions of KSFs and combined like KSFs across alternative futures to enable understanding of which KSFs provided substantial value across multiple futures. VES algorithms calculated the weight of each KSF in each future and further refined weighting by including the impact by the weighting of a particular future. Importantly, many of the KSFs played a significant role in multiple alternative futures. A KSF that weighs highly in all four futures is a "no regret" activity and is worthy of leadership attention.

The following list identifies the highest scoring KSF's.

1) Rapid Cyber Fielding (5.77)

2) Adaptable Flexible HR System (5.63)

3) Resilient Infrastructure w/ Enclaves (4.94)

4) Create Professional Cyber Career Field (4.37)

5) Global AI Enabled Maritime Domain Awareness (3.27)

6) Utilize AI/Autonomous Decision-Making (3.12)

7) Clear, Robust National/International Cyber Standards (2.78)

8) Cyber Use of Force Continuum (2.65)

9) Increased Tolerance for Innovation (2.63)

10) Mission Enabling AI/Autonomous Systems (2.60)

11) Cyber Center of Excellence (2.37)

12) Enhanced Operational Cyber Training (2.27)

13) Cyber Mission teams (1.91)

The numbers next to the KSFs indicate the score that each one received. Workshop participants scored each KSF during the workshop. The score was calculated using a VES derived algorithm. A high score indicates that the Coast Guard should address the challenge as it will help the Coast Guard succeed across multiple futures.

Several observations are pertinent in the output of these 13 highest-ranking KSFs. First, one must highlight that these KSFs tend to be highly ranked across multiple futures. Secondly, the KSFs cover multiple areas of emphasis including culture, organization, training, and investment.[2] Some of the highest ranked KSF's involve the creation of a Professional Cyber career field. In support of this, the workshop teams recommended adopting a flexible human resources (HR) system. Such a system would enable recruitment of more senior personnel with targeted skill sets and would enable such personnel to depart and reenter the service as required. Of note, workshop participants rated the flexible HR system higher than development of a professional cyber career field because of the additional salutary effects to a range of other USCG mission areas. Other linked KSFs include development of cyber mission teams for operational deployment or placement as required and a robust operational cyber training program. VES participants believe a cyber operational training program would likely have even greater positive impact than denoted by its overall workshop derived score. Simply put, operational weaknesses and strengths would be uncovered by aggressive exercises and training resulting in significant feedback to mission teams and other cyber support personnel. No organization has unlimited management and the rigorous analysis enabled the team to determine a natural breakpoint after KSF 13.

A key concern of the workshop teams was the security of the Maritime Transportation System. As a result of concern for this system and for other key infrastructure, the workshop participants rated development of resilient infrastructure with secure enclaves and non-networked redundancies highly. This concern also played a role in the recommendation to emphasize Artificial Intelligence (AI) and autonomous decision-making to speed response to threat-

---

2   Of note, high tolerance for innovation and integrating new technologies requires a focus on culture within the U.S. Coast Guard. Changing culture often requires sustained leadership focus and is generally hard. The development of clear national and international cyber standards and norms is also a challenging long term activity.

ening cyber activities. Workshop participants also emphasized the importance of rapid cyber fielding to enable timely response of cyber capabilities. Teams recommended this rapid fielding not just to protect critical infrastructure, but also to enable appropriate protection of USCG operational units and to extend USCG response operations.

Several workshop teams recommended a "Cyber Center of Excellence" to select and oversee requisite capabilities and training. Also of note, the idea of having a cyber element in the use of force continuum needs further review. The concept would enable the USCG to use cyber in support for activities that could result in use of force. For example, cyber could be used where appropriate to stop a drug running fast boat before more lethal methods are employed.

In many of the futures examined, the teams assessed a growing importance of Maritime Domain Awareness (MDA). Use of cyber, and big data analytics could potentially enable improved execution of this task with reduced use of expensive operational assets. Team members also emphasized the importance of AI linked with autonomous systems both to reduce use of USCG operational assets but also to provide support for other key mission areas such as marine environmental protection where unmanned underwater vehicles (UUV) and autonomous capabilities are likely to grow in importance.

VES then examined the links between KSFs in order to determine interdependencies. Highly- ranked KSFs that support lots of other KSFs but require little support themselves tend to be good early investments. "Central" KSFs support many, and require little support; these KSFs are heavily supported but also have many dependencies. "Reactive" KSFs requires a great deal of support but supports fewer KSFs.

VES then compared desirable early areas of emphasis (i.e., KSFs) with the workshop's recommendations of 13 key KSFs. The result was seven KSFs recommended for early consideration or adoption. Of these we recommend three of the 'Driving' KSFs for initial consideration. Also of note, workshop participants placed a high emphasis on protection of the maritime transportation system (MTS). The "Resilient Infrastructure" KSF ranked highly as a result, but could require significant coordination with cyber professionals. However, we moved this KSF up temporally because of the importance placed on MTS by workshop participants.

VES then reviewed the overall results to determine if the USCG should consider one or more lower ranked KSFs as a "bet", even if only beneficial in one future. We felt that the "innovation pool" concept is worthy of consideration for inclusion with the thirteen other highest ranked KSFs. In this concept, the USCG would consider pooling small amounts of R&D dollars with either private sector organizations or government organizations in support of software and other innovations that could have outsize impact on USCG missions. Organizations like In-Q-Tel and the newly created Defense Innovation Unit Experimental (DIUx) enable the intelligence community and the defense community to gain access to cutting edge private sector ideas while leveraging pooled resources.[3]

COMPARISON OF RESULTS WITH COAST GUARD STRATEGIC DOCUMENTS

VES conducted a review of workshop output and conclusions and compared it with the recently released (June 2015) USCG Cyber strategy. We found the KSFs were entirely consistent and nested within the three specific strategic priorities noted by the Coast Guard (defending cyber space, enabling operations, and protecting infrastructure). We also compared the KSFs with the seven long term support factors developed by the Coast Guard. Thirty-two of the 37 KSFs linked to one of the seven support factors (see table, following page). Five additional KSFs touched on additional capabilities that cyber could enable but also touched on broader USCG capabilities. In short, the overall workshop results fit extremely well with developed and articulated U.S. Coast Guard Cyber strategy.

VES also compared workshop results with the Evergreen IV Strategic Needs and other recently released strategy documents (USCG's Arctic Strategy, USCG Living Marine Resources Ocean Guardian, 2014 USCG's Western Hemisphere Strategy). The workshop results synchronized well across the board with a variety of key strategic needs articulated by Evergreen. For example, Evergreen discusses the need for a fully integrated Maritime Domain Awareness (MDA) capability; workshop participants echoed this with a KSF recommending examination of AI-enabled MDA. Evergreen specifically recognized the importance of Talent Management and Individual Technology Specialization; two of the highest-ranked workshop KSFs involve a more flexible and adaptable HR system and a Professional Cyber Career field. Evergreen highlighted the importance of a Secure C4IT system and specifically noted the importance of "preventing unauthorized actors from infiltrating automated systems with the Maritime Transportation System." The generated KSFs provide several key concepts that are strongly aligned and supportive of this Evergreen recommendation, and independently determined. Other technical and culture change KSFs aligned with needs outlined in the other USCG strategic documents as enablers.

---

3   For more information see In-Q-Tel: https://www.iqt.org/ and DIUx: https://www.diux.mil/.

USCG Cyber Strategy Support Factor	Workshop Key Success Factors
Recognition of cyber space as an operational domain	• Cyber use of force continuum • Update decision-making framework for cyber
Developing cyber guidance and defining mission space	• Clear robust National and International standards in cyber space CG wide insider threat identification mitigation program • Unique government only cyber mission
Leveraging partnerships to build knowledge, resource capacity, and an understanding of MTS cyber vulnerabilities	• Partner with industry for protection of Undersea infrastructure • Collaborative protection of key drilling assets • Improved interoperability with partners in cyber degraded environment • Secure interoperability between MILSPEC and Industry standards • Embassy liaisons and International training teams • Automated protocols for coordinated stakeholder response to cyber incidents on MTS
Sharing of real-time information	• Mission enabling AI/autonomous systems • Global AI enabled Maritime Domain Awareness • Utilize AI/autonomous decision-making to focus on key events • On scene cyber analysis capabilities
Organizing for success	• Cyber mission teams with effective combined capabilities to respond to maritime cyber threats • Corporate knowledge strategy • COOP cyber plan/standards • Increased tolerance for innovation and integrating new technologies
Building a well-trained cyber workforce	• Adaptable flexible HR System • Create professional cyber career field/specialty • Continuous technical/cyber professional development • Enhanced operational cyber training/exercises for units
Making thoughtful future cyber investments	• Innovation investment pool/capability • CG/DARPA
	• Cyber center of excellence • Redundant non-networked backups to enhance resilience and recovery • Agile mission support for cyber product lines • Rapid cyber fielding capability • Highly autonomous and flexible units • Resilient infrastructure—secure enclaves in a single unified infrastructure with non-networked redundancies

Figure 2: Strategic linkages (adapted from Ventus Solutions and Evergreen 2015)

CHAPTER 33

Conclusions

Evergreen/VES analyzed a variety of plausible cyber futures and the potential impact of those futures on the Coast Guard. The team assessed the potential impact of these cyber futures as significant. We caution that no one can predict the future. Indeed, the methodology presented here is designed to "bound" the future and enable an organization to "future proof" strategy. However, participants noted with concern that in several of the futures the pace of technology would provide many challenges (and opportunities) to organizations such as the Coast Guard. As a result, workshop participants proposed several significant cultural, organizational, training, and investment areas of emphasis which should enable the Coast Guard to continue to perform its critical missions despite these uncertainties. The team recommended seven of these areas for early emphasis and consideration within existing USCG assessment processes.

References

Christensen, Clayton. 2011. *The Innovator's Dilemma*. New York: Harper Business.

Schoemaker, Paul J. H., and V. Michael Mavaddat. 2000. "Scenario Planning for Disruptive Technologies." In *Wharton On Managing Emerging Technologies*, eds. George S. Day and Paul S. Schoemaker. Hoboken, NJ: John Wiley & Sons, 206–241.

United States Coast Guard. 2015. *United States Coast Guard Cyber Strategy*. Washington, DC, June. http://www.uscg.mil/seniorleadership/DOCS/cyber.pdf.

Ventus Solutions and Evergreen. 2015. "US Coast Guard Future of Cyber Project: Final Brief: Workshop Review, Alternative Futures and Analysis." PowerPoint (February 2016).

White House. 2015. *National Security Strategy*. Washington, DC, February. https://www.whitehouse.gov/sites/default/files/docs/2015_national_security_strategy.pdf.

APPENDIX A:
KEY SUCCESS FACTORS AND DEFINITIONS

KSF	Definition
Rapid Cyber Fielding	Augment existing acquisitions process; greater agility to meet timely software and hardware needs.
	Ensure procurements are interoperable with current systems and upgradeable.
Adaptable Flexible HR System	A workforce that integrates active duty, civilian, and other cyber professionals, allowing flexible entry/exit, and meets defined competencies. Streamline hiring practices using non-traditional means of acquiring and retaining cyber professionals through DCO, reserve, contractor and temporary options.
Resilient Infrastructure with Enclaves in a single unified infrastructure with non-networked redundancies.	Provide defensive capability for assured data integrity within core USCG systems such as the MTS (Assured Data Integrity).
	Creation of hardened protected nodes that can also work offline.
Create a Professional Cyber Career Field	Establish a career identity; define a portable civilian, enlisted, and officer career path with dedicated training programs, qualifications, and structures aligned with industry standards and credentials.
Global AI Enabled Maritime Domain Awareness	Aggregated information to create a comprehensive COP (Common Operating Picture) for the maritime domain. Utilize segmented, real-time data to influence mission operations with enhanced decision-making.
Utilize AI/Autonomous Decision-Making	Create AI-enhanced decision –making systems that augment and support all levels of strategic and tactical decisions. AI-enabled mission execution and planning; utilize big data mining, crowd sourcing and algorithm-based data-driven decision-making tools to filter out "white noise" and summarize information for human consumption and informed decision-making.
Clear, Robust National/International Cyberspace Standards	Clarify existing cyber legal authorities and seek new authorities where necessary to ensure a robust international and domestic legal architecture that establishes the USCG as the preeminent authority in the maritime related cyber domain. This includes establishing

CHAPTER 33

Cyber Use of Force Continuum	international cyber related norms for sovereign nations to prevent and respond to maritime incidents; and establishing norms for the use of cyber force against non-state actors implemented through US domestic law and regulations. Clearly define authorities and jurisdictions to conduct offensive and defensive cyber operations against non-state actors. Recognition of cyber as a tool on the force continuum: (e.g. use of cyber tools to stop UAS drug running).
Increased Tolerance For Innovation and Integrating New Technologies	Increased risk tolerance for innovation and integrating new technologies (automation and outsourcing of systems), with expectation of efficiency gains. Change management culture in regards to decision-making unencumbered by the status quo.
Mission Enabling AI/Autonomous Systems	Adoption of emergent AI technologies and resilient systems (including UAVs, UUVs, etc.) to help perform missions and improve decision-making. Utilize all source capabilities including big data/social media to allocate resources; ensure understanding of actual SAR mission vs. diversion; improve real-time asset allocation. Incorporate unmanned systems where practical to reduce resource requirements and ensure operational efficiency.
Cyber Center of Excellence	Hub for cooperation, collaboration, and communication between CG/DHS/interagency/academia/NGOs/allies. National and international industry recognized cyber expertise focused on technologies, authorities, and enforcement; building partnerships; directing innovation investments.
Enhanced Operational Cyber Training/Exercises for units	Incorporate cyber into exercise plans, policies and procedures. Create cyber-com deployed training teams to educate, assist, evaluate, and inspect cyber readiness.
Cyber Mission Teams	Codified Coast Guard cyber mission teams that leverage OGA/Industry cyber capabilities to respond to maritime cyber threats and inspect, assure, and protect networks and network functions both at sea and ashore.

581

Made in the USA
Middletown, DE
11 September 2017